Carrier-Mediated Dermal Delivery

Carrier-Mediated Dermal Delivery

Applications in the Prevention and Treatment of Skin Disorders

edited by
Andreia Ascenso
Sandra Simões
Helena Ribeiro

Pan Stanford Publishing

Published by

Pan Stanford Publishing Pte. Ltd.
Penthouse Level, Suntec Tower 3
8 Temasek Boulevard
Singapore 038988

Email: editorial@panstanford.com
Web: www.panstanford.com

British Library Cataloguing-in-Publication Data
A catalogue record for this book is available from the British Library.

Carrier-Mediated Dermal Delivery: Applications in the Prevention and Treatment of Skin Disorders

ISBN 978-981-4745-58-1 (Hardcover)
ISBN 978-1-315-36447-6 (eBook)

Printed in the USA

Contents

Preface

We are honored to present to the readers the book entitled *Carrier-Mediated Dermal Delivery: Applications in the Prevention and Treatment of Skin Disorders*, which is based on the systematic revision of the most recent findings. This book contains several contributions on new approaches for the management of skin aging and photocarcinogenesis and topical formulations based on nanocarrier systems for skin disorders. These chapters discuss the structure and skin morphology in detail. Cosmeceuticals, laser and photodynamic therapy, and melatonin-based treatments are presented as important strategies for photoaging management. Photodynamic therapy and melatonin can be also used in the context of photocarcinogenesis. Therefore, the inclusion of this strong antioxidant in sunscreen products could be a promising approach. The book discusses the safety and efficacy of sunscreen products as well.

Topical formulations, including emulsions (conventional formulations and emulsions stabilized by solid particles), nail films, and nanocarriers used for different actives delivery, are reviewed concerning certain skin and nail diseases context (e.g., acne, psoriasis, atopic dermatitis, fungal diseases, leishmaniasis, skin cancer). Finally, several nanocarriers are introduced, such as lipid vesicles (from the first generation of conventional liposomes until the more recent deformable vesicles), liquid crystalline nanodispersions, and gelatin and solid lipid nanoparticles. Their composition, formulation process, characterization, and examples of topical applications are discussed in detail for each system. In fact, these nanocarrier systems can be useful as topical and/or transdermal delivery systems attending to a higher skin penetration and permeation profiles, besides contributing to improving technological drawbacks (e.g., formulation stability) and increasing the therapeutic index.

Although this is a quite broad topic, the most important (nano)pharmaceutical formulations are presented in the book. Future perspectives are also discussed in some chapters. This book will be a useful reference for researchers and professionals interested in nanotechnology in the skin delivery context.

Part 1

New Approaches for Management of Skin Aging and Photocarcinogenesis

Chapter 1

New Trends in Anti-Aging Skin Care

Joice Lana[a] and Andreia Ascenso[b]

[a]*Faculdade de Farmácia, Universidade de Lisboa, Av. Prof. Gama Pinto, 1649-003 Lisboa, Portugal*
[b]*Instituto de Investigação do Medicamento (iMed.ULisboa), Faculdade de Farmácia, Universidade de Lisboa, Av. Prof. Gama Pinto, 1649-003 Lisboa, Portugal*

joice_lana@outlook.com

1.1 Introduction

Life expectancy has never been so high in developed countries according to the World Health Organization. Nowadays, the desire to maintain a young image and healthy appearance is omnipresent in both genders. Skin aging is generally the main concern with a marked social impact. In fact, the skin properties and functions based on maintaining the organism homeostasis and protecting it from external harmful agents [1–3] will decrease over time.

Several techniques have been developed that aim to prevent, slow, or revert the skin aging process with diverse successful outcomes alone or in combination with others. Those techniques can be surgical, such as eyelid surgery and facial lifting or non-surgical procedures, for example, botulinum toxin injection, filler injection, laser treatment, dermabrasion, and chemical peelings [3].

Carrier-Mediated Dermal Delivery: Applications in the Prevention and Treatment of Skin Disorders
Edited by Andreia Ascenso, Sandra Simões, and Helena Ribeiro

ISBN 978-981-4745-58-1 (Hardcover), 978-1-315-36447-6 (eBook)
www.panstanford.com

Some of these non-surgical techniques will be mentioned in this review, especially injectable and laser-based treatments.

Although the skin nervous system has not been fully understood and somewhat underexplored, the possibility of manipulating neuropeptides to improve the skin appearance by delaying or treating its aging has becoming a recent research area with a significant potential. On the other hand, laser therapy is an excellent and safe technique. Both advantages and limitations of these techniques will be addressed as well as future perspectives in this context.

1.2 Skin

The integumentary system includes the skin and accessory structures (hair, nails, and glands). The skin or the integument covers the body externally, being its largest organ (about 15% to 20% of total body weight in adults). It possesses several functions, such as (i) **protective,** as a physical barrier against potential pathogens and chemicals as well as thermal, mechanical, and ultraviolet radiation damages; (ii) **sensorial,** as sensory skin receptors are able to detect heat, cold, pressure, touch, and pain, allowing skin to monitor the environment and regulate interactions with it; (iii) **thermoregulatory,** through water loss mechanisms or insulation ability; (iv) **endocrine** and **metabolic,** by secreting hormones, cytokines, and growth factors and synthesizing vitamin D (indispensable for calcium metabolism and bone formation); (v) **excretory,** through the skin pores and glands [4–6].

1.2.1 Structure

The skin is formed by three main layers: the epidermis, an epithelial layer of ectodermal origin; the dermis, a layer of mesodermal connective tissue; and the hypodermis, or subcutaneous tissue, a layer of loose connective tissue that connects the skin to the muscles or bones. Additionally, skin can have different thickness, being thick on the palms and soles and thin on the rest of the body [4–6].

1.2.1.1 Epidermis

Epidermis, the most superficial layer of the skin, is mainly formed by stratified squamous keratinized epithelial tissue. The living cells receive nutrients and excrete waste products by diffusion through the epidermis layer and the capillaries of the dermis [4, 5].

Different types of cells are present in the epidermis layer:

- **Keratinocytes,** which produce keratin, a protein that contributes for cells strength;
- **Melanocytes**, the melanin producers. These cells are accumulated in the *stratum basale* (1 melanocyte for every 5–6 basal keratinocytes), being attached to the basal lamina. Melanin synthesis starts with a reaction catalyzed by tyrosinase that converts tyrosine into 3,4-dihydroxyphenylalanine (DOPA) that will be transformed and polymerized in different forms of melanin. Melanin is stored in cell vesicles until they form melanosomes that are transported and taken up by keratinocytes. This molecule will protect the DNA (deoxyribonucleic acid) of living cells from the ionizing mutagenic effects of ultraviolet radiation (UVR);
- **Langerhans cells**, the antigen-presenting cells with an important role in the skin immune system. These cells are mostly present in the *stratum spinosum*;
- **Merkel cells,** which are tactile epithelial cells associated with the nerve endings, being responsible for perceiving light, touch, and superficial pressure. These cells are present in a higher percentage in the sensitive skin [4, 5].

Regarding the **epidermis renewal**, new keratinocytes are produced in the deepest layer of epidermis, pushing the older cells to the surface, and changing their shape and chemical composition along the way (keratinization process). During this process, the cells will die and the outer layer of dead cells will provide resistance to abrasion, forming a barrier with relative permeability [4].

The human epidermis is renewed in 15 to 30 days, depending on several factors (age, region of the body, etc.). During the process of keratinization, the number of cell layers in each stratum is influenced by the body location. The deepest stratum, **stratum basale** or **germinativum**, rest on the basal lamina and is formed by a single layer of keratinocyte steam cells (basophilic cuboidal or columnar cells) that go through mitotic divisions. **Hemidesmosomes** join these cells to the basal lamina and **desmosomes** bind the cells together on their lateral and upper surfaces. The **stratum spinosum**, or the spinous layer, lies on the stratum basale. Additional keratin fibers and lamellar bodies are formed inside the larger and polyhedral keratinocytes of this layer. Following these two strata, there is the **stratum granulosum** formed by flattened diamond-shaped cells. Keratohyalin granules are produced in this layer. These granules are basophilic masses rich in cystine and histidine proteins, the precursors of filaggrin, which aggregates the keratin filaments on the stratum corneum (SC). It also facilitates the lipids' release by lamellar bodies from the cells creating an impermeable layer around it. At this stage, the cells die. The **stratum lucidum** is above the stratum granulosum being constituted by some layers of dead cells with indistinct borders. It is normally present in thick skin and absent in thin skin. The most superficial layer is the **stratum corneum** formed by dead cells, which have a hard protein envelope filled with keratin (a mixture of keratin fibers and keratohyalin) providing structural strength. The released lipids from lamellar bodies are responsible for the skin permeability. At the end of the keratinization process, the cells are fully keratinized or cornified (squames) and shed from the epidermal surface as the desmosomes and lipid-rich cell envelopes break down [4–6].

The **basement membrane** or **basal lamina** separates the epidermis from the dermis. This irregular junction and projections (dermal papillae) merge with invaginating epidermal ridges, which contribute to a stronger adhesion of these two layers [4, 5].

1.2.1.2 Dermis

The **dermis** provides most of the structural strength of the skin. Its connective tissue is formed by collagen, elastic and reticular fibers. Several cells and structures can be found in dermis, such as fibroblasts/fibrocytes/myofibroblasts, adipocytes, macrophages, monocytes, mast cells, Langerhans cells, T lymphocytes, dendritic cells, nerve endings, hair follicles, smooth muscle cells, and glands, besides the presence of blood and lymphatic vessels. It is divided into two sub-layers (i) the **papillary layer**, which includes the dermal papillae, made up of loose connective tissue with collagen types I/III and collagen type VII, which connects the dermis to the epidermis fibrils and to the basal lamina, respectively; (ii) the **reticular layer**, which is a dense and irregular connective tissue mainly constituted by elastic fibers (more fibers than cells compared with the papillary layer) and collagen type I surrounded by proteoglycans rich in dermatan sulfate. Dermis thickness differs with the region of the body, reaching a maximum of 4 mm on the back [4–7].

Two major plexuses are formed by nutritive vessels on the dermis. The **microvascular subpapillary plexus** lies between the papillary and reticular dermis layers, and its extensions form a capillary matrix below the epidermis. The **deep plexus** lies between the dermis and the subcutaneous layer and is formed by larger vessels. These plexuses are connected by blood vessels [5, 7].

1.2.2 Innervation

The skin innervation is represented by a neural matrix formed by cholinergic and adrenergic nerves and sensory fibers. This dense innervation extends to the superficial layers of the epidermis, including the stratum corneum. Immunolabeling (with an antibody against protein gene product 9.5) is the best method for the visualization of skin innervation [8]. The skin possesses an enormous diversity of receptors, channels, neurotransmitters, and modulators, including various types of sensory receptors, motor nerve endings at the blood vessels, pili muscles, and sweat glands. Nevertheless, there is also contact between nerve fibers and keratinocytes, melanocytes, Langerhans cells, mastocytes, and dendritic and endothelial cells. The nerve endings differ within different body parts, reflecting their distinct functions [6–9].

1.2.2.1 Sensory receptors

Sensory receptors transduce and transmit pain (nociception) and can be free or encapsulated nerve endings (or fibers). Sensory axons are morphologically categorized as myelinated (A-fibers) or unmyelinated (C-fibers). Myelination degree increases the axonal signal transmission velocity. A-α, -β, and -δ fibers are highly, moderately, and poorly myelinated with a conduction velocity ranging from 70–120 m/s (highly myelinated) to 4–30 m/s (poorly myelinated). C-Fibers are unmyelinated and small fibers having a low conducting velocity, approximately 0.5–2 m/s [4, 5, 7, 8, 10].

Free nerve endings include (i) *Merkel cells,* which are associated with the recognition of continuous light touch and texture, consisting of A-fibers; (ii) *nerve terminations* in the papillary dermis, also extended to lower epidermal layers; these nerve terminations respond to extreme temperatures, pain and itching, consisting in A-fibers and C-fibers; (iii) *root hair fibers complex* localized around the hair follicles bases and *Haarschiebe touch domes*; these last structures are specialized in pressure reception, localized near the hair follicles and their neurite plexus, containing both A and C-fibers [4–6, 10].

Encapsulated nerve endings comprise (i) *Meissner corpuscles,* which are sensory axons (A and C-fibers) winding among Schwann cells located in the dermal papillae, mainly responsible for perceiving light touch or low frequency stimuli; their number decreases gradually after puberty; (ii) *Pacinian corpuscles,* which are *myelinated nerve endings (A-fibers)* encircled by a capsule (formed by flattened Schwann cells and collagen) and deeply located in the reticular dermis and hypodermis. It mostly detects pressure changes and vibrations applied on the skin; (iii) *Rufinni or bulbous corpuscles,* which surround A-fibers, have a thin collagenous fusiform capsule anchored to the surrounding connective tissue. It typically responds to tension and torque applied to the skin and intravenous; (iv) *Krause end bulbs* that have a very thin collagenous capsule penetrated by a nerve A-fiber, being generally associated with low frequency vibrations [4–6, 10].

1.2.2.2 Non-sensory receptors

Non-sensory receptors are expressed in **diverse types of skin cells**, and their signal transduction lead to diverse manifestations. Some non-sensory receptors are as follows:

- **Cholinergic receptors**, found in sweat glands and ducts, keratinocytes, sebocytes, fibroblasts and melanocytes. Their activation on keratinocytes stimulates adhesion, motility and differentiation. Furthermore, acetylcholine may be a promoter in sebocytes differentiation, sebum production and hyperpigmentation. In addition to the presence of the cholinergic receptors in a non-neuronal manner, cholinergic neurons (sympathetic and parasympathetic) supply innervation to blood and lymphatic vessels, pili muscles, hair follicles and glands play a crucial role in the regulation of the body temperature [7, 10–12].

- **Adrenergic receptors** of several α and β subtypes and are expressed in melanocytes, keratinocytes, and eccrine epithelial cells. Their activation leads to keratinocyte differentiation stimulation; α and β receptors are found in dermal blood vessels, and their stimulation leads to vasoconstriction and decreasing of the vascular permeability [7, 10].
- **Corticotropin-releasing hormone (CRH) and urocortin receptors** expressed in dermal fibroblasts, endothelial cells, hair follicle, smooth muscle of blood vessels, keratinocytes, melanocytes, and mast cells. CRH can act as a pro- or anti-inflammatory agent, and antinociceptive and wound-healing accelerator. Urocortin and CRH are able to inhibit the proliferation of keratinocytes and induce their differentiation as well as stimulate or inhibit melanoma cell proliferation depending on culture conditions [7, 10, 13].
- **Melanocortin receptors (MCR)** activated by the adreno-corticotropic hormone (ACTH) and some melanocyte-stimulating hormones (MSH). These receptors have been detected in melanocytes, keratinocytes, monocytes, sebo-cytes, fibroblasts, and endothelial, epithelial, Langerhans, and dermal immune cells. They are responsible for differ-ences in skin color. The stimulation of melanogenesis and its switching to eumelanogenesis from pheomelanogenesis is the most recognized phenotypic effect of ACTH and MSH peptides [7, 10].
- **Opioid receptors**, m-opioid receptors identified in keratinocytes, hair follicles, sebaceous glands and sweat glands, and z-opioid receptors identified in epidermal keratinocytes. Opioid peptides can be divided into endorphins, enkephalins, and dynorphins. Generally,

the activation of opioid receptors leads to inhibition of neuronal excitability. Met- and leu-enkephalins can inhibit the differentiation of human keratinocytes *in vitro*. β-Endorphin and enkephalins have antinociceptive and immunomodulatory properties [7, 10].

- **Growth hormone receptor (GHR)** detected in the human skin epidermis, hair follicle, eccrine glands, dermal fibroblasts, keratinocytes, adipocytes, melanocytes, and in Schwann and muscle cells. GH may directly modulate keratinocyte and fibroblast function [7, 14].
- **Prolactin (PRL) and luteinizing hormone/chori-ogonadotropin receptors (LH/CG-R)** also expressed in the skin. In humans, hyperprolactinemia has been associated with hirsutism [7].
- **Neurokinin receptors (NKR)** activated by substance P (SP) or neurokinins A and B (NKA and NKB), leading to the stimulation of keratinocytes, fibroblasts, and endothelial cell proliferation and neovascularization. In addition, they are also related with the modulation of pro-inflammatory processes. Their expression has been detected in keratinocytes, endothelial cells, mast cells, fibroblasts, and Merkel and Langerhans cells [7, 10].
- **Calcitonin gene-related peptide receptors (CGRP-R)** present in mast cells, keratinocytes, melanocytes, smooth muscle, blood vessels, and Merkel and Langerhans cells. CGRP has several functions, such as increasing the vascular permeability, producing dermal edema, stimulating endothelial cell, keratinocyte, and melanocyte proliferation, and regulating cytokine production [7, 10].
- **Vasoactive intestinal peptide and pituitary adenylate cyclase-activating polypeptide receptors (PACAP/VIP-R or PVR)** found in sweat glands, keratinocytes, smooth muscle, and immune and endothelial cells. VIP stimulates the keratinocyte proliferation and sweat production, while PACAP is correlated with inflammation and neurotransmission in the skin. Through these receptors, peptide histidine-methionine (PHM) and GH-releasing factor (GFR) lead to keratinocytes proliferation [7, 10].
- **Neurotrophin (NT) receptors**, which are receptor proteins of the tyrosine kinase (Trk) and p75 panneurotrophin

(p75NTR) families expressed in epidermal and follicular keratinocytes, epidermal melanocytes, specialized dermal fibroblasts, mast cells, immunocytes, and cutaneous sensorial nerves. Nerve growth factor (NGF) stimulates dendrite formation. NTs take part in apoptosis, cutaneous nerve growth, and development [7, 10].

- **Parathyroid hormone (PTH) and/or PTH-related protein (PTHrP)** receptors are expressed in dermal fibroblasts. When stimulated, these receptors respond by producing cytokine and keratinocyte growth factor (KGF) [7].
- **Receptors for the active form of vitamin D [1,25-dihydroxyvitamin D (1,25-(OH)2D or calcitriol] (vitamin D receptor (VDR))** expressed in epidermal and follicular keratinocytes. Epidermal synthesis of vitamin D3 from 7-dehydrocholesterol is dependent on sun exposure [15]. Vitamin D is involved in keratinocyte differentiation, cell proliferation, and hair growth [7]. Additionally, it has a significant role in calcium homeostasis and metabolism, as is well known.
- **Glucocorticoid receptors (GR)** expressed in epidermal and follicular keratinocytes, epithelial cells from glands, sebocytes, melanocytes, immune cells, dermal fibroblasts, and smooth muscle, while **mineralocorticoid receptors (MR)** have been identified in epidermal keratinocytes, hair follicle, and sweat and sebaceous glands. They are linked to those cell functions exerting anti-inflammatory effects [7, 16].
- **Androgen (AR) and estrogen (ER) receptors** disseminated in the skin. AR promotes hair growth and sebum secretion depending on their location. ER are located in epidermal epithelial cells, hair follicle, sebaceous, glands, melanocytes, and fibroblasts [7].
- **Thyroid hormone receptors (THR)** activated by triiodothyronine (T3), which participates in the epidermal differentiation, increasing its response to growth factors. THR also has a role in the functioning of sebaceous and other glands, hair growth, and production of aminoglycans. Changes on the skin in the presence of hyper- or hypothyroidism reveal the functions of this receptor [7].

1.2.2.3 Neuropeptides

Skin neuromodulators are now starting to be seen as possible targets to prevent and treat skin aging. Neuropeptides are neurotransmitters and chemical messengers present in the skin. Their location in skin can be quite variable. Accordingly, Schulze *et al.* studied the distribution of some neuropeptides (VIP, NPY, CGRP, SP, NKA and CR) through different skin layers of both face and forearm by immunohistochemical detection. These authors have reported the following observations: VIP was present in dermis and subcutis layers of both face and forearm; NPY was only detected on the forearm subcutis; CGRP and SP were identified in all skin layers on both sites; NKA was only absent on the face epidermis while CR was absent on the epidermis of both sites [10, 17].

Additionally, neuropeptides can be produced by **resident and circulating skin cells** (e.g., gastrin-releasing peptide, somastostatin, NPY, atrial natriuretic peptide, PHM/PHI (peptide histidine isoleucine), galanin, neurokinins, substance P, neurotensin, CGRP, VIP, bradikinin, cholecystokinin, endothelins, CRH, urocortin, α-, β-, and γ-MSH, ACTH, β endorphin, enkephalins), and/or **in nerve endings** (e.g., substance P, neurokinins, neurotensin, CGRP, VIP, somatostatin, NPY, atrial natriuretic endothelins, α-, β-, and γ-MSH, β endorphin, CRH, urocortin, dynorphin, enkephalins) [7].

There is a linear reduction of epidermis innervation and content of NGF with **skin aging**, while neuropeptide levels and their receptor expression may increase or decrease. For example, epidermal resident cells suffer an age-related density decrease leading to more or less marked consequences on skin structure and appearance. As already stated, neuropeptides bind to skin receptors and produce several cellular responses related to skin functions. Consequently, the alteration of their levels is particularly relevant for skin cells survival, maintenance, and regeneration [18–20]. The list of some neuropeptides and their effect on the skin intrinsic aging is given below:

- **Corticotropin-releasing hormone (CRH)**: neuropeptide hormone that coordinates neuroendocrine and behavioral responses to stress. Aging leads to the dysregulation of CRH system subjecting the skin to a continuous stress environment [10, 18]. Stopping this over-activation of the stress environment would be beneficial to skin health.

- **Somatostatin**: neuropeptide related to Alzheimer's disease, suggesting that lower values of somatostatin with aging may have some neurodegenerative effects (including those on the skin) [18, 21]. Thus, the maintenance of normal values may be a way to prevent this neurodegenerative effect.
- **Melanocortins** (pro-opiomelanocortin or POMC, derivatives —ACTH, α–, β–, and γ–MSH, γ–lipotropin, and β–endorphins): neuropeptides with anti-inflammatory properties, regulating cortisol levels and melanogenesis process, also involved in pigmentary responses (with CRH) and cornification of the skin. POMC system is greatly disturbed with aging. On one hand, peptides levels seem to increase. On the other hand, receptors for ACTH (MCR-2) and β-endorphin (I-opioid receptor 1) decrease with skin aging. However, it has been reported that *Achillea millefolium* extract is able to upregulate their expression. Consequently, this non-endogenous MCR agonist is able to increase skin thickness, which is quite beneficial for skin aging management. [20, 22, 23].
- **Endothelin**: induces collagen synthesis and is involved in tissue remodeling and fibrogenesis [10]. Thus, it could be also for used for the anti-aging process.
- **Pituitary adenylate cyclase-activating polypeptide (PACAP)**: involved in neuroprotective, regenerative, and immunomodulatory functions. However, high levels of this peptide are associated to neurogenic inflammation and lesions (e.g., psoriasis). In addition, an intravenous injection of PACAP leads to a long-lasting flush [10]. Consequently, this peptide should be kept at its normal levels instead of stimulating its synthesis.
- **Substance P (SP)**: a stress-related neuropeptide that can induce cytokines release, the major mediators in skin inflammatory processes and itch. SP is highly increased in sun-aged skin, promoting fibroblast and keratinocyte proliferation. Additionally, it is also involved in wound healing due to fibroblast proliferation with simultaneous collagen deposition and angiogenesis. However, SP can

enhance the virulence of bacteria normally present on the skin, which can limit its applicability to anti-aging methods [9, 10, 18, 24–26].

- **Neuropeptide Y (NPY)**: takes part in the skin defense mechanism owing to its antimicrobial activity. However, NPY should be used cautiously considering that it is highly expressed in melanoma [18].
- **Calcitonin gene-related peptide (CGRP):** expression decreases with age, being associated with the reduced survival and proliferation rate of keratinocytes. In addition, this peptide is able to stimulate melanocytes proliferation, which can be also changed in the aged skin. Moreover, CGRP has anti-inflammatory activity as well [18, 20, 27].

Neuropeptides levels and/or their effects can be conditioned by respective receptors and peptidases levels, which in turn could be affected by aging. For example, the level of neutral endopeptidase-24.11 (NEP, EC 3.4.24.11), which hydrolyzes bioactive regulatory peptides in the skin, significantly increases with aging. There is also an increase of elastase-type endopeptidase activity, leading to the degradation of the extracellular matrix [28, 29]. Regarding photoaging, there is an increase on the densities of dermal and intraepidermal nerve fibers, neuropeptidergic sensory nerve fibers in the dermis, neuropeptides levels (e.g., SP and CGRP), and NGF. The severity of the photodamage is closely related to the subjacent epidermal innervation changes [30].

Neuropeptides (mainly SP and CGRP) have been successfully used to accelerate and modulate **wound healing** in a dose-dependent manner, and since this process is delayed with aging, this could be another application of neuropeptides. In addition, they can be also potential **topical painkillers** (e.g., capsaicin excites C-fibers, releasing tachykinins and CGRP, and then those fibers become desensitized due to the lack of the same neuropeptides) [25, 31–34].

Moreover, neuropeptide modulation can be also used to treat other **skin disorders**, such as sensitive skin. Accordingly, an *in vitro* study with Sensiline® complex showed its influence on neuropeptides (mainly SP) levels in sensitive skin [35].

Incorporation of neuromediators in pharmaceutical formulations is quite difficult due to their fragile nature. However, agonists or antagonists of those neuromediators can be easily formulated in order to control their release (e.g., botulinum toxin and capsacine), and prevent their degradation or modulate their synthesis [9]. Therefore, it is possible to determine the neuromediators effect on skin functions by modulating its levels, leading to **new approaches** toward skin disorder (including skin aging) and disease management.

1.2.3 Skin Changes with Aging

The skin suffers several **structural and functional modifications** with aging in both neuronal and non-neuronal structures.

In general, the **epidermis** becomes thinner due to a progressive decline on the renewal rate of epidermal cells. Furthermore, repair mechanisms slow down and the dermal-epidermal junction becomes smoother, leading to an increase of skin fragility and decrease of nutrient transfer between the dermis and epidermis layers. Additionally, corneocytes have a tendency to accumulate on the skin surface, creating a rough appearance and texture [3–5, 36, 37].

Normally, there is a reduction of melanocytes function except on more exposed regions, such as the face and hands, where there is a production of age spots correspondent to melanin accumulation. Therefore, older skin is more susceptible to sunburn and skin cancer owing to a thinner epidermis and deficiency of melanocytes. Simultaneously, the absence of melanin production (since melanocytes from the bulb are lost) leads to gray or white hair [3, 4, 37].

In the **dermis** layer, the quantity of collagen and elastic fibers is decreased due to a reduction in fibroblast percentage and its synthetic ability and an increased expression of enzymes able to degrade the collagen matrix. Collagen fibers also become denser. These changes combined with a loss of adipose subcutaneous tissue and lymphatic deregulation lead to skin sagging, loss of elasticity and wrinkling (with volume loss). Gravity and disappearance of substance of facial bones and cartilage as well as muscular contractions influence the wrinkle appearance and degree of expression. Additionally, there is a loss of fibers in sweat glands and Meissner corpuscles. The activity of both sweat and sebaceous glands is also lower. The progressive reduction of blood supply affects the deterioration of these appendages [3–5, 36–39].

The skin immune system is also affected by the aging process. Langerhans cells may decrease up to 50% in late adulthood, lowering the skin immune surveillance level, and leading to a higher risk of skin cancer and infections [37].

Besides the **growth factors**, skin aging is also affected by **hormones**. The activity of several hormones declines with time, mainly estrogen, testosterone, and dehydroepiandrosterone (DHEA), but also melatonin, cortisol, thyroxine, and growth hormone [3].

The modifications already referred are **intrinsic modifications** dependent on time and individuals genetic. In addition, numerous **extrinsic factors** can also affect and accelerate skin aging, such as ultraviolet radiation from sunlight, ionizing radiation, severe physical and physiological stress, pollution, severe weather, alcohol intake, overeating and tobacco smoking. For example, if the skin is usually exposed to sunlight, the normal elastic fibers can be replaced with an elastic-like material, the number of collagen fibers lowers, keratinocytes division is compromised, and the lymphatic vessels become also damaged [3, 4, 37, 38]. Photoaging degree can be classified according to Glogau wrinkle scale (Table 1.1).

Table 1.1 Glogau wrinkle scale [40–42]

Type	Age reference (Years)	Photoaging degree
I	20 – 30	Early
II	30 – 40	Early to moderate
III	50 – 60	Advanced
IV	60 and over	Severe

1.3 Rejuvenation Procedures

The rejuvenation procedures have as a major goal the **reversion** or **delay** the dermal and epidermal signs of aging. Therefore, collagen, elastin, and glycosaminoglycans are the main targets of these procedures, which can have a preventive or therapeutic approach [43, 44].

Skin aging prevention may be achieved by essentially using daily skin care products (cleaning, moisturizing, among others) and a **sunscreen** (besides the sun avoidance and the use of protective clothing).

Skin aging treatment can be divided into **invasive or non-invasive procedures. Topical application of anti-aging agents** is usually one of the most common non-invasive procedures. These agents are predominantly antioxidants and cell regulators. The use of **antioxidants** (vitamins, mainly C, B_3, and E, and botanical compounds such as flavonoids, carotenoids, and polyphenols) will reduce and neutralize free radicals, diminishing collagen destruction, and repairing oxidized membranes. On the other hand, the use of **cell regulators** (vitamin A and its derivatives/retinols, peptides and growth factors) will stimulate the production of collagen and elastic fibers and also act on collagen metabolism [37, 43, 44].

At least, **invasive procedures** are based on the removal of the damaged epidermis, as described in Table 1.2).

Table 1.2 Invasive rejuvenation procedures

Rejuvenation technique	Compounds/methods
Injectable	Hyaluronic acid
	Autologous fat
	Autologous platelet-rich plasma
	Botulinum toxin
Skin Resurfacing	Chemical peels
	Dermabrasion
	Ablative Lasers
	Ablative Fractional Lasers
	Non-ablative procedures

1.3.1 Injectable Techniques

Microinjections in the superficial dermis can contribute to **recover an ideal physiologic environment** by improving the synthesis of collagen, elastin and hyaluronic acid, cell activity, and skin hydration. These microinjections can contain one or more biocompatible active components [43, 45].

Dermal fillers are products that are injected into the skin in order to improve its physical features by soft tissue augmentation, one of the most common minimal invasive procedures of skin rejuvenation. These fillers can be categorized according to their source and permanence in the tissue. Regarding their origin, fillers can be **autologous**, i.e., from the person himself (e.g., fat tissue) or **heterologous**, such as collagen derived from human, porcine or bovine tissue cultures, animal or synthetic hyaluronic acid, synthetic or pseudo-synthetic implants, polymers, etc. According to the time that fillers remain in the tissue, they can be **temporary** (results visible only for a few months), **semi-permanent** (results visible for 1 to 2 years) or **permanent** (results visible for more than 2 years). Permanent fillers are usually non-biodegradable unlike non-permanent fillers, which are eliminated eventually through digestion or metabolism [43, 46–49]. The use of temporary fillers is more common in rejuvenation procedures since skin aging is a dynamic process and time adaptation is inevitable. These fillers are associated with a low incidence of secondary effects and complications. Nevertheless, the fillers can produce redness, inflammation, pain, bruising, edema, erythema, presence of visible material in the form of papules or nodules, tissue necrosis (a rare complication due to alterations on the blood flow), infections, hypersensitivity reactions, etc. [43, 46–48].

Hyaluronic acid (HA) is a glycosaminoglycan disaccharide composed of an alternating and repeating unit of D-glucuronic acid and *N*-acetyl-D-glucosamine. It is an important component of the interstitial matrix of the dermis layer and has become the "gold-standard" for skin rejuvenation through injectable technique. It is a space filler, lubricant, cell regulator (on proliferation and locomotion) and promotes the stabilization of the connective tissue. HA also increases skin hydration due to its hydrophilic ability, and activates fibroblasts, leading to skin augmentation/rejuvenation. As a temporary filler, its effect can last from 3 to 12 months, depending on the type of HA used [43, 45, 47, 49].

The distribution of subcutaneous fat between compartments becomes more evident with aging in the form of the loss of fat volume in facial skin. **Autologous fat** can be used as a safe and natural filler since no rejection or allergic reaction would be expected. Although this material can be easily obtained from thighs, abdomen, or buttocks, it requires an operation room to be extracted. The duration of its effect is unpredictable; it can last from months to several years [45, 47].

Autologous platelet-rich plasma (PRP) from the fresh whole blood contains a high concentration of platelets and various growth factors that can regulate several processes, such as cell migration, attachment, proliferation, and differentiation. It can promote collagen synthesis and stimulate fibroblasts activation, also leading to skin rejuvenation [43].

Botulinum toxin (BTX) is a neurotoxin produced by different strains of *Clostridium botulinum.* There are seven subtypes, neurotoxins A–G, from which only A, B, and F subtypes are available for clinical use, the A subtype being the most potent. Although BTX can slow down the skin aging process, it cannot discontinue this process. The BTX mechanism of action consists of blocking the presynaptic release of acetylcholine resulting in temporary chemical denervation at the neuromuscular junction and leading to striated muscular flaccid paralysis. This process occurs through different steps as follows: (a) The neurotoxin binds to a specific receptor on the presynaptic cholinergic neuron; (b) the toxin/receptor complex suffers internalization through endocytosis into nerve terminals, and (c) the formed vesicle is lysed, preventing the acetylcholine release from inside the cell. A significant wrinkle reduction will be obtained through this transitory and reversible paralysis state. Although BTX effects are temporary and localized, repeated injections may lead to a long-term effect. Nevertheless, the side effects are related with BTX's mechanism of action. Therefore, BTX is not indicated under conditions that may be exacerbated by the toxin (e.g., pre-existing neuromuscular disorders, psychiatric disorders, local infection, etc.). Complications are mild and may include pain, edema, erythema, ecchymosis, headache, and short-term hypoesthesia [43, 44, 50, 51].

The **ideal dermal filler** should present several characteristics, including biocompatibility; absence of immunogenicity; being not carcinogenic, infectious, teratogenic neither non-migratory; easy to obtain and store; inexpensive; removable (if required); with reproducible results, etc. [44, 49].

These techniques have been shown to be effective individually. Anyway, the effectiveness of combinations has been studied in order to obtain better and prolonged results with fewer side effects. The combination of two or more complementary techniques may be the answer to achieve an optimal, adaptable, and durable result for skin aging management without compromising the skin health or the patient comfort.

1.3.1.1 Skin resurfacing techniques: Chemical peelings and dermabrasion

There are several skin resurfacing methods, such as chemical peelings, dermabrasion, and laser treatment as follows:

Chemical peelings consist of a chemical ablation of the exposed tissue. Chemical exfoliating agents are applied to the skin destroying portions of the epidermis and/or dermis through epidermolysis, protein precipitation or tissue denaturation, which will lead to the activation of skin regeneration and repair mechanisms. Peelings can reach different skin depths, depending on the substance used, its concentration, pH of the formulation as well as the type and condition of the skin, mode and time of application, etc. Accordingly, they can reach the epidermal layers (**superficial peels**), the papillary dermis or upper reticular dermis (**medium-depth peels**) or even to the mid/lower reticular dermis (**deep peels**) [43, 44, 52, 53].

Some skin modifications may be observed as a uniform distribution of melanocytes and melanin grains, a homogeneous thickness of the basal membrane, a fresh sub-epidermal band of collagen and network of elastic fibers [43, 44, 54].

Although all skin phototypes (Fitzpatrick scale) respond to all peel types, phototypes I to III are less predisposed to scar or suffer pigment changes while phototypes IV and V have a greater risk of post-treatment dyspigmentation [44, 52].

Dermabrasion is the process of uniform mechanical abrasion of the skin, including the epidermis and upper papillary dermis layers, and consequently removes or reduces superficial wrinkles. It can be performed by using a serrated wheel, diamond embedded fraises, wire brush or sterilized sandpaper as a cutting tool, which is attached to a rotating handpiece electrically powered. However, this technique is highly dependent of the operator, requiring specialized skills and experience since any inaccuracy may result in major scarring. The healing time can be extended to one month. Once more, patients with darker skin may experience dyspigmentation. This technique tends to be replaced with or used simultaneously with resurfacing lasers [52, 55, 56].

1.3.1.2 Skin resurfacing techniques: Laser and light therapy

Light possesses particle and wave properties. Whereas wave properties comprise light behavior in space and large interfaces, such as skin and air, particle properties involve the tissue interactions on a molecular level. Lasers have a well-defined energy with the capability of ablating selected tissue while preserving the surrounding tissue. They may differ in wavelength, intensity, and duration of action, being the wavelength the responsible for the heat generation and, consequently, the tissue destruction. The extreme control of these parameters (mainly intensity and monochromaticity) permits a higher degree of precision compared with a non-laser source [52, 57].

The transformation of light or electrical energy into heat is the basic mechanism of photothermal and electrothermal skin rejuvenation procedures, respectively. The energy of a laser is proportional to its frequency and inversely proportional to its wavelength [57].

The light source produces a photon, which transfers its energy to a chromophore. The energy absorbed by the chromophore leads to an excited state, and to leave that state the chromophore may dissipate the energy as heat or fluorescence (reemission of light). It is possible to be selective toward the heating process by considering specific wavelengths and pulse duration as different chromophores may absorb at diverse wavelengths bands. The primary absorber is the water present in deeper skin layers. Water absorption peaks appear at 980, 1,480, 2,940, and 10,600 nm. The optical properties of the skin will determine the laser penetration, absorption, and internal dosimetry. Lasers act by selective photothermolysis, targeting water in the skin and stimulating collagen synthesis, and thus preventing skin aging [51, 55, 57, 58].

In fact, lasers have become quite popular for skin aging management. Ablative lasers were first used for this dermatologic approach. Then, fractional photothermolysis was introduced in order to obtain lesser side effects and lower risks [55, 58]. There was also an evolution from laser generation working at a continuous wavelength to recent pulsed lasers.

In particular, CO_2 and erbium:yttrium aluminum garnet (Erb:YAG) lasers, which operate at high wavelengths of 10,600 nm and 2,940 nm, respectively, are the main lasers used for skin rejuvenation. CO_2 laser is mainly absorbed by skin water, which favors its application for resurfacing techniques. However, Erb:YAG laser is around 10 times more selectively absorbed by water than CO_2 laser, leading to an extremely rapid heating with minimal damage to the surrounding tissue. In addition, it is more superficial and promotes the skin re-epithelialization earlier. These Erb:YAG lasers are extremely effective for all skin types, quite flexible, and produce from pure ablation craters to rapid superficial abrasion. Although several studies have compared these two laser types, no statistical differences were observed regarding the wrinkle reduction in most cases. Nevertheless, CO_2 laser seems to be more efficient for collagen synthesis, while Erb: YAG has milder effects. On the other hand, an improved outcome and decreased healing time has been shown with combined laser techniques [51, 52, 58].

Ablative lasers are hypothetically more effective than non-ablative lasers once the ablative technique removes the epidermis, causing a more extensive regeneration and prolonged effect, while the non-ablative technique keeps it intact. Consequently, dermal histological changes are more perceptible after the first procedure. On the other hand, the non-ablative approach presents lesser side effects [41, 58].

1.3.1.2.1 Ablative laser resurfacing

Ablative laser resurfacing consists of the controlled ablative removal of the superficial skin layers, based on the selective photothermolysis. An insignificant injury to the adjacent tissue can be observed, and the wound healing may occur from days to weeks. Whereas the laser penetration into tissue will depend on the selective absorption of skin water, the instantaneous tissue effects will depend on the laser potency, irradiation speed and treatment area. Possible thermal damage may occur due to an extended period of laser–tissue interaction. Nevertheless, the skin will be renewed by the re-epithelialization process, and damaged collagen, elastic fibers, and epithelial cells will be replaced with their normal homonyms [44]. There are two types of ablative laser, as already mentioned:

- **Carbon dioxide lasers (CO_2 resurfacing):** CO_2 lasers consist of unfractionated fully ablative laser emitting at 10,600 nm wavelength that deliver peak fluences above the ablation threshold of the cutaneous tissue and shorter tissue-dwell time. Although CO_2 lasers emitting a continuous wavelength were initially used, pulsed lasers are now more common. In fact, the first CO_2 laser generation (continuous wavelength lasers) emitted a quite high energy dispended for heating instead of ablating tissue. Therefore, this laser was limited by its risk of scarring, variable level of thermal damage, delayed healing, among other side effects. On the contrary, the last laser generation provides a precise control of tissue vaporization, hemostasis, and less residual thermal damage [41, 44]. CO_2 lasers ablate 20 to 60 μm of tissue per pass at the ablation threshold for cutaneous tissue (5 J/cm^2), resulting in a thermal damage zone up to 150 μm [41].
- **Erbium:yttrium-aluminum garnet (lasers Er:YAG)** were developed after CO_2 lasers, consisting of a laser emitting at 2,490 nm wavelength. These lasers can be used for more superficial resurfacing processes since they ablate 5 to 15 μm of skin tissue per pass at the ablation threshold resulting in a thermal damage zone smaller than 15 μm. This minimum thermal damage can be justified by their coefficient of water absorption, which is approximately 16 times higher than the CO_2 lasers. They also cause less pigmentation changes in people with higher skin phototypes (III and above) [41, 44].

Ablative lasers are exceptionally versatile tools and can lead to surprising results with just one treatment. However, their side effects may limit their use, such as erythema (which can be maintained for several months mainly after treatment with CO_2 laser), permanent hypopigmentation, transitory hyperpigmentation, prolonged healing time and post-operative period, possible infections (bacterial, viral, or even fungal), and scarring [41, 43, 44, 51–53, 58, 59].

This technique is primarily contraindicated in people with several clinical conditions, including abnormal wound healing (e.g., keloids); vascular diseases; abnormal decrease of adnexal structures of skin (as occurs after radiation therapy); deep peels or scars; active infections or immunosuppression; isomorphic diseases (e.g., vitiligo); uncontrolled hypertension; diabetes; and other significant medical conditions that may be compromised by the procedure or compromise the procedure itself [41].

The ideal laser skin resurfacing would provide skin vaporization with minimal side effects, relying considerably on both depth of ablation and energy fluency. Recent resurfacing methods allow the choice of different systems, and the treatment can be adjusted individually in order to improve its safety and efficacy [60].

1.3.1.2.2 Ablative fractional laser resurfacing

Fractional ablative lasers (CO_2 and erbium types) have been developed to diminish some side effects of fully ablative lasers. In this technique, only skin fractions are ablated, and the laser beams just damage or remove an array of microscopic columns of skin tissue leading to dermal controlled thermal damage. In addition, fractional ablative lasers deeply penetrate into the skin. This technique leads to the enhancement of elastin and collagen synthesis and to a shorter healing time since the healthy skin that surrounds the ablated zone will help in the healing process. Fractional ablative lasers differ from fully ablative forms by the pixelation of the laser beam [41, 44, 59].

Although the possible side effects may include those already mentioned for the fully ablative method (particularly if there is an excessive heating), their severity is much moderate. Additionally, immediate side effects may include burning sensation, discomfort, and redness, which disappear after a few days. On the other hand, this method is mainly contraindicated in people with a history of keloid scarring, immunosuppression, vitiligo, psoriasis, vasculitis, active infections, prior radiation, or recent oral retinoid treatment [41, 44, 59].

1.3.1.2.3 Non-ablative procedures

Non-ablative resurfacing methods have been developed in order to overcome the morbidity and complications (mainly permanent hypopigmentation and scarring) associated to ablative techniques. These methods employ visible, near-infrared, and mid-infrared wavelengths and a skin cooling system to protect it. In general, less promising results may be obtained compared with ablative approaches. Nevertheless, the risk/benefit ratio between the different techniques should be taken into account for each case [41, 42, 44, 61]. Their **mechanism of action** is based on non-ablative fractional photothermolysis, in which the dermis is inflamed and thermally denatured in order to stimulate the healing process also improved by the surrounding healthy skin. During this process, collagen is synthesized and melanin is released leading to a pigmentary skin redistribution [42, 51, 61].

In contrast to fully ablative resurfacing methods in which the thermal damage is homogeneous at a specific depth, in this case (non-ablative fractional photothermolysis and fractional ablative photothermolysis) the thermal damage only occurs in microscopic skin columns called **microthermal zones** (MTZs). The extension and size of these columns are defined by the energy used (i.e., higher energy will provide wider and deeper columns). The MTZs are encircled with healthy tissue that will help the healing process, as already mentioned. This process is safer and quicker than the ablative skin resurfacing method. It is used to improve skin texture and dyspigmentation, being able to target both epidermis and dermis layers. Even though the results can be pronounced, it involves multiple treatments to achieve them. Side effects are usually mild and temporary, mainly mild redness, swelling, blistering, herpes activation, acne flare up, etc. Pigmentary changes might be observed in darker skin phototypes when compared with treatment with other wavelengths [44, 59, 61, 62].

Non-ablative **infrared lasers** were developed to rejuvenate the skin with higher safety. Mid-infrared laser heating can target dermal collagen by dermal water heating and stimulation of collagen synthesis concomitant with epidermal protection (cooling). Repigmentation is linked to melanocytes migration and proliferation, sideways with the release of cytokines and inflammatory mediators. Although this process has fewer and less severe side effects, they can still be present, mainly scarring and temporary hyperpigmentation due to an aggressive cooling [41, 44, 61].

Non-ablative fractional lasers also comprise **erbium** (1,410; 1,440; 1,540 and 1,550 nm), **neodymium-doped** (1,320), **diode laser** (1,450), and **thulium** (1,927 nm) [42, 61].

Pulsed dye lasers (PDL) which use oxyhemoglobin as the principal chromophore and intense pulsed light (IPL) are other non-ablative rejuvenation procedures based on selective photothermolysis. Additionally, less aggressive histological changes are observed after treatment with different IPL devices. The flash lamps of these devices produce a broad spectrum of light heating and denaturating the target tissue and, consequently, accelerating the epidermal turnover. Thus, new collagen in the papillary and reticular dermis is observed as well as an increase of the fibroblasts number. Although IPL absorption spectrum of skin chromophores is not monochromatic, a filter addition allows the adaption to specific applications. Nevertheless, the effectiveness is weaker compared with ablative results [42, 43, 57, 63].

In summary, the effects of different laser resurfacing methods are shown in Fig. 1.1.

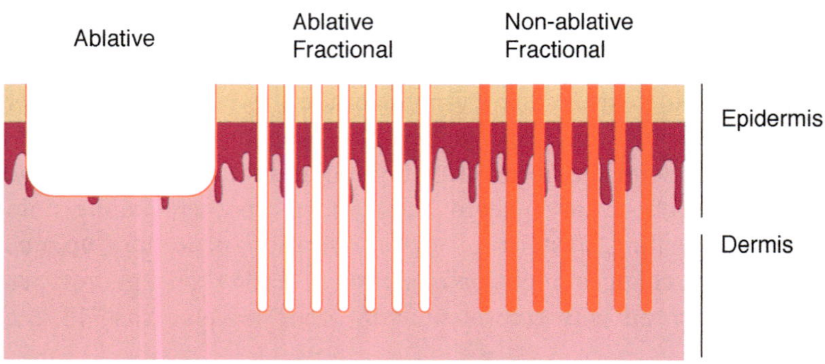

Figure 1.1 Effect of different laser resurfacing methods.

1.3.2 Cosmeceuticals

Cosmetic ingredients were once considered inert. Cosmeceuticals are topical products on the borderline between cosmetics and drugs, presenting a cosmetic effect associated to a physiological mechanism. The search for better and effective skin products with real bioactivity is increasing since consumers start getting interested in understanding the technology and mechanisms behind these products design. Therefore, there is a good acceptance to cosmeceuticals. It is important to use not only rejuvenation methods but also cosmeceuticals to delay and prevent skin aging.

Their application will result in significant reduction of lines and wrinkles, redness, and discoloration as well as an improved skin texture [64–66].

This new field is growing and new possibilities are being explored. Cosmeceuticals may have different active constituents as follows:

- **Vitamin A and derivatives** (retinoids): They increase skin hydration, collagen levels (due to increased synthesis and reduced metabolism), and epidermal proliferation and differentiation (by normalizing keratinization). Additionally, they also act as antioxidants and participate in immune response regulation. As a result, the skin becomes thicker leading to diminished wrinkle fine lines. Deep wrinkles require longer treatment periods (weeks to months) to obtain visible results [65, 67–69].
- **Vitamin B_3**: It is a precursor of endogenous enzymes cofactors, nicotinamide adenine dinucleotide (NAD) and NAD phosphate (NADP), participating in several reactions. It is associated with several effects, including skin color, anti-inflammatory activity, increase of collagen synthesis, barrier function improvement due to the stimulation of ceramide synthesis and an increase of the stratum corneum thickness. Photoprotection associated to antioxidant ability is also suggested for this vitamin [65, 67–69].
- **Provitamin B_5:** It is a precursor of vitamin B_5 and its effects are linked to that vitamin's role. It improves the skin barrier, hydration (humectant ability) and wound healing, leading to an enhanced skin elasticity and appearance [65, 67].
- **Vitamin C and derivatives:** They inhibit tyrosinase and show antioxidant activity against radical oxygen species (ROS), modulating the UV radiation-induced damage. These derivatives have been used in skin lightening and anti-aging products since they also play an essential role in collagen and elastin synthesis [65, 67–69].
- **Vitamin E and derivatives**: These are lipophilic antioxidants that can prevent skin damage by reducing lipid peroxidation caused by ROS. Furthermore, they reduce UV- induced edema and erythema, increase skin hydration, control collagen breakdown, and diminish wrinkles and tumor formation [65, 67–69].

- **Vitamin K**: It prevents or even treats some vascular alterations caused by skin aging [67, 69].
- **Vitamin D**: Although it has not been studied in the context of skin aging yet, its mechanism of promoting epidermal differentiation suggests that this molecule could be a possible active component of skin products for some age-related abnormalities [70].
- **Hydroxy acids:** They exfoliate the skin improving its texture and color distribution. Thus, the skin becomes more homogenous and hydrated. For example, glycolic acid (α-hydroxy acid) is able to improve skin texture, fine wrinkling, and hyperpigmentation [65, 70].
- **Sugar amines**: Examples are hexose amines glucosamine and *N*-acetyl glucosamine (NAG), which are precursors of hyaluronic acid. They can reduce hyperpigmentation, increase hydration levels, and exfoliate the skin, leading to the improvement of fine wrinkles [65].
- **Ceramide**s: These are the lipids present in the stratum corneum that decrease with age. Thus, skin supplementation with these lipids will lead to an improvement of the barrier function [65].
- **Metals** (magnesium, zinc, copper, selenium, etc.): These are cofactors of several enzymes. Their mechanism to prevent/treat skin aging is based on antioxidant properties and collagen synthesis [65, 70].
- **Peptides:** A few examples are acetyl-glutamate-glutamate-methionine-glutamine-arginine-arginine, which inhibits the release of neuromediators, having a botulinum toxin-like effect; palmitoyl-lysine-threoninethreonine-lysine-serine and tripeptide copper glycine-histidine-lysine, which are fragments of human dermal collagen stimulating the collagen production and synthesis of dermal matrix components, respectively [65, 71].

Vitamins can be suggested for skin care products to prevent and treat skin aging. The effects of vitamins A, B_3, C, and E have more scientific support in this field [69]. However, vitamin formulations can present some technological drawbacks regarding their chemical instability and possible inactivation. Enzymes formulations can also present some problems and short-lived catalytic activity in the skin. Alternatives are being studied to improve or contour those drawbacks, such the use of nanocarriers (liposomes, lipid and polymeric nanoparticles, cyclodextrins, etc). In addition, extremozymes obtained from *Thermus thermophilus* are able to stand extreme environmental conditions and have shown detoxification properties and UVR and heat resistance. Additionally, their ability to protect the skin from ROS, peroxidation, and UVA-induced damage has been revealed. Curiously, this activity increases proportionally under hot climate or solar exposure [64].

Other several active ingredients have antioxidant activity and are able to regulate epidermal protection from solar radiation, such as astaxanthin, carotenoids, cynaropicrin, fucoxanthin, myricetin, resveratrol from botanical extracts, etc. [72]. Many of these substances are incorporated in sunscreens to protect and treat the skin against UVR damage [73, 74].

1.3.3 Procedures Comparison

Skin rejuvenation treatments must be adapted to each patient considering the skin type, wrinkle degree, expectations, side effects, etc. However, the **combination** of those treatments could be necessary leading to the inevitable comparison between them.

In general, the main goal is to protect the skin remaining its functions and avoiding further aging-induced damage (Fig. 1.2). **Dermabrasion**, a superficial skin resurfacing method, is suitable to treat superficial wrinkles. **Cosmeceuticals** can be used at any degree of wrinkle even though results may not be visible for deep wrinkles since they require an amount of functioning cells that may not be present. On the other hand, **dermal fillers** are usually more useful for more visible wrinkles. **Chemical peels** and **lasers** therapies are extremely flexible and adaptable to each situation, since may also provide a simple superficial resurfacing or a deep aggressive treatment depending on the characteristics and mechanisms of each procedure (Table 1.3).

Although techniques that are effective for more advanced wrinkle degrees can be also effective for less severe cases, milder methods developed for more early stages of aging may not have a visible effect for deeper wrinkles.

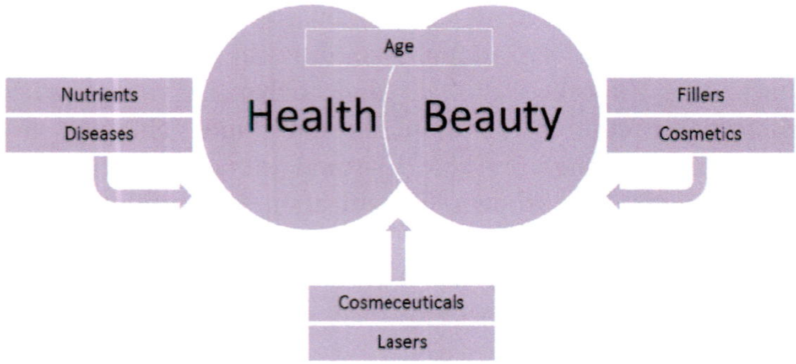

Figure 1.2 Health and beauty interaction.

Table 1.3 Most suitable techniques with respect to the degree of the wrinkle

<table>
<tr><th>Wrinkle Degree</th><th colspan="6">Most Suitable Techniques</th></tr>
<tr><td>Superficial</td><td rowspan="3">Skin protection & aging prevention</td><td>Dermabrasion</td><td rowspan="3">Cosmeceuticals</td><td></td><td rowspan="3">Chemical peelings</td><td rowspan="3">Lasers</td></tr>
<tr><td>Medium</td><td rowspan="2"></td><td rowspan="2">Fillers</td></tr>
<tr><td>Deep</td></tr>
</table>

There is no literature until this date regarding a standardized comparison between all types of procedures mentioned along this work. In general, the studies report the absence or presence of the effects on wrinkle reduction or a relation of superiority or inferiority between treatments However, the outcome extrapolation is not particularly accurate since those studies are performed at different conditions and including different populations. Nevertheless, some studies compare similar or more distinct approaches regarding their efficacy.

For example, Ooe *et al.* studied the comparison between four methods with different levels of invasiveness [63]: fractional erbium:YAG laser (2,940 nm), IPL, tretinoin (carboxylic acid form of vitamin (A)), and a nutraceutical formulation (amino acids: arginine, leucine and glutamine combined with 11 vitamins). Significant improvement in wrinkles evaluated by the Lemperle wrinkle scale was observed after IPL, tretinoin, and nutritional therapy. In addition, WAF (wrinkle area fraction) was also measured showing significant effects on wrinkle improvement for IPL and tretinoin treatment. The change of skin quality was also controlled against a ratio (immediate recovery to total deformation), and the outcomes showed a significant improvement with fractional erbium:YAG laser, IPL and tretinoin. Furthermore, the satisfaction level was equivalent for all treatments. Regarding the downtime, only erbium:YAG laser and IPL required 3 days. Finally, these authors concluded that there was no correlation between the depth of skin penetration and the wrinkle improvement. Accordingly, minimally invasive procedures were effective and should be preferred first. In addition, they also suggested the combination of different types of non-invasive procedures, such as tretinoin and the amino acid supplement [63].

Regarding the different laser treatments mentioned in this review, there is no statistically relevant data yet proving a superior effectiveness of a particular type of full ablative lasers. Nevertheless, there is sufficient data to assume that ablative approaches are more effective than non-ablative ones, as concluded from a study comparing three different non-ablative lasers (1,320 nm Nd:YAG, 1,450 nm diode and 1,540 nm erbium:glass), and CO_2 and erbium:YAG ablative lasers. Although non-ablative lasers presented good results improving the skin texture, tone, and elasticity, these effects were not still comparable to those of ablative methods. In addition, it was also concluded that none of the non-ablative methods showed superiority over the others [75]. Additionally, a study comparing two ablative fractional lasers (CO_2 and erbium:YAG) for peri-orbital rhytides suggested that both had similar results as well as post-operative healing time [76]. Furthermore, another study comparing an ablative (CO_2 laser) with a non-ablative method (IPL) for perioral rhytides presented significantly better results for CO_2 laser but with more side effects. Nevertheless, patients were more satisfied with the ablative method outcomes [77]. It should be noted that these results obtained for specific areas may need to be confirmed regarding other skin areas. El-Domyati *et al.* compared the fractional and ablative erbium:YAG lasers, observing that both laser types had similar effects on the skin (increase of epidermal thickness and dermal collagen, etc.) even though more marked effects were obtained with the ablative method. Indeed, the outcomes from multiple treatments of the fractional method were comparable to one session of the ablative method [78].

Regarding dermal fillers, Baumann *et al.* studied the comparison between HA smooth-gel dermal fillers with injectable bovine collagen for the nasolabial fold. The results revealed that the improvement on wrinkles provided by HA gels had a longer lasting effect than bovine collagen. Additionally, HA gels were also preferred by patients [79].

In general, the combination of procedures is already showing promising results, such as diode laser/radiofrequency, CO_2 laser/dermabrasion, CO_2 laser/autologous platelet-rich plasma, autologous platelet-rich plasma/fractional laser therapy, and pulsed light/non-ablative fractional laser [62, 80–83].

As a suggestion, a future comparable study between the best different methods for each wrinkle level could be designed (e.g., the best non-ablative laser for superficial wrinkles vs. the best chemical peeling). Further investigation could be also based on the combined treatment of non-ablative fractional laser for a primary approach with immediate results, followed by a neuropeptide-based cosmeceutical as a maintenance routine in order to maintain stable levels of beneficial neuropeptides in the skin. In fact, lasers are able to trigger neuropeptide production [84].

1.4 Conclusions and Future Perspectives

Time and environmental aggressions result in visible alterations on the skin surface, leading to intrinsic and extrinsic aging, respectively. Nevertheless, people are becoming more interested about their appearance and looking for more effective anti-aging procedures in order to prevent and/or delay this process. In fact, there are already many available treatments and a growing list of new approaches to contour skin aging. However, very high or unreal expectative should be avoided considering that it is still not possible to revert this natural phenomena.

A good looking, healthy, and functional skin is the result of the interaction of many systems (endocrine, nervous, and immune) leading to a perfect balance. Once this balance is disturbed, it must be restored even considering the adaptive response. In general, the main goal is to improve the skin health and appearance.

Regarding the existing treatments, non-ablative fractional lasers will probably gain more ground on this business. However, more aggressive techniques will never cease to exist owing to the requirements of some patients and certain situations that demand those types of treatment. The ideal anti-aging treatment is a personalized treatment in which patient expectations are considered.

New approaches to target the skin nervous system, particularly neuropeptides, are also being explored. Cosmeceuticals are an important evolution of cosmetics with active ingredients that can actually protect, restore, and rejuvenate skin. Cosmeceuticals containing active molecules capable of modulating neuropeptides in the skin will revolutionize daily-use anti-aging creams.

In summary, although the existent procedures used for skin rejuvenation are effective and considerably safe, improvement and innovation strongly supported by knowledge, alongside with sustainable and comparative results, are the future issues in this field. The search for safer and more effective procedures should never stop. In addition, personalized and combined techniques with different action mechanisms seem to be the key for a brighter future in skin rejuvenation context.

List of abbreviations

ACTH: Adrenocorticotropic hormone
AR: Androgen receptors
BTX: Botulinum toxin
cAMP: Cyclic adenosine monophosphate
CGRP: Calcitonin gene-related peptide
CGRP-R: Calcitonin gene-related peptide receptors
CRH: Corticotropin-releasing hormone
CR: Calretinin
DHEA: Dehydroepiandrosterone
DNA: Deoxyribonucleic Acid
DOPA: 3,4-dihydroxyphenylalanine
ER: Estrogen receptors
Erb:YAG: Erbium: yttrium-aluminum garnet
GFR: Growth hormone releasing factor
GH: Growth hormone
GHR: Growth hormone receptor
GR: Glucocorticoid receptors
GRP: Gastrin-releasing peptide
GRPR: Gastrin-releasing peptide receptor

HA: Hyaluronic acid
IPL: Intense pulsed light
IV: Intravenous
KGF: keratinocyte growth factor
LH/CG-R: Luteinizing hormone/choriogonadotropin receptors
MCR: Melanocortin receptors
MR: Mineralocorticoid receptors
mRNA: Messenger ribonucleic acid
MSH: Melanocyte-stimulating hormones
MTZs: Microthermal zones
NAD: Nicotinamide adenine dinucleotide
NADP: Nicotinamide adenine dinucleotide phosphate
NAG: *N*-acetyl glucosamine
NGF: Nerve growth factor
NK: Neurokinin
NKA: Neurokinin A
NKB: Neurokinin B
NKR: Neurokinin receptors
NPY: Neuropeptide Y
NT: Neurotrophin
PACAP: Pituitary adenylate cyclase-activating polypeptide
PACAP/VIP-R or PVR: Pituitary adenylate cyclase-activating polypeptide and vasoactive intestinal peptide receptors
PDL: Pulsed dye lasers
PHM: Peptide histidine-methionine
PHI: Peptide histidine isoleucine
POMC: Pro-opiomelanocortin
PRL: Prolactin
PRP: Autologous Platelet-Rich Plasma
PTH: Parathyroid hormone
PTHrP: Parathyroid hormone related protein
ROS: Radical oxygen species
SC: Stratum corneum
SP: Substance P
T3: Triiodothyronine
THR: Thyroid hormone receptors
Trk: Tyrosine kinase
UV: Ultraviolet
UVA: Ultraviolet A

UVB: Ultraviolet B
UVR: Ultraviolet radiation
VDR: Vitamin D receptor
VIP: Vasoactive intestinal peptide

References

1. Rawlings A, Harding C. Moisturization and skin barrier function. *Dermatologic Therapy*. 2004; 17: 43–48.
2. Boelsma E, Hendriks HF, Roza L. Nutritional skin care: Health effects of micronutrients and fatty acids. *The American Journal of Clinical Nutrition*. 2001; 73: 853–864.
3. Puizina-Ivi N. Skin aging. *Acta Dermatoven APA*. 2008; 17: 47.
4. VanPutte C, Seeley R. *Seeley's Anatomy & Physiology*. New York, NY: McGraw-Hill; 2014.
5. Junqueira LCU, Mescher AL. *Junqueira's Basic Histology: Text and Atlas*. 13 ed: McGraw-Hill Medical; 2013.
6. Ross MH, Pawlina W. *Histology: A Text and Atlas: With Correlated Cell and Molecular Biology*. Philadelphia: Wolters Kluwer/Lippincott Williams & Wilkins Health; 2011.
7. Slominski A, Wortsman J. Neuroendocrinology of the Skin 1. *Endocrine Reviews*. 2000; 21: 457–487.
8. Oaklander AL, Siegel SM. Cutaneous innervation: Form and function. *Journal of the American Academy of Dermatology*. 2005; 53: 1027–1037.
9. Misery L. Les nerfs à fleur de peau. *International Journal of Cosmetic Science*. 2002; 24: 111–116.
10. Roosterman D, Goerge T, Schneider SW, Bunnett NW, Steinhoff M. Neuronal control of skin function: The skin as a neuroimmunoendocrine organ. *Physiological Reviews*. 2006; 86: 1309–1379.
11. Donadio V, Nolano M, Provitera V, Stancanelli A, Lullo F, Liguori R, *et al.* Skin sympathetic adrenergic innervation: An immunofluorescence confocal study. *Annals of Neurology*. 2006; 59: 376–381.
12. Kurzen H, Wessler I, Kirkpatrick C, Kawashima K, Grando S. The non-neuronal cholinergic system of human skin. *Hormone and Metabolic Research = Hormon-und Stoffwechselforschung = Hormones et Metabolisme*. 2007; 39: 125–135.

13. Slominski A, Wortsman J, Pisarchik A, Zbytek B, Linton EA, Mazurkiewicz JE, *et al.* Cutaneous expression of corticotropin-releasing hormone (CRH), urocortin, and CRH receptors. *The FASEB Journal.* 2001; 15: 1678–1693.

14. Oakes S, Haynes K, Waters M, Herington A, Werther G. Demonstration and localization of growth hormone receptor in human skin and skin fibroblasts. *The Journal of Clinical Endocrinology & Metabolism.* 1992; 75: 1368–1373.

15. Crissey SD, Ange KD, Jacobsen KL, Slifka KA, Bowen PE, Stacewicz-Sapuntzakis M, *et al.* Serum concentrations of lipids, vitamin D metabolites, retinol, retinyl esters, tocopherols and selected carotenoids in twelve captive wild felid species at four zoos. *The Journal of Nutrition.* 2003; 133: 160–166.

16. Leiferman KM, Schroeter A, Kirschner MK, Spelsberg TC. Characterization of the glucocorticoid receptor in human skin. *Journal of Investigative Dermatology.* 1983; 81: 355–360.

17. Schulze E, Witt M, Fink T, Hofer A, Funk R. Immunohistochemical detection of human skin nerve fibers. *Acta Histochemica.* 1997; 99: 301–309.

18. Elewa R, Makrantonaki E, Zouboulis CC. Neuropeptides and skin aging. *Hormone Molecular Biology and Clinical Investigation.* 2013; 16: 29–33.

19. Chang Y-C, Lin W-M, Hsieh S-T. Effects of aging on human skin innervation. *Neuroreport.* 2004; 15: 149–153.

20. Legat FJ, Wolf P. Photodamage to the cutaneous sensory nerves: Role in photoaging and carcinogenesis of the skin? *Photochemical & Photobiological Sciences.* 2006; 5: 170–176.

21. Makrantonaki E, Schönknecht P, Hossini AM, Kaiser E, Katsouli M-M, Adjaye J, *et al.* Skin and brain age together: The role of hormones in the ageing process. *Experimental Gerontology.* 2010; 45: 801–813.

22. Pain S, Altobelli C, Boher A, Cittadini L, Favre-Mercuret M, Gaillard C, *et al.* Surface rejuvenating effect of Achillea millefolium extract. *International Journal of Cosmetic Science.* 2011; 33: 535–542.

23. Schiller M, Brzoska T, Böhm M, Metze D, Scholzen TE, Rougier A, *et al.* Solar-simulated ultraviolet radiation-induced upregulation of the melanocortin-1 receptor, proopiomelanocortin, and α-melanocyte-stimulating hormone in human epidermis *in vivo.* *Journal of Investigative Dermatology.* 2004; 122: 468–476.

24. Mijouin L, Hillion M, Ramdani Y, Jaouen T, Duclairoir-Poc C, Follet-Gueye M-L, *et al.* Effects of a skin neuropeptide (substance p) on cutaneous microflora. *PLoS One.* 2013; 8: e78773.

25. Kant V, Gopal A, Kumar D, Bag S, Kurade NP, Kumar A, *et al.* Topically applied substance P enhanced healing of open excision wound in rats. *European Journal of Pharmacology.* 2013; 715: 345–353.

26. Chen Y, Lyga J. Brain-Skin Connection: Stress, Inflammation and skin aging. *Inflammation & Allergy Drug Targets.* 2014; 13: 177.

27. Hara M, Toyoda M, Yaar M, Bhawan J, Avila EM, Penner IR, *et al.* Innervation of melanocytes in human skin. *The Journal of Experimental Medicine.* 1996; 184: 1385–1395.

28. Kletsas D, Caselgrandi E, Barbieri D, Stathakos D, Franceschi C, Ottaviani E. Neutral endopeptidase-24.11 (NEP) activity in human fibroblasts during development and ageing. *Mechanisms of Ageing and Development.* 1998; 102: 15–23.

29. Robert L, Labat-Robert J, Robert AM. Physiology of skin aging. *Pathologie Biologie.* 2009; 57: 336–341.

30. Toyoda M, Nakamura M, Nakada K, Nakagawa H, Morohashi M. Characteristic alterations of cutaneous neurogenic factors in photoaged skin. *British Journal of Dermatology.* 2005; 153: 13–22.

31. Shipton EA. Skin matters: Identifying pain mechanisms and predicting treatment outcomes. *Neurology Research International.* 2013; 2013: 13.

32. Cheret J, Lebonvallet N, Buhe V, Carre J, Misery L, Le Gall-Ianotto C. Influence of sensory neuropeptides on human cutaneous wound healing process. *Journal of Dermatological Science.* 2014; 74: 193–203.

33. Khalil Z, Helme R. Sensory peptides as neuromodulators of wound healing in aged rats. *The Journals of Gerontology Series A: Biological Sciences and Medical Sciences.* 1996; 51: B354–B61.

34. Lotti T, Hautmann G, Panconesi E. Neuropeptides in skin. *Journal of the American Academy of Dermatology.* 1995; 33: 482–496.

35. Costa A, Eberlin S, Polettini AJ, Pereira AFdC, Pereira CS, Ferreira NMC, *et al.* Neuromodulatory and anti-inflammatory ingredient for sensitive skin: *In vitro* assessment. *Inflammation & Allergy-Drug Targets (Formerly Current Drug Targets-Inflammation & Allergy).* 2014; 13: 191–198.

36. Lauria G, Holland N, Hauer P, Cornblath DR, Griffin JW, McArthur JC. Epidermal innervation: Changes with aging, topographic location, and in sensory neuropathy. *Journal of the Neurological Sciences.* 1999; 164: 172–178.

37. Mccullough JL, Kelly KM. Prevention and treatment of skin aging. *Annals of the New York Academy of Sciences*. 2006; 1067: 323–331.

38. Pessa JE, Nguyen H, John GB, Scherer PE. The Anatomical Basis for Wrinkles. *Aesthetic Surgery Journal*. 2014; 34: 227–234.

39. Jenkins G. Molecular mechanisms of skin ageing. *Mechanisms of Ageing and Development*. 2002; 123: 801–810.

40. Durai PC, Thappa DM, Kumari R, Malathi M. Aging in elderly: Chronological versus photoaging. *Indian Journal of Dermatology*. 2012; 57: 343–352.

41. Ibrahimi OA, Fitzpatrick RE, Goldman MP, Kilmer SL. *Skin Resurfacing with Ablative Lasers. Lasers and Energy Devices for the Skin*: CRC Press; 2013. pp. 110–161.

42. Carniol P, Sadick NS. *Clinical Procedures in laser Skin Rejuvenation*: CRC Press; 2007.

43. Ganceviciene R, Liakou AI, Theodoridis A, Makrantonaki E, Zouboulis CC. Skin anti-aging strategies. *Dermato-Endocrinology*. 2012; 4: 308–319.

44. Avram M. *The Color Atlas of Cosmetic Dermatology*. 2nd ed: McGraw-Hill Professional; 2011.

45. Athre RS. Facial filler agents. *Operative Techniques in Otolaryngology-Head and Neck Surgery*. 2007; 18: 243–247.

46. Wollina U, Goldman A. Dermal fillers: Facts and controversies. *Clinics in Dermatology*. 2013; 31: 731–736.

47. Sánchez-Carpintero I, Candelas D, Ruiz-Rodríguez R. Dermal fillers: Types, indications, and complications. *Actas Dermo-Sifiliográficas (English Edition)*. 2010; 101: 381–393.

48. Alijotas-Reig J, Fernández-Figueras MT, Puig L. Inflammatory, immune-mediated adverse reactions related to soft tissue dermal fillers. *Seminars in Arthritis and Rheumatism*: Elsevier; 2013; 43: 241–258.

49. Buck DW, Alam M, Kim JY. Injectable fillers for facial rejuvenation: A review. *Journal of Plastic, Reconstructive & Aesthetic Surgery*. 2009; 62: 11–18.

50. Nanda S, Bansal S. Upper face rejuvenation using botulinum toxin and hyaluronic acid fillers. *Indian Journal of Dermatology, Venereology, and Leprology*. 2013; 79: 32.

51. Gilchrest BA, Krutmann J. *Skin Aging*: Springer; 2006.

52. Friedman S, Lippitz J. Chemical peels, dermabrasion, and laser therapy. *Disease-a-Month*. 2009; 55: 223–235.

53. Hassan KM, Benedetto AV. Facial skin rejuvenation: Ablative laser resurfacing, chemical peels, or photodynamic therapy? *Facts and Controversies. Clinics in Dermatology.* 2013; 31: 737–740.

54. Landau M. Chemical peels. *Clinics in Dermatology.* 2008; 26: 200–208.

55. Hession MT, Graber EM. Atrophic acne scarring: A review of treatment options. *The Journal of Clinical and Aesthetic Dermatology.* 2015; 8: 50.

56. Zisser M, Kaplan B, Moy RL. Surgical pearl: Manual dermabrasion. *Journal of the American Academy of Dermatology.* 1995; 33: 105–106.

57. Ross EV, Anderson RR. Laser–tissue interactions. *Lasers and Energy Devices for the Skin*: CRC Press; 2013. pp. 1–30.

58. Borges J, Cuzzi T, Mandarim-de-Lacerda CA, Manela-Azulay M. Fractional Erbium laser in the treatment of photoaging: Randomized comparative, clinical and histopathological study of ablative (2940 nm) vs. non-ablative (1540 nm) methods after 3 months. *Anais Brasileiros de Dermatologia.* 2014; 89: 250–258.

59. Shai A, Maibach HI, Baran R. *Handbook of Cosmetic Skin Care*: Martin Dunitz, London; 2001.

60. Papadavid E, Katsambas A. Lasers for facial rejuvenation: A review. *International Journal of Dermatology.* 2003; 42: 480–487.

61. Beasley KL. Nonablative fractional lasers. *Lasers and Energy Devices for the Skin*, CRC Press; 2013. pp. 178–191.

62. Chan CS, Saedi N, Mickle C, Dover JS. Combined treatment for facial rejuvenation using an optimized pulsed light source followed by a fractional non-ablative laser. *Lasers in Surgery and Medicine.* 2013; 45: 405–409.

63. Ooe M, Seki T, Miura T, Takada A. Comparative evaluation of wrinkle treatments. *Aesthetic Plastic Surgery.* 2013; 37: 424–433.

64. Lintner K, Mas-Chamberlin C, Mondon P, Peschard O, Lamy L. Cosmeceuticals and active ingredients. *Clinics in Dermatology.* 2009; 27: 461–468.

65. Bissett DL. Common cosmeceuticals. *Clinics in Dermatology.* 2009; 27: 435–445.

66. Zhang L, Falla TJ. Cosmeceuticals and peptides. *Clinics in Dermatology.* 2009; 27: 485–494.

67. Lupo MP. Antioxidants and vitamins in cosmetics. *Clinics in Dermatology.* 2001; 19: 467–473.

68. Manela-Azulay M, Bagatin E. Cosmeceuticals vitamins. *Clinics in Dermatology.* 2009; 27: 469–474.

69. Zussman J, Ahdout J, Kim J. Vitamins and photoaging: Do scientific data support their use? *Journal of the American Academy of Dermatology*. 2010; 63: 507–525.

70. Cunningham WJ. Cosmeceuticals in Photoaging. Cosmeceuticals and Active Cosmetics: Drugs vs Cosmetics. 2005: 261.

71. Lupo MP. Cosmeceutical peptides. *Dermatologic Surgery*. 2005; 31: 832–836.

72. Lorencini M, Brohem CA, Dieamant GC, Zanchin NI, Maibach HI. Active ingredients against human epidermal aging. *Ageing Research Reviews*. 2014; 15: 100–115.

73. Lintner K. *Antiaging Actives in Sunscreens. Sunscreens: Regulations and Commercial Development*, 3rd ed., Marcel Dekker, New York. 2005: 673–695.

74. Mas-Chamberlin C, Mondon P, Lamy F, Peschard O, Lintner K. Strategies of antiaging actives in sunscreen products. *Cosmetics and Toiletries*. 2006; 121: 45.

75. Alster TS, Lupton JR. Are all infrared lasers equally effective in skin rejuvenation. *Seminars in cutaneous medicine and surgery*: No longer published by Elsevier; 2002. p. 274–279.

76. Karsai S, Czarnecka A, Jünger M, Raulin C. Ablative fractional lasers (CO_2 and Er: YAG): A randomized controlled double-blind split-face trial of the treatment of peri-orbital rhytides. *Lasers in Surgery and Medicine*. 2010; 42: 160–167.

77. Hedelund L, Bjerring P, Egekvist H, Haedersdal M. Ablative versus non-ablative treatment of perioral rhytides. A randomized controlled trial with long-term blinded clinical evaluations and non-invasive measurements. *Lasers in Surgery and Medicine*. 2006; 38: 129–136.

78. El-Domyati M, Abd-El-Raheem T, Abdel-Wahab H, Medhat W, Hosam W, El-Fakahany H, *et al.* Fractional versus ablative erbium: Yttrium-aluminum-garnet laser resurfacing for facial rejuvenation: An objective evaluation. *Journal of the American Academy of Dermatology*. 2013; 68: 103–112.

79. Baumann LS, Shamban AT, Lupo MP, Monheit GD, Thomas JA, Murphy DK, *et al.* Comparison of smooth-gel hyaluronic acid dermal fillers with cross-linked bovine collagen: A multicenter, double-masked, randomized, within-subject study. *Dermatologic Surgery*. 2007; 33: S128–S35.

80. Sadick NS, Trelles MA. Nonablative wrinkle treatment of the face and neck using a combined diode laser and radiofrequency technology. *Dermatologic Surgery*. 2005; 31: 1695–1699.

81. Fezza JP. Laserbrasion: The combination of carbon dioxide laser and dermasanding. *Plastic and Reconstructive Surgery*. 2006; 118: 1217–1221.

82. Na J-I, Choi J-W, Choi H-R, Jeong J-B, Park K-C, Youn S-W, *et al.* Rapid healing and reduced erythema after ablative fractional carbon dioxide laser resurfacing combined with the application of autologous platelet-rich plasma. *Dermatologic Surgery*. 2011; 37: 463–468.

83. Shin MK, Lee JH, Lee SJ, Kim NI. Platelet-rich plasma combined with fractional laser therapy for skin rejuvenation. *Dermatologic Surgery*. 2012; 38: 623–630.

84. Hochman B, Pinfildi CE, Nishioka MA, Furtado F, Bonatti S, Monteiro PK, *et al.* Low-level laser therapy and light-emitting diode effects in the secretion of neuropeptides SP and CGRP in rat skin. *Lasers in Medical Science*. 2014; 29: 1203–1208.

Chapter 2

Melatonin: A "Guardian" of the Genome and Cellular Integrity for Prevention of Photocarcinogenesis

Patrícia Manteigas[a] **and Andreia Ascenso**[b]

[a]*Faculdade de Farmácia, Universidade de Lisboa, Av. Prof. Gama Pinto, 1649-003 Lisboa, Portugal*
[b]*Instituto de Investigação do Medicamento (iMed.ULisboa), Faculdade de Farmácia, Universidade de Lisboa, Av. Prof. Gama Pinto, 1649-003 Lisboa, Portugal*

patricia.manteigas1@gmail.com

2.1 Introduction

A recent study published in the Skin Cancer Foundation has revealed a significant rise in melanoma cases among people aged 18 to 39 years old. The prevalence of 800% among young women and 400% among young men sounded like alarm bells.

The skin is our more visible organ and represents the main human barrier against external aggressions. Consequently, it is particularly susceptible to deleterious effects of UV radiation overexposure, including the photoaging, photoimmune suppression, and photocarcinogenesis. Although the human skin is provided by an endogenous system composed of numerous anti-oxidant

Carrier-Mediated Dermal Delivery: Applications in the Prevention and Treatment of Skin Disorders
Edited by Andreia Ascenso, Sandra Simões, and Helena Ribeiro

ISBN 978-981-4745-58-1 (Hardcover), 978-1-315-36447-6 (eBook)
www.panstanford.com

molecules, this sophisticated organ could be not efficient enough to counteract the activity of several offenders.

Melatonin as a pleiotropic agent can counteract the generation of free radicals and, subsequently, decrease the harmful consequences of oxidative stress phenomenon. In addition, this active agent may allow the enhancement of the immune surveillance due to its anti-inflammatory, anti-apoptotic, and anti-oxidant properties playing a crucial role in the genome stability. Last, melatonin as a circadian pacemaker is also able to prevent the circadian desynchrony and to restore the circadian rhythm. This multiplicity of actions and the variety of biological effects of melatonin justify its potential in a range of clinical and wellness-enhancing uses.

Notwithstanding a proper behavior by avoiding the excessive intermittent or cumulative solar exposure and using sunscreen products, the topical supplementation of skin with anti-oxidants like melatonin is gaining increasing importance in order to counteract the UV-induced depletion of those molecules in skin layers.

In fact, we believe that the incorporation of this "Swiss army knife" in sunscreen formulation through a successful topical delivery system may be the key strategy to prevent UV-radiation–induced skin damage, such as pre-cancerous and cancerous skin lesions. Therefore, this review aims to highlight the potential use of melatonin for inhibition and/or repair of oxidative stress and other UV-induced damage in the photocarcinogenesis context.

2.2 What Is Melatonin?

Melatonin[1] is an evolutionally phylogenic old molecule [1, 2]. It was discovered by Aaron B. Lerner and his colleagues at Yale University School of Medicine in 1958 [3–5]. It is a ubiquitous indolamine (*N*-acetyl-5-methoxytryptamine) derived from the amino-acid tryptophan being a widely occurring neurotransmitter-like compound primarily derived from the pineal gland [2, 4, 6]. Melatonin has pleiotropic bioactivities, working not only as a

[1]Melatonin: "mela" from melanin and "tonin" from serotonin [1] Reiter RJ, Tan DX, Galano A. Melatonin: exceeding expectations. *Physiology* (Bethesda, Md). 2014; 29: 325–333.

neurotransmitter but also as a hormone, cytokine and biological-response modifier.

Melatonin is a small-size molecule with an amphiphilic nature (Table 2.1)—the two characteristics that promote its permeability to all cellular compartments allowing it to be present in cytosol, membrane, mitochondria, and nuclear structures [6].

Table 2.1 Chemical properties of melatonin [7]

Chemical structure	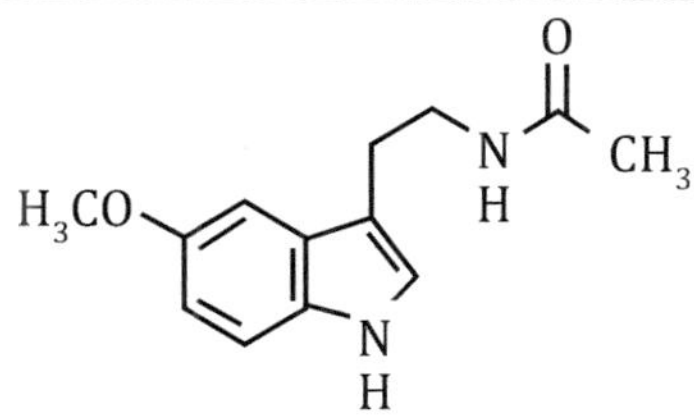
Chemical group	Serotonin
IUPAC name	*N*-[2-(5-Methoxy-1H-indol-3-yl)ethyl]acetamide
Molecular formula	$C_{13}H_{16}N_2O_2$
Molar mass (g/mol)	232.28
LOG P	~0.48/1.6
Water solubility	2 g/L (at 20°C); 5 g/L (at 50°C)

This hormone production follows a circadian light-dependent rhythm of secretion, i.e., its production is influenced by the detection of light and dark by our retina. During the day, melatonin levels are low because of the light intensity, and by contrast its level peaks occur during the night being responsible for inducing physiological changes that promote sleep, such as decreased body temperature and respiration rate [3].

Melatonin is a functionally and versatile molecule, which is involved in numerous aspects of body's biological and physiologic regulation. Its primary role and the first to be described was the regulation of circadian and circannual cycles—"**circadian pacemaker**" [8–10]. The role of endogenous melatonin in circadian rhythm disturbances and sleep disorders is well established but this chronobiotic is also responsible for the sexual maturation, energy expenditure or body-mass regulation [6, 11]. More recently, some studies found that melatonin and its derivatives were potent free radical scavengers and broad-spectrum anti-oxidants. Accordingly, it was suggested that this indolic compound may inhibit tumors proliferation and modify immunity, the stress response, and certain aspects of the aging process [12]. Moreover, melatonin production gradually declines with age, and its loss is associated with several age-related problems, including Alzheimer's disease and cardiovascular diseases [3, 13–15].

The **multiplicity of actions** (as shown in Fig. 2.1) and variety of biological effects of melatonin justify its potential in a range of clinical and wellness-enhancing uses.

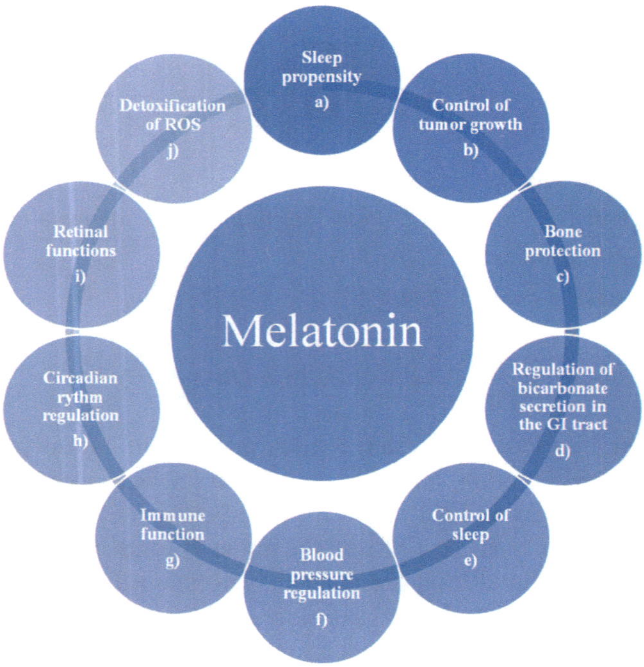

Figure 2.1 Melatonin: a pleiotropic molecule. (a) [16–18] (b) [4, 19] (c) [20] (d) [21] (e) [18] (f) [22, 23] (g) [24–26] (h) [27] (i) [28] (j) [29].

2.3 Melatoninergic System

2.3.1 Synthesis of Melatonin in the Skin

Originally, melatonin was believed to be synthesized exclusively in **pineal gland** of vertebrates. Not long after, some studies have shown that many **extrapineal tissues** and organs have the capacity to synthesize it. To date, melatonin synthesis has been also identified in the thymus, brain, retina, Harderian gland, lens, ciliary body, airway epithelium, bone marrow, immune cells, placenta, GI tract, gonads and skin [30–34]. It is believed that mammalian skin contains the entire molecular and biochemical machinery necessary to perform the process [6]. However, it is important to state that the only above-mentioned site that seems to exhibit a circadian rhythm of melatonin production is the **retina** [35].

The **melatonin-biosynthesis pathway in skin,** similar to what happens in other organs, is divided into four stages (Fig. 2.2), initiated by the uptake of the essential amino acid L-tryptophan by pineal parenchymal cells [6, 31, 36]. After this, L-tryptophan is converted to another amino acid, 5-hydroxytryptophan due the action of tryptophan hydroxylase enzyme (TPH), which is dependent on (6R) 5,6,7,8-tetrahydrobiopterin (6-BH4) [37]. There are two isoforms of tryptophan hydroxylase identified as TPH1 and TPH2. The first one is expressed in many peripheral tissues, including the skin, whereas TPH2 is expressed predominantly in the central nervous system [38]. Thus, TPH1 is the responsible enzyme for the production of melatonin at skin level [33]. The second step of melatonin synthesis involves the decarboxylation of 5-hydroxytryptophan to serotonin by the aromatic amino acid decarboxylase enzyme (AAD). In the third step, serotonin is acetylated to *N*-acetylserotonin by arylalkylamine *N*-acetyltransferase (AANAT). Further, it is methylated to melatonin via hydroxyindole-*O*-methyltransferase (HIOMT).

During the melatonin synthesis process, it is considered the existence of three limiting stages [36]. These steps are the result of TPH, AANAT, and HIOMT enzymatic action. The idea that AANAT was essential to the biosynthesis process persisted for several decades, but recent findings suggest that the skin has the capability to convert serotonin to melatonin by an AANAT-independent pathway. In other words, evidence has shown that HIOMT might be a rate-limiting enzyme for melatonin biosynthesis rather than AANAT [39].

Ribelayga *et al.* observed that AANAT failed to promote melatonin production in Siberian hamster pineal gland while the increased activity of HIOMT was positively related to this synthesis [40].

Moreover, Liu and Borjigin evaluated whether AANAT is a rate-limiting enzyme of melatonin biosynthesis using a genetic mutant rat model [41]. Their data demonstrated that melatonin levels were not influenced by AANAT activity neither NAS (product of the AANAT activity). In addition, these authors also found that melatonin synthesis was limited by the HIOMT activity rate, and the recent identification of missense mutations in HIOMT from patients who suffered from autism spectrum disorders also supports the idea that HIOMT is the rate-limiting enzyme in the melatonin production as these patients displayed lower levels of melatonin secretion [42].

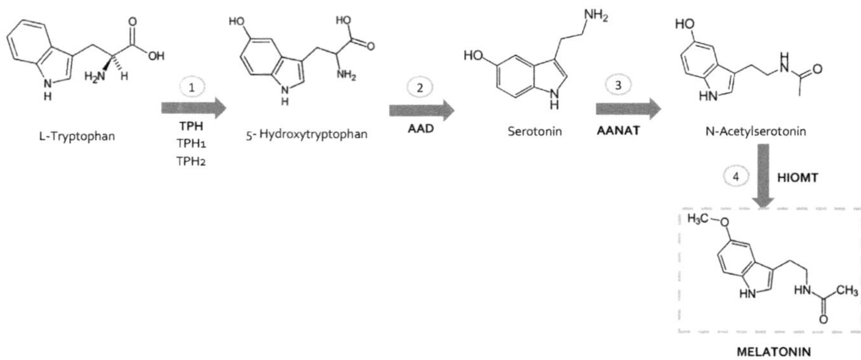

Figure 2.2 Scheme of melatonin synthesis in the skin.

2.3.2 Melatonin and Its Metabolites

Melatonin is metabolized through enzymatic as well as chemical reactions in a total of three major metabolic pathways: classic, indolic and kynuric (Fig. 2.3) [36]. All these pathways are operative in skin [43]. In this review, several derivatives were identified, including, 5-MTT, 3-OHM, AFMK, AMK, 6-hydroxymelatonin sulfate (epidermis major metabolite), 2-hydroxymelatonin, 1-nitromelatonin, and nitrosomelatonin [31, 37]. Due to its biological significance, three of these metabolites are more widely studied, namely: 3-OHM, AFMK and AMK [2, 31, 44]. 2-Hydroxymelatonin, 3-OHM, 1-nitromelatonin, and nitrosomelatonin are the resultant products of melatonin interaction with ROS/RNS.

The **classical pathway** for melatonin degradation starts with circulating melatonin being metabolized mainly in the liver by the CYP450 family enzymes (CYP1A2, CYP1A1 and, to a lesser extent, by CYP1B1) to 6-hydroxymelatonin (the main product of this pathway) [36, 45]. Then, it becomes more polarized by the addition of either sulfate (6-sulfatoxymelatonin) or glucuronide (extremely limited) facilitating its excretion process. Although it was initially thought that melatonin was catabolized to 6(OH)M exclusively in the liver, recent evidence indicates that some 6(OH)M may be formed at extrahepatic sites, namely in the skin [31]. The metabolism in extrahepatic tissues exhibits substantial differences, such as the prevalence of the classical, indolic, or kynuric pathways depending on the tissue. In the brain, a substantial fraction of melatonin is metabolized to kynuramine derivatives, for example [46].

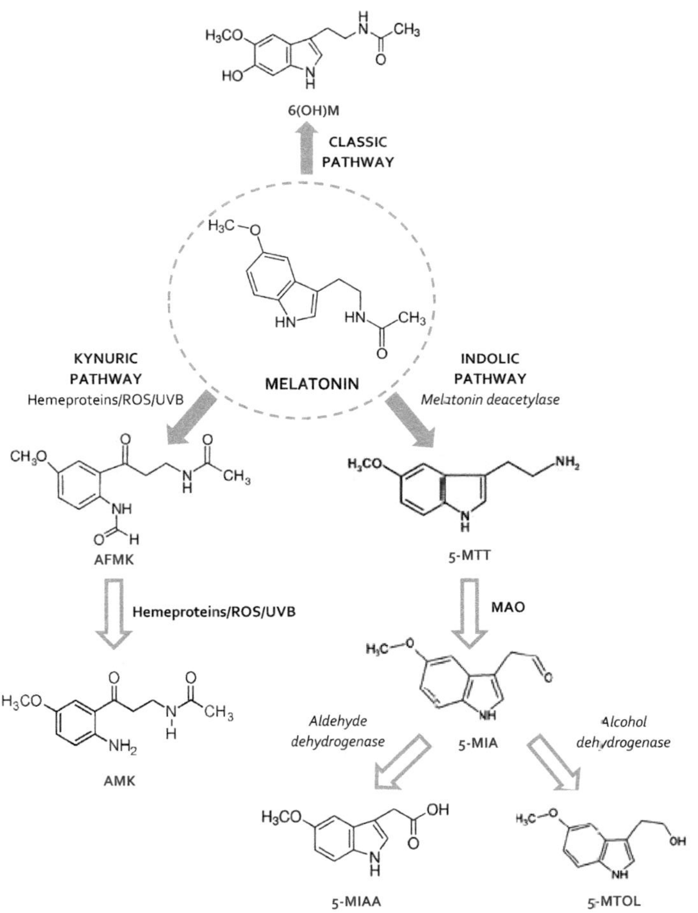

Figure 2.3 Three possible pathways of melatonin metabolism.

The **indolic** is the main pathway of melatonin degradation in skin and involves the metabolization by melatonin deacetylase to 5-methoxytryptamine (5-MTT)—minor derivative [45].

Subsequently, monoamine oxidase transforms 5-MTT to 5-methoxyindolacetaldehyde (5-MIA), which is enzymatically metabolized to either 5-methoxyindole acetic acid (5-MIAA) by aldehyde dehydrogenase or 5-methoxytryptophol (5-MTOL) by alcohol dehydrogenase [36, 45].

The **kynuric pathway** metabolizes the melatonin through two forms: enzymatic and non-enzymatic, the last one being involved in the action of free radicals or ultraviolet B (UVB) radiation. Moreover, **non-enzymatic reactions** with free radicals, in particular the superoxide anion and the hydroxyl radicals, represent a significant aspect of melatonin's biological role, which will be explained in detail in the chapter—melatonin as a "guardian" for prevention of photocarcinogenesis [47]. In **enzymatic reactions**, melatonin is metabolized by indoleamine 2,3 dioxygenase to produce *N*-acetyl-*N*-formyl-5-methoxykynuramine (AFMK). AFMK is further metabolized by arylamine formamidase to *N*-acetyl-5-methoxykynuramine (AMK). Posteriorly, AMK can produce two different metabolites: AAMC or MQA by reacting with either NO^+, NO, or HNO or with carbamoyl phosphate, H_2O_2 and Cu^{2+}, respectively [2, 6, 31].

Moreover, the metabolism of melatonin is cell-type dependent and expressed in all three main cell populations of human skin—**keratinocytes**, **melanocytes**, and **fibroblasts**. The skin cell type-specific metabolism, which determines the biological effects of melatonin, has not been established until Kim *et al.*'s work in 2013 [45]. These authors have performed several experiments with normal epidermal keratinocytes and melanocytes, dermal fibroblasts as well as immortalized epidermal keratinocytes, and melanoma cells [45]. The main conclusions of this multidimensional experimental study were as follows: (a) identification of 6(OH)M and AFMK as the main metabolites of the melatonin degradation; (b) identification of the indolic pathway as the predominant route of melatonin metabolism; (c) importance of the kynuric pathway in organs exposed to environmental stress such as skin, in contrast to the systemic metabolism, in which AFMK may be below or at the limit of detection; (d) differences in the rate of melatonin metabolism dependent on cell type [45].

Probably, this metabolization exhibits species-, site-, and tissue compartment-dependent differences, which are modulated by several factors, such as environmental UVR or local adrenergic stimulation levels and hemeproteins [48].

Melatonin has been frequently associated with a protective action against oxidative stress both *in vitro* and *in vivo* [44, 47,49]. This protective action is also shared by its metabolites constituting an anti-oxidant chain, also known as the **melatonin's cascade**, even at low concentrations [2, 31]. Its particular metabolism makes it one of the most potent physiological scavengers of hydroxyl radicals found to date [47].

2.3.3 Melatonin Receptors in the Skin

As seen above, some or even much of this multitasking agent's activity may be attributed to its metabolites [32, 39, 50, 51]. Furthermore, the pleiotropic activity of melatonin and its respective derivatives is mediated by receptor dependent and receptor independent mechanisms.

In mammals, two classes of receptors were described: the G-protein coupled receptor superfamily and the quinone reductase enzyme family [52–54].

MT_1 and **MT_2** are considered the true melatonin receptor system in mammals and belong to the superfamily of G-protein coupled receptors containing the typical seven transmembrane domains [50, 55]. They are also described as high-affinity melatoninergic receptors.

MT_1 is the dominant receptor in the human skin being located in the following regions: *stratum granulosum*, *stratum spinosum*, eccrine sweat glands, upper and inner root sheath, and the endothelium of blood vessels [56]. By contrast, MT_2 is only found in the inner hair sheath, eccrine sweat glands, and in the endothelium of blood vessels [36]. MT_1 and MT_2 were distinguished pharmacologically with the use of agonists as 2-[^{125}I]iodomelatonin [57]. However, these receptors share a close pharmacological profile and a homology at the amino acid level of around 60% [54].

After the receptor stimulation, it occurs the activation of either G_i or G_o, depending on the receptor type. Subsequently, the activation of second messengers (adenylate cyclase, phospholipases C and A2), potassium and calcium channels and guanyl cyclase, mediates the intracellular signaling [6].

Physiologically, the activation of these two receptors triggers a distinctive type of effects and cellular responses. Moreover, the expression of MT_1 and/or MT_2 is affected by several variables, such as environmental factors (UVB), genetic heritage and certain diseases, including skin cancer [58]. Some evidence suggests that exposure to UVB induces up-regulation of MT_1 expression in normal neonatal epidermal melanocytes, while down-regulates it in melanoma lines [58]. Further, Kim *et al.* have revealed that the modulatory effect of melatonin on proliferation or differentiation of keratinocytes is strongly connected to the high expression of MT_1 and MT_2 in mammalian skin [45].

There is a third melatonin binding site initially described as **MT_3** [51]. However, this biological target of melatonin has been posteriorly regarded as a cytosolic enzyme—quinone reductase II [59]. Quinone reductases (QR) may be linked with detoxifying properties and related to the protection against oxidative stress [60]. The hypothesis of MT_3 to be more closely similar to the enzyme concept instead of the receptor term was supported by the observance of the temperature-dependency or by the very rapid ligand association/dissociation kinetics. While only a small number of natural QR_2 co-substrates have been identified, recent studies have shown that melatonin may be one of them [61].

Besides the receptors described above, melatonin also binds to **nuclear receptors** of the retinoic acid sub-family or orphan receptors (**ROR/RZR**). This occurs at relatively higher concentrations than those required for membrane-bound melatonin-receptor activation [6]. All human skin cells express the nuclear receptor ROR, which has different isoforms as ROR-1,-2,-3 or RZR1 depending on the cell type. While dermal fibroblasts express ROR-1 and -2, melanocytes or epidermal keratinocytes express RZR1 [58].

In addition, other melatonin interactions have also been described, such as **calmodulin**, **calreticulin**, and **microtubular proteins** acting as melatonin sensors within the cells [55].

In summary, the multiplicity of actions of this promising molecule involves on one hand a diversity of targets such as membrane bound or nuclear receptors and, on the other hand, certain actions which do not require receptors such as direct scavenging of free radicals as described below [32, 50].

2.3.4 Mechanism of Action

The melatonin biosynthesis pathway and the receptors involved in its broad-spectrum activity have been well characterized. This bioactive substance stands out by the variety of mechanisms that it employs to modulate the physiology and molecular biology of cells [62].

As already described, melatonin is not synthesized in a single organ nor exerts its effects on a specific target organ. For this reason, it is not regarded as a hormone, in the classical sense, but rather as a cell protector [63].

Melatonin exerts different **receptor dependent** and **receptor-independent** actions mediated by **paracrine**, **autocrine,** and **anti-oxidant** mechanisms [64, 65]. It is important to state that these diverse actions are not only attributable to melatonin since the entire melatoninergic system (melatonin and its metabolites) acts to preserve the physical and functional integrity of the organism [37, 58].

Melatonin and its derivatives have several abilities by acting as potent anti-oxidants, activating cytoprotective pathways, counteracting oxidative stress or modulating metabolic and enzymatic activities [6]. Although behind different effects are different mechanisms, in this review the beneficial effects of melatonin in a specific and extremely important organ—the **skin** will be focused.

Several studies have reported the action of melatonin against the harmful effects of UVR. *In vitro* assays have revealed the effects of melatonin topical application which were consistent with its role as a protector of the skin integrity. Melatonin was responsible for an increase in cell viability, mitochondrial membrane potential and DNA synthesis. Furthermore, it caused a decrease in ROS generation, caspase activity (family of endoproteases that provide critical links in cell regulatory networks controlling inflammation and cell death), and apoptosis process [66–73].

Regarding its **anti-oxidant** activity, melatonin acts as a powerful radical scavenger of ROS. Free radicals comprise two groups of toxic species: reactive oxygen species (ROS) and reactive nitrogen species (RNS) (Table 2.2). ROS and RNS production is dependent on the intracellular metabolism. In addition, a number of external agents can promote ROS generation, such as ionizing radiation, ultraviolet light, environmental toxins, inflammatory, and cytokines.

Table 2.2 ROS and RNS toxic species

ROS	RNS
• $^1O_2$ (singlet oxygen)	• $NO^•$ (nitric oxide)
• $O_2^{•-}$ (superoxide)	• $ONOO^-$ (peroxynitrite)
• H_2O_2 (hydrogen peroxide)	• ONOOH (peroxynitrous acid)
• $^•OH$ (hydroxyl radical)	• $NO_2^•$ (nitric dioxide)

In other words, both ROS and RNS can function as double-edged "sword" as they can be either harmful or beneficial to biological systems, depending on the site of generation and their respective concentrations [74].

In high levels, ROS/RNS are often associated with the oxidative stress which occurs when the levels of free radicals overwhelm the cellular defense system designed to metabolize them [2]. This last process can involve cell death, senescence, or tumorogenesis. Radical scavengers as melatonin neutralize directly free radicals and stimulate anti-oxidant enzymatic systems to convert toxic species to non-toxic molecules. Melatonin detoxifies oxidants by several mechanisms, including single electron transfer, hydrogen transfer, and radical adduct formation. Furthermore, melatonin enhances the expression of anti-oxidative enzymes such as superoxide dismutase (SOD), catalase (CAT) and glutathione peroxidase (GPx) [64]. It should be noted that melatonin also up-regulates the expression of genes involved in detoxifying both ROS/RNS and suppresses the activity of genes involved in the generation of those damaging species. Most recently, melatonin has been investigated regarding its ability to counteract UVR-induced enhanced lactate dehydrogenase (LDH) and UVB-induced acidification of cytosol [36].

The direct radical scavenging actions of melatonin are receptor independent, while the regulation of gene expression may involve an interaction of melatonin with its receptors, mainly, including MT1 and MT2. In skin cells, receptor-independent melatonin actions also might be mediated partly by the QRII and by the up-regulation of the main anti-oxidants enzymes [6, 64].

Concerning to its **anti-apoptotic** effect, melatonin has proved to down-regulate effector caspases (caspase-3, -7, and -9) and to reduce PARP (an indirect marker of DNA damage) activation [75].

Finally, different studies suggested the **anti-inflammatory** role of melatonin. This last property was demonstrated by inhibition of lipid peroxidation and nitric oxide formation and by the interference with arachidonic acid metabolism, preventing its conversion to pro-inflammatory substances [36].

In summary, the melatoninergic system has diverse actions (Table 2.3), all of which could contribute to counteract external or internal stresses. Consequently, it preserves the biological integrity and maintains homeostasis of the organism, including the skin.

Table 2.3 Melatonin as a pleiotropic agent [36, 64, 75]

Antioxidant
• Neutralizes directly ROS/RNS (through hydrogen and electron transfers and through the formation of radical aducts);
• Stimulates antioxidant enzymes (SOD; CAT; GPx);
• Up-regulates genes responsible for the detoxification;
• Suppresses the activity of genes involved in the generation of ROS/RNS;
• Prevents the acidification of cytosol;
Anti-inflamatory
• Inhibits the lipid peroxidation;
• Avoids the nitric oxid formation;
• ↓ iNOS; COX-2; TNF-α; adhesion molecules;
• ↓ NFkB pathway;
Anti-apoptotic
• ↓ Caspases (-3, -7, -9);
• ↓ PARP activation;
• ↓ Bax gene expression;
• ↓ Bcl-2 gene expression.

2.4 Photocarcinogenesis

The **skin** is our passport for a life on dry land [76]. Although it may seem strange to consider the skin as an organ, it actually represents the body's largest (1.5–2 m^2), heaviest (16% of body weight) and most visible organ [77, 78]. Skin is responsible for important processes, such as absorption, excretion, secretion, thermoregulation, immune surveillance, and sensation [79]. Moreover, it protects vital internal organs against external harmful agents [76, 80]. Environmental pollution, poor nutrition, smoking, severe physical and psychological stress, alcohol intake, ionizing, and UV radiations are the main responsible elements for skin injury, being estimated that among all these agents, UVR contributes up to 80% [81].

Anatomically, the architecture of human skin is made up of three structural layers: **epidermis** (the external layer), **dermis** (which contains hair roots, nervous cells, sweat glands, blood and lymph vessels), and **hypodermis** (the deeper layer consisting of fatty tissue that connects the dermis to underlying skeletal components) [76, 80, 82–84]. Although **keratinocytes** are the main cells of epidermis, representing 95% of epidermal population, melanocytes and Langerhans cells (LCs) are also included in this population [80, 85]. **Melanocytes** localized in the *stratum basale* are the only entities able to synthesize melanin in melanosomes which is transferred through melanocytic dendrites to the adjacent keratinocytes [83, 86]. The photoprotective role of this skin pigment is evidenced by the observance of the greater extent of DNA damage in the upper layers of the epidermis (with lower melanin levels) [87]. **Langerhans cells** are considered the first immunological barrier, once they have the ability to promote appropriate adoptive and adaptive immune responses [80, 82, 88]. These cells are down-regulated by the UVR [87, 89]. This phenomenon is based on local and systemic immunosuppression, which is known as a major risk factor for the photocarcinogenesis induction [87, 90].

The sunlight has obvious **benefits** in terms of the human health, being responsible for different processes, such as synthesis of vitamin D, reduction of blood pressure or mental health improvement through neurotransmitters regulation [82]. Its importance becomes even more unquestionable if we consider the implications of its deficiency. In fact, there is already evidence that suggests that an insufficient sunlight exposure could result in certain disorders like osteomalacia and increased risk of cardiovascular disease, rheumatoid arthritis, diabetes, depression, or multiple sclerosis [82].

However, the daily solar exposure should be controlled since humans are also susceptible to sun damage. According to Skin Cancer Foundation Statistics, the incidence of both **non-melanoma** (NMSC) and **melanoma skin** (MSC) **cancers** has been increasing over the past decades. All around the world, between 2 and 3 million NMSC (the most prevalent form of human neoplasia) [91] and 132,000 MSC are diagnosed each year [92]. Squamous cell carcinoma (SCC) and basal cell carcinoma (BSC) are the two types of NMSC. The transformation of keratinocytes and melanocytes both present in the epidermis layer may contribute to the emergence of NMSC and MSC, respectively [92–94]. Although each type of tumor is linked to different cells, all of them have a common etiology. Intermittent or cumulative UV exposure generally results in basal cell carcinoma, while squamous cell carcinoma is often associated to a lifelong cumulative UV exposure [95, 96]. Patients who develop melanoma are normally associated to an intense and intermittent UV exposure [95, 97].

Although many factors may contribute to the development of skin tumors, including ionizing radiations, viruses, genetic factors, and immunologic status, the UVR from the sunlight or from artificial sources is considered the most important etiological factor [64, 90, 98]. The skin, as the main barrier acting against external aggressions, is particularly susceptible to deleterious effects of UVR overexposure [99]. These effects include the photocarcinogenesis, photoaging, and photoimmune suppression processes [90, 99].

Different range of UVR interacts with various types of molecules causing distinct consequences in skin. Therefore, it is important to identify its action spectrum. The UV light represents the portion between visible light and X-rays of the electromagnetic spectrum, and can be classified according to the International Convention into three major components—UVA (320–400 nm), UVB (290–320 nm), and UVC (200–290 nm), as evidenced by Fig. 2.4 [64, 93, 98]. The energy of each part of the UVR is inversely proportional to its wavelength, **UVC** being the most energetic range and not favorable to life on earth. Fortunately, it is filtered by the ozone layer [100]. Regarding **UVB**, it represents approximately 5% of UVR that reaches the earth [94]. It has pleiotropic effects on the human skin (especially on the epidermis layer), including apoptosis, inflammation, local and systemic immunosuppression or mutagenic phenomena that could contribute to the photocarcinogenesis process [90].

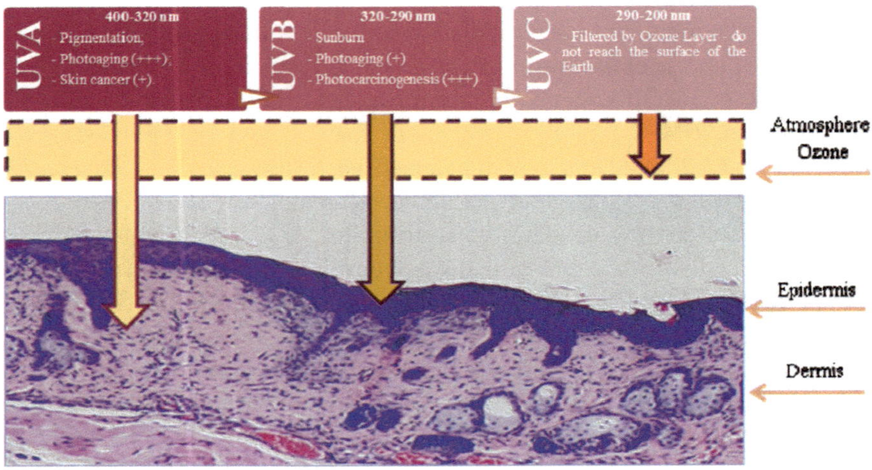

Figure 2.4 UVR penetration into different skin layers and its deleterious effects [104].

Finally, **UVA** rays represent 95% of the solar UVR that reaches the earth surface [98]. Their energy is relatively lower, and therefore, its harmful effect on skin was underestimated for many years. However, recent studies have proven that UVA is capable of generating ROS by photosensitization reactions which could result in DNA, proteins, and/or lipids damage [101, 102]. Furthermore, UVA is considered the major source of oxidative stress in human skin [103].

While erythema is mainly caused by UVB radiation, photoaging, pigmentation changes, photoimmune suppression, and photocarcinogenesis can be attributed to both [94]. These effects may be explained by the different penetration degrees in the skin depend on the UVR wavelength [101]. As shown in Fig. 2.4, UVB is limited to the epidermis layer affecting mostly keratinocytes but also melanocytes which could lead to NMSC and MSC, respectively, as already mentioned. On the other hand, UVA rays penetrate into the upper layers of the dermis being more responsible for the photoaging process [99].

Photocarcinogenesis is a multistep phenomenon (initiation, promotion and progression) wherein the skin cells acquire genomic mutations induced by UV irradiation [93, 94]. This phenomenon is inseparable of two other processes, namely, oxidative stress and photoimmune suppression.

UVR interacts with the skin being absorbed by target-molecules known as chromophores [94]. Different **chromophores** have been identified, such as melanin and melanosomes; proteins like porphyrins, riboflavin, tyrosine, histidine, or collagen; cholesterol; folate; *trans*-urocanic acid and DNA [82, 105]. In addition, the UVR induces the generation of photoproducts at the molecular level which are extremely cytotoxic and mutagenic [100]. The most cytotoxic **DNA photoproducts** include 75% of cyclobutane-pyrimidine dimers (CPDs) and 25% of 6–4 photoproducts (6-4PPs) both responsible for the distortion of the DNA helix [80, 100, 106, 107]. However, CPDs species are more important since these species are more mutagenic than 6-4PPs besides being removed quite slowly [107, 108]. The CPDs formation occurs especially in the single stranded DNA and at the flexible ends of poly (dA:dT) tracts. UVB damage occurs essentially on adjacent sites of pyrimidine bases Thymine (T) and Cytosine (C) on the same side of the DNA-chain resulting in the four-membered ring. These **TC** to **TT** and **CC** to **TT transitions** are unique on neighboring dypirimidine bases representing specific markers/fingerprints of the UVR implication in photocarcinogenesis [107–109]. In general, both types of these bulky photoproducts are removed by a complex process of **nucleotide excision repair (NER)**.

The NER encompasses approximately 30 proteins (e.g., XPC; XPA; RNA Polymerase II; DNA Polymerase; ERRCC1) and operates through 4 steps to repair the DNA damage: (a) damage recognition, including two mechanisms, the entire genome repair and the transcription coupled repair; (b) DNA unwinding; (c) DNA incision, and (d) DNA excision and synthesis [107].

Apart from these specific UV-induced damage aforementioned, UV radiation triggers other types of effects. It is also able to induce oxidative damage and purine photoproducts, DNA crosslinks and single-strand breaks. However, these last deleterious events are not specific markers of UV action, and may be attributable to other damaging agents [105].

Unfortunately, the restoration of normal DNA sequence, structure, and chemistry is not always possible. Nevertheless, the cell has also other protective mechanisms which do not require the removal of DNA damage [110]. These processes are called **DNA damage tolerance (DDT)** mechanisms, and include the translesion DNA synthesis (**TLS**) and the homology-dependent repair (**HDR**) [111]. Although DDT mechanisms are not a repair pathway per se, they constitute an important cellular strategy by increasing its survival and preventing the genomic instability [112].

If the photolesions are not recognized or incorrectly repaired and, if DNA repair mechanisms fail before its replication, certain cancer relevant genes suffer mutations and a cascade of deleterious events takes place [94, 113, 114]. There are two types of cancer genes: the **suppressor genes** which regulate the DNA damage, cell cycle control and apoptosis; and the **proto-oncogenes** (e.g., RAS, WNT, MYC, ERK, and TRK) which are responsible for the organ growth and the tissue repair [94].

One of the most important tumor suppressor genes is ***p53*-gene**, which is often called the "guardian of the genome" and is located on chromosome 17p [94, 101, 107, 115, 116]. In fact, the mutation of *p53*-gene is the most dangerous UV damage by disabling the cells to inhibit the cancer development [94]. UV acute exposure up-regulates the transcription of this crucial gene resulting in an increase of *p53* protein production. This protein permits to restore the damaged DNA by stopping the cell cycle in G1 phase prior to its incorrect replication in next phases [94]. If the damage is too extensive and this repair mechanism is not possible, the cell has another important repairing mechanism which is represented by the **apoptosis** process. This programmed cell death occurs through the action of the tumor suppressor gene (*p53*-gene) or due to the activation of death receptors and the subsequent activation of the caspases cascade [94].

Although the carcinogenic effect of UVB has been widely described, the role of UVA rays has only recently been reported [117]. Accordingly, recent studies showed that UVA radiation induces the ROS production and promotes a suppression of the local and systemic immunological response, which are indirectly associated with the carcinogenesis process [94].

Oxidative stress has a key role on the indirect pathway of UV-induced damage. As seen before, ROS could be normal and relatively harmless considering that these species become reactive only under certain conditions. For instance, H_2O_2 needs transition metals to promote Fenton reactions [118]. However, when the levels of free radicals overwhelm the anti-oxidant system, they can become very dangerous entities as a major source of oxidative injury in all aerobic organisms [74, 118]. While the majority of ROS has a short half-life and the damage is confined to a local level, other oxidants like H_2O_2, could lead to DNA damage at distant locations due to their longer half-life [118].

Two typical DNA lesions arise from the interaction of ROS with this molecule [105]. The first comes from the oxidation of the Guanine (G) base resulting in 8-oxo-7,8-dihydroguanine (**8-oxo-dG**) and 8-hydroxy-2-deoxyguanosine (**8-OH-dG**) [119, 120]. The second is represented by the formation of 2,6-diamino-4-hydroxy-5-formamidopyrimidine (**FapydG**), which is the most predominant guanine-derived modification under hypoxic conditions [118]. In addition, other base alterations can occur. The production of substances like 5,6-dihydroxy-5,6 dihydrothymine (**thymine glycol**) and 5,6-dihydroxy-5,6-dihydrocytosine (**cytosine glycol**) is the result of the reaction between $^{\cdot}$OH and pyrimidine bases [118]. It should be mentioned that thymine glycol and 8-oxo-dG are known to be markers of oxidative stress in carcinogenesis process [118]. The **base excision repair** (BER) is another repair mechanism of lesions that result from exposure to endogenous or exogenous ROS avoiding the accumulation of unrepaired 8-oxo-dG and 8-OH-dG photoproducts [121, 122]. Besides ROS, RNS are also able to induce DNA lesions [120]. The increment of nitric oxide (NO) levels can be due to UV-induced up regulation of NO synthetases and to UVA decomposition of endogenous NO sources. **Nitrotyrosine** and **8-nitroguanine** are two species resulting from the interaction between RNS and proteins or DNA, respectively [120]. The last interaction occurs commonly with Guanine, once it has the lowest oxidation potential being the most susceptible base [102]. The genotoxicity process is represented in Fig. 2.5.

In fact, RNS in combination with ROS may lead to even more genotoxic products and DNA injuries and, ultimately, resulting in mutagenesis and carcinogenesis events [120].

Fortunately, the cell has defense machinery represented by enzymatic and non-enzymatic anti-oxidants that help to protect against the UV-induce oxidative damage [102, 123]. Beyond the endogenous anti-oxidant protectors as SOD, GSH, or catalase, **free radical scavengers**, such as **melatonin,** may be a key factor for the prevention and/or inhibition of the photocarcinogenesis process. In fact, this is one of the mechanisms by which melatonin is identified as a guardian of the genome.

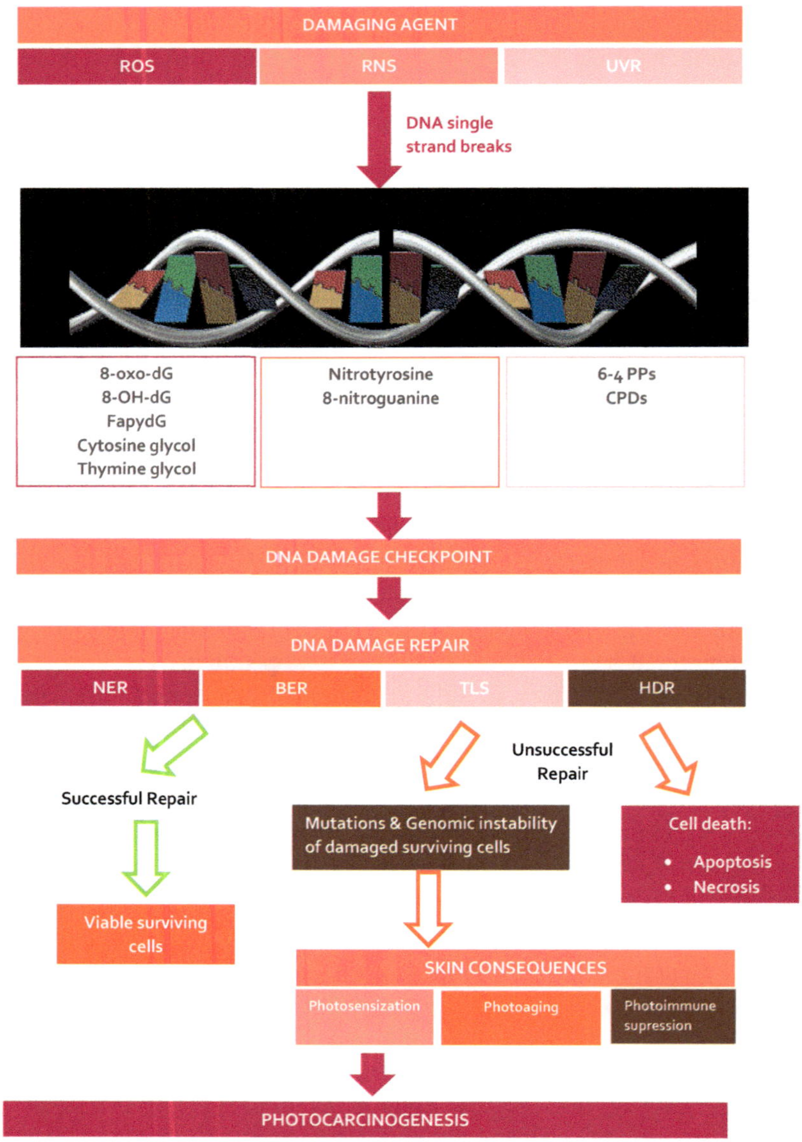

Figure 2.5 Theoretic scheme of genotoxicity associated to photocarcinogenesis.

Besides the anti-oxidant mechanism, the human body is constantly trying to maintain the homeostasis through its immune system [118]. **Photoimmune suppression** is considered another major risk factor for skin cancer development and can result from a single exposure to UVR during a normal occupational or recreational exposure [90, 124]. This effect is usually mediated by several mechanisms, as follows: (a) release of immunosuppressive cytokines (IL-10; TGF-β), (b) decrease of immunostimulatory cytokines (IL-12, -23), (c) migration of UV damaged Langerhans cells to lymph nodes, and (d) defects induction in antigen presentation [125]. Some authors, including Ullrich [90], suggest that photoimmune suppression is a secondary effect of the cell defense mechanisms to guarantee the genomic stability.

In any case, it is known that both processes, i.e., oxidative stress and photoimmune suppression, not only favor the mutational changes but also improve the genomic instability [90, 126].

2.4.1 Genomic Instability and Its Impact on Photocarcinogenesis

Genomic instability is a dynamic process considered as a key driver of many types of human cancer, including the skin cancer [112, 127, 128]. The loss of genome integrity involves a continuous and accumulated modification of the cellular genome [128].

The genomic instability can emanate from either exogenous or endogenous sources [128]. Numerous **exogenous agents** are well established, such as the cigarette smoke, the occupational exposition to heavy metals, the infrared radiation (IR), the UVR and, the most recently described, the light at night (LAN) [128]. **LAN** is associated with night shift work and involves the circadian disruption. In other words, this phenomenon is translated by the suppression of circadian **melatonin** production, and it was recognized as a possible carcinogen in 2007 by the World Health Organization (WHO) [128, 129].

It is widely established that both UVA and UVB radiation cause genomic instability in mammalian cells. As stated above, UVB is a complete carcinogen agent, promoting direct damage to the DNA molecule as well as generating ROS in the skin [126]. UVA is more linked to the generation of ROS, being suspected to play an important role on the induction of genomically unstable malignant melanoma [130, 131].

Among **endogenous factors**, there are the ROS, mitochondrial dysfunction, replication errors and transposable elements—"entities that can rearrange the genomic material of their hosts in the process of their mobilization" [128].

After the initial insult, the genomic instability can be manifested by different ways [128]. The affected cells may experience several harmful situations which can occur at the nucleotide level and/or at the gross chromosomal level [127]. The highpoint of these mutations can be a **gene mutation** (single base-pair substitution or deletions); **chromosomal rearrangement** (insertions, deletions, translocations or inversions); **cellular transformation**; **cell death** and, ultimately, **carcinogenesis** [128, 132].

Unfortunately, some of these situations represent **persistent changes**, such as the genomic instability induced by the UVR, which can be perpetuated for many cell generations [133, 134]. In fact, Dahle *et al.* have reported that UVR induces an enhancement of mutation rates and chromosomal instability in many cell generations after the radiation insult [134–136]. This can be explained by the fact that UV-light could initiate a process in a cell that can be spread to other cells through a cascade of cellular events. The mechanisms that underline this continuous instability include several factors, such as

(a) persistent induction of DNA damage by the increased ROS level;
(b) mutation in DNA repair enzymes;
(c) cell-to-cell gap junction communication;
(d) secreted factors from unstable cells.

These last two factors are responsible for what is known as the **bystander effect**. The concept arose in the 1970s from a group of immunologists, and it is defined as the induction of biological alterations in non-stimulated cells by signals transmitted from stimulated neighboring cells [137]. This phenomenon was initially tied up with ionizing radiation even though recent studies have evidenced that it is also valid for other cell stressors, namely UVR [137, 138]. Besides the direct mechanisms, the bystander effect is fivefold higher for UVB than for UVA radiation [134].

According to Widel, the cellular consequences of the bystander effect can include reduction of cell survival, induction of apoptosis and genomic instability, chromosomal aberrations, micronuclei formation and delayed cell death [137]. These biological adverse events can arise even in non-irradiated cells via intercellular communication or/and via the medium [134, 137, 138]. Since the mediators of the bystander mechanism are not well established yet, further studies are required to better elucidate them. However ROS, RNS and pro-inflammatory cytokines such as IL-1, IL-6, IL-8, interferon gamma (IFN), and tumor necrosis factor (TNF)-alpha are probable candidates. ROS may be considered as the main inducers of this phenomenon [136, 137]. Such hypothesis is evidenced by the reduction of bystander effect when the irradiated and bystander cells are treated with free radical scavengers like ascorbic acid or DMSO [139, 140].

The significance of the bystander effect on health remains to be evaluated, and its *in vivo* occurrence has not been clearly defined as well [137]. Nonetheless, the amplification of the deleterious genomic consequences by this phenomenon is quite evident [137, 141] especially when considering the photoimmune suppression process. As stated before, this event is considered a major risk factor for photocarcinogenesis and is not restricted to the irradiated zone. It is also able to affect the surrounding tissues even the entire organism due the immunosuppressive cytokines or the defective antigen presentation [90].

Melatonin as a pleiotropic agent can counteract the generation of free radicals, and subsequently, decrease the level of bystander effect-mediators. In addition, this active agent may allow the enhancement of the immune surveillance due to its anti-inflammatory, anti-apoptotic and anti-oxidant properties. All these contributions could not only represent a reduction in the genomic instability but also a decline in its perpetuation, which justifies the melatonin potential on photocarcinogenesis prevention.

2.4.2 Circadian Cycle Connection between Cell Physiology and Photocarcinogenesis

The concept of clock entrainment or **circadian[2] rhythm** has emerged as a result of the natural survival instinct [142, 143]. This term is defined by the American Heritage® Science Dictionary

[2](Latin *circa diem*, meaning "about a day").

as "a daily cycle of biological activity based on a 24 h period and influenced by regular variations in the environment, such as the alternation of night and day" [144]. The clock core machinery is controlled by genes responsible for cycling negative feedback loops, being synchronized by light detected by the retina, and subsequently, transmitted to the hypothalamic suprachiasmatic nuclei (SCN) in the brain [128, 145, 146]. Typically, the circadian system consists of three major constituents: (a) a **central clock** (CC), (b) an **entrainment pathway**(s), and (c) **CC-responsive peripheral tissues** [128]. One of the master clock characteristics is the self-sustainment, i.e., it has an auto-regulatory activity [128]. Nonetheless, it is entrained by external stimuli being the environmental light/dark cycle the most powerful external stimulus. Curiously, it is the daily periodicity of the light/dark cycle that synchronizes the CC-driven oscillation of melatonin production in the pineal gland [128]. Similar to melatonin synthesis, there are several peripheral CC-responsive tissues, namely, liver, intestine, heart, adipose tissue, or retina [143, 147–149].

The role of the circadian rhythm is unquestionable, and it is represented by the control of different behavioral and physiological systems, including the body temperature, renal activity, locomotor activity, sleep-wake cycles, cardiovascular activity, endocrine regulation of energy metabolism and gastrointestinal tract motility [145]. At cellular level, it orchestrates changes in xenobiotic metabolism and detoxification, cell cycle events, DNA repair, apoptosis, and angiogenesis [150].

This system has such importance that it still remains a hot topic in science research. Regarding this, light-mediated perturbation of circadian rhythms has been associated with three major events: (a) melatonin suppression, (b) chronodisruption, and (c) sleep deprivation [128]. Each of these events, individually or in combination, may be associated with increased morbidity and mortality [145]. In fact, both *in vitro* and *in vivo* data suggest the implication of circadian desynchrony in several pathologic conditions, including **tumorigenesis** and **progression of cancer** [146]. Besides the photoimmune suppression and genomic instability, the desynchrony of rhythmicity is also presumed to lead to an uncontrolled cell proliferation status and seems to increase the photocarcinogenesis process [151–153]. To understand better this interconnection, we can also note that

NER, which has an undoubted relevance in the maintenance of genome integrity, depends on the circadian rhythm. In other words, this excision repair mechanism has as rate-limiting factor, the XPA protein, which is controlled by the circadian clock what makes that genomic instability, circadian rhythm and photocarcinogenesis are three inseparable concepts [154].

Accordingly, some epidemiological studies clarify the link between disruption of circadian rhythms and the emergence of tumors [146]. However, it still lacks scientific evidence about how and why the circadian rhythm affects the cell proliferation *in vivo* and its disruption may conduce to tumorigenesis [145].

Melatonin as a chronobiotic molecule has been intensively investigated. It is responsible for resynchronize the CC by providing information about light/darkness from the retinohypothalamic system [128]. Furthermore, the synthesis of this molecule follows a circadian rhythm which has been related to the core circadian machinery genes [146].

Moreover, it has been suggested that DNA damage itself can disrupt the circadian clock, therefore promoting its maintenance [128]. In this sense, by restoring the circadian rhythm and preventing the circadian desynchrony, exogenous melatonin has been claimed as a novel strategy for prevention of the genome instability and consequently of the skin cancer.

2.5 Melatonin as a "Guardian" for Prevention of Photocarcinogenesis

2.5.1 Endogenous and Exogenous Anti-Oxidants as Skin Defenders

The avoidance of an excessive intermittent or cumulative exposure to UV rays and the use of a proper sunscreen are the main weapons to reduce skin aging and the risk of skin cancer (**primary approach**) [90]. In addition, anti-oxidants could represent a **secondary approach** for the maintenance of skin integrity by attenuating the harmful effects of free radicals and by enhancing the efficiency of DNA repair systems [155].

Human skin is provided by an **endogenous anti-oxidant system** composed by numerous molecules, such as melanin, GSH, catalase, SOD, uric acid and ubiquinol[3] [156]. As a general rule, the epidermis comprises higher concentrations of anti-oxidants than the dermis layer [155]. However, this sophisticated organ is not always sufficiently "strong" to counteract the activity of its several offenders [155]. Furthermore, UV exposure leads to a depletion of skin anti-oxidants, which in turn seems to trigger skin diseases [155, 157].

In this context, the **exogenous supplementation** and the topical application of anti-oxidants might be a protective strategy against skin oxidative injuries by preventing the harmful consequences of ROS action [158–161].

Several *in vitro* studies have convincingly evidenced the potential benefits of sunscreens supplemented with anti-oxidants [157, 162]. However, the extrapolation of these data to *in vivo* conditions is not always easy [98]. Surprisingly, anti-oxidants can also act as **pro-oxidants**, the reason why Sing *et al.* affirmed that anti-oxidants can pass from "Miracle Molecules" to "Marvelous Molecules", depending on the presence of $O_2^{\bullet-}$ and transition metals (like iron or copper) [81, 163]. These compounds may interfere with tumorigenesis in a dichotomous way: either suppressing it by preventing oxidative stress to DNA molecule or promoting it by permitting survival of cells that are metabolically impaired [155]. Therefore, it is important to recognize that the treatment of skin with exogenous anti-oxidants after UV exposure could be not always associated with a protective effect [81, 98, 157].

2.5.2 Melatonin Protects against Skin Photodamage

Presently, the topical treatment of skin with anti-oxidants like **melatonin** is gaining increasing importance among dermatologists [72, 81, 164, 165]. In fact, melatonin has a prominent role among a wide variety of anti-oxidants or other phytochemical compounds (coenzyme Q, lycopene, selenium, bioflavonoids, genistein, silymarin, retinoic acid, or proanthocyanidins from grape seeds), once it combines its **radical scavenger activity** with its **anti-inflammatory** and **anti-carcinogenic** properties [62, 146, 157, 164].

In addition, some general facts of melatonin on skin cells should be reported to comprehensively characterize it as a **guardian** for prevention of photocarcinogenesis.

[3]An electron-rich (reduced) form of coenzyme Q10.

Firstly, it should be reminded that apart from being a melatonin target, human skin is also a synthetic and metabolic organ of this multifunctional agent [36]. Moreover, it is important to remember that beneficial effects of melatonin by preventing cells death occur a step before "the point of no return" and may be amplified by its derivatives through the melatoninergic cascade [36]. Furthermore, melatonin exerts **different effects on distinct UV-irradiated cell types**. For example, melatonin has proved to increase the fibroblasts viability by avoiding the formation of polyamines, the accumulation of malondialdehyde (a mutagenic and carcinogenic substance), and by decreasing the apoptosis process [58]. Regarding the main population of epidermis, melatonin has been shown to counteract the typical UV-induced reduction of cell viability when applied at high doses (10^{-3} *M*). In both cell types, melatonin has exerted its radical scavenger activity by suppressing the generation of ROS as well as increasing the rate of cell survival.

Additionally, it is convenient to emphasize that the time of application (prior or after UV exposure) and the melatonin dose may be critical for determining its efficacy. Thus, relevant *in vitro* and clinical studies were included in Tables 2.4 and 2.5.

Regarding the **melatonin dose**, it is important to consider what is expected of this molecule also called as "Swiss army knife" [1]. Accordingly, the doses required for receptor-mediated melatoninergic effects (immune and circadian modulation, anti-oxidant enzyme regulation, etc.) are typically lower than those required for receptor-independent actions (like free radical scavenging) [1]. According to Reiter *et al.*, the idea that melatonin acts exclusively as a free radical scavenger in pharmacological concentrations is not correct particularly in photocarcinogenesis context. Moreover, and as evidenced below, the melatoninergic benefits are dose-dependent [45, 166]. As reviewed by Scheuer *et al.*, the topical application of melatonin conduced to a dose-dependent decrease of erythema degree. Thus, a formulation containing 2.5% of melatonin proved to be more effective in preventing UV-induced erythema than a formulation with 1% of the same active. However, it is important to note that the ideal concentration of melatonin is not well defined yet and further studies will be needed to assess the required dose to achieve an effective photoprotection [166].

Table 2.4 *In vitro* studies regarding the effect of melatonin against UV damage

Aim of the study	Cell line & UV irradiation	Melatonin dose	Time application	Main conclusions
Kim *et al.* studied the metabolism of melatonin and biological activity of intermediates of melatoninergic pathway in human skin cells [45]	Keratinocytes (HaCaT) melanocytes, fibroblasts, melanoma cells	10^{-10}–10^{-3} M	Skin cells were exposed to a wide range of melatonin doses in a time-dependent manner (1 or 24 h)	• Melatonin had a differential effect on keratinocytes proliferation depending on the conditions: *in vitro* culture in monolayer vs. spatially restricted proliferation in the epidermis of intact human skin • Metabolites showed selective anti-proliferative effects on human primary epidermal keratinocytes • Exposure of normal keratinocytes to melatonin and its metabolites led to a slight but significant inhibition of DNA synthesis • Melatonin showed inhibition of DNA synthesis at low (10^{-10} M) and relatively high (10^{-5} M) concentrations. Any effect was observed at intermediate levels
Salucci S. *et al.* investigated the potential protective effect of melatonin against UVB-induced apoptosis in keratinocytes [170]	Keratinocytes (HaCaT) Exposed to UVB radiation $t = 10$ min $t = 30$ min	100 μM	24 h before UVB exposure	• A massive DNA damage was inhibited by melatonin administration. This anti-oxidant was able to: significantly reduce cell death after UVB exposure (living cells: Mel. 71.8% vs. 39% without Mel.); prevent or delay ultrastructural apoptotic patterns appearance as well as DNA fragmentation and reduce caspases activation (mainly caspase-9) • Melatonin seemed to mostly counteract the mitochondrial pathways by preventing apoptotic cell death

Aim of the study	Cell line & UV irradiation	Melatonin dose	Time application	Main conclusions
				• Melatonin showed major anti-}apoptotic properties representing a potential tool in the treatment of UVB-induced skin disorders
Slominsky *et al.* studied the potential of melatonin as a therapeutic and/ or protective agent for application in UVR-induced skin pathology [36]	Keratinocytes (HaCaT) exposed to 25, 50 or 75 mJ/cm^2 UVB; Fibroblasts exposed to 140 mJ/cm^2 UVB	10^{-4} or 10^{-3} M 10^{-7} and 10^{-9} M	Skin cells were pre-incubated for 1 h with melatonin before UVR exposure	• Melatonin was active in UV-irradiated keratinocytes, maintaining the mitochondrial membrane potential and inhibiting the consecutive activation of the intrinsic apoptotic pathway. This led to a reduction of PARP activation, an indirect marker of DNA damage • Melatonin enhanced expression of anti-oxidative enzymes both at the RNA and protein levels • Melatonin and its derivatives enhanced the DNA repair capacity of UVB-induced pyrimidine photoproducts or cyclobutane pyrimidine dimers generation in human keratinocytes • Melatonin could protect dermal fibroblasts against UVB-induced damage. (The protective effect against UVA remained to be confirmed in human dermal fibroblasts) • The strong protective effects of Melatonin against the UVR-induced skin damage observed at its high (pharmacological) concentrations indicated that these are mainly mediated through its potent and diverse direct radical scavenging actions as well as to its metabolic and anti-oxidative enzyme stimulatory effects.

(*Continued*)

Table 2.5 Human studies regarding the effect of melatonin against UV damage

Aim of the study	Human Study/ Volunteers	Melatonin Dose	Time Application	Main Conclusions
Scheuer *et al.* have reviewed several human studies regarding the protective effects and cellular mechanisms of melatonin against UV radiation damage [98]	This review has evaluated the results of four human studies (randomized clinical trials) and sixteen experimental studies. Eligibility criteria: - Healthy volunteers - Randomized clinical trials - Topical application of melatonin in various doses	0.5–2.5%	After or Before UVR exposure	• Topical application of Melatonin led to a dose-dependent inhibition of erythema and prevent the following UV-induced cellular damage • decrease in cell viability; increased generation of ROS; mitochondrial membrane damage and subsequent activation of the intrinsic pathway of apoptosis; decrease in DNA synthesis; amount of oxidative damaged DNA • To answer the question if melatonin can be used as a sun protective agent, further studies are needed to evaluate its protective effect by using natural sunlight. Further studies should also investigate whether application of melatonin on large skin areas will result in adverse reactions or sedative/cognitive side effects
Scheuer *et al.* performed an interventional study (Phase I) to investigate the sun protective effect of melatonin topically applied before sun exposure [166]	Inclusion criteria: - Healthy volunteers of both sexes (18–90 years) - Skin type 1–3 according to Fitzpatrick skin type scale - No sun exposure on skin area tested in the study 4 weeks prior to the study	Melatonin cream (0.5%; 2.5%; 12.5%)	After randomized and received treatment, the volunteers were exposed to strong midday sun (UV-index ~9) for 40 min	• Although this study has been completed, the respective results have not been posted yet in ClinicalTrials report

Concerning the **time of application**, it is well established that the application of melatonin after UVR has no skin protective effects [167]. This fact was corroborated by Howes *et al.* through a clinical study conducted with 16 healthy volunteers. The authors studied the effects of topical melatonin, and concluded that melatonin did not protect against sunburn or immune suppression when applied after solar-simulated UV exposure [168]. On the contrary, its photoprotective role is widely recognized once applied prior to UV exposure [169]. Dreher *et al.* justify this time-dependent response based on the fact that UV-induced skin damage is a rapid event, which implies relevant anti-oxidants concentrations in the appropriate target both at the beginning and during oxidative insult [169].

Accordingly and as summarized in Table 2.3, melatonin exerts its role as a skin guardian due to its multitasking nature, and consequently, this active molecule seems to exceed expectations in terms of cancer prevention, namely skin cancer [1].

2.5.3 Cosmetic and Therapeutic Perspectives

Clinical and cosmetic applications of melatonin were firstly based on the evidence of its benefits in the plant kingdom [32]. In fact, the presence of high levels of melatonin in most plants led to its use in traditional Chinese medicine. Those botanical extracts are usually used to treat pathologies or disorders which could be associated with free radical injury, such as certain skin diseases or age-related phenomenon [171]. Curiously other studies evidenced that plants exposed to strong UV irradiation, such as Mediterrean flora, present high levels of melatonin, which is consistent with its role on the prevention of UV-induced damage [172].

Although melatonin is produced in humans as extensively described, it could be also supplemented especially for certain diseases and/ or disorders. Regarding the route of administration, melatonin undergoes extensive first-pass degradation when taken orally, whereas its ability to penetrate into the stratum corneum and to form a depot when topically applied justifies the option for this last route [83, 173, 174].

Oral melatonin supplementation is especially used for phase shift and sleep disorder treatments. There are also several studies that recently claimed its oncotherapeutic role [175, 176].

In terms of **oral toxicity**, Bella *et al.* have investigated the safety use of melatonin over 42 years. These authors showed that a daily oral dose of 20–40 mg (pharmacological doses) up to a maximum of 1000 mg I.V. was perfectly well tolerated by the patients without toxicity or significant side effects, except the temporary drowsiness reported at the beginning of the treatment [176]. Moreover, interventional human studies with a melatonin diary intake of 1–6.6 g during 30–45 days followed by a battery of biochemical tests, have not detected any potential toxicity [176]. Accordingly, the Scientific Committee on Consumer Safety of the European Commission considers an acute oral LD50-value of melatonin > 3200 mg/kg body weight (high dose), which justifies the fact that melatonin is not considered as acutely harmful [177]. Nevertheless, a higher inspection of these nutraceutics by the competent regulatory authorities should be assured.

The melatonin incorporation in novel **topical** pharmaceutical and cosmetic delivery systems is a recent hot topic of formulation development [83, 178]. Nowadays, topical melatonin is mainly incorporated in **hair care products** indicated for androgenetic alopecia. In addition, recent clinical studies have tried to include this hormone in **sunscreen** (clinical trial—phase I) and **anti-aging creams** [83, 179].

Additionally, it is important to note that melatonin is a photosensitive substance that is particularly important during the development of topical formulations (mainly sunscreens) [180]. The strategy to encapsulate this radical scavenger on a safe and effective manner may be based on nanodelivery systems such as ethosomes, liposomes, and Pickering emulsions [181, 182]. Recently, we have successfully developed and characterized an innovative and "green" sunscreen formulation based on Pickering emulsions concept, i.e., surfactant-free emulsions stabilized by solid nanoparticles, natural oils and physical UV filters associated to melatonin as a key strategy for preventing the UV-induced skin damage (submitted data).

Concerning the **dermal toxicity**, no specific studies have been performed yet. However, Sugden *et al.* investigated the toxicity of melatonin applied subcutaneously in mice and rats determining an LD50-value (24 h) > 1600 mg/kg body weight [177].

2.6 Conclusion

The skin is an organ with an unquestionable importance for the organism protection and integrity. Unfortunately, this protective role could be overwhelmed by solar/UV overexposure. In this context, the prevalence of skin cancers is increasing at an alarming rate around the world, which justifies the urgent adoption of effective protective strategies.

The intracutaneous melatoninergic system is activated in response to external offenders as UVR. Consequently, melatonin may be a pivot agent for maintaining the functional integrity of the skin protecting it from UV damage. Furthermore, it is important to remember that beneficial and multiple effects of melatonin may be amplified by its several derivatives through the melatoninergic cascade. In fact, its particular metabolism contributes to consider melatonin as one of the most potent physiological scavengers of free radicals found to date.

The multiple clinical applications could be based from its primary role as a circadian pacemaker to its promising action as an oncotherapeutic agent. Oral and/or topical formulations are already used for an effective and safe melatonin supplementation.

Melatonin can be also considered as a new and effective compound which could be included in sunscreens and/or anti-aging formulations. However, further studies are still needed regarding its photostability, effective dose, and dermal toxicity.

In a near future, we believe that these constraints will be overcome, and melatonin will take its *place under the sun* in the cosmetic and pharmaceutical market.

List of Abbreviations

3-OHM: Cyclic 3-hydroxymelatonin
5-MIA: 5-Methoxyindolacetaldehyde
5-MIAA: 5-Methoxyindole acetic acid
5-MTOL: 5-Methoxytryptophol
5-MTT: 5-Methoxytryptamine
6(OH)M: 6-Hydroxymelatonin
6-BH4: (6R) 5,6,7,8-Tetrahydrobiopterin
6-4PPs: 6-4 Photoproducts
8-oxo-dG: 8-Oxo-7,8-dihydroguanine

8-OH-dG: 8-Hydroxy-2-deoxyguanosine
FapydG: 2,6-Diamino-4-hydroxy-5-formamidopyrimidine
AAD: Amino acid decarboxylase
AANAT: Arylalkylamine *N*-acetyltransferase
AAMC: 3-Acetyamidomethyl-6-methoxycinnolinone
AMK: *N*-Acetyl-5-methoxykynuramine
AFMK: *N*1-Acetyl-*N*2-formyl-5-methoxykynuramine
BER: Base excision repair
BSC: Basal cell carcinoma
C: Cytosine
CC: Central Clock
Cytosine glycol: 5,6-Dihydroxy-5,6-dihydrocytosine
CAT: Catalase
CYP: Cythocrome
CPDs: Cyclobutane-Pyrimidine Dimers
DDT: DNA damage tolerance mechanisms
DNA: Deoxyribonucleic Acid
GI: Gastro Intestinal
GPx: Glutathione Peroxidase
G: Guanine
HDR: Homology-dependent repair
HIOMT: Hydroxyindole-*O*-methyltransferase
IL: Interleukin
IR: Infrared radiation
LAN: Light at night
LC: Langerhans cells
MAO: Monoamine oxidase
MQA: *N*-[2-(6-Methoxyquinazolin-4-yl)-ethyl]-acetamide
MSC: Melanoma skin cancer
$\mathbf{MT_1}$: Melatonin surface receptor 1
$\mathbf{MT_2}$: Melatonin surface receptor 2
$\mathbf{MT_3}$: Melatonin receptor 3/quinone reductase II
NAS: *N*-Acetylserotonin
NER: Nucleotide excision repair
NMSC: Non-melanoma skin cancer
PARP: Poly(ADP-ribose) polymerase
$\mathbf{QR_2}$: Quinone reductase II
RNS: Reactive nitrogen species
ROR: Nuclear receptors of the tyrosine kinase-like orphan receptor
ROS: Reactive oxygen species

RZR: Retinoid Z receptor
SCC: Squamous cell carcinoma
SOD: Superoxide dismutase
T: Thymine
$\mathbf{T_{1/2}}$: Half-life
TGF: Transforming growth factor
Thymine glycol: 5,6-Dihydroxy-5,6 dihydrothymine
$\mathbf{TiO_2}$: Titanium dioxide
TLS: Translesion DNA synthesis
TPH: Tryptophan hydroxylase
UVA: Ultraviolet A (320–400 nm)
UVB: Ultraviolet radiation type B (280–320 nm)
UVR: Ultraviolet radiation
WHO: World Health Organization
ZnO: Zinc oxide

References

1. Reiter RJ, Tan DX, Galano A. Melatonin: Exceeding expectations. *Physiology (Bethesda, Md)*, 2014; 29: 325–333.
2. Zhang H-MZY. Melatonin: A well-documented antioxidant with conditional pro-oxidant actions. *J Pineal Res*, 2014; 57: 131–146.
3. Britannica E. Melatonin. *Encyclopædia Britannica Online*.
4. Mihandoost E, Shirazi A, Mahdavi SR, Aliasgharzadeh A. Can melatonin help us in radiation oncology treatments? *Bio Med Res Int*, 2014; 2014: 578137.
5. Vijayalaxmi, Reiter RJ, Tan DX, Herman TS, Thomas CR. Jr. Melatonin as a radioprotective agent: A review. *Int J Radiat Oncol Biol Phys*, 2004; 59: 639–653.
6. Slominski A, Tobin DJ, Zmijewski MA, Wortsman J, Paus R. Melatonin in the skin: Synthesis, metabolism and functions. *Trends Endocrinol Metab TEM*, 2008; 19: 17–24.
7. DrugBank. Melatonin (DB01065).
8. Marczynski TJ, Yamaguchi N, Ling GM, Grodzinska L. Sleep induced by the administration of melatonin (5-methoxyn-acetyltryptamine) to the hypothalamus in unrestrained cats. *Experientia*, 1964; 20: 435–437.
9. Reiter RJ. Melatonin: The chemical expression of darkness. *Mol Cell Endocrinol*, 1991; 79: C153–C158.

10. Reiter RJ. The melatonin rhythm: Both a clock and a calendar. *Experientia*, 1993; 49: 654–664.
11. Sánchez-Hidalgo M., Guerrero JM, Villegas I., Packham G., de la Lastra C. A. Melatonin, a natural programmed cell death inducer in cancer. *Curr Med Chem*, 2012; 19: 3805–3821.
12. Malhotra S, Sawhney G, Pandhi P. The therapeutic potential of melatonin: A review of the science. *Med Gen Med*, 2004; 6: 46.
13. Liu RY, Zhou JN, van Heerikhuize J, Hofman MA, Swaab DF. Decreased melatonin levels in postmortem cerebrospinal fluid in relation to aging, Alzheimer's disease, and apolipoprotein E-epsilon 4/4 genotype. *J Clin Endocrinol Metab*, 1999; 84: 323–327.
14. Sakotnik A, Liebmann PM, Stoschitzky K, Lercher P, Schauenstein K, Klein W, *et al.* Decreased melatonin synthesis in patients with coronary artery disease. *Eur Heart J*, 1999; 20: 1314–1317.
15. Yaprak M, Altun A, Vardar A, Aktoz M, Ciftci S, Ozbay G. Decreased nocturnal synthesis of melatonin in patients with coronary artery disease. *Int J Cardiol*, 2003; 89: 103–107.
16. Wurtman RJ, Zhdanova I. Improvement of sleep quality by melatonin. *Lancet*, 1995; 346: 1491.
17. Lavie P. Melatonin: Role in gating nocturnal rise in sleep propensity. *J Biol Rhythms*, 1997; 12: 657–665.
18. Zisapel N. Circadian rhythm sleep disorders: Pathophysiology and potential approaches to management. *CNS Drugs*, 2001; 15: 311–328.
19. Blask DE, Dauchy RT, Sauer LA. Putting cancer to sleep at night: The neuroendocrine/circadian melatonin signal. *Endocrine*, 2005; 27: 179–188.
20. Cardinali DP, Ladizesky MG, Boggio V, Cutrera RA, Mautalen C. Melatonin effects on bone: Experimental facts and clinical perspectives. *J Pineal Res*, 2003; 34: 81–87.
21. Bubenik GA. Gastrointestinal melatonin: Localization, function, and clinical relevance. *Dig Dis Sci*, 2002; 47: 2336–2348.
22. Doolen S, Krause DN, Dubocovich ML, Duckles SP. Melatonin mediates two distinct responses in vascular smooth muscle. *Eur J Pharmacol*, 1998; 345: 67–69.
23. Scheer FA, Van Montfrans GA, van Someren EJ, Mairuhu G, Buijs RM. Daily nighttime melatonin reduces blood pressure in male patients with essential hypertension. *Hypertension*, 2004; 43: 192–197.
24. Guerrero JM, Reiter RJ. Melatonin-immune system relationships. *Curr Top Med Chem*, 2002; 2: 167–179.

25. Carrillo-Vico A, Guerrero JM, Lardone PJ, Reiter RJ. A review of the multiple actions of melatonin on the immune system. *Endocrine*, 2005; 27: 189–200.

26. Esquifino AI, Pandi-Perumal SR, Cardinali DP. Circadian organization of the immune response: A role for melatonin. *Clin Appl Immunol Rev*, 2004; 4: 423–433.

27. Armstrong SM. Melatonin and circadian control in mammals. *Experientia*, 1989; 45: 932–938.

28. Iuvone PM, Tosini G, Pozdeyev N, Haque R, Klein DC, Chaurasia SS. Circadian clocks, clock networks, arylalkylamine *N*-acetyltransferase, and melatonin in the retina. *Prog Retin Eye Res*, 2005; 24: 433–456.

29. Reiter RJ, Tan DX, Maldonado MD. Melatonin as an antioxidant: Physiology versus pharmacology. *J Pineal Res*, 2005; 39: 215–216.

30. Rezzani R RL, Favero G, Damiani G, Paganelli C, Reiter RJ. Attenuation of ultraviolet A-induced alterations in NIH3T3 dermal fibroblasts by melatonin. *Br J Dermatol*, 2014; 170: 382–391.

31. Tan DX, Manchester LC, Terron MP, Flores LJ, Reiter RJ. One molecule, many derivatives: A never-ending interaction of melatonin with reactive oxygen and nitrogen species? *J Pineal Res*, 2007; 42: 28–42.

32. Pandi-Perumal SR, Srinivasan V, Maestroni GJ, Cardinali DP, Poeggeler B, Hardeland R. Melatonin: Nature's most versatile biological signal? *FEBS J*, 2006; 273: 2813–2838.

33. Slominski A WJ, Tobin DJ. The cutaneous serotoninergic/Melatoninergic system: Securing a place under the sun. *FASEB J*, 2005; 19: 176–194.

34. Carrillo-Vico A, Lardone PJ, Fernandez-Santos JM, Martin-Lacave I, Calvo JR, Karasek M, *et al.* Human lymphocyte-synthesized melatonin is involved in the regulation of the interleukin-2/interleukin-2 receptor system. *J Clin Endocrinol Metab*, 2005; 90: 992–1000.

35. Tosini G, Davidson AJ, Fukuhara C, Kasamatsu M, Castanon-Cervantes O. Localization of a circadian clock in mammalian photoreceptors. *FASEB J*, 2007; 21: 3866–3871.

36. Slominski AT, Kleszczyński K, Semak I, Janjetovic Z, Żmijewski MA, Kim T-K, *et al.* Local melatoninergic system as the protector of skin integrity. *Int J Mol Sci*, 2014; 15: 17705–17732.

37. Fischer TW, Sweatman TW, Semak I, Sayre RM, Wortsman J, Slominski A. Constitutive and UV-induced metabolism of melatonin in keratinocytes and cell-free systems. *FASEB J*, 2006; 20: 1564–1566.

38. Zhang X, Beaulieu JM, Sotnikova TD, Gainetdinov RR, Caron MG. Tryptophan hydroxylase-2 controls brain serotonin synthesis. *Science*, 2004; 305: 217.

39. Reiter RJ, Tan DX, Terron MP, Flores LJ, Czarnocki Z. Melatonin and its metabolites: New findings regarding their production and their radical scavenging actions. *Acta Biochim Pol*, 2007; 54: 1–9.

40. Ribelayga C, Pevet P, Simonneaux V. HIOMT drives the photoperiodic changes in the amplitude of the melatonin peak of the Siberian hamster. *Am J Physiol Regulatory, Integr Comp Physiol*, 2000; 278: R1339–R1345.

41. Borjigin J, Zhang LS, Calinescu AA. Circadian regulation of pineal gland rhythmicity. *Mol Cell Endocrinol*, 2012; 349: 13–19.

42. Chattoraj A, Liu T, Zhang LS, Huang Z, Borjigin J. Melatonin formation in mammals: *In vivo* perspectives. *Rev Endocr Metab Disord*, 2009; 10: 237–243.

43. Radomir S, Andrzej TS. Synthesis and metabolism of melatonin in the skin and retinal pigment epithelium. In *Melatonin in the Promotion of Health*, 2nd ed, CRC Press; 2011. pp. 69–80.

44. Galano A, Medina ME, Tan DX, Reiter RJ. Melatonin and its metabolites as copper chelating agents and their role in inhibiting oxidative stress: A physicochemical analysis. *J Pineal Res*, 2015; 58: 107–116.

45. Kim TK, Kleszczynski K, Janjetovic Z, Sweatman T, Lin Z, Li W, *et al.* Metabolism of melatonin and biological activity of intermediates of melatoninergic pathway in human skin cells. *FASEB J*, 2013; 27: 2742–2755.

46. Hirata F, Hayaishi O, Tokuyama T, Seno S. *In vitro* and *in vivo* formation of two new metabolites of melatonin. *J Biol Chem*, 1974; 249: 1311–1313.

47. Hardeland R, Reiter RJ, Poeggeler B, Tan DX. The significance of the metabolism of the neurohormone melatonin: Antioxidative protection and formation of bioactive substances. *Neurosci Biobehav Rev*, 1993; 17: 347–357.

48. Slominski AT, Zmijewski MA, Skobowiat C, Zbytek B, Slominski RM, Steketee JD. Sensing the environment: Regulation of local and global homeostasis by the skin's neuroendocrine system. *Adv Anat Embryol Cell Biol*, 2012; 212: v, vii, 1–115.

49. Galano A, Tan DX, Reiter RJ. Melatonin as a natural ally against oxidative stress: A physicochemical examination. *J Pineal Res*, 2011; 51: 1–16.

50. Slominski A, Wortsman J, Tobin DJ. The cutaneous serotoninergic/melatoninergic system: Securing a place under the sun. *FASEB J*, 2005; 19: 176–194.

51. Slominski RM, Reiter RJ, Schlabritz-Loutsevitch N, Ostrom RS, Slominski AT. Melatonin membrane receptors in peripheral tissues: Distribution and functions. *Mol Cell Endocrinol*, 2012; 351: 152–166.

52. Reiter RJ. Melatonin: Clinical relevance. *Best Pract Res Clin Endocrinol Metab*, 2003; 17: 273–285.

53. Dubocovich ML, Markowska M. Functional MT1 and MT2 melatonin receptors in mammals. *Endocrine*, 2005; 27: 101–110.

54. Witt-Enderby PA, Bennett J, Jarzynka MJ, Firestine S, Melan MA. Melatonin receptors and their regulation: *Biochem Struct Mech Life Sci*, 2003; 72: 2183–2198.

55. Luchetti F, Canonico B, Betti M, Arcangeletti M, Pilolli F, Piroddi M, *et al.* Melatonin signaling and cell protection function. *FASEB J*, 2010; 24: 3603–3624.

56. Ebisawa T, Karne S, Lerner MR, Reppert SM. Expression cloning of a high-affinity melatonin receptor from Xenopus dermal melanophores. *Proc Natl Acad Sci USA*, 1994; 91: 6133–6137.

57. Nosjean O, Nicolas J-P, Klupsch F, Delagrange P, Canet E, Boutin JA. Comparative pharmacological studies of melatonin receptors: MT1, MT2 and MT3/QR2. Tissue distribution of MT3/QR2. *Biochem Pharmacol*, 2001; 61: 1369–1379.

58. Slominski AFT, Zmijewski MA, Wortsman J, Semak I, Zbytek B, Slominski M, Tobin DJ. On the role of melatonin in skin physiology and pathology. *End Rev*, 2005; 27: 137–148.

59. Nosjean O, Ferro M, Coge F, Beauverger P, Henlin JM, Lefoulon F, *et al.* Identification of the melatonin-binding site MT3 as the quinone reductase 2. *J Biol Chem*, 2000; 275: 31311–31317.

60. Madeo J, Zubair A, Marianne F. A review on the role of quinones in renal disorders. *Springer Plus*, 2013; 2: 139.

61. Tan DX, Manchester LC, Terron MP, Flores LJ, Tamura H, Reiter RJ. Melatonin as a naturally occurring co-substrate of quinone reductase-2, the putative MT3 melatonin membrane receptor: Hypothesis and significance. *J Pineal Res*, 2007; 43: 317–320.

62. Cutando A, Lopez-Valverde A, Arias-Santiago S, J DEV, RG DED. Role of melatonin in cancer treatment. *Anticancer Res*, 2012; 32: 2747–2753.

63. Tan DX, Manchester LC, Hardeland R, Lopez-Burillo S, Mayo JC, Sainz RM, *et al.* Melatonin: A hormone, a tissue factor, an autocoid, a paracoid, and an antioxidant vitamin. *J Pineal Res*, 2003; 34: 75–78.

64. Kleszczynski K, Hardkop LH, Fischer TW. Differential effects of melatonin as a broad range UV-damage preventive dermato-endocrine regulator. *Dermatoendocrinology*, 2011; 3: 27–31.

65. Simonneaux V, Ribelayga C. Generation of the melatonin endocrine message in mammals: A review of the complex regulation of melatonin synthesis by norepinephrine, peptides, and other pineal transmitters. *Pharmacol Rev*, 2003; 55: 325–395.

66. Fischer TW, Scholz G, Knoll B, Hipler UC, Elsner P. Melatonin reduces UV-induced reactive oxygen species in a dose-dependent manner in IL-3-stimulated leukocytes. *J Pineal Res*, 2001; 31: 39–45.

67. Fischer TW, Scholz G, Knoll B, Hipler UC, Elsner P. Melatonin suppresses reactive oxygen species induced by UV irradiation in leukocytes. *J Pineal Res*, 2004; 37: 107–112.

68. Fischer TW, Zmijewski MA, Wortsman J, Slominski A. Melatonin maintains mitochondrial membrane potential and attenuates activation of initiator (casp-9) and effector caspases (casp-3/casp-7) and PARP in UVR-exposed HaCaT keratinocytes. *J Pineal Res*, 2008; 44: 397–407.

69. Fischer TW, Zbytek B, Sayre RM, Apostolov EO, Basnakian AG, Sweatman TW, *et al.* Melatonin increases survival of HaCaT keratinocytes by suppressing UV-induced apoptosis. *J Pineal Res*, 2006; 40: 18–26.

70. Luchetti F, Canonico B, Curci R, Battistelli M, Mannello F, Papa S, *et al.* Melatonin prevents apoptosis induced by UV-B treatment in U937 cell line. *J Pineal Res*, 2006; 40: 158–167.

71. Kleszczynski K, Tukaj S, Kruse N, Zillikens D, Fischer TW. Melatonin prevents ultraviolet radiation-induced alterations in plasma membrane potential and intracellular pH in human keratinocytes. *J Pineal Res*, 2013; 54: 89–99.

72. Izykowska I, Cegielski M, Gebarowska E, Podhorska-Okolow M, Piotrowska A, Zabel M, *et al.* Effect of melatonin on human keratinocytes and fibroblasts subjected to UVA and UVB radiation *in vitro*. *In vivo*, 2009; 23: 739–745.

73. Nickel A, Wohlrab W. Melatonin protects human keratinocytes from UVB irradiation by light absorption. *Arch Dermatol Res*, 2000; 292: 366–368.

74. Fransen M, Nordgren M, Wang B, Apanasets O. Role of peroxisomes in ROS/RNS-metabolism: Implications for human disease. *Biochim Biophys Acta*, 2012; 1822: 1363–1373.

75. Fischer TW, Kleszczynski K, Hardkop LH, Kruse N, Zillikens D. Melatonin enhances antioxidative enzyme gene expression (CAT, GPx, SOD), prevents their UVR-induced depletion, and protects against the formation of DNA damage (8-hydroxy-2′-deoxyguanosine) in *ex vivo* human skin. *J Pineal Res*, 2013; 54: 303–312.

76. Madison KC. Barrier function of the skin: "la raison d'etre" of the epidermis. *J Invest Dermatol*, 2003; 121: 231–241.

77. Swann G. The skin is the body's largest organ. *J Vis Commun Med*, 2010; 33: 148–149.

78. Slominski A, Wortsman J. Neuroendocrinology of the skin. *Endocr Rev*, 2000; 21: 457–487.

79. Schafer M, Werner S. The cornified envelope: A first line of defense against reactive oxygen species. *J Invest Dermatol*, 2011; 131: 1409–1411.

80. Lee C-H, Wu S-B, Hong C-H, Yu H-S, Wei Y-H. Molecular mechanisms of UV-induced apoptosis and its effects on skin residential cells: The implication in UV-based phototherapy. *Int J Mol Sci*, 2013; 14: 6414–6435.

81. Poljšak B, Dahmane R. Free radicals and extrinsic skin aging. *Dermatol Res Pract*, 2012; 2012: http://dx.doi.org/10.1155/2012/135206.

82. Moan J, Juzeniene A. Solar radiation and human health. *J Photochem Photobiol B*, 2010; 101: 109–110.

83. Kleszczynski KFT. Melatonin and human skin aging. *Dermatol Endocrinol*, 2012; 4: 245–252.

84. D'Orazio J, Jarrett S, Amaro-Ortiz A, Scott T. UV radiation and the skin. *Int J Mol Sci*, 2013; 14: 12222–12248.

85. Pincelli C, Marconi A. Keratinocyte stem cells: Friends and foes. *J Cell Physiol*, 2010; 225: 310–315.

86. Ando H, Niki Y, Yoshida M, Ito M, Akiyama K, Kim J-H, *et al.* Involvement of pigment globules containing multiple melanosomes in the transfer of melanosomes from melanocytes to keratinocytes. *Cell Logistics*, 2011; 1: 12–20.

87. Brenner M, Hearing VJ. The protective role of melanin against UV damage in human skin. *J Photochem Photobiol B*, 2008; 84: 539–549.

88. Koch S, Kohl K, Klein E, von Bubnoff D, Bieber T. Skin homing of Langerhans cell precursors: Adhesion, chemotaxis, and migration. *J Allergy Clin Immunol*, 2006; 117: 163–168.
89. Kripke ML. Immunology and photocarcinogenesis. New light on an old problem. *J Am Acad Dermatol*, 1986; 14: 149–155.
90. Ullrich SE. Photoimmune suppression and photocarcinogenesis. *Front Biosci A J Virtual Library*, 2002; 7: d684–d703.
91. Holloway A, Simmonds M, Azad A, Fox JL, Storey A. Resistance to UV-induced apoptosis by beta-HPV5 E6 involves targeting of activated BAK for proteolysis by recruitment of the HERC1 ubiquitin ligase. *Int J Cancer J Int Du Cancer*, 2014.
92. Foundation SC. Skin Cancer Facts. Skin Cancer Foundation; 2015.
93. Matsumura Y, Ananthaswamy HN. Molecular mechanisms of photocarcinogenesis. *Front Biosci A J Virtual Library*, 2002; 7: d765–d783.
94. Gruber F, Zamolo G, Kastelan M, Massari LP, Cabrijan L, Peharda V, *et al.* Photocarcinogenesis–molecular mechanisms. *Coll Antropol*, 2007; 31: 101–106.
95. Armstrong BK, Kricker A. The epidemiology of UV induced skin cancer. *J Photochem Photobio B*, 2001; 63: 8–18.
96. Leiter U, Garbe C. Epidemiology of melanoma and nonmelanoma skin cancer–the role of sunlight. *Adv Exp Med Biol*, 2008; 624: 89–103.
97. Gandini S, Sera F, Cattaruzza MS, Pasquini P, Picconi O, Boyle P, *et al.* Meta-analysis of risk factors for cutaneous melanoma: II. Sun exposure. *Eur J Cancer*, 2005; 41: 45–60.
98. Scheuer C, Pommergaard HC, Rosenberg J, Gogenur I. Melatonin's protective effect against UV radiation: A systematic review of clinical and experimental studies. *Photodermatol Photoimmunol Photomed*, 2014; 30: 180–188.
99. Polefka TG, Meyer TA, Agin PP, Bianchini RJ. Effects of solar radiation on the skin. *J Cosmet Dermatol*, 2012; 11: 134–143.
100. Sinha RP, Hader DP. UV-induced DNA damage and repair: A review. *J Photochem Photobiol B*, 2002; 1: 225–236.
101. Pustisek N, Situm M. UV-radiation, apoptosis and skin. *Coll Antropol*, 2011; 35: 339–341.
102. Halliday GM. Inflammation, gene mutation and photoimmunosuppression in response to UVR-induced oxidative damage contributes to photocarcinogenesis. *Mutat Res*, 2005; 571: 107–120.
103. Narayanapillai S, Agarwal C, Tilley C, Agarwal R. Silibinin is a potent sensitizer of UVA radiation-induced oxidative stress and apoptosis

in human keratinocyte HaCaT cells. *J Photochem Photobiol B*, 2012; 88: 1135–1140.

104. Ascenso A SA, Euletério C, Praça F, Bentley MV, Marques HC, *et al.* *In vitro* and *in vivo* topical delivery studies of tretinoin-loaded ultradeformable vesicles. *Eur J Pharm Biopharm*, 2014; 88: 48–55.

105. Menezes AC, Raposo S, Simoes S, Ribeiro H, Oliveira H, Ascenso A. Prevention of photocarcinogenesis by agonists of 5-HT1A and antagonists of 5-HT2A receptors. *Mol Neurobiol*, 2015; 53: 1145–1164.

106. You YH, Szabo PE, Pfeifer GP. Cyclobutane pyrimidine dimers form preferentially at the major p53 mutational hotspot in UVB-induced mouse skin tumors. *Carcinogenesis*, 2000; 21: 2113–2117.

107. Budden T, Bowden NA. The role of altered nucleotide excision repair and UVB-induced DNA damage in melanomagenesis. *Int J Mol Sci*, 2013; 14: 1132–1151.

108. Anna B, Blazej Z, Jacqueline G, Andrew CJ, Jeffrey R, Andrzej S. Mechanism of UV-related carcinogenesis and its contribution to nevi/melanoma. *Exp Rev Dermatol*, 2007; 2: 451–469.

109. Douki T, Cadet J. Individual determination of the yield of the main UV-induced dimeric pyrimidine photoproducts in DNA suggests a high mutagenicity of CC photolesions. *Biochemistry*, 2001; 40: 2495–24501.

110. Friedberg EC, Walker GC, Siede W, Wood RD. *DNA Repair and Mutagenesis: American Society for Microbiology*; 2006.

111. Izhar L, Ziv O, Cohen IS, Geacintov NE, Livneh Z. Genomic assay reveals tolerance of DNA damage by both translesion DNA synthesis and homology-dependent repair in mammalian cells. *Proc Natl Acad Sci U S A*, 2013; 110: E1462–E1469.

112. Ghosal G, Chen J. DNA damage tolerance: A double-edged sword guarding the genome. *Trans Cancer Res*, 2013; 2: 107–129.

113. Murphy GM. Ultraviolet radiation and immunosuppression. *Br J Dermatol*, 2009; 161: 90–95.

114. Black HS, deGruijl FR, Forbes PD, Cleaver JE, Ananthaswamy HN, deFabo EC, *et al.* Photocarcinogenesis: An overview. *J Photochem Photobiol B*, 1997; 40: 29–47.

115. Aloni-Grinstein R, Shetzer Y, Kaufman T, Rotter V. p53: The barrier to cancer stem cell formation. *FEBS Lett*, 2014; 588: 2580–2589.

116. Rao B, Lain S, Thompson AM. p53-Based cyclotherapy: Exploiting the "guardian of the genome" to protect normal cells from cytotoxic therapy. *Br J Cancer*, 2013; 109: 2954–2958.

117. Besaratinia A, Synold TW, Chen HH, Chang C, Xi B, Riggs AD, *et al.* DNA lesions induced by UV A1 and B radiation in human cells: Comparative analyses in the overall genome and in the p53 tumor suppressor gene. *Proc Natl Acad Sci USA*, 2005; 102: 10058–10063.

118. Kryston TB, Georgiev AB, Pissis P, Georgakilas AG. Role of oxidative stress and DNA damage in human carcinogenesis. *Mutat Res*, 2011; 711: 193–201.

119. Kawachi Y, Xu X, Taguchi S, Sakurai H, Nakamura Y, Ishii Y, *et al.* Attenuation of UVB-induced sunburn reaction and oxidative DNA damage with no alterations in UVB-induced skin carcinogenesis in Nrf2 gene-deficient mice. *J Invest Dermatol*, 2008; 128: 1773–1779.

120. Song EJ, Gordon-Thomson C, Cole L, Stern H, Halliday GM, Damian DL, *et al.*, 1-alpha, 25-Dihydroxyvitamin D3 reduces several types of UV-induced DNA damage and contributes to photoprotection. *J Steroid Biochem Mol Biol*, 2013; 136: 131–138.

121. Huang XX, Scolyer RA, Abubakar A, Halliday GM. Human 8-oxoguanine-DNA glycosylase-1 is downregulated in human basal cell carcinoma. *Mol Gen Metab*, 2012; 106: 127–130.

122. Legrand M, Chan CL, Jauert PA, Kirkpatrick DT. Analysis of base excision and nucleotide excision repair in Candida albicans. *Microbiology*, 2008; 154: 2446–2456.

123. Wolfle U, Haarhaus B, Schempp CM. The photoprotective and antioxidative properties of luteolin are synergistically augmented by tocopherol and ubiquinone. *Planta Med*, 2013; 79: 963–695.

124. Fartasch M, Diepgen TL, Schmitt J, Drexler H. The relationship between occupational sun exposure and non-melanoma skin cancer: Clinical basics, epidemiology, occupational disease evaluation, and prevention. *Deutsches Arzteblatt Int*, 2012; 109: 715–720.

125. Aubin F. Mechanisms involved in ultraviolet light-induced immunosuppression. *Eur J Dermatol EJD*, 2003; 13: 515–523.

126. Strozyk E, Kulms D. The role of AKT/mTOR pathway in stress response to UV-irradiation: implication in skin carcinogenesis by regulation of apoptosis, autophagy and senescence. *Int J Mol Sci*, 2013; 14: 15260–15285.

127. Abdel-Rahman WM. Genomic instability and carcinogenesis: An update. *Curr Gen*, 2008; 9: 535–541.

128. Belancio VP, Blask DE, Deininger P, Hill SM, Jazwinski SM. The aging clock and circadian control of metabolism and genome stability. *Front Gen*, 2014; 5: 455.

129. Bonde JP, Hansen J, Kolstad HA, Mikkelsen S, Olsen JH, Blask DE, *et al.* Work at night and breast cancer–report on evidence-based options for preventive actions. *Scand J Work Environ Health*, 2012; 38: 380–390.

130. Cahill DP, Kinzler KW, Vogelstein B, Lengauer C. Genetic instability and darwinian selection in tumours. *Trends Cell Biol*, 1999; 9: M57–M60.

131. Lengauer C, Kinzler KW, Vogelstein B. Genetic instabilities in human cancers. *Nature*, 1998; 396: 643–649.

132. Huang L, Snyder AR, Morgan WF. Radiation-induced genomic instability and its implications for radiation carcinogenesis. *Oncogene*, 2003; 22: 5848–5854.

133. Suzuki K, Ojima M, Kodama S, Watanabe M. Radiation-induced DNA damage and delayed induced genomic instability. *Oncogene*, 2003; 22: 6988–6993.

134. Dahle J, Kvam E, Stokke T. Bystander effects in UV-induced genomic instability: Antioxidants inhibit delayed mutagenesis induced by ultraviolet A and B radiation. *J Carcinog*, 2005; 4: 11.

135. Abdel-Rahman WM, Mecklin JP, Peltomaki P. The genetics of HNPCC: Application to diagnosis and screening. *Crit Rev Oncol/Hematol*, 2006; 58: 208–220.

136. Dahle J, Kvam E. Induction of delayed mutations and chromosomal instability in fibroblasts after UVA-, UVB-, and X-radiation. *Cancer Res*, 2003; 63: 1464–1469.

137. Widel M. Bystander effect induced by UV radiation; why should we be interested? *Postepy Hig Med Dosw*, 2012; 66: 828–837.

138. Little JB. Genomic instability and bystander effects: A historical perspective. *Oncogene*, 2003; 22: 6978–6987.

139. Harada T, Kashino G, Suzuki K, Matsuda N, Kodama S, Watanabe M. Different involvement of radical species in irradiated and bystander cells. *Int J Radiat Biol*, 2008; 84: 809–814.

140. Konopacka M, Rzeszowska-Wolny J. The bystander effect-induced formation of micronucleated cells is inhibited by antioxidants, but the parallel induction of apoptosis and loss of viability are not affected. *Mutat Res*, 2006; 593: 32–38.

141. Whiteside JR, McMillan TJ. A bystander effect is induced in human cells treated with UVA radiation but not UVB radiation. *Radiat Res*, 2009; 171: 204–211.

142. Panda S, Hogenesch JB, Kay SA. Circadian rhythms from flies to human. *Nature*, 2002; 417: 329–335.

143. Reppert SM, Weaver DR. Coordination of circadian timing in mammals. *Nature*, 2002; 418: 935–941.

144. Obeng MK. The American Heritage Stedman's Medical Dictionary, *J Natl Med Assoc*, 2003; 95: 634–635.

145. Froy O. Circadian rhythms, aging, and life span in mammals. *Am J Physiol*, 2011; 26: 225–235.

146. Kovacic P, Somanathan R. Melatonin and circadian rhythm: Aging, cancer, and mechanism. *Open J Prev Med*, 2014; 4: 16.

147. Froy O, Chapnik N. Circadian oscillation of innate immunity components in mouse small intestine. *Mol Immunol*, 2007; 44: 1954–1960.

148. Lee C, Etchegaray JP, Cagampang FR, Loudon AS, Reppert SM. Posttranslational mechanisms regulate the mammalian circadian clock. *Cell*, 2001; 107: 855–867.

149. Young ME. The circadian clock within the heart: Potential influence on myocardial gene expression, metabolism, and function. *Am J Physiol Heart Circ Physiol*, 2006; 290: H1–H16.

150. Levi F, Okyar A, Dulong S, Innominato PF, Clairambault J. Circadian timing in cancer treatments. *Ann Rev Pharmacol Toxicol*, 2010; 50: 377–421.

151. Anea CB, Zhang M, Stepp DW, Simkins GB, Reed G, Fulton DJ, *et al.* Vascular disease in mice with a dysfunctional circadian clock. *Circulation*, 2009; 119: 1510–1517.

152. Davis S, Mirick DK. Circadian disruption, shift work and the risk of cancer: A summary of the evidence and studies in Seattle. *Cancer Causes Control*, 2006; 17: 539–545.

153. Jung-Hynes B, Reiter RJ, Ahmad N. Sirtuins, melatonin and circadian rhythms: Building a bridge between aging and cancer. *J Pineal Res*, 2010; 48: 9–19.

154. Partch CL, Green CB, Takahashi JS. Molecular architecture of the mammalian circadian clock. *Trends Cell Biol*, 2014; 24: 90–99.

155. Godic A, Polj, #x161, ak B, Adamic M, Dahmane R. The role of antioxidants in skin cancer prevention and treatment. *Oxidative Med Cell Longevity*, 2014; 2014: 6.

156. Shindo Y, Witt E, Packer L. Antioxidant defense mechanisms in murine epidermis and dermis and their responses to ultraviolet light. *J Invest Dermatol*, 1993; 100: 260–265.

157. Pandel R, #x17e, Polj, #x161, ak B, Godic A, *et al.* Skin photoaging and the role of antioxidants in its prevention. *ISRN Dermatol*, 2013; 2013: 11.

158. Afaq F, Adhami VM, Ahmad N, Mukhtar H. Botanical antioxidants for chemoprevention of photocarcinogenesis. *Front Biosci A J Vir Library*, 2002; 7: d784–d792.

159. Wang ZY, Agarwal R, Bickers DR, Mukhtar H. Protection against ultraviolet B radiation-induced photocarcinogenesis in hairless mice by green tea polyphenols. *Carcinogenesis*, 1991; 12: 1527–530.

160. Ahmad N, Mukhtar H. Green tea polyphenols and cancer: Biologic mechanisms and practical implications. *Nutr Rev*, 1999; 57: 78–83.

161. Liebler DC, Burr JA. Effects of UV light and tumor promoters on endogenous vitamin E status in mouse skin. *Carcinogenesis*, 2000; 21: 221–225.

162. Poljsak B, Dahmane R, Godic A. Skin and antioxidants. *J Cosmet Laser Ther*, 2013; 15: 107–113.

163. Carocho M, Ferreira IC. A review on antioxidants, prooxidants and related controversy: Natural and synthetic compounds, screening and analysis methodologies and future perspectives. *Food Chem Toxicol*, 2013; 51: 15–25.

164. Ganceviciene R, Liakou AI, Theodoridis A, Makrantonaki E, Zouboulis CC. Skin anti-aging strategies. *Dermato Endocrinol*, 2012; 4: 308–319.

165. Nguyen G, Torres A. Systemic antioxidants and skin health. *J Drugs Dermatol*, 2012; 11: e1–e4.

166. Cecilie Scheuer HH. The sun protective effect of melatonin. *Clinical Trials.gov*; 2013.

167. Fischer T, Bangha E, Elsner P, Kistler GS. Suppression of UV-induced erythema by topical treatment with melatonin. Influence of the application time point. *Biol Signals Receptors*, 1999; 8: 132–135.

168. Howes RA, Halliday GM, Damian DL. Effect of topical melatonin on ultraviolet radiation-induced suppression of Mantoux reactions in humans. *Photodermatol Photoimmunol Photomed*, 2006; 22: 267–269.

169. Dreher F, Denig N, Gabard B, Schwindt DA, Maibach HI. Effect of topical antioxidants on UV-induced erythema formation when administered after exposure. *Dermatology*, 1999; 198: 52–55.

170. Salucci S, Burattini S, Curzi D, Buontempo F, Martelli AM, Zappia G, *et al.* Antioxidants in the prevention of UVB-induced keratynocyte apoptosis. *J Photochem Photobiol B*, 2014; 141: 1–9.

171. Chen G, Huo Y, Tan DX, Liang Z, Zhang W, Zhang Y. Melatonin in Chinese medicinal herbs. *Life Sci*, 2003; 73: 19–26.

172. Hardeland R, Pandi-Perumal SR. Melatonin, a potent agent in antioxidative defense: Actions as a natural food constituent, gastrointestinal factor, drug and prodrug. *Nutr Metab*, 2005; 2: 22.

173. Lane EA, Moss HB. Pharmacokinetics of melatonin in man: First pass hepatic metabolism. *J Clin Endocrinol Metabol*, 1985; 61: 1214–1216.

174. Fischer TW, Greif C, Fluhr JW, Wigger-Alberti W, Elsner P. Percutaneous penetration of topically applied melatonin in a cream and an alcoholic solution. *Skin Pharmacol Physiol*, 2004; 17: 190–194.

175. Skene D, Lockley S, Arendt J. Use of melatonin in the treatment of phase shift and sleep disorders. In *Tryptophan, Serotonin, and Melatonin* (Huether G, Kochen W, Simat T, Steinhart H, eds), Springer US; 1999. pp. 79–84.

176. Di Bella G, Mascia F, Gualano L, Di Bella L. Melatonin anticancer effects: Review. *Int J Mol Sci*, 2013; 14: 2410–2430.

177. Safety SCoC. Opinion on melatonin, European Commission, Brussels 2010.

178. Rivara S, Pala D, Bedini A, Spadoni G. Therapeutic uses of melatonin and melatonin derivatives: A patent review (2012–2014). *Exp Opin Ther Patents*, 2015; 25: 425–441.

179. Scheuer C. Evaluation of percutaneous dissolvement of melatonin when used as sunscreen; a randomized, placebo controlled, double-blind crossover study on healthy volunteers. In: GCP CU, editor. Denmark 2014.

180. Xu X-R, Li X-Y, Li X-Z, Li H-B. Degradation of melatonin by UV, UV/H2O2, Fe^{2+}/H_2O_2 and UV/Fe^{2+}/H_2O_2 processes. *Separation Purif. Technol.*, 2009; 68: 261–266.

181. Parfenyuk E, Alyoshina N, Antsiferova YS, Sotnikova NY. Drug Delivery Nanosystems as a Promising Area of Modern Chemistry and Medicine. Silica Nanoparticles as Potential Drug Carriers. Silica Nanoparticles as Drug Delivery System for Immunomodulator GMDP (Biomedical & Nanomedical Technologies-Concise Monograph Series): ASME Press; 2012.

182. Mehanna MM, Motawaa AM, Samaha. MW. Pharmaceutical particulate carriers: Lipid-based carriers. *Natl J Physiol Pharm Pharmacol*, 2012; 2: 10–22.

Chapter 3

Safety and Efficacy of Sunscreen Formulations Containing Carrier or Non-Carrier-Based UV-Filters

Carolina Gomes Benevenuto and Lorena Rigo Gaspar

Faculdade de Ciências Farmacêuticas de Ribeirão Preto, Universidade de São Paulo, Ribeirão Prato, SP, Brazil

lorena@fcfrp.usp.br

Several parameters should be assessed during the formulation phase in order to provide safe and efficient sunscreens with a high cosmetic quality. Product safety is guaranteed by reduced adverse reactions mainly caused by fragrances, preservatives, emulsifiers, and UV-filters.

Both raw materials and finished products must undergo screening with safety tests using *in vitro* assays to determine irritation, cytotoxicity, corrosivity, phototoxicity, and allergenicity. When acceptable results are obtained (i.e., absence of any risks), these materials and/or products can be further analyzed by human *in vivo* tests. *In vitro* and *in vivo* efficacy studies should be also conducted considering the type of product. Although *in vitro* tests are useful to evaluate isolated events, it is necessary to extrapolate them to the human skin since skin

Carrier-Mediated Dermal Delivery: Applications in the Prevention and Treatment of Skin Disorders
Edited by Andreia Ascenso, Sandra Simões, and Helena Ribeiro

ISBN 978-981-4745-58-1 (Hardcover), 978-1-315-36447-6 (eBook)
www.panstanford.com

penetration does not occur in these studies. Thus, both *in vitro* and topical studies are necessary to guarantee the efficacy of a cosmetic product.

A careful selection of *in vitro* tests is required to evaluate sunscreens since most UV-filters are poorly solubilized in culture medium, and some physical UV-filters (nanoparticles) have high adsorption capacity and optical activity, which may influence some conventional assays. The most commonly used physical UV-filter is titanium dioxide that is found either in fine particles and in nanoparticles. TiO_2 nanoparticles are widely manufactured and applied to sunscreen formulations. Notably, these nanoparticles have different physicochemical and bioactivity properties compared to fine particles. Although most authors consider that TiO_2 nanoparticles do not penetrate the skin, some studies have reported that some skin damage such as UVB-induced burning could enhance the penetration of TiO_2 nanoparticles. However, it is still accepted that TiO_2 nanoparticles do not achieve systemic availability after cutaneous exposure.

Therefore, this chapter aims to focus on the safety and efficacy of sunscreen formulations containing carrier- or non-carrier-based UV-filters.

3.1 UV Radiation and the Skin

Of the UV radiation that reaches the Earth, 90–99% is UVA (320–400 nm) and only 1–10% is UVB (280–320 nm). UVC (200–280 nm) is filtered by the ozone layer and does not reach the Earth [1, 2]. UVA radiation represents the longest wavelength (lowest frequency and energy), is able to penetrate into the deeper layers of the skin, and is responsible for the induction of skin photoaging and tanning, as well as immediate and persistent skin pigmentation. UVB radiation represents a lower wavelength than UVA (higher frequency and energy), is able to penetrate the superficial layers of the skin due to the presence of higher amount of UVB than UVA chromophores, and is responsible for sunburn, skin cancer, and photoaging [3]. Considering that both UVA and UVB radiation has enough energy to affect the skin, exposure to UV can damage the skin. Among the acute effects, the most important are erythema (inflammation) and tanning (increase in melanogenesis) [4]. The chronic effects of UV exposure involve a gradual deterioration of structures and skin function due to cumulative DNA damage and chronic inflammation [4, 5]. In addition, chronic exposure to UV radiation leads to skin aging, wrinkles, uneven pigmentation, loss of elasticity, and changes in skin barrier function [6].

Whereas UVB is erythemogenic and carcinogenic, UVA can induce photoaging, damage to DNA, RNA and proteins, and also reactive oxygen species (ROS) production, such as formation of singlet oxygen, superoxide anion and the hydroxyl radical, which damage DNA and cell membranes and promote carcinogenesis and some changes associated with photoaging [7]. The first association between the development of skin cancer and sunlight was described in 1894 [8]. Since then, many correlation studies have been conducted in order to associate the incidence of skin cancer with cumulative exposure to UV radiation [9].

The skin contains various chromophores and UV absorption can lead to a series of photochemical reactions, which in turn promote photo-induced inflammatory responses. Among skin chromophores are the nitrogenous bases, which absorb UVB, and amino acids, including tyrosine and tryptophan, which are the most common aromatic amino acids. These two amino acids also absorb UVB. Urocanic acid is the main chromophore in the stratum corneum, which absorbs UV with an absorption peak at 275 nm [9]. When a skin chromophore absorbs UV radiation, it can generate free radicals, which are chemically reactive species that have been related to the etiopathogenesis of many diseases [9].

The harmful effects of UV radiation on the skin are caused by direct cell damage and by alterations in immune function. UV radiation produces DNA damage (formation of cyclobutane pyrimidine dimers), gene mutation, immunosuppression, oxidative stress, and inflammatory responses, which have an important role in photoaging, and skin cancer [10]. Therefore, although the pathogenesis of cancer is multifunctional, UV radiation has been described as a major contributor and a potent carcinogen [11–13].

On this basis, the tumor suppressor gene p53 is of fundamental importance in carcinogenesis since it allows not only DNA repair, but also cell apoptosis when DNA damage is too large [14]. Cells containing mutant p53 can be found in skin with chronic sun damage, as well as in actinic keratosis, basal cell carcinoma and squamous cell carcinoma [15]. This can correlate UV radiation to the production of mutations in the tumor suppressor gene p53, which can lead to skin cancer initiation [16]. Immunosuppression is also considered an important event in skin carcinogenesis [17]. UV exposure can impair the immune system by reducing antigen presentation, by inducing the production of immunosuppressive cytokines, and by modulating hypersensitivity reactions [18].

Nowadays, 65–90% of melanoma skin cancer cases are associated with exposure to UV radiation [19]. While UVA radiation has an important role in the carcinogenesis of skin stem cells [10], UVB can induce DNA damage, which can lead to an inflammatory response and tumorigenicity [16]. The development of basal cell carcinoma, squamous cell carcinoma and malignant melanoma is often associated with painful sunburn [20]. UV exposure is known to be the most important risk factor for the development of actinic keratosis, which can be a precursor of squamous cell carcinoma [20].

Intense intermittent exposures are related to an increased risk for the development of basal cell carcinoma and melanoma when compared to chronic mild exposure, even though the total dosage of UV radiation in both types of exposure is the same. Moreover, the risk of squamous cell carcinoma is closely associated with chronic UV exposure, but not with intermittent exposure [21].

3.2 UV-Filters and Photoprotection

The first commercial sunscreen, an emulsion containing benzyl salicylate and benzyl cinnamate, became available in the US in 1928 [22]. In Germany, an ointment containing sulfonic acid benzimidazole was available in 1933 [23]. In France, a formulation containing benzyl salicylate was available in 1936 [24].

The number of patents increased: In 1943, p-aminobenzoic acid (PABA) was patented, but it was commercialized as a sunscreen much later [25]. In 1944, one of the substances known today as the first widely used UV-filter, the red veterinary petrolatum, was produced in the US [23] and was extensively used in World War II [26]. However, this red and sticky substance had limited efficacy and had to be applied as a thick layer for the soldiers not to develop sunburn. Other substances were later patented: in 1948 PABA esters were patented [27], in 1962 benzophenone, *3-benzyl-4-hydroxy-6-methoxy-benzenesulfonic acid*, was introduced [28], in 1977 the first water-resistant sunscreen became available [29], and in 1979, UVA dibenzoylmethane-derived filters were introduced, followed by the micronized inorganic particles titanium dioxide and zinc oxide in 1989 [30].

Today, the European Commission considers that sunscreen products should be sufficiently effective against UVB and UVA radiation to ensure a high protection of public health. In order to achieve this protection, a sunscreen product should provide a minimum UVB and UVA protection. An increased sun protection factor (SPF) (i.e., mainly UVB protection) should include an increase in UVA protection as well, at least 1/3 of increase in UVB protection, with preferentially a critical wavelength of at least 370 nm, to ensure broad protection [31].

UV-filters can be divided into inorganic (or physical) filters, which reflect or scatter UV light, and organic (or chemical) UV-filters, which absorb radiation [32]. They can be divided into different classes according to their chemical structures and among these are aminobenzoates, salicylates, cinnamates, benzophenones, camphor derivatives, triazine derivatives, benzotriazoles, cyanoacrylates, and dibenzoylmethanes.

On the other hand, the main representatives of physical filters are titanium dioxide (TiO_2) and zinc oxide (ZnO).

3.3 Chemical UV-Filters (Organic UV-Filters)

The aminobenzoates (PABA derivatives) are among the first UV-filter molecules used. This molecule has an electron-releasing group (NR_2) that is substituted *para* to an electron-accepting group (COOR), which permits efficient electron delocalization [33]. Many studies have demonstrated the photoallergic potential of PABA in guinea pigs [34] and in humans [35] due to its photoactivation [36] and photodegradation [37]. For this reason, the use of PABA sunscreens was discontinued and the "PABA-free" claim is still used today.

Salicylates are *ortho*-disubstituted molecules with a spatial arrangement that permits internal hydrogen bonding within the molecule itself [33]. The main representatives of this class are ethyl-hexyl salicylate and homosalate, which are considered weak, but photostable UV absorbers. Ethyl-hexyl salicylate is also used as a solvent for other UV-filters, especially for avobenzone and these UV-filters rarely appear alone in a formulation [35].

Salicylates are generally well tolerated by the skin, and sensitization reactions, photocontact and skin irritation induced by these compounds are rare [38]. Studies have shown that only 1% of ethyl-hexyl salicylate can penetrate the human skin after application [39, 40]. There are indications of an estrogen-like effect *in vitro*, but not *in vivo* [41]. Thus, salicylates have a favorable toxicological profile for use in sunscreens [35].

Cinnamates are aromatic compounds disubstituted with both an electron-releasing group (OCH_3) and an electron-accepting group (ester group further conjugated with a double bond). This allows the extended delocalization of electrons enabling these molecules to absorb at 310 nm [33]. The cinnamates are subject to photo *cis-trans* isomerization (*cis*-Z isomer, *trans*-E isomer). While the *E* isomer is predominant and has a $\lambda_{máx}$ = 310 nm, with a molar absorption coefficient (ε) of 19,500 $dm^3mol^{-1}cm^{-1}$, the *cis* isomer, which is formed under sunlight exposure, has a $\lambda_{máx}$ of 312 nm with a lower ε, 10,000 $dm^3mol^{-1}cm^{-1}$. Thus, sunlight exposure results in an equilibrated proportion of *E* and *Z* isomers and a consequent decreased absorption of UV light [33].

Octyl methoxycinnamate is the most widely used UV-filter [35]. The topical application of octyl methoxycinnamate is well tolerated [35] and the incidence of photodermatitis related to this substance is very low [42]. *In vitro* skin permeation studies have shown that only 0.2–0.3% of the applied UV-filter penetrates the viable epidermis and these values can be overestimated when compared to human studies [39, 43, 44]. Although some studies have reported an *in vivo* estrogen-like effect of octyl methoxycinnamate [45], this effect was not observed in humans [43].

Unlike most other UV-filters that possess an ester or amide linkage, benzophenones are aromatic ketones. They present an electron-releasing group in either *ortho* or *para* position or both, which results in two λ_{max} at 286 nm (UVB) and 324 (UVA). Benzophenone-3 has been described as the UV-filter responsible for most cases of photodermatitis [46] and can be systemically absorbed [47]. Some studies have also reported estrogen-like [48] and antiandrogenic effects of benzophenone-3 [49]. However, these results were based on *in vitro* studies and market experience with benzophenones has been favorable to its use in sunscreens [35].

Camphor derivatives protect against UVB radiation in the 290–300 nm region. As reported for cinnamates, they can reach equilibrium of isomer formation after sunlight exposure and also exhibit good photostability [50].

Avobenzone is the main representative of the dibenzoylmethanes due to its UVA absorption, with a λ_{max} at 357 nm [51]. Although avobenzone has an appropriate UVA absorption spectrum and consequently is found in several photoprotective cosmetic formulations, this molecule may present photoreactivity when exposed to UV [52–55]. This photoreactivity occurs due to the fact that avobenzone undergoes keto-enol tautomerization (AVO^K–AVO^E). In a sunscreen formulation, avobenzone exists predominantly in the enol form, which absorbs in the UVA range. The relative amounts of the two isomers are solvent dependent but the equilibrium always lies in the direction of the enol tautomer, which will consequently impart high sunscreen efficiency. However, under UV radiation, the keto-enol equilibrium is shifted toward the keto form of avobenzone, which absorbs in the UVC range from 260 to 280 nm and can lead to the majority of the harmful effects of avobenzone, such as its irreversible photodegradation and ROS formation [51, 56].

Among the high molecular weight UV-filter molecules (over 500 Daltons) are bemotrizinol (MW = 629 Dalton) and ethylhexyl triazone (MW = 823 Dalton). Their high molecular weight reduces their skin penetration [33] and they are also used to stabilize or even replace avobenzone due to their high photostability [53, 55].

The benzotriazole derivatives are good UVB and UVA absorbers, with a λ_{max} at 305 nm. They are heterocyclic compounds containing nitrogen atoms. One representative of this class is benzotriazolyl dodecyl p-cresol.

Octocrylene is the main representative of the cyanoacrylate class. Until recently, there was no evidence of contact allergy to octocrylene; however, new studies have shown its photoallergenic potential, leading to contact dermatitis [57].

3.3.1 Nanocarrier-Based Chemical UV-Filters

In order to improve the performance and decrease the side-effects of UV-filters, some research groups have been focusing on the encapsulation of these [58, 59]. Luppi *et al.* [58] described nanoparticles of benzophenone-3 encapsulated in modified poly(vinyl alcohol) (PVA) with fatty acids (FAs), which prevented the movement of benzophenone-3 into the skin, limiting its percutaneous absorption.

Substituted PVA, at two different degrees of substitution (40% and 80%), with saturated fatty acids (myristic, palmitic, stearic, and behenic), can be employed for the preparation of nanomatrices for sunscreen delivery. These systems can prevent the movement of benzophenone-3 towards the skin, thereby limiting its percutaneous absorption.

Other studies have shown that chemical UV-filters encapsulated with nanoparticles can have a synergistic effect in terms of UV protection efficacy [60, 61], reduction of photodegradation [62, 63], improvement of photostability *in vitro* [64], and increased SPF *in vitro* [59].

3.4 Physical UV-Filters (Inorganic UV-Filters)

While there are more chemical than physical UV-filters on the market, the use of inorganic UV-filters, i.e., TiO_2 and ZnO, has many advantages when compared to the chemical UV-filters, such as enhancement of the protection level in the longer UVA range (>370 nm) [65], photostability of the molecules, and a low photoallergic potential [66]. However, among the main disadvantages of this class are the few available molecules, the opaque and white appearance of the skin and the poor dispersibility in the formulation [66].

Among the main forms of TiO_2, the rutile crystal form has a high refractive index, can absorb UV radiation and is used in most cases. The anatase form of TiO_2 has properties similar to those of the rutile form, but has lower UV radiation absorption and a higher tendency to photocatalysis.

3.4.1 Nanocarrier-Based Physical UV-Filters

In order to improve the appearance and effectiveness of physical UV-filters, many efforts have been devoted to decreasing the size of these molecules to the nanoscale since this can lead to changes in physical and chemical properties. Many authors have also been concerned about the biological properties of these UV-filters at the nanoscale. The TiO_2 nanoparticle, for example, presents a high surface-area to mass ratio [67] and a high redox potential, which can change the biological behavior of this UV-filter.

UV-filter nanoparticles are used in formulations to improve the efficacy [68] and the performance of the product since they promote a more efficient UV protection [69] in sunscreens. At the nanoscale (60–120 nm), TiO_2 reflects and scatters UV light in the most efficient manner [70]. Nanotechnology has improved inorganic UV-filters and consequently enhanced their acceptance and use in sunscreens [69]. However, particle size is related to its photoprotective effects. As described by Forestier [65], small TiO_2 particles are more efficient at scattering shorter light wavelengths, i.e., the absorption is dislocated from UVA to UVB absorption, which sometimes is not desired. Egerton and Tooley [71] also showed that titanium dioxide 145 nm particles are more effective in the UVA region than 35 and 50 nm particles.

ZnO is another commonly used UV-filter nanoparticle, which exists as aggregates of primary particles ranging from 30 to 150 nm in size. These molecules started to be used because their nanoparticles can protect against UV radiation more efficiently than their corresponding micron-sized particles [69].

In order to improve the dispersion of physical UV-filters in sunscreen formulations, inert materials such as aluminum oxide or silicon oils are frequently used to coat the particles. These substances can change the behavior of the nanoparticles, such as solubility, biological reactivity, penetration and distribution [72].

There is an effort to cover nanoparticle UV-filters with substances that decrease skin penetration in order to avoid the biological effects of nanoparticle UV-filters. In addition, considering the photocatalytic effect, TiO_2 is coated with aluminum oxide or silica in order to prevent oxygen radical formation when used in sunscreens [73].

3.4.2 Side Effects of Carrier- and Non-Carrier-Based Physical UV-Filters

Many authors have reviewed the side effects of physical UV-filters, especially TiO_2, concerning their pulmonary effects, such as inflammation, pulmonary damage, fibrosis and carcinogenicity [74]. However, these effects are related to systemic toxicity and to respiratory exposure, while topical exposure is normally considered safe for the consumer.

Some authors have also described toxicity of nanoparticled physical UV-filters: some TiO_2 nanoparticles have shown DNA damage potential [75] and should be considered genotoxic or photogenotoxic [76–78].

The side effects of UV-filters are dependent on their penetration into the skin. For nanoparticles, their penetration depends on characteristics such as their size and coating. The coating of nanoparticles can change their solubility, as well as their penetration into the skin and consequently their safety [69, 72, 79], and should be studied case by case.

Regarding nanoparticle size, it was found that TiO_2 nanoparticles with a diameter of more than 20 nm can be considered safe since they do not reach viable cells in the epidermis. In addition, the toxicological profile of TiO_2 did not raise any safety concerns regarding its application (SCCNFP/0005/98) [80] to non-flexed (not occluded) and unburned human skin. Consequently, the exclusion of particles of less than 20 nm in the production process would improve product safety for this particular compound.

Some authors have shown that UV-filter nanoparticles do not penetrate living cells and *in vitro* and *in vivo* (animals and humans) assays have revealed that penetration of nanoparticled UV-filters is limited only to the stratum corneum [81–90]. Monteiro-Riviere *et al.* [91] observed that despite the slightly enhanced penetration of TiO_2 or ZnO nanoparticles in UVB sunburnt skin, there was no evidence of systemic absorption of these nanoparticles.

The German NANODERM project [92] is an extensive investigation of dermal penetration of TiO_2 nanoparticles. In 2007, the authors published a final report containing tests performed on many skin sources such as porcine and healthy human skin from biopsies and explants, human foreskin transplanted to SCID-mice and psoriatic skin. They used different formulations with nanoparticled TiO_2, including commercial sunscreens. In all but a few cases, Ti was detected on top of the stratum corneum (topmost 3–5 corneocyte layers); the Ti-spots detected in viable epidermis were identified as preparation artifacts. They also found that frequently the nanoparticles were aggregated and no diffusive transport was observed. Despite the deep penetration into hair follicles, the nanoparticles were not found inside vital tissue. They also supposed that desquamation could explain the clearance of nanoparticles found in the stratum corneum and that the sebum excretion would explain the clearance pathway for nanoparticles in hair follicles.

In July 2013, the Scientific Committee on Consumer Safety (SCCS) [79] delivered an opinion on TiO_2 nanoparticles. They concluded that TiO_2 nanoparticles in sunscreens applied to healthy skin, intact or sunburnt, is unlikely to lead to systemic exposure, acute toxicity via dermal application, skin and eye irritation, skin sensitization or reproductive effects. They also pointed out that the potential genotoxicity of TiO_2 nanomaterials is not conclusive. Thus, they concluded that the use of TiO_2 nanoparticles at a concentration up to 25% in topically applied cosmetic products, for example sunscreens, should not pose any significant risk to the consumer. However, they did not recommend the use of TiO_2 nanoparticles in spray products, such as sprayable sunscreens, since that might lead to exposure of the consumer's lungs to TiO_2 nanoparticles by inhalation.

Regarding ZnO nanoparticles, the SCCS of the European Commission issued an opinion in 2012 [93] considering many toxicological aspects of ZnO. They concluded that in *in vivo* (human volunteers) and *in vitro* (human, porcine and nude mouse skin) tests, there was no evidence that ZnO nanoparticles can penetrate the intact or compromised skin and there is only minimal absorption of ZnO nanoparticles by sunscreen application. They did not observe differences between coated/ uncoated nano ZnO, or between normal/damaged skin. They assumed that penetration into the skin, if any, is caused by Zn ions released from ZnO nanoparticles. Since nano ZnO-containing cosmetic formulations are likely to contain a small proportion of solubilized zinc, the amount of absorbed zinc could be considered insignificantly small compared to the large zinc pool already present in the body.

3.5 Assessment of Sunscreen Performance

3.5.1 Efficacy of Carrier- and Non-Carrier-Based UV-Filters

Due to the understanding of the potentially unfavorable long-term side effects of solar irradiation acquired during the last two decades, there is a universal call for harmless and effective sunscreens. During the First World War, Hausser and Vale, in Germany, conducted the first study of the detailed action spectrum of erythema and pigmentation of human skin. They showed that these effects depend on UV radiation, mainly due to shorter wavelengths (<320 nm). In 1922, they published the erythema and pigmentation action spectrum for human skin using a monochromator and an artificial mercury lamp [23].

In Copenhagen, during the Second International Congress on Light in 1932, Coblentz proposed the division of UV into three regions: UVA (315–400 nm), UVB (280–315 nm), and UVC (<280 nm) [23]. Later, in Germany, in 1956, Schulze proposed a test to be used for commercially available UV-filters, adding a protective factor to them [23]. The protective factor was obtained by dividing the exposure time required to induce erythema in sunscreen-protected skin by the time required to induce erythema in unprotected skin. This was done by applying a series of increasing doses of UV (40% increase) to both protected and unprotected skin. In 1974, Greiter, in Austria, improved and popularized the concept of SPF [23].

In 1978, the US Food and Drug Administration (FDA) published guidelines for formulating and evaluating sunscreens in the Federal Register [23]. This method has been accepted internationally. At that time, filters were used not only to prevent sunburn but also to prolong the exposure time for tanning. In this context, researchers have become concerned that sunscreens could create a false sense of security and cause people to sustain damage to their skin [23].

In 1994, Cosmetics Europe (former European Cosmetic, Toiletry and Perfumery Association—Colipa), introduced new techniques to characterize and specify the emission spectrum of the UV source and designated skin types based on skin color. In 2000, COLIPA, together with Japanese and South African authorities, discussed the harmonization of the SPF evaluation method. A joint agreement about the international test method SPF was reached in October 2002 and updated in 2006 by the European, Japanese, American and South African industries [94]. In December 2010, the European Standardisation Organisation (CEN) published the Standard EN ISO 24444:2010 Cosmetics—Sun protection test methods—*In vivo* determination of the SPF. This standard test method replaces the earlier reference method (International Sun Protection Factor Test Method (2006) and is considered as the reference method within the EU to determine the Sun Protection Factor (SPF) [95].

In 2007, the FDA published the Final Proposed Rules on sunscreens [96], and one of the proposed changes was changing the acronym "SPF" from "sun protection factor" to "sunburn protection factor," which avoids a false sense of security; an increase of SPF from SPF30+ to SPF50+ was also proposed. In addition, for the first time, the FDA also proposed evaluation and labeling of UVA protection of sunscreens, using a combination of *in vitro* tests (spectrophotometric) and *in vivo* persistent pigment darkening (PPD).

In 2011, the FDA announced new requirements for sunscreens currently sold over-the-counter as follows: sunscreens approved by FDA's broad spectrum test procedure may be labeled as "Broad Spectrum SPF [value]" on the front label; the claim "reduce the risk of skin cancer and early skin aging" can only be used for broad spectrum sunscreens with an SPF higher than 15; the claims "waterproof," "sweatproof" and the term "sunblock" cannot be used; the "water resistance" claim must indicate if the product remains effective while swimming or sweating, for 40 or 80 min.

Today, these concepts designed by various regulatory agencies are well known. The SPF determined *in vivo* is known as a universal indicator of the effectiveness of sunscreens against sunburn, defined as the "UV energy required to produce an MED on protected skin divided by the UV energy required to produce an MED on unprotected skin." Minimum erythemal dose (MED) is "The quantity of erythema-effective energy required to produce the first perceptible, redness reaction with clearly defined borders."

Since the SPF only addresses erythema, which is caused by UVB radiation, and is not able to predict UVA protection, a similar assay performed with UVA radiation was created to evaluate tanning. Nowadays the *in vivo* test most widely used and recommended by regulatory agencies is called Persistent Pigment Darkening (PPD), which measures the pigmentation of volunteers' skin 2 h after UVA irradiation. The minimal pigmenting dose (MPD), which is similar to MED, is the quantity of erythema-effective energy required to produce the first perceptible pigment darkening. However, this test may have some problems related to prolonged irradiance times and exposure to high doses of UVA radiation [23].

Although the *in vitro* SPF determination is still not satisfactory, the *in vitro* determination of the various UVA indices is now considered a valuable method. The UVA/UVB ratio and the critical wavelength [97], which are relative measurements of the absorption spectrum of the sunscreen, can be determined. Regarding the critical wavelength, the sunscreens are grouped into five categories, with the category of higher UVA being achieved with a critical wavelength exceeding 370 nm [97]. The UVA/UVB ratio is widely used for the measurement of photostability since it is a simple, fast, reproducible and inexpensive index [39, 53, 94]. For sunscreen registration, the accepted methods are SPF evaluation and PPD in humans [31].

In silico SPF calculation, i.e., BASF Sunscreen Simulator [98], is also a useful tool for the development of sunscreen products since it provides the estimation of the SPF as well as various common UVA-Metrics. This tool helps the formulator to plan the composition of active ingredients in a sunscreen formulation, but it does not replace the actual formulation work and *in vivo* SPF testing of the final product.

Scientific findings have shown that some biological damage to the skin can be prevented and/or reduced if the ratio of the protective factor, as measured in the PPD test, is at least 1/3 of the factor measured under the sun protection factor testing method. Thus, the European Commission recommended a UVA/UVB ratio of 1/3 [31].

The International Standardization Organization (ISO) is also dedicated to the determination of relevant UVA protection parameters [99]. They consider the *in vivo* UVA-PF method to be the second method to become official since there is a consensus about the PPD method [100].

3.5.1.1 Efficacy-characterization of carrier-based UV-filters

The characterization of nanostructures can provide evidence for the correct formation of the desired shapes and also contributes to a better understanding of their formation and functionality [101]. The characterization includes particle size, shape, and particle surface charge determinations [102] and is important to guarantee the performance of the nanoparticles [103].

Among the available techniques, there are dynamic light scattering (DLS) measurements, soft X-ray tomography, cryogenic light microscopy, transmission electron microscopy (TEM), and scanning electron microscopy (SEM) [104]. As described by Silva *et al.* [104], each method has some limitations, and may lead to errors of characterization. Therefore, the formulator must combine several techniques in order to better characterize nanostructures.

3.5.2 Safety of Carrier- and Non-Carrier-Based UV-Filters

According to the SCCS, the safety evaluation of the finished product is the responsibility of the "responsible person" through the safety assessor of a company and SCCS only evaluates ingredients based on the dossier sent by the cosmetic company.

In general, tests addressing acute toxicity (if available), irritation and corrosivity (skin and eye), skin sensitization, dermal/percutaneous absorption, repeated dose toxicity, and mutagenicity/genotoxicity of cosmetic ingredients are considered to be the minimal base set requirements for safety evaluation. However, photo-induced toxicity data are required when the substance present in a cosmetic product is expected or intended to be used on sunlight-exposed skin, i.e., in sunscreens.

Although it is inconceivable that tests in human volunteers would replace animal tests, it is known that tests in animals and alternative methods are of limited predictive value with respect to the human situation. Therefore, a skin compatibility test with human volunteers, confirming that there are no harmful effects when applying a cosmetic product for the first time to human skin or mucous membranes, may be needed scientifically and ethically. Consequently, before finished cosmetic products are tested in humans, the toxicological profiles of the substances must be provided. For safety assessment, cosmetic products usually are tested in small populations to confirm their skin and mucous membrane compatibility, as well as their cosmetic acceptability (= fulfilment of in-use expectations). The general ethical and practical aspects related to human volunteer compatibility studies on finished cosmetic products are described in SCCNFP/0068/98 [105] and SCCNFP/0245/99 [106].

For many years, animal testing was the only tool able to assess the safety of raw materials and finished cosmetic products before human use. However, many efforts have been made to develop alternative methods to animal use as a safe way to evaluate cosmetics.

The European Community, since the publication of the 3Rs concept (reduction, refinement, replacement) by Russell and Burch in 1959, has been gradually inserting alternative methods to animal experimentation for safety assessment of cosmetic products. According to the seventh amendment of the European Directive published in February 2003, deadlines were established for banning animal testing for cosmetic and/or finished product (testing ban) ingredients, as well as banning the marketing of products in the European Community (or its ingredients) tested on animals (marketing ban). Therefore, some tests on animals can no longer be performed in Europe since 2004.

According to the seventh amendment, a full ban of animal use for cosmetic testing was established in March 2013 by the European Union, despite the fact that there was no alternative test completely validated for many safety purposes such as skin sensitization. Therefore, there is a need to develop methods able to assess the safety parameters of cosmetic products, especially products as new and unknown as nanoparticles.

One of the most common tests used to assess the phototoxic potential of UV-filters is the 3T3 NRU PT (3T3 Neutral Red Uptake Phototoxicity Test). Since 2000, animal tests for the assessment of phototoxic potential are no longer allowed in the European Union since the acute phototoxic potential is considered to be better evaluated by *in vitro* methods, such as the 3T3 NRU PT. The 3T3 NRU PT is a phototoxicity test based on neutral red uptake by 3T3 murine fibroblasts. It is considered to be the main test for this assessment since it presents high sensitivity and specificity and has also been validated [107]. In this test, Balb/c 3T3 cells cultured on two 96-well microplates are pre-incubated with eight different concentrations of the chemical. One plate is then exposed to UVA radiation and the other is kept in the dark for the determination of cell viability in the presence and absence of radiation for later determination of the cyto- and phototoxicity of the chemical [107].

Among the tests used to assess the safety of nanomaterials, there is no well-defined procedure to estimate their potential toxicity [108] and there is no sufficient information on the safety of nanomaterials, such as toxicity information and physicochemical properties. Thus, many tests have been developed or adapted to cover this new class of materials. Among the alternative methods used to assess the toxicological parameters of nanoparticles, cell culture is one of the most important for both carrier- and non-carrier-based cosmetic products.

Besides the 3T3 NRU phototoxicity test, the tests that assess cytotoxicity are also important, especially to predict the toxicological parameters of nanomaterials. One of the most used tests is the MTT assay, which is based on the conversion of MTT (3-[4,5-dimethylthiazol-2-yl]-2,5 diphenyl tetrazolium bromide) to formazan crystals by living cells [109] by mitochondrial activity, which is related to cell viability.

Many nanoparticles can interact with dyes, leading to loss of accuracy and precision [110–112]. Some studies have shown that single-walled carbon nanotubes could interact with some dyes such as MTT [113, 114]. Other interferences can occur; some nanoparticles can absorb proteins on their surface, which can affect cytokine release quantification [115], and nanoparticles can agglomerate in culture medium, which can change the stability of nanoparticle dispersion. Thus it is important to ensure the stability of nanoparticle dispersion in the test medium, i.e., DMEM or PBS, and to determine that there is no interaction with the vital dye.

Many studies have focused on the development of more realistic models to determine the toxic effects of NPs [116, 117, 112]. Thus, the formulator must: (a) perform a more adequate test for the specific nanoparticle and (b) perform more than one test in order to compare results, cover many aspects, obtain accuracy, and minimize test limitations.

Among the advanced technologies, the use of reconstructed skin models represents an interesting alternative to the use of experimental animals for the evaluation of a wide variety of toxicological studies involving corrosiveness, cytotoxicity, phototoxicity, irritation, melanogenesis, skin penetration, skin sensitization, drug metabolism, and mechanism of action [118–121]. The development of reconstructed skin models was initiated by Rheinwald and Green [122] and has been progressively refined [123, 124] and processed in order to be commercially available [125].

Nanoparticle characterization does not only provides evidence for the correct formation of human-designed structures, but can also contribute to a better understanding of their structure and functional capabilities.

3.5.3 Stability and Photostability of Carrier- and Non-Carrier-Based UV-Filters

The stability and photostability of sunscreens are deeply related to their safety and efficacy. For example, a photo-unstable sunscreen can impair the SPF, i.e., lose efficacy, which can lead to consumer-unprotected skin. In addition, an unstable formulation can generate some unexpected reaction products, such as degradation products, which can compromise the safety for the consumer.

Thus, the assessment of photostability is necessary for the development of a new sunscreen and for the determination of its stability. It can also be used as a tool for quality control of such formulations, since the produced degradation products may lead to toxic or allergic effects on the skin [126]. On the other hand, the stability of nanocarrier-based UV-filters should be assessed in order to guarantee that the nanoparticles will remain in the formulation until the stated expiration date and will be able to protect the skin against UV radiation.

Stability studies are needed to check if there is any unstable UV-filter that needs to be stabilized, as well as to predict the effects of the environment on chemical stability, shelf life and the appropriate storage conditions. Moreover, these tests add to the process the choice of the components and the type of formulation [127–130] in order to guarantee a safe sunscreen formulation for use.

The reactions that occur with UV-filters may lead to the loss of sunscreen effectiveness and to the undesirable formation of toxic or allergenic products. In this respect, formulators attempt to develop photostable associations of UV-filters, photoprotective formulations that remain stable after exposure to UV radiation.

Thus, many studies of photostability have been conducted, generally using in chemico techniques such as spectrophotometry [50, 131–133], or separation techniques such as high-performance liquid chromatography (HPLC) [50, 53–55, 132, 133] and gas chromatography [134].

The different techniques should be performed in parallel because, while spectrophotometric analysis may be affected by optical artifacts, such as photoproducts that also absorb UV radiation, the separation techniques do not allow the quantification of the photo-induced effects or changes, or of their effects on photoprotection [135].

3.6 Conclusions

It is well accepted that the use of adequate broad spectrum UV-filters prevents photoaging and skin cancer. In addition to new high molecular weight UV-filter organic molecules, inorganic nanoparticles are improving sunscreen efficacy by allowing broader UVA protection, which is not achieved by employing only combinations of usual organic UVB-filters. Thus, the use of inorganic UV-filters is important to satisfy the European

Commission's recommendations regarding UVA/UVB protection rates (UVA-PF should be at least 1/3 of the SPF).

In recent years, an improvement of safety assessment and quality of analytical techniques has been observed worldwide and, currently it is well accepted that nanomaterials are similar to normal chemicals/substances in that some may be toxic and some may not. Consequently, the safety of nanoparticles, including TiO_2 and ZnO, should be analyzed case by case due to different coatings.

The European Union Scientific Committee on Emerging and Newly Identified Health Risks concluded that the topical use of TiO_2 and ZnO in cosmetic products does not pose a risk to humans. However, the inhalation of these nanoparticles must be avoided once they may lead to lung and systemic exposure.

Finally, despite recent advances in safety and analytical techniques, they still must be improved and adapted for the detection, characterization, and analysis of nanoparticles in order to be standardized and validated for assessing hazards and exposure risks of any nanomaterial that is in contact with human beings.

References

1. Miller S. A., Hamilton S. L., Wester U. G., CYR W. H. An analysis of UVA emissions from sunlamps and the potential importance for melanoma. *Photochemistry and Photobiology,* 68 (1998) pp. 63–70.
2. Pastila R., Leszczynski D. Ultraviolet-A radiation induces changes in cyclin G gene expression in mouse melanoma B_{16}-F_1 cells. *Cancer Cell International,* 7 (2007) p. 7.
3. Narayanan D. L., Saladi R. N., Fox J. L. Ultraviolet radiation and skin cancer. *International Journal of Dermatology,* 49 (2010) pp. 978–986.
4. Matsumura Y., Ananthaswamy H. N. Toxic effects of ultraviolet radiation on the skin. *Toxicology and Applied Pharmacology,* 195 (2004) pp. 298–308.
5. Wlaschek M., Tantcheva-Poor I., Naderi L., MA W., Schneider L. A., RAZI-Wolf Z., Schuller J., Scharffetter-Kochanek K. Solar UV irradiation and dermal photoaging. *Journal of Photochemistry and Photobiology B Biology,* 63 (2001) pp. 41–51.

6. Schroeder P., Haendeler J., Krutmann J. The role of near infrared radiation in photoaging of the skin. *Experimental Gerontology*, 43 (2008) pp. 629–632.

7. Pathak M. A., Carraro C. Reactive oxygen species in cutaneous photosensitivity reactions in porphyrias, PUVA photochemotherapy, and melanin pigmentation. In Hayaishi O., Imamura S., Myachi Y. (eds.) *The Biological Role of Reactive Oxygen Species in the Skin*. Tokyo University Press, Tokyo (1988) pp. 75–84.

8. Gamage D. A., Fawzy A. A., Frason, R. C. Preferential hydrolysis of peroxidized phospholipid by liposomal phospholipase. *Biochimica et Biophysica Acta*, 958 (1988) pp. 116–124.

9. Tedesco A. C., Martínez L., González S. Photochemistry and photobiology of actinic erythema: Defensive and reparative cutaneous mechanisms Brazilian. *Journal of Medical and Biological Research*, 30 (1997) pp. 561–575.

10. Meeran S. M., Punathil T., Katiyar S. K. Interleukin-12-deficiency exacerbates inflammatory responses in UV irradiated skin and skin tumors. *Journal of Investigative Dermatology*, 128 (2008) pp. 2716–2727.

11. Preston D. S., Stern R. S. Nonmelanoma cancers of the skin. *The New England Journal of Medicine*, 327 (1992) pp. 1649–1662.

12. Diepgen T. L., Mahler V. The epidemiology of skin cancer. *British Journal of Dermatology*, 61 (2002) pp. 1–6.

13. Gilchrest B. A., Eller M. S., Geller A. C., Yaar M. The pathogenesis of melanoma induced by ultraviolet radiation. *The New England Journal of Medicine*, 340 (1999) pp. 1341–1348.

14. Hornung R. L. Photoprotection. In *Harper's Textbook of Pediatric Dermatology* (Irvine A. D., Hoeger P. H., Yan A. C., eds.), 3rd ed. Blackwell Publishing Ltd. (2011).

15. Marrot L., Meunier J. R. Skin DNA photodamage and its biological consequences. *Journal of the American Academy of Dermatology*, 58 (2008) pp. 139–148.

16. Benjamin C. L., Ananthaswamy H. N. p53 and the pathogenesis of skin cancer. *Toxicology and Applied Pharmacology*, 224 (2007) pp. 241–248.

17. Brenner M., Hearing V. J. The protective role of melanin against UV damage in human skin. *Photochemistry and Photobiology*, 84 (2008) pp. 539–549.

18. Beissert S., Schwarz T. Mechanisms involved in ultraviolet light-induced immunosuppression. *Journal of Investigative Dermatology Symposium Proceedings*, 4 (1999) 61–64.
19. Glanz K., Buller D. B., Saraiya M. Reducing ultraviolet radiation exposure among outdoor workers: State of the evidence and recommendations. *Environmental Health*, 6 (2007) pp. 22.
20. Kennedy C., Bajdik C. D., Willemze R., De Gruijl F. R., Bouwes Bavinck J. N. The influence of painful sunburns and lifetime sun exposure on the risk of actinic keratoses, seborrheic warts, melanocytic nevi, atypical nevi, and skin cancer. *Journal of Investigative Dermatology*, 120 (2003) pp. 1087–1093.
21. Westerdahl J., Olsson H., Ingvar C. At what age do sunburn episodes play a crucial role for the development of malignant melanoma. *European Journal of Cancer*, 30A (1994) pp. 1647–1654.
22. Shaath N. A. Evolution of modern sunscreen chemicals. In (Lowe N. J., Shaath N. A., eds.), *Sunscreens, Development, Evaluation, and Regulatory Aspects*. New York, Marcel Dekker (1990) pp. 3–35.
23. Lim H. W., Draelos Z. D. *Clinical Guide to Sunscreens and Photoprotection*. Informa Healthcare USA, Inc. New York. (2009) 322p.
24. Rebut D. The sunscreen industry in Europe: Past, present and future. In (Lowe N. J., Shaath N. A., eds.) *Sunscreens, Development, Evaluation, and Regulatory Aspects*. New York: Marcel Dekker (1990) pp. 161–171.
25. Mackie B. S., Mackie L. E. The PABA story. *Australasian Journal of Dermatology*, 40 (1999) pp. 41–53.
26. MacEACHERN W. N., JILLSON O. F. A practical sunscreen "Red Vet Pet". *Archives of Dermatology*, 89 (1946) pp. 147–150.
27. Kumler W. D., Daniels T. C. Sunscreen compounds. *Journal of the American Pharmaceutical Association. Scientific Edition*, 37 (1948) pp. 474–476.
28. Urbach F. The historical aspects of sunscreens. *Journal of Photochemistry and Photobiology B Biology*, 64 (2001) pp. 99–104.
29. Thomas L., Lim H. W. Sunscreens. *Journal of Drugs in Dermatology*, 2 (2003) pp. 174–177.
30. Roelandts R. Advances in sunscreen technology: Choosing the sunscreen to suit. *Current Opinion in Dermatology*, 2 (1995) pp. 173–177.
31. Commission Recommendation of 22 September 2006 on the efficacy of sunscreen products and the claims made relating

thereto (notified under document number C(2006) 4089) (Text with EEA relevance) (2006/647/EC). *Official Journal of the European Union.* L 265/39 Volume 49. Brussels, 22 September (2006).

32. Draellos Z. D. *Dermatologia Cosmética: Produtos e Procedimentos.* Editora Santos Grupo Gen (2012).
33. Shaath N. A. Ultraviolet filters. *Photochemistry and Photobiology Sciences,* 9 (2010) pp. 464–469.
34. Gerberick G. F., Ryan C. A. Contact photoallergy testing of sunscreens in guinea pigs. *Contact Dermatitis,* 20 (1989) pp. 251–259.
35. Nash J. F. Human safety and efficacy of ultraviolet filters and sunscreen products. *Dermatologic Clinics,* 24 (2006) pp. 35–51.
36. Allen J. M., Gossett C. J., Allen S. F. Photochemical formation of singlet molecular oxygen ($^1O^2$) in illuminated aqueous solutions of p-aminobenzoic acid (PABA). *Journal of Photochemistry and Photobiology B Biology,* 32 (1996) pp. 33–37
37. Gasparro F. P. UV-induced photoproducts of paraaminobenzoid acid. *Photodermatology,* 2 (1985) pp. 151–157.
38. Dromgoole S. H., Maibach H. I. Sunscreening agent intolerance: Contact and photocontact sensitization and contact urticaria. *Journal of the American Academy of Dermatology,* 22 (1990) pp. 1068–1078.
39. Chatelain E., Gabard B., Surber C. Skin penetration and sun protection factor of five UV filters: Effect of the vehicle. *Skin Pharmacology and Applied Skin Physiology,* 16 (2003) pp. 28–35.
40. Walters K. A., Brain K. R., Howes D., James V. J., Kraus A. L., Teetsel N. M., Toulon M., Watkinson A. C., Gettings S. D. Percutaneous penetration of octyl salicylate from representative sunscreen formulations through human skin in vitro. *Food and Chemical Toxicology,* 35 (1997) pp. 1219–1225.
41. Morohoshi K., Yamamoto H., Kamata R., Shiraishi F., Koda T., Morita M. Estrogenic activity of 37 components of commercial sunscreen lotions evaluated by in vitro assays. *Toxicology In Vitro,* 19 (2005) pp. 457–469.
42. Darvay A., White I. R., Rycroft R. J., Hawk J. L., Mcfadden J. P. Photoallergic contact dermatitis is uncommon. *British Journal of Dermatology,* 145 (2001) pp. 597–601.
43. Janjua N. R., Mogensen B., Andersson A. M., Petersen J. H., Henriksen M., Skakkebaek N. E., Wulf H. C. Systemic absorption of the sunscreens benzophenone-3, octyl-methoxycinnamate, and 3-(4-methyl-benzylidene) camphor after whole-body topical application

and reproductive hormone levels in humans. *Journal of Investigative Dermatology*, 123 (2004) pp. 57–61.

44. Freitas J. V., Praça F. S., Bentley M. V., Gaspar L. R. Trans-resveratrol and beta-carotene from sunscreens penetrate viable skin layers and reduce cutaneous penetration of UV-filters. *International Journal of Pharmaceutics*, 484 (2015) pp. 131–137.
45. Seidlova-Wuttke D., Christoffel J., Rimoldi G., Jarry H., Wuttke W. Comparison of effects of estradiol (E2) with those of octylmethoxycinnamate (OMC) and 4-methylbenzylidene camphor (4MBC)—2 filters of UV light on several uterine, vaginal, and bone parameters. *Toxicology and Applied Pharmacology*, 214 (2006) pp. 1–7.
46. Schauder S., Ippen H. Contact and photocontact sensitivity to sunscreens. Review of a 15-year experience and of the literature. *Contact Dermatitis*, 37 (1997) pp. 221–232.
47. Jiang R., Roberts M. S., Collins D. M., Haller V., Steinmann B., Lichtensteiger W. Absorption of sunscreens across human skin: An evaluation of commercial products for children and adults. *British Journal of Clinical Pharmacology*, 48 (1999) pp. 635–637.
48. Suzuki T., Kitamura S., Khota R., Sugihara K., Fujimoto N., Ohta S. Estrogenic and antiandrogenic activities of 17 benzophenone derivatives used as UV stabilizers and sunscreens. *Toxicology and Applied Pharmacology*, 203 (2005) pp. 9–17.
49. MA R., Cotton B., Lichtensteiger W., Schlumpf M. UV filters with antagonistic action at androgen receptors in the MDA-kb2 cell transcriptional-activation assay. *Toxicology Sciences*, 74 (2003) pp. 43–50.
50. Deflandre A., Lang G. Photostability assessment of sunscreens. Benzylidene camphor and dibenzoylmethanederivatives. *International Journal of Cosmetic Science*, 10 (1988) pp. 53–62.
51. Lhiaubet-Vallet V., Marin M., Jimenez O., Gorchs O., Trullasb C., Miranda M. A. Filter–filter interactions. Photostabilization, triplet quenching and reactivity with singlet oxygen, *Photochemistry and Photobiology Sciences*, 9 (2010) pp. 552–554.
52. Gaspar L. R., Tharmann J., Maia Campos P. M., Liebsch M. Skin phototoxicity of cosmetic formulations containing photounstable and photostable UV-filters and vitamin A palmitate. *Toxicology In Vitro*, 1 (2013) pp. 418–425.
53. Benevenuto C. G., Guerra L. O., Gaspar L. R. Combination of retinyl palmitate and UV-filters: Phototoxic risk assessment based

on photostability and *in vitro* and *in vivo* phototoxicity assays. *European Journal of Pharmaceutical Sciences*, 68 (2015) pp. 127–136.

54. Freitas J. V., Lopes N. P., Gaspar L. R. Photostability evaluation of five UV-filters, trans-resveratrol and beta-carotene in sunscreens. *European Journal of Pharmaceutical Sciences*, 78 (2015) pp. 79–89.

55. Kawakami CM, Gaspar LR. Mangiferin and naringenin affect the photostability and phototoxicity of sunscreens containing avobenzone. *Journal of Photochemistry and Photobiology B Biology*, 151 (2015) pp. 239–247.

56. Mturi G. J., Martincigh B. S., Photostability of the sunscreening agent 4-tert-butyl-4′-methoxydibenzoylmethane (avobenzone) in solvents of different polarity and proticity. *Journal of Photochemistry and Photobiology A: Chemistry*, 200 (2008) pp. 410–420.

57. Avenel-Audran M., Dutartre H., Goossens A., Jeanmougin M., Comte C., Bernier C., Benkalfate L., Michel M., Ferrier-Lebouëdec M. C., Vigan M., Bourrain J. L., Outtas O., Peyron J. L., Martin L. Octocrylene, an emerging photoallergen. *Archieves of Dermatology* 146 (2010) pp. 753–757.

58. Luppi B., Cerchiara T., Bigucci F., Basile R., Zecchi V. Polymeric nanoparticles composed of fatty acids and polyvinylalcohol for topical application of sunscreens. *Journal of Pharmacy and Pharmacology*, 56 (2004) pp. 407–411.

59. Marcato P. D., Caverzan J., Rossi-Bergmann B., Pinto E. F., Machado D., Silva R. A., Justo G. Z., Ferreira C. V., Durán N. Nanostructured polymer and lipid carriers for sunscreen. Biological effects and skin permeation. *Journal of Nanoscience and Nanotechnology*, 11 (2011) pp. 1880–1886.

60. Wissing S. A. Müller R. H. Cosmetic applications for solid lipid nanoparticles (SLN). *International Journal of Pharmaceutics*, 254 (2003) 65–68.

61. Xia Q., Saupe A., Müller R. H., Souto E. B. Nanostructured lipid carriers as novel carrier for sunscreen formulations. *International Journal of Cosmetic Sciences*, 29 (2007) pp. 473–482.

62. Perugini P., Simeoni S., Scalia S., Genta I., Modena T., Conti B., Pavanetto F. Effect of nanoparticle encapsulation on the photostability of the sunscreen agent, 2-ethylhexyl-p-methoxycinnamate. *International Journal of Pharmaceutics*, 246 (2002) pp. 37–45.

63. Paese K., Jaeger A., Poletto F. S., Pinto E. F., Rossi-Bergmann B., Pohlmann A. R., Guterres S. S., Semisolid formulation containing a nanoencapsulated sunscreen: Effectiveness, *in vitro* photostability

and immune response. *Journal of Biomedical Nanotechnology*, 5 (2009) pp. 240–246.

64. Do Nascimento D. F., Silva A. C., Mansur C. R., Presgrave R. F., Alves E. N., Silva R. S., Ricci-Júnior E., De Freitas Z. M., Dos Santos E. P. Characterization and evaluation of poly(epsilon-caprolactone) nanoparticles containing 2-ethylhexyl-p-methoxycinnamate, octocrylene, and benzophenone-3 in anti-solar preparations. *Journal of Nanoscience and Nanotechnology*, 12 (2012) pp. 7155–7166.
65. Forestier S. Rationale for sunscreen development. *Journal of the American Academy of Dermatology*, 58 (2008) pp. 133–138.
66. Burnett M. E., Wang S. Q. Currentt sunscreen controversies: A critical review. *Photodermatology, Photoimmunology & Photomedicine*, 27 (2011) pp. 58–67.
67. Oberdörster G., Oberdörster E., Oberdörster J. Nanotoxicology: An emerging discipline evolving from studies of ultrafine particles. *Environmental Health Perspectives*, 113 (2005) pp. 823–839.
68. The European Consumer Organisation. Nano-materials in cosmetic products: Definition needs to effectively protect consumers. http://www.beuc.eu/publications/2012-00537-01-e.pdf (accessed in 15 february 2016).
69. Contado C. Nanomaterials in consumer products: A challenging analytical problem. *Frontiers in Chemistry*, 3 (2015) 48.
70. Nasir A., Friedman A., Wang S. (eds.) *Nanotechnology in Dermatology* (2013), Springer Science & Business Media.
71. Egerton T. A., Tooley I. R. UV absorption and scattering properties of inorganic-based sunscreens. *International Journal of Cosmetic Science*, 34 (2012) pp. 117–122.
72. Teubl B. J., Schimpel C., Leitinger G., Bauer B., Fröhlichc E., Zimmer A., Roblegg E. Interactions between nano-TiO_2 and the oral cavity: Impact of nanomaterial surface hydrophilicity/hydrophobicity. *Journal of Hazardous Materials*, 286 (2015) pp. 298–305.
73. Jansen R., Osterwalder U., Wang S. Q., Burnett M., Lim H. W. Photoprotection: Part II. Sunscreen: Development, efficacy, and controversies. *Journal of the American Academy of Dermatology*, 69 (2013) pp. 867.e1–14.
74. Chen T., Yan J., Li Y. Genotoxicity of titanium dioxide nanoparticles. *Journal of Food and Drug Analysis*, 22 (2014) pp. 95–104.
75. Wamer W. G., Yin J. J., Wei R. R. Oxidative damage to nucleic acids photosensitized by titanium dioxide. *Free Radical Biology & Medicine*, 23 (1997) pp. 851–858.

76. Nakagawa Y., Wakuri S., Sakamoto K., Tanaka N. The photogenotoxicity of titanium dioxide particles. *Mutation Research*, 27 (1997) pp. 125–132.

77. Dunford R., Salinaro A., Cai L., Serpone N., Horikoshi S., Hidaka H., Knowland J. Chemical oxidation and DNA damage catalyzed by inorganic sunscreen ingredients. *FEBS Letters*, 418 (1997) pp. 87–90.

78. Hidaka H., Kobayashi H., Koike T., Sato T., Serpone N. DNA damage photoinduced by cosmetic pigments and sunscreen agents under solar exposure and artificial UV illumination. *Journal of Olec Science*, 55 (2006) pp. 249–261.

79. Scientific Committee on Consumer Safety. Opinion on Titanium Dioxide (nano form) COLIPA n° S75 SCCS/1516/13. European Union (2013).

80. SCCNFP. Opinion of the Scientific Committee on Cosmetic Products and Non-Food Products Intended for Consumers Concerning Titanium Dioxide Colipa No S 75, European Commission (2000) Brussels, Belgium.

81. Tan M. H., Commens C. A., Burnett L., Snitch P. J. A pilot study on the percutaneous absorption of microfine titanium dioxide from sunscreens. *Australasian Journal of Dermatology*, 37 (1996) pp. 185–187.

82. Pirot F., Millet J., Kalia Y. N., Humbert P. *In vitro* study of percutaneous absorption, cutaneous bioavailability and bioequivalence of zinc and copper from five topical formulations. *Skin Pharmacology*, 9 (1996) pp. 259–269.

83. Lansdown A. B., Taylor A. Zinc and titanium oxides: Promising UV-absorbers but what influence do they have on the intact skin? *International Journal of Cosmetic Science*, 19 (1997) pp. 167–172.

84. Dussert A. S., Gooris E., Hemmerle J. Characterization of the mineral content of a physical sunscreen emulsion and its distribution onto human stratum corneum. *International Journal of Cosmetic Science*, 19 (1997) pp. 119–129.

85. Lademann J., Weigmann H., Rickmeyer C., Barthelmes H., Schaefer H., Mueller G., Sterry W. Penetration of titanium dioxide microparticles in a sunscreen formulation into the horny layer and the follicular orifice. *Skin Pharmacology and Applied Skin Physiology*, 12 (1999) pp. 247–256.

86. Pflucker F., Wendel V., Hohenberg H., Gärtner E., Will T., Pfeiffer S., Wepf R., Gers-Barlag H. The human stratum corneum layer: An effective barrier against dermal uptake of different forms of

topically applied micronised titanium dioxide. *Skin Pharmacology and Applied Skin Physiology*, 14 (2001) pp. 92–97.

87. Schulz J., Hohenberg H., Pflücker F., Gärtner E., Will T., Pfeiffer S., Wepf R., Wendel V., Gers-Barlag H., Wittern K. P. Distribution of sunscreens on skin. *Advanced Drug Delivery Reviews*, 54 (2002) pp. S157–S163.
88. Cross S. E., Innes B., Roberts M. S., Tsuzuki T., Robertson T. A., Mccormick P. Human skin penetration of sunscreen nanoparticles: In-vitro assessment of a novel micronized zinc oxide formulation. *Skin Pharmacology and Physiology*, 20 (2007) pp. 148–154.
89. Gamer A. O., Leibold E., Van Ravenzwaay B. The *in vitro* absorption of microfine zinc oxide and titanium dioxide through porcine skin. *Toxicology in vitro*, 20 (2006) pp. 301–307.
90. Mavon A., Miquel C., Lejeune O., Payre B., Moretto P. *In vitro* percutaneous absorption and *in vivo* stratum corneum distribution of an organic and a mineral sunscreen. *Skin Pharmacology and Physiology*, 20 (2007) pp. 10–20.
91. Monteiro-Riviere N. A. Wiench K., Landsiedel R., Schulte S., Inman A. O., Riviere J. E. Safety evaluation of sunscreen formulations containing titanium dioxide and zinc oxide nanoparticles in UVB sunburned skin: An *in vitro* and *in vivo* study. *Toxicological Sciences*, 123 (2011) pp. 264–280.
92. Nanoderm Quality of Skin as a Barrier to ultra-fine Particles. QLK4-CT-2002-02678. Final Report (2007). http://www.uni-leipzig.de/~nanoderm/Downloads/Nanoderm_Final_Report.pdf (assessed in 15 february 2016).
93. SCCS opinion in 2012 SCCS (Scientific Committee on Consumer Safety) (2012). SCCS/1489/12-Opinion on Zinc oxide (Nano Form) COLIPA S 76. Brussels.
94. Herzog B. Prediction of sun protection factors by calculation of transmissions with a calibrated step film model. *Journal of Cosmetic Science*, 53 (2002) pp. 11–26.
95. The Standard EN ISO 24444:2010 Cosmetics—Sun protection test methods—*In vivo* determination of the sun protection factor (SPF) http://www.iso.org/iso/catalogue_detail.htm?csnumber=46523 (assessed in 15 February 2015).
96. Department of Health and Human Services Food and Drug Administration 21 CFR Parts 347 and 352 [Docket No. 1978N–0038] (formerly Docket No. 78N–0038) RIN 0910–AF43 Sunscreen Drug

Products for Over-the Counter Human Use; Proposed Amendment of Final Monograph.

97. Diffey, B. L. A method for broad spectrum classification of sunscreens. *International Journal of Cosmetic Science*, 16 (1994) pp. 47–52.

98. Basf Homepage: https://www.sunscreensimulator.basf.com (accessed 26 January 2016).

99. ISO/NP 24443. *In vitro* determination of UVA protection (2012) http://www.iso.org/iso/catalogue_detail?csnumber=46522__(accessed 15 February 2016).

100. Osterwalder U., Herzog B. Sun protection factors: World wide confusion. *British Journal of Dermatology*, 161 (2009) pp. 13–24.

101. Sander B., Golas M. M. Visualization of bionanostructures using transmission electron microscopical techniques. *Microscopy Research and Technique*, 74 (2011) pp. 642–666.

102. Xu, R. Progress in nanoparticles characterization: Sizing and zeta potencial measurement. *Particuology*, 6 (2008) pp. 112–115.

103. Kumart S. A., Khan M. I. Heterofunctional nanomaterials: Fabrication, properties and applications in nanobiotechnology. *Journal of Nanoscience and Nanotechnology*, 10 (2010) pp. 4124–4134.

104. Silva R. C., Muehlmann A. L., Silva J. R., Azevedo R. B., Lucci C.M. Influence of nanostructure composition on its morphometric characterization by different techniques. *Microscopy Research and Technique*, 77 (2014) pp. 691–696.

105. Sccnfp: Sccnfp/0068/98 Guidelines on the use of human volunteers in compatibility testing of finished cosmetic products, adopted by the SCCNFP during the plenary session of 23 June 1999.

106. Sccnfp: Sccnfp/0245/99 Opinion concerning basic criteria of the protocols for the skin compatibility testing of potentially cutaneous irritant cosmetic ingredients or mixtures of ingredients on human volunteers, adopted by SCCNFP during plenary session of December 1999.

107. Liebsch M., Spielmann H., Pape W., Krul C., Deguercy A., Eskes, C. UV-induced effects. *Alternatives to Laboratory Animals*, 33 (2005) pp. 131–146.

108. Silbergeld E. K., Contreras E. Q., Hartung T., Hirsch C., Hogberg H., Jachak A. C., Jordan W., Landsiedel R., Morris J., Patri A., Pounds J. G., De Vizcaya Ruiz A., Shvedova A., Tanguay R., Tatarazako N., Van Vliet E., Walker N. J., Wiesner M., Wilcox N., Zurlo J. T[4] workshop report. Nanotoxicology: "The end of the beginning"—signs on the

roadmap to a strategy for assuring the safe application and use of nanomaterials. *ALTEX*, 28 (2011) pp. 236–241.

109. Mosmann, T. Rapid colorimetric assay for cellular growth and survival: Application to proliferation and cytotoxicity assays. *Journal of Immunological Methods*, 65 (1983) pp. 55–63.

110. Lewinski N., Colvin V., Drezek R. Cytotoxicity of Nanoparticles. *Small*, 4 (2008) pp. 26–49.

111. Hillegass J. M., Shukla A., Lathrop S. A., Macpherson M. B., Fukagawa N. K., Brooke T., Mossman B. T. Assessing nanotoxicity in cells in vitro. wiley interdisciplinary reviews. *Nanomedicine and Nanobiotechnology*, 2 (2010) pp. 219–231.

112. Sharifi S., Behzadi S., Laurent S., Forrest M. L., Stroeve P., Mahmoudi M. Toxicity of nanomaterials. *Chemical Society Reviews*, 41 (2012) pp. 2323–2343.

113. Casey A., Herzog E., Davoren M., Lyng F. M., Byrne H. J., Chambers G. Spectroscopic analysis confirms the interactions between single walled carbon nanotubes and various dyes commonly used to assess cytotoxicity. *Carbon*, 7 (2007) 1425–1432.

114. Wörle-Knirsch J. M., Pulskamp K., Krug H. F. Oops they did it again! carbon nanotubes hoax scientists in viability assays. *Nano Letters*, 6 (2006) pp. 1261–1268.

115. Riccardi C. S., Santos M. L., Clift A. C. G., *Engineered Nanomaterials, Nanotoxicology Issues, Nanosafety and Regulatory Affairs*. Cultura Acadêmica (2015).

116. Clift M. J. D., Gehr P., Rothen-Rutishauser B. Nanotoxicology: A perspective and discussion of whether or not *in vitro* testing is a valid alternative. *Archives of Toxicology*, 85 (2011) pp. 723–731.

117. Hartung T., Sabbioni E. Alternative *in vitro* assays in nanomaterial toxicology. wiley interdisciplinary reviews. *Nanomedicine and Nanobiotechnology*, 3 (2011) pp. 545–573.

118. Augustin C., Collombel C., Damour O. Use of dermal equivalent and skin equivalent models for identifying phototoxic compounds *in vitro*. *Photodermatololy Photoimmunology Photomedicine*, 13 (2007) pp. 27–36.

119. Portes P., Grandidier M. H., Cohen C., Roguet R. Refinement of the Episkin protocol for the assessment of acute skin irritation of chemicals: Follow-up to the ECVAM prevalidation study. *Toxicology in vitro*, 16 (2002) pp. 765–770.

120. Kejlová K., Jírová D., Jírová H., Kandárová H., Weidenhoffer Z., Kolárová H., Liebsch M. Phototoxicity of bergamot oil assessed by *in vitro* techniques in combination with human patch tests. *Toxicology in vitro*, 21 (2007) pp. 1298–1303.

121. Dos Santos G. G., Spiekstra S. W., Sampat-Sardjoepersad S. C., Reinders J., Scheper R. J., Gibbs S. A potential *in vitro* epidermal equivalent assay to determine sensitizer potency. *Toxicology in vitro* 25 (2011) pp. 347–357.

122. Rheinwald J. G., Green H. Formation of a keratinizing epithelium in culture by a cloned cell line derived from a teratoma. *Cell*, 5 (1975) pp. 317–330.

123. Poumay Y., Dupont F., Marcoux S., Leclercq-Smekens M., Hérin M., Coquette A. A simple reconstructed human epidermis: Preparation of the culture model and utilization in *in vitro* studies. *Archives of Dermatological Research*, 296 (2014) pp. 203–211.

124. Carlson M. W., Alt-Holland A., Egles C., Garlick J. A. Three-dimensional tissue models of normal and diseased skin. *Current Protocols in Cell Biology*, Chapter 19 (2008) Unit 19.9.

125. Poumay Y., Coquette A. Modelling the human epidermis in vitro: Tools for basic and applied research. *Archives of Dermatological Research*, 298 (2007) pp. 361–369.

126. Granger K. L., Brown P. R. The chemistry and HPLC analysis of chemical sunscreen filters in sunscreen and cosmetics. *Journal of Liquid Chromatography and & Related Technologies*, 24 (2001) pp. 2895–2924.

127. Rieger M. M. Photostability of cosmetic ingredients on the skin. *Cosmetics & Toiletries*, 112 (1997) pp. 65–72.

128. Schrader A., Jakupovic J., Baltes W. Photochemical studies on trans-3-methylbutyl 4-methoxycinnamate. *Journal of the Society of Cosmetic Chemists*, 45 (1994) pp. 43–52.

129. Nohynek G. J., Schaefery H. Benefit and risk of organic ultraviolet filters. *Regulatory Toxicology and Pharmacology*, 33 (2001) pp. 285–299.

130. Butt S. T., Christensen T. Toxicity and phototoxicity of chemical sun filters. *Radiation Protection Dosimetry*, 91 (2000) pp. 283–286.

131. Kammeyer A., Westerhof W., Bolhuis P. A., Ris A. J., Hische E. A. The spectral stability of several sunscreening agents on stratum corneum sheets. *International Journal of Cosmetic Science*, 9 (1987) pp. 125–136.

132. Berset G., Gozenbach H., Christ R., Martin R., Deflandre A., Mascotto R. E., Jolley J. D. R., Lowell W., Pelzer R. Stiehm T. Proposed protocol for determination of photostability Part I: Cosmetic UV filters. *International Journal of Cosmetic Science*, 18 (1996) pp. 167–177.

133. Marginean Lazar G., Fructus A. E., Baillet A., Bocquet J. L., Thomas P., Marty J. P. Sunscreens' photochemical behaviour: *In vivo* evaluation by the stripping method. *International Journal of Cosmetic Science*, 19 (1997) pp. 87–101.

134. Roscher N. M., Lindemann M. K. O., Kong S. B., Cho C. G., Jiang P. Photodecomposition of several compounds commonly used as sunscreen agents. *Photochemistry and Photobiology*, 80 (1994) pp. 417–421.

135. Stokes R., Diffey B. *In vitro* assessment of sunscreen photostability: The effect of radiation source, sunscreen application thickness and substrate. *International Journal of Cosmetic Science*, 21 (1999) pp. 341–351.

Chapter 4

Topical Photodynamic Therapy for Skin Diseases: Current Status of Preclinical and Clinical Research, Nanocarriers and Physical Methods for Photosensitizer Delivery

Fabíola Silva Garcia Praça,[a] Patrícia Mazureki Campos,[a] Josimar O. Eloy,[a] Raquel Petrilli,[a] Maria Vitória Lopes Badra Bentley,[a] and Wanessa Silva Garcia Medina[b]

[a]*Faculdade de Ciências Farmacêuticas de Ribeirão Preto, Universidade de São Paulo, Ribeirão Preto, SP, Brazil*
[b]*Faculdades Integradas Padre Albino- FIPA, Catanduva, SP, Brazil*

wasigame@gmail.com

4.1 Introduction

Photodynamic therapy (PDT) with systemic application is effective for superficial cutaneous tumors treatment; however, this comes with the risk of prolonged skin phototoxicity. PDT based on topical application has shown promising results in the treatment of superficial skin disorders avoiding the prolonged phototoxicity.

Carrier-Mediated Dermal Delivery: Applications in the Prevention and Treatment of Skin Disorders
Edited by Andreia Ascenso, Sandra Simões, and Helena Ribeiro

ISBN 978-981-4745-58-1 (Hardcover), 978-1-315-36447-6 (eBook)
www.panstanford.com

This non-invasive technique has specificity for the target tissue and allows the treatment of multiples lesions simultaneously due to its low toxicity and good cosmetic results. The reactive end products of topical PDT result in rapid cell cytotoxicity, in particular to the mitochondria, which can lead to apoptosis and/or necrosis of diseased cells. For this purpose, the photosensitizer must to be delivered into skin deep layers to achieve promising results. This chapter will provide an overview of our experience with PDT covering the fundamental aspects, nanocarrier products, physical methods for enhanced penetration, potentials and limitations of using topical PDT in dermatology. Furthermore, we aim to provide an overview of the studies published to date based on oncological and non-oncological dermatology clinical trials and preclinical orientations.

4.2 Photodynamic Therapy and Mechanism of Photosensitization

Photodynamic therapy is able to cause cell death by both necrosis and apoptosis, and due to the ease of access to the skin, it can be easily performed in the treatment of cutaneous diseases. Three components are essential for the efficacy of PDT: a photosensitizing agent, a specific light wavelength, and molecular oxygen [1]. PDT was approved by the FDA for use in dermatologic conditions such as actinic keratosis and non-melanoma skin cancers but off-label uses continue to increase [2]. In recent years, the dermatological photodynamic therapy expanded the spectrum of application to other cutaneous diseases, including inflammatory processes, infectious diseases and skin diseases that are resistant to multiple treatments [3]. The mechanism of photosensitization (Fig. 4.1) involves the effective interaction of both photosensitizer and appropriated light wavelengths, which turns the photosensitizer active from a ground state to an excited state causing the generation of free radicals and radical ions (type I) as well as reactive oxygen species (Type II) which are singlet and triplet oxygen, both of them leading to oxidative damage and cell death [4]. It is not an easy task distinguishing from type I and II reaction in biological systems. However, it is generally accepted that generation of singlet oxygen species (type II reaction) predominates during PDT. Both type I and II reaction of photosensitization have been exhaustively reviewed by several researchers [5–9]. Basically, a successful PDT treatment is related to the ability of the photosensitizer to damage cells followed by apoptosis or necrosis effect. Moreover, it was demonstrated that the cell death pathway is dependent on several factors including cell type, photosensitizer dose administered, incubation time of photosensitizer in target tissue and PDT light dose. Most photosensitizers target mitochondria more efficiently than other cell components and because of this, there is a consensus that the first damage of PDT occurs based on mitochondrial cell death, with signaling pathways leading to apoptosis. The apoptosis induced by PDT was tested using cancer cell lines by Miller and colleagues (2007). Different cell lines were exposed to doses of photosensitizer and the phototoxicity

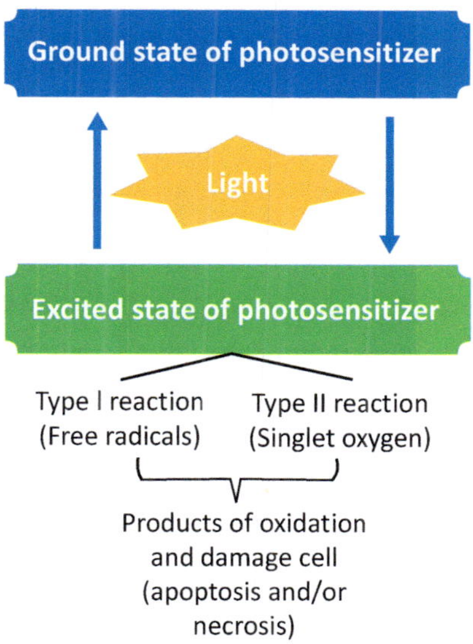

Figure 4.1 Mechanism of action of photodynamic therapy. Photosensitizer excitation begins in the ground state, following the absorption of light, which turns to an excited state and its returns to a ground state, when it can directly react with the substrate by proton or electron transfer occurring the generation of free radicals and radical ions (type I reaction). Alternatively, it can react with molecular oxygen to produce singlet oxygen (type II reaction). Both types of reactions are able to release oxidation products and damage cells by apoptosis and necrosis.

mechanism was measured after transfection of apoptosis inhibitor (Bcl-2). The loss of mitochondrial membrane was an early step found, allowing the release of cytochrome c and other pro-apoptotic factors [10]. In a previous report, Luo and Kessel (1997) showed that organelles such as the mitochondria and lysosomes were mainly affected using lower PDT doses resulting in the induction of apoptosis as opposed to necrosis while at higher PDT doses severe membrane damage resulting in necrosis was observed [11].

4.3 Most Commonly Used Photosensitizers for Dermatological Diseases

The photosensitizers (PSs) used in PDT are chemical compounds with a chromophore agent administrated to treat skin diseases by several routes (e.g., topical, oral, and intravenous). After PS accumulation in the target skin lesion, it is exposed to light of specific wavelengths and befalls the PS activation with generation of reactive oxygen species. Prerequisites for an ideal PS include chemical purity, chemical and physical stability, rapid accumulation in tumor tissue, activation at therapeutic wavelengths with optimal tissue penetration, and rapid clearance from the body [9]. The most extensively investigated PSs in dermatological PDT studies are both pro-drugs that require conversion to porphyrin such as aminolevulinic acid (ALA), methyl aminolevulinic acid (MAL), besides phthalocyanine class and chorine derivatives. Other PSs, which have been studied less extensively, include texaphyrins, hypericin, methylene blue, Nile blue, porphycenes, and indole-3-acetic acid among others. However, they will not be included in this review chapter [9, 12–14].

4.3.1 Topical Photodynamic Therapy with Aminolevulinic Acid and Methyl Aminolevulinic

ALA and MAL are considered endogenous amino acid photosensitizers because in target cells they are metabolized into protoporphyrin IX, a photosensitizing porphyrin generated in mitochondria. Since ALA has hydrophilic nature and its use is restricted to superficial skin disease, the ester of ALA (MAL) enhances lipophilicity and thus can be expected to penetrate deeper into the skin. Most cells in the human body can metabolize ALA or MAL into protoporphyrin IX while neoplastic tissue can produce excess of protoporphyrin IX as compared to the surrounding normal tissue. As described by Peng and colleagues (2001), porphyrins accumulate mostly in sebaceous glands and epidermis, whereas they accumulate preferentially in dysplastic cells and hyperproliferative tissue after application of MAL [15]. In addition, the normal epidermis showed decreased PSs accumulation than surrounding skin lesions reducing PDT photo toxicity to local healthy tissue confirming the PS high selectivity for the tumor.

Both ALA and MAL can be activated with 635 nm light. The effectiveness of ALA and MAL in topical PDT has so far been proven in several studies [16–20] and very similar results were observed in comparative studies [21–22]. However, ALA-PDT resulted in more prolonged and severe adverse effects after treatment using blue light source while MAL-PDT followed by red light is present in most of the clinical trial protocols [23–25]. These findings suggest that the appropriated light source can improve the efficacy and safety of dermatological PDT. Usually photosensitizers absorbing in the far red or near infrared regions have been used for the treatment of deep skin pathologies because the light can penetrate deeply into the skin. On the other hand, photosensitizers absorbing in the blue region may be suitable for the treatment of superficial skin malignances once the light reaches only the stratum corneum [26]. A series of commercial products containing ALA or MAL are available in United States, Brazil, and Europe.

Levulan® is a commercial topical solution containing 20% of ALA and recommended to be administrated following a blue light illumination. Since 1999 it is approved by the FDA for the treatment of minimally to moderately thick actinic keratosis of the face or scalp [27]; however, it is commercialized only in the United States. The most common side effects described by the manufacturer include scaling/crusting, hypo/hyper-pigmentation, itching, stinging, and/or burning, erythema and edema. Severe stinging and/or burning at one or more lesions being treated was reported by at least 50% of patients at some time during the treatment (http://www.dusapharma.com/levulan-product-information.html).

Metvix® is a cream containing 16% of MAL (160 mg/g), which was approved by the FDA in 2004, thus some years after Levulan®, for the same use but it is commercialized in United States, Brazil, and Europe. Metvix® cream layer should be applied about 1 mm thick on the lesion and 10 mm to surrounding normal skin using a spatula followed by an occlusive dressing for 3 h. After this, the area needs to be cleaned using saline solution and immediately exposed to red light. Multiple lesions may be treated during the same treatment session. The usual side effects of Metvix® include painful and burning skin sensation typically beginning during illumination or soon after and lasting for a few hours (http://www.metvix.com/home/about-metvix/overview).

Ameluz® is an innovative topical nanoemulsion-based gel formulation (BF-200) containing 5-aminolevulinic acid (78 mg/g) approved in Europe for use in actinic keratosis and available in Austria, Denmark, Germany, Netherlands, Norway, Sweden, and United Kingdom. The high efficacy of Ameluz® was demonstrated in the pivotal clinical phase III study with 122 patients with actinic keratosis treated by PDT with Ameluz® BF-200 ALA or a registered MAL cream formulation compared to the placebo gel. PDT with Ameluz® BF-200 ALA was superior to placebo PDT with respect to patient complete lesion complete clearance rate such as 81% vs. 22%, respectively [28].

A novel self-adhesive ALA patch is still waiting for approval by the FDA, but it is available outside the United States. The ALA patch–PDT efficacy was evaluated in a clinical trial by Hauschild and colleagues (2009). Twelve months after the use of a single ALA patch-PDT, it was able to treat the majority of mild to moderate actinic keratosis lesions with an excellent cosmetic outcome [29].

4.3.2 Topical Photodynamic Therapy with Phthalocyanines Class and Chlorine Derivatives

A high number of phthalocyanines are currently available and differentiate by the modifying central metal ligand (usually zinc, silicon, or aluminum used to increase singlet oxygen production) and type of side chains [26]. The first structure of phthalocyanine was elucidated in 1935 by Robertson via X-ray crystallography (*apud* by Allen *et al.*, 2001) [4]. In addition, phthalocyanines have emerged as more appropriate PSs for PDT of pigmented lesion because these are active in the red region (650 to 850 nm) and thus avoid the visible light absorption scattering process caused by melanin pigment and allows deeper tissue penetration with higher tumor specificity [30].

Allen and colleagues (2001) discussed that the presence of zinc as the central metal ion in phthalocyanine-based photodynamic therapy (PC-PDT) induces an oxidative stress with short triplet lifetime, high triplet quantum yields, and high singlet oxygen quantum yields, which increase its photo activity and induce apoptosis in several malignant and non-malignant cell lines [4]. These findings support PC-PDT clinical trials and clinical orientations to dermatological diseases shown in Tables 4.1 and 4.2. High photosensitizing activity was also found for water-soluble sulfonated zinc phthalocyanines used against *S. cerevisiae* [31], *C. albicans* and multi drug resistant *C. albicans*, suggesting that the presence of cationic charge is necessary for inactivation of these microorganisms.

Furthermore, silicon and aluminum PC showed excellent clinical response for actinic keratosis, Bowen's disease, skin cancer, mycosis fungoides (http://clinicaltrials.gov.br/ct2/show/NCT00103246) and cutaneous neoplasms [33–34]. Medina and colleagues (2009) evaluated the tissue penetration activity and efficacy of a tetrasulfonate zinc phthalocyanine with and without illumination at 670 nm to cells in order to evaluate the intrinsic toxicity and phototoxic action, respectively [35]. The light and PSs doses used were lower than those usually applied in clinical practice. The authors concluded that low intrinsic toxicity and intense phototoxic effect on brain mitochondria were observed when sulfonated phthalocyanine was used [35]. Tetrasulfonate or silicon zinc phthalocyanine has numerous advantages for the use in dermatological PDT, such as the activation by larger wavelength, which allows increased tissue penetration; minimal systemic toxicity and minimal skin photosensitization due to a faster clearance from tissues and circulation [10, 35].

Chlorins are reduced forms of porphyrins that are attracting attention in PDT field due to the pronounced absorption at 650–900 nm and because of the ability to sensitize singlet oxygen in high quantum yields and low remaining toxicity in the dark [36–39]. Furthermore, chlorin absorption bands in the UV, blue region and red visible, the latter usually in the range of red at 640–700 nm, can be modified by chelated metal ions [40, 41]. As a consequence of their low green region absorption, chlorins exhibit the green color that is related to the name in greek "chloros" [40]. Basically, a chlorin consists of a large heterocyclic aromatic compound with a core of three pyrroles and one reduced pyrrole ring that are coupled by methine linkages. One important example is the chlorophyll presence in plants that consists of a naturally occurring magnesium chlorin [42]. Since the 1980s, chlorins have been used in PDT and many derivatives were developed [42]; however, only the m-tetra(hydroxyphenyl) chlorin (m-THPC or termoporfirin, marketed as Foscan®) has been approved in Europe for the treatment of head and neck squamous cell carcinomas [43]. Gupta and colleagues (1999) were the first to study the effect of topical application of m-THPC using PDT for basal cell carcinoma and Bowen's disease in humans. Authors observed the pathological clearance in nine out of the 28 patients under treatment [44]. With the purpose of improved efficacy and decreased side effects, topical delivery of Foscan® delivery systems can be applied [45] and will be further discussed in the section "Nanocarriers for improved dermal skin delivery of photosensitizers in PDT".

Chlorin e6, also named phytochlorin, is another important class of chlorins, which is derived from chlorophyll A [9]. Among its main applications we can mention skin cancer [46–48] and acne treatment [49]. In the study of Dube and colleagues (2006), chlorin p6, also a chlrophyll A derivative, was applied for the treatment of squamous cell carcinoma in the hamster cheek pouch model using both systemic and topical application. Authors observed that for large tumors (>8 mm) topical application was more effective than intraperitoneal treatment which was justified by the direct absorption of chlorin e6 by the skin layers [50]. In another pilot study, topical application of mono-L-aspartyl chlorin e6 (NPe6) was performed in balb-c mice skin after tape stripping. It was observed an orange to red fluorescence at the epidermis, dermis and also at the muscular layer when 10 mg/mL was applied. Additionally, a dose-dependent pattern was found and fluorescence intensity peak was shown 1 h post-application [51].

Another chlorin derivative that is already under clinical trials is radachlorin, also known as bremachlorin, that consists of a composition of 3 chlorophyll a derivatives [52]. In this phase II clinical studies, both radachlorin solution for intravenous injection and a gel for topical application were administered and activated by light at 662 nm aiming the treatment of basal cell carcinoma. Intravenous administration at a dose of 1.0–1.2 mg/kg or 4 PDT sessions using topical gel with 1 week interval and light exposure at 300 J/cm^2 and 400 J/cm^2, respectively, revealed efficacy of approximately 80%.

4.4 PDT Applied to Skin Diseases

Owing to systemic side effects, the interest for the PDT skin cancer treatment remained decreased until 1990 when Kennedy *et al.* carried out a clinical trial of ALA-induced Pp IX photodynamic therapy, and they found that the complete response rate of the first 80 basal cell carcinomas treated was 90% when examined 2–3 months post-treatment, while 7.5% showed a partial response (tumor area decreased more than 50%) and 2.5% corresponded to failure. It was also used to treat six lesions having a histological diagnosis of either *in situ* or early invasive squamous cell carcinoma. All six showed a complete response. Furthermore, cosmetic results were excellent with desirable patient acceptance [53]. On the other hand, the use of PDT for non-cancer skin treatment, such as antimicrobial skin lesion has been known for over 100 years. Antimicrobial PDT or photodynamic inactivation (PDI) is an alternative treatment against multidrug resistant infections [54].

Table 4.1 Dermatologic application of photodynamic therapy

Photosensitizer used on PDT	Dermatologic condition	References
(ALA) Aminolevulinic acid	Non-melanoma skin cancer	[16]
	Nodular basal cell carcinoma	[55]
	Cutaneous T cell lymphoma	[17, 56]
	Actinic keratoses	[57–59]
	Psoriasis	[60–62]

(*Continued*)

Table 4.1 (*Continued*)

Photosensitizer used on PDT	Dermatologic condition	References
	Leishmaniasis cutaneous	[63]
	Paget's disease Mycosis fungoides	[64–68]
	Molluscum contagiosum	[69]
	Warts	[70]
	Acne	[27, 71–73]
	Sebaceous skin disease	[69, 74–75, 88]
	Photorejuvenation	[55, 76–77]
(MAL) Methyl aminolevulinic	Leishmaniasis cutaneous	[78]
	Paget's disease	[66]
	Actinic keratoses	[18–20]
	Acne	[23–24]
	Nodular basal cell carcinoma	[20, 79–80]
	Queyrat disease	[81]
Phthalocyanines	Lymphoma, non-melanoma skin cancer and precancerous conditions	[82]
	Non-melanoma skin cancer	[83–84]
	Recurrent cutaneous T-cell lymphoma and recurrent mycosis fungoides	[82]
	Cutaneous T cell Lymphoma	[10]
	Psoriasis	[86]
	Non-melanoma skin cancer	[87]
	Actinic keratoses, Bowen's disease, squamous cell carcinoma, basal cell carcinoma, or mycosis fungoides	[33]
Chlorins	Melanoma skin cancer	[47–48, 89]
	Squamous cell carcinoma	[50]
	Bacteria-induced inflammation	[49]

4.4.1 Skin Cancer Treatment

Melanoma basal cell carcinoma and squamous cell carcinoma belong to non-melanoma skin cancer group which is the most common skin cancer worldwide and is able to be treated with PDT [90]. Actinic keratosis and squamous cell carcinoma *in situ*, also called Morbus Bowen, are precursors of squamous cell carcinoma and therefore are also normally treated with topical PDT.

4.4.1.1 Actinic keratosis and squamous cell carcinoma *in situ*

AKs are proliferations of transformed neoplastic keratinocytes confined to the epidermis while the squamous cell carcinoma *in situ* can reach the dermis [90–91]. These are likely to occur on sun exposed sites such as face, dorsum of the hands, scalp, often with multiple injuries. The photosensitizers of first generation ALA and MAL are approved by the FDA and the European regulatory authorities, respectively for actinic keratosis and basal cell carcinoma. The protocol provides the application of ALA in injuries (only individual lesions) followed by exposure to blue light (10 J/cm^2) [92, 93]. In a clinical trial several light sources, photosensitizers and various therapeutic regimes have been used successfully [94].

Alexiades *et al.* 2003 showed the efficiency of ALA treatment using pulsed dye laser at 595 nm during 8 months. Forty-one individuals with AKs were subjected to a single treatment, with topical application of ALA (20%) and incubation period of 3 to 18 h, followed by pulsed laser exposure. In this study, the control group received only laser treatment (without ALA). In addition to efficacy, safety, the duration of treatment and recovery period were also assessed. The authors concluded that the treatment of AKs using pulsed dye laser is safe and effective, causing minimal patient discomfort, in addition, excellent cosmetic results were reached in the post-treatment [95, 96].

A randomized controlled clinical trial of PDT using MAL to treat AK in transplant recipients was evaluated by Dragieva *et al.* (2004). The results showed that MAL-PDT is safe and effective. After the end of treatment, all 129 injuries were either resolved or were reduced in number and size [97]. Meanwhile, the PDT is not recommended for invasive squamous cell carcinoma but has been reported in the treatment of other skin proliferative disorders (e.g., vascular malformations), and other skin tumor Kaposi's sarcoma, Paget's disease, cutaneous B cell lymphoma [2] and cutaneous T-cell lymphoma (CTCL) [85]. CTCL is a malignant tumor derived from T-helper lymphocytes and a variety of treatment options are available, such as topical corticosteroids, topical nitrogen mustard, psoralen plus UVA (PUVA), radiation therapy, excision, carbon dioxide laser, and photodynamic therapy in patients with localized lesions that are resistant to standard treatments [3]. Topical ALA cream was applied to two patients with CTCL lesions (stage I and stage III) for 16 h and the kinetics of ALA-induced protoporphyrin IX (PP) accumulation were investigated before, during, and 1 hour after photoirradiation (580–720 nm). Authors argued that different patients exhibited different patterns of fluorescence kinetics and according to lesion thickness the PP fluorescence intensity was diminished during photoirradiation process. Absence of PP fluorescence was observed in the stage I lesion after the treatment; however, PP fluorescence reappeared in the thick stage III MF lesions. Complete response was achieved in the stage I lesion with a single irradiation light of 170 J/cm^2, while for stage III lesion it was required fractionated irradiation with a total light dose of 380 J/cm^2. Thus, the ALA-PDT treatment were effective for CTCL showing a high response for both stage I and stage III MF lesions.

Fernandez-Guarino *et al.* 2007, concluded that parameters of PDT treatment for CTCL have to be optimized, given that it inactivates but does not eliminate the lymphocytes in the plaque and that the periods of remission are highly variable and follow-up is necessary to monitor possible recurrence [3]. In addition, early detection of basal cell carcinoma, actinic keratosis and seborrheic keratosis by fluorescence imaging using PDT and ALA as biomarkers was reported by Andrade and colleagues (2014) (Fig. 4.2). The authors argued that ALA-mediated photodynamic detection is a potential auxiliary tool for skin diagnosis. Significant discrimination of normal and tumor cell as well as better delineation of the lesion margins (which is important for effective treatment of skin cancer lesions) were demonstrated [99].

4.4.1.2 Non-melanoma carcinoma

These are also proliferations of transformed neoplastic keratinocytes confined to the epidermis, whereas it extends more deeply including dermis. In this way, better results were reported using MAL-PDT compared to ALA-PDT. This situation is likely due to greater lipophilicity, selectivity, and penetration rates of MAL [93].

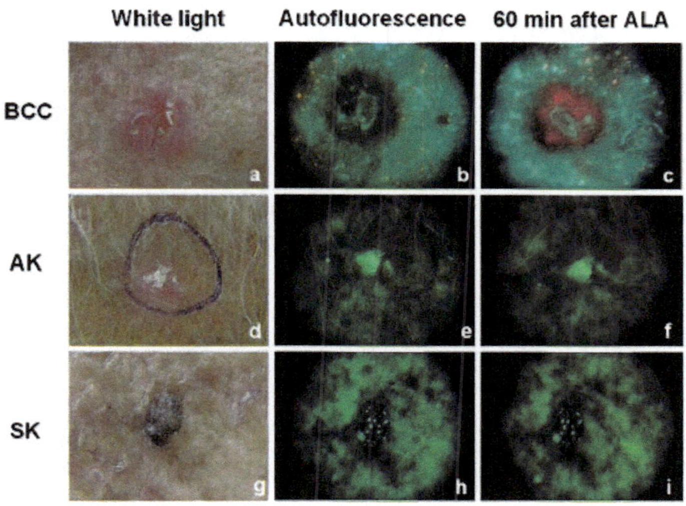

Figure 4.2 Skin lesions images using white light illumination (left column), autofluorescence (central column) and 1 h after ALA application (right column). The lines show typical images for: basal cell carcinoma (top), actinic keratosis (center), and seborrheic keratosis (bottom). Reproduced with permission from Andrade *et al.*, 2014 [99].

However, scientific orientation showed the response of ALA-PDT in the treatment of skin cancer is satisfactory in 85% to 100% and 10% to 64% in superficial and nodular carcinoma, respectively [93]. MAL-PDT effect on skin lesion was successfully confirmed for difficult-to-treat basal cell carcinoma [80], nodular carcinoma [100], and queyrat erythroplasia [81]. A MAL-PDT multicenter clinical study with large number of patients and follow-up of five years after the end of treatment showed cure rate of 95% and 73% for superficial and nodular carcinoma, respectively. The recurrence rate was 22% for superficial cancer similar to conventional cryotherapy treatments [100]. Ongoing clinical trials of PDT for skin cancer are presented in Table 4.2.

Table 4.2 Examples of FDA ongoing clinical trials for skin diseases treated with photodynamic therapy

Clinical trial	Identifier number	Clinical phase	Current status
Evaluation of efficacy of PDT in basal cell carcinoma	NCT00985829	Phase I	This study has been completed
Short term effects of photodynamic therapy in basal cell carcinoma	NCT01015898	Phase I	This study has been completed
Safety and efficacy study for the field-directed treatment of actinic keratosis with photodynamic therapy	NCT01966120	Phase III	This study has been completed
Protocols for painless photodynamic therapy (PDT of Actinic keratoses)	NCT02124733	Phase III	This study is ongoing, but not recruiting participants
Safety study of photodynamic therapy using phthalocyanine injection in treating patients with malignant tumors	NCT01043016	Phase I	The recruitment status of this study is unknown because the information has not been verified recently
Photodynamic therapy using blue light or red light in treating basal cell carcinoma in patients with basal cell nevus syndrome	NCT02258243	Phase II	This study is not yet open for participant recruitment
Pilot study evaluating the efficiency and the tolerance of the PDT in the treatment of Epidermal Dysplasia for patients affected by hereditary DEB	NCT02004600	Pilot study	This study is not yet open for participant recruitment

Clinical trial	Identifier number	Clinical phase	Current status
Fractional laser assisted daylight photodynamic therapy versus daylight photodynamic for treatment of actinic keratoses	NCT01898936	Not informed	This study is currently recruiting participants
A clinical trial evaluating the role of systemic antihistamine therapy in the reduction of adverse effects associated with topical 5 aminolevulinic acid photodynamic therapy	NCT02451579	Phase I	This study is currently recruiting participants
A photodynamic therapy for treatment of actinic keratoses	NCT00558688	Phase I	This study has been completed
Study of the efficacy of daylight activated photodynamic therapy in the treatment of cutaneous leishmaniasis	NCT00840359	Phase II	This study is enrolling participants by invitation only
IM versus 5-FU versus IMI versus MAL-PDT in treatment of actinic keratosis	NCT02281682	Phase IV	This study is currently recruiting participants
Study of methyl aminolevulinate photodynamic therapy with and without Er:YAG laser in actinic cheilitis	NCT02198469	Phase I	This study has been completed.
Actinic keratosis treatment with Metvix in combination with light	NCT02373371	Phase III	This study is not yet open for participant recruitment.
Evaluation of efficacy of photodynamic therapy in basal cell carcinoma with 6 months follow-up	NCT00988455	Phase I Phase II	This study has been completed.

(*Continued*)

Table 4.2 (*Continued*)

Clinical trial	Identifier number	Clinical phase	Current status
Photodynamic therapy with levulan and blue light for treatment of actinic cheilitis	NCT02409732	Phase IV	This study is not yet open for participant recruitment.
Safety and efficacy study for the treatment of non-aggressive basal cell carcinoma with photodynamic therapy	NCT02144077	Phase III	This study is currently recruiting participants.
Photodynamic therapy using methyl-5-Aminolevulinate Hydrochloride Cream in Determining Pain Threshold in patients with skin cancer	NCT01292668	Phase I	This study has been completed.
Photodynamic therapy compared to adapalene 0.1% gel plus doxycycline in the treatment of acne vulgaris	NCT01245946	Phase II	This study has been completed.
Treatment of actinic keratoses of the face with imiquimod 3.75% cream followed by photodynamic therapy	NCT01203878	Phase IV	This study has been terminated.
Photodynamic therapy using silicon phthalocyanine 4 in treating patients with actinic keratosis, Bowen's disease, skin cancer, or stage I or stage II mycosis fungoides	NCT 00103246	Phase I	This study has been completed.

Source: Data retrieved from http://clinicaltrials.gov.

4.4.2 Other Skin Disease Treatments

PDT applications for non-oncologic dermatology are based on (1) photosensitizing skin appendages; (2) immunomodulatory action through lymphocyte activation and (3) microbial inactivation by irradiation of visible light. Thus, both topical ALA and MAL-PDT are considered in practical clinical an alternative for the treatment of selected skin inflammatory diseases (e.g., psoriasis) [1], localized infections by virus, bacteria and fungi [101] and other skin disorders including those caused by photodamage [102].

4.4.2.1 Viral lesions treated with PDT

Several clinical trials of PDT have been conducted against viral infections, especially for cutaneous warts also known as *verrucae vulgaris* and *condyloma accuminata* (genital warts) caused by human papilloma virus (HPV) [54]. A clinical trial of 31 patients with cutaneous warts treated with ALA-PDT showed complete response and no significant side effects [103]. Ichimura *et al.* showed that PDT treatment was effective not only in improving the cytological and histological characteristics but also for eradicating cervical HPV [104]. Similar results were found when used for the treatment of cervical intraepithelial neoplasia [105], inoperable vulvar Paget's disease [106]; molluscum contagiosum lesion (sexually transmitted) successfully treated with ALA-PDT in HIV-positive individuals [69, 107]; condyloma in the vulva, vagina, and penis [108–110].

Recent clinical data of anogenital condylomata acuminata treated with topical PDT was provided by Nucci *et al.*, 2010. Thirteen out of 14 cases were successfully treated and the best results were likely to be achieved with 16% to 20% gel formulation of ALA and red light dose of 100 to 150 J/cm^2 [111]. It is important to highlight that the effectiveness of photosensitizer compounds against viruses is governed by physical and chemical parameters such as virus type and photosensitizer structure [112].

Mark Wainwright presented an excellent review article on these aspects and described several photosensitizers that are able to inactivate both enveloped and non-enveloped viruses and are active against all stages of the virus life. Thus, the influence of molecular charge and lipophilicity/hydrophilicity balance (log *P*) as well as the degree of ionization (pKa) of photosensitizers are important to ensure that it is able to enter the infected cell before attacking the virus [112]. For example, amphiphilic phthalocyanine derivatives were more effective against cell-associated virus than were hydrophilic analogues, while anionic phthalocyanines are thought to act against the viral envelope. In this way, considerable efforts have been made with both anionic and cationic phthalocyanines to improve the PDT treatment effectiveness [113–115].

4.4.2.2 Bacterial lesion treated with PDT

Several problems with conventional bacterial treatments, including antibiotic resistance, create a demand for PDT. The most common bacterial infection of skin is acne, caused by *Propionibacterium acnes* (*P. acnes*), found in the sebaceous glands of patients. Common therapies for acne include topical antibacterial therapies and phototherapy without photosensitizer [101]. The skin photoactivation by phototherapy using blue, red, yellow light and pulsed laser devices can produce a significant improvement over the time once the *P. acnes* is capable of producing endogenous porphyrins [101]. Moreover, Hongcharu and colleagues (2000) reported for the first time a clinical trial of combined photosensitizer and illumination by ALA-PDT treatment for inflammatory acne disease [116].

PDT acne treatment mechanism occurs by damage into sebaceous glands concomitantly with light activation of porphyrin (produced from *P. acnes* and ALA) resulting in photodynamic killing of *P. acnes* [116]. PDT after ALA and MAL applied to the skin were safe and effective for acne treatment since *P. acnes* is capable of producing endogenous porphyrins and even more porphyrins can be accumulated in the presence of ALA or MAL, ensuring the effectiveness of PDT treatment. Also, no significant differences in the response rate between ALA-PDT and MAL-PDT was found [21]. Greater clinical effect was observed by Taylor *et al.* using short time contact of topical ALA and MAL applied on skin before the illumination with non-coherent light source [117]. Furthermore, an alternative mode of action for ALA-PDT other than direct damage to sebaceous glands or photodynamic killing of *P. acnes* was suggested by Pollock *et al.*, 2004 [73].

Hui-Lin Ding and colleagues (2011) showed a case report using a repeated weekly short-cycle ALA-PDT to treat severe facial acne lesions refractory to systemic retinoid and antibiotics. Freshly prepared 3% ALA cream was applied to acne lesions 3 h before light irradiation. Mild erythema and edema occurred immediately after the treatment and lasted 3–5 days. Mild exfoliation and crusts occurred 2–4 days post-treatment and spontaneously subsided within 2–3 days. No severe exfoliation, erosion, or purpura was seen during or after the treatment. Since most of the lesions were completely cleared after 3 sessions, no further treatment was required. The results of this case report, however, suggest that repeated weekly short-cycle ALA-PDT of low drug/light dose could be an alternative treatment for facial acne refractory to conventional therapies [118].

Moreover, there is number of clinical trials that successfully treated acne with both ALA and MAL–PDT using different light sources and protocols [21, 119–122] and most of them considered that neutral or anionic photosensitizers are able to bind and inactivate the Gram positive bacteria but only bind to the outer membrane of Gram-negative bacteria because the latter has physical and functional barrier provided of two cell envelope membranes, which influence the response of PDT [123]. Wardlaw and colleagues showed a greater PDT susceptibility for Gram-positive bacteria than Gram-negative using topical ALA treatment of skin wound caused by *Staphylococcus intermedius, Streptococcus canis, Pseudomonas aeruginosa, and Escherichia coli. Staphylococcus intermedius* and *S. canis* had 60–70% cell death and *Pseudomonas aeruginosa* also responded significantly, but to a lesser extent with about 30% cell death, while *E. coli* did not have any significant bacterial death with PDT treatment [123].

4.4.2.3 Fungal skin lesions treated with PDT

Unlike bacterial therapy, antifungal standard treatments are limited due to the reduced number of drug substances, several side effects, prolonged and expensive drug treatment in addition to high incidence of drug resistant fungi [124]. Advantageously, the use of PDT for fungal skin lesion treatment reported did not cause resistance to antifungal and mutagenic effects on fungi cells. A review of superficial fungal infections of the skin treated by photodynamic therapy was provided by Calzavara-Pinton *et al.* 2005 in order to bring attention to this application of PDT that is cost-effective, highly selective and prevents the occurrence of drug resistant [26].

In this book chapter, we intended to show the general principles of photosensitization of fungi and discuss the most common photosensitizers used for this application. As discussed above, the photosensitizer structure and spectrum of light used for PDT of skin diseases is a critical issue. Cationic photosensitizers appear to be more effective for microorganism treatment and are the predominant type used against fungal skin lesions. Fungi cells consist in much more complex targets than bacteria. It is enveloped by a thick external wall composed of a mixture of glucan, mannan, chitin, and lipoproteins and separated from the plasma membrane by a periplasmic space, which makes it more similar to mammalian cells than bacteria cells [124]. Furthermore, photosensitizers with absorption peak at red or near infrared regions may be suitable for the treatment of dermatophytes (which can invade both the stratum corneum and hair follicles) due to the deep penetration into tissue while the photosensitizers absorbing at blue region are more suitable for *Candida* species (which can superficially invade the tissue) [26].

Mycosis is an infection caused by direct fungal invasion of the skin. Interdigital mycosis of the feet is the most frequently observed condition of fungal infection in men [125]. ALA-PDT in the treatment of interdigital mycosis of the feet caused by *Candida* or *Trichophyton* species were clinically effective in an open pilot study using nine patients and 20% of topical ALA under occlusive dressing, followed, 4 h later, by the irradiation of 75 J/cm^2 of broadband red light. Four weeks after the treatment, patients had a final clinical follow-up and laboratory examination. Clinical and microbiological recovery was identified in most patients. However, after 4 weeks, recurrences were seen in four patients and overall tolerability was good [125].

4.4.2.4 Other microbiological lesions treated with PDT

Pityriasis versicolor, tinea cruris, papillomas, molluscum contagiosum, leishmaniose cutanea are microbiological lesions also treated with PDT [126]. Cutaneous leishmaniose is an infection depending on the host genetic background and the leishmaniose parasite type. Recently, Akilov *et al.*, 2007, reported the efficacy of PDT for cutaneous leishmaniasis. The parasites were inoculated intradermally into each ear of BALB/c mice and ALA-PDT was performed 3 weeks after infection. Infected sites were illuminated and mice were sacrificed after 5 days of PDT and the parasites load was assessed. *In vivo*, authors observed that PDT with ALA resulted in significant reduction of the parasite loads; however, it led to tissue destruction [127].

Shol *et al.*, 2007 reported excellent results of PDT treatment in patient with facial cutaneous leishmania proved to be resistant to various other therapeutic regimens. The PDT treatment using Metvix® cream (Photocure, Odlo, Norway) as photosensitizer consisted of two applications of light administered at one week interval, in addition a third application was conducted after four weeks to ensure treatment success. The lesions healed rapidly after PDT with good cosmetic results [128]. Finally, it was reported the successful treatment of a 69-year-old patient with a relapse of long-standing cutaneous leishmaniasis using intralesional ALA-PDT [129].

4.4.2.5 Other inflammatory lesions treated with PDT

Psoriasis, characterized to be a chronic inflammatory skin disease, has an estimated prevalence worldwide from 0.6% to 4.8% and approximately 2.5% and 1.3% of these cases are predominant in caucasian and African populations, respectively [1]. Current therapies for psoriasis include corticosteroids, cyclosporine, topical vitamin D analogs, phototherapy such as ultraviolet B, psoralen plus ultraviolet A and others. PDT is an innovative light treatment for psoriasis but some disadvantages of this therapy have been reported, such as a very painful sensation during and after the treatment, besides the treatment cost and the partial clearance of lesions [1]. Schleyer and colleagues (2006), evaluated an intrapatient comparison study on 12 patients in order to determine the topical ALA-PDT as a suitable treatment for chronic plaque-type psoriasis. However, the authors achieved disappointing results and low treatment tolerability. All patients reported that the PDT caused pain or burning during irradiation, which lasted for several hours after treatment [130]. Avoiding similar results, a clinical randomized of topical ALA-PDT carried out by Radakovic-Fijan and colleagues (2005), applied 1% of ALA in 29 patients because of their observation of very painful sensations when 10% to 20% ALA was used during PDT of psoriasis.

The authors concluded that satisfactory response in localized psoriasis using a smaller doses was achieved [62]. Recently, a good clinical response was observed in most patients treated with ALA-PDT in a clinical and immunohistochemical evaluation. The topical ALA-PDT applied in 12 patients caused an improvement in the psoriasis but several discomforts were related [131].

Different clinical studies compared the efficacy of both topical PDT and dithranol in patients with chronic plaque psoriasis [132, 133]. Nine patients were treated on symmetrical plaques twice a week for 2 weeks with PDT and were treated daily with 0.5% dithranol. Patients were evaluated before and 20 days after the treatment. The results showed that 56% of the patients reported that PDT caused less pain and irritation when compared with dithranol. However, treated plaque thickness improved in 57% and 78% of the PDT group and dithranol group, respectively [133].

Rosacea is another common skin disease of unknown cause and frequently treated with conventional antibiotics, which are tetracycline and metronidazole [134]. The rosacea symptoms involve redness area and inflamed skin lesions particularly on the nose, forehead, cheeks and chin and it can affect women more than men between the 30 and 50 years old. Nybaek and Jemec (2005) reported clinical results of the combined PDT and standard treatment using tetracycline and metronidazole. The treated skin cleared in three of the four patients. Treatment with light alone did not appear to have any effect. In one of the patients, there was no relapse during a 9-month follow-up. In two patients, remissions lasted 3 months [134]. Others studies in a series of rosacea patients using MAL-PDT showed good results in 10 out of 17 patients, and fair results in other 4 patients suggesting an apparent effect of MAL-PDT on rosacea [135].

4.4.2.6 Other dermatological applications of PDT

4.4.2.6.1 Photorejuvenation, cosmetic enhancement and hair loss

As results of a consensus conference of PDT, the experts established consensus statements for pretreatment, post-treatment, ALA contact time, light sources, and number of sessions for ALA PDT to treat superficial basal cell carcinoma, actinic keratosis, acne, rhinophyma, rosacea, sebaceous skin, cosmetic enhancement, and photorejuvenation. In this evaluation, it was concluded that ALA PDT offers many advantages, being a safe and effective tool for use in dermatology treatment [136].

In clinical practice, the human skin with photoaging damage was treated using ALA-PDT and red light. The results showed general improvement of skin color, collagen and elastic fibers for 92.3% of the cases. However, the melanose was not totally removed. These results suggested better efficacy of ALA-PDT for lower photo-type human skin [183]. Cosmetic outcomes using PDT were reported as good to excellent [76, 77, 137].

Gold in 2003 observed, after full face ALA-PDT treatments, improvements of about 90% in crow's feet, 100% in tactile skin roughness, 90% in mottled hyperpigmentation and 70% in telangiectasias [76].

An industry-sponsored multicenter study was published in 2007 by Rhodes *et al.* The authors compared 5-year lesion recurrence rates in nodular basal cell carcinoma treated with topical MAL-PDT or simple excise surgery. Excellent cosmetic outcome: 87% versus 54% was achieved for MAL-PDT and excision surgery, respectively [32].

Similar cosmetics results were reported single-center and open-label investigatory studies of photodamaged skin using topical ALA-PDT, in which the subjects have reported improvements in their fine winkles, skin roughness and overall hyperpigmentation for about 67%, 33%, and 67% of the patients, respectively [77] and excellent cosmetics rates higher than 90% for nonhyperkeratotic actinic keratosis lesions [138].

Another recent report was a pilot study to compare the effectiveness of ALA-PDT and red light alone for photoaging therapy. For this purpose, patients with photoaging skin were evaluated. The treated forearm sites using ALA-PDT or red light alone were examined by dermoscopy and the transepidermal water loss (TEWL), before and after treatment, as well as the changes in stratum corneum (SC) hydration was assessed. In addition, microscopic examination of collagen and elastin was performed (Fig. 4.3). The results showed better skin rejuvenation effect using ALA-PDT treatment when compared to ref light alone (TEWL decreased and SC hydration increased). Furthermore, the signs of typical solar elastosis damage were enhanced in the groups [139].

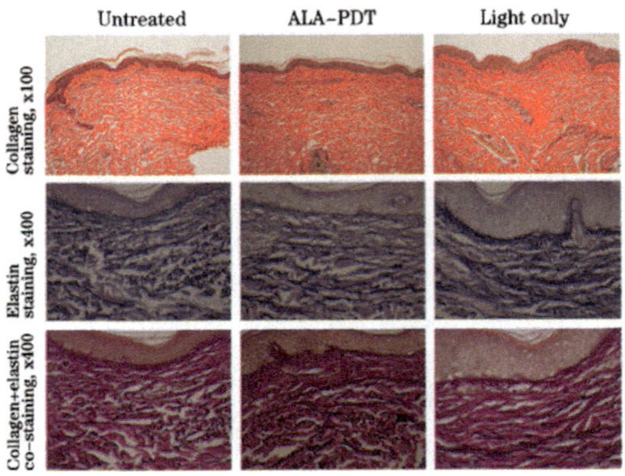

Figure 4.3 Representative microscopic images of collagen and elastin staining. After PDT or red light illumination, the epidermis became slightly thicker but more prominent in the ALA-PDT group. The photoaging signs were improved and the content of collagen fibers increased in both groups but the PDT group showed more collagen depositions. Improvement in the denaturation of elastic fibers with more obvious increase in the content of elastic fibers was confirmed in the PDT group in addition the distribution of collagen and elastic fibers in the dermal layers appeared more uniform, regular, and tighter after ALA-PDT. Image reproduced with permission from Jie Ji *et al.* (2014) [139].

ALA-PDT was also evaluated for hair loss. The authors applied ALA 5, 10% and 20% in patient with alopecia areata 3 h before the red light application. The results showed no significant hair growth after 20 twice-weekly treatment sessions. A significant increase in erythema and pigmentation was observed for the three concentrations of ALA [60]. In this way, PDT is a common tool for dermatology worldwide both for clinical practice and for preclinical studies, and perhaps better performance can be reached using appropriated protocols established depending on a case.

4.5 Nanocarriers and Physical Methods for Improved Dermal Skin Delivery of Photosensitizers in PDT

Skin naturally exerts hindrance against molecule penetration playing a role as a protective layer, also against mechanical and radiation-induced injuries. Skin has other functions such as immunologic, metabolic, sensory, temperature regulation and excretory through sebaceous and sweat glands [140]. The skin barrier is particularly provided by the outermost layer with lipophilic properties, the stratum corneum (SC), which is the limiting factor of dermal absorption. However, other numerous factors can influence the permeation of substances through the skin, such as contact area, exposure duration, integrity of SC and thickness of the epidermis. After SC, there is a hydrophilic compartment represented by viable epidermis and dermis, formed by stratified epithelium and conjunctive tissue, respectively [141].

The SC is a selective skin layer that hampers drug permeation according to physicochemical characteristics of molecules such as lipophilicity, molecular weight, and concentration. Related to PSs, the lipophilic molecules are absorbed more easily, whereas hydrophilic molecules are barred by SC [9]. So, there is a need of drug incorporation into effective delivery systems, addition of penetration enhancers into the formulation or use of physical methods (ionthophoresis and sonophoresis), which also will determine the skin permeation and overcome the skin barrier to access viable skin [142].

Another important issue to consider is the cutaneous absorption in diseased skin, mainly for conditions that alter stratum corneum such as wounds and inflammation or those related to disturbed epidermal differentiation as thickened stratum corneum in non-melanoma skin cancer, for example [143]. Dermal permeation of substances is enhanced in compromised skin compared to normal skin [144]. Within this context, Gattu and Maibach (2011) provided examples of these conditions such as mycosis fungoides, psoriasis and skin atopic dermatitis patients, concluding that damaged or diseased skin display different patterns in penetration [145].

4.5.1 Nanocarriers used in Topical PDT

Nanotechnology represents an advance for PDT use, because nanoparticles may offer large payload of different types of PSs, possess multiple ligands and promotes controlled release [146]. For topical application, drug delivery systems based on nanotechnology can deliver efficiently PSs molecules into viable epidermis and enable clinical use of PDT [83, 147–149]. Features required for this kind of development are the use of biocompatible and biodegradable materials, maintenance of small size in the range of nanometers with large surface area to mass ratio and high reactivity (Fig. 4.4). Furthermore, it should display safety, interact with biological systems, improve solubility of hydrophobic molecules, protect against degradation and increase the bioavailability of PSs [146].

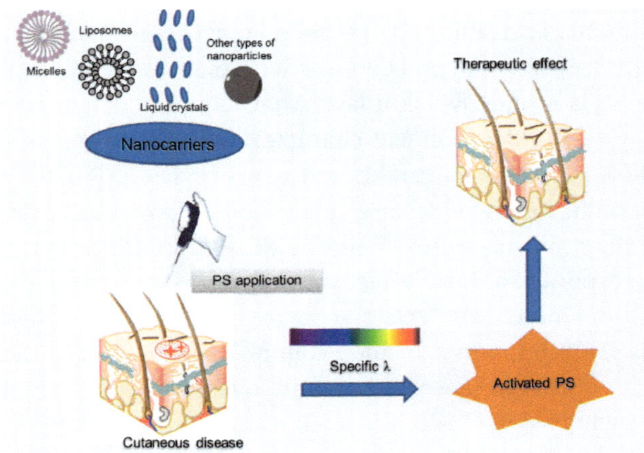

Figure 4.4 Schematic presentation of nanocarriers used to skin cancer treatment and diagnosis based on PDT.

4.5.1.1 Polymeric carriers

The most common polymeric nanoparticles are polymer-PS conjugates, PS-loaded polymeric nanocarriers and polymeric micelles containing PS, which could be applied for passive or active delivery strategies [150]. Also, the highly branched polymer known as dendrimer can be used as PS carrier, which possesses well-defined structure with high capability of drug payload [151].

Regarding polymeric nanoparticles, the main advantages lie in their versatility, physical robustness, surface properties, high drug loading and diversified composition of polymeric matrices, which could be controlled for polymer degradation and drug release kinetics [152].

Within polymeric carriers, there are biodegradable polymer nanoparticles, which have advantages such as biodegradability, ability of controlled release, high drug loading and allowing incorporation of site-specific moieties. For this kind of particle, there is a large variety of available materials and manufacturing processes, which is interesting for application in PSs because they can accommodate PSs with several degrees of hydrophobicity, molecular weight and pH. Also, biodegradable polymeric nanoparticles can improve PDT response compared to conventional formulations [153, 154].

Da Silva *et al.* (2013) investigated the effect of Poly(D,L lactic-co-glycolic acid) (PLGA) based nanoparticles for PpIX application on the skin using fluorescence microscopy. Penetration and distribution of PpIX was analyzed after 8 and 24 h post-application and revealed greater fluorescence intensity in deeper skin layers after 24 h [155]. *In vivo* studies using polymeric nanoparticles for topical PDT were performed by Wang *et al.* (2015). PLGA nanoparticles loaded with 5-ALA were prepared, characterized and applied topically in SKH-1 hairless mice with cutaneous SCC. Animals treated with nanoparticles in the absence of PDT presented tumor growth similar to the negative control group. In contrast, when nanoparticles were applied and PDT was performed, a significant tumor reduction was observed a week after the second treatment ($P \sim 0.0001$). As a result, a reduction of approximately 68% tumor sizes after four treatments was observed [156].

4.5.1.2 Lipid-based carriers

Among lipid-based carriers, the most commonly reported are liposomes and lipid nanoparticles, which embrace solid lipid nanoparticles and nanostructured lipid carriers, as well as nanoemulsions and liquid crystalline nanodispersion, among others. In general, all these lipid-based carriers increase solubilization of water-insoluble substances, dissolve or suspend the drug into lipid matrix [157].

Liposomes are vesicles formed by bilayers of phospholipids that can accommodate hydrophilic and hydrophobic PSs. For example, lipophilic PSs such as phthalocyanine derivatives are solubilized into the bilayer region and hydrophilic PSs such as aminolevulinic acid solubilize in the inner core of liposomes [152]. Liposomes are regarded as versatile and superior compared to other lipid-based carriers, once they can have controllable sizes according to preparation procedures [158] as well as are able to decrease concentration of PSs and lower light doses [154]. Improvement of 5-ALA skin penetration from topical application using liposomes was reported in the last decade [149, 159]. For example, liposomes containing lipid composition similar to mammalian stratum corneum such as ceramide type III, cholesterol, cholesteryl sulfate and palmitic acid, were able to promote effective PSs delivery into the skin [149]. However, malleable, flexible, and deformable liposomes such as ethosomes, flexosomes, and invasomes, respectively, are the new generation of liposomes developed for enhanced skin delivery (Fig. 4.5).

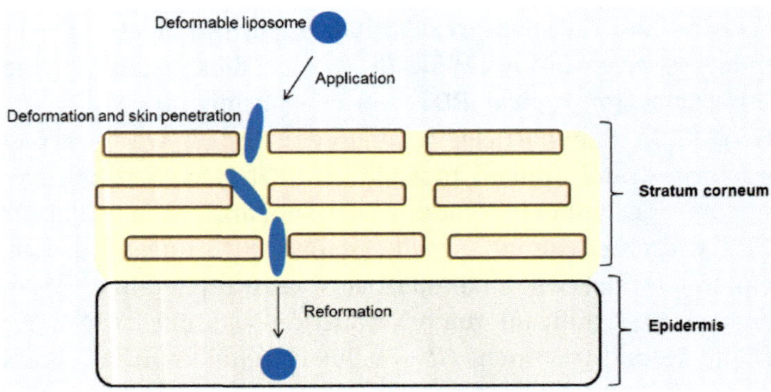

Figure 4.5 New generation of deformable liposome able to adapt its shapes and volume when passing through the stratum corneum.

Deformable liposomes were capable of adapting their shape and volume when passing through the SC [192] (Santana *et al.*, 2011) and have been used as 5-ALA carriers [160, 161] and *meta*-tetrahydroxyphenyl chlorine (*m*THPC) carriers [162–165]. There are fundamental differences between them, such as composition; for example, ethosomes are composed of phospholipids, water, and high concentration of ethanol since invasomes contain terpenes, phospholipids, and ethanol, which make the vesicle deformable [9].

Solid lipid nanoparticles are formed by a solid lipid matrix, present average diameter around 200 nm, physical stability, are able to protect labile molecules and provide biological tolerability. Nanostructured lipid carriers are formed by a mixture of solid and liquid lipids in the core; as well their size is around the nanometric range. This latter carrier has better drug loading and retention during storage compared to solid lipid nanoparticles.

Because of their lipid composition they interact with skin and due to their adhesiveness (lipids and small size) they can form a film on the skin and generate occlusive effect, being good candidates for TPDT [154, 166]. Lima and colleagues (2013) prepared and characterized hypericin-loaded solid lipid nanoparticles to decrease aggregation of hypericin, increase PS solubility and PDT efficacy. They showed lifetime prolongation of the triplet state and better cell uptake in epithelial and melanoma cell lines [167].

Liquid crystalline phases are lipid-structured systems, in which the lipid composition is formed by polar lipids such as glycerol monooleate, organizing itself in contact with water. Their ability to control PSs skin delivery [148, 168] makes them particularly attractive carriers for skin cancer treatment using PDT. Specific reverse hexagonal and cubic phase can co-exist in equilibrium with an excess of water forming stable colloidal dispersions in nanometer size with potential use for drug delivery and has already been used to encapsulate PS for PDT application [83, 147, 148]. Furthermore, this system has promoted better *in vitro* and *in vivo* penetration into skin layers of a chlorin derivative and phthalocyanines, which is essential for topical PDT [147, 148].

4.5.1.3 Inorganic nanoparticles

There are several examples of inorganic nanoparticles employed in PDT, such as gold nanoparticles, silica-based nanoparticles and iron oxide nanoparticles, but their potential toxicity should be taken into account and studied case by case [169].

Gold nanoparticles, due their small size, chemical inertness, and improved targeting, can have enhanced permeation into tumor tissue and vasculature [169]. Although gold nanoparticles have not been evaluated for the *in vivo* efficacy in topical PDT, they could be potentially applied for this purpose. For instance, in the study by Navarro and colleagues (2013), a thiol-terminated charged block copolymer was grafted on the gold nanoparticles surface to favor steric hindrance and these nanoparticles were labeled with the photoactive molecule dibromobenzene, a PS with residual fluorescence, for fluorescence imaging and efficient PDT on $B_{16}F_{10}$ melanoma cells [170].

Silica-based nanoparticles present several advantages, such as the large surface area and pore volume, enabling high drug loading, with various precursors and synthesis methods available, bringing flexibility to carry PSs. Moreover, they are compatible with biological systems and allow functionalization [169]. An example was the encapsulation of silicon phthalocyanine 4 (Pc4), resulting in improved aqueous solubility and stability, being able to deliver Pc4 with more efficacy compared to free drug towards melanoma cells [171]. Magnetic nanoemulsions were studied to deliver Foscan®, a chlorin derivative, through the skin [172]. In this study, authors synthesized magnetic nanoemulsions through the combination of colloidal nanoparticles with magnetic fluids. These inorganic nanoparticles were able to increase Foscan® retention in the stratum corneum and viable epidermis *in vitro* after 12 h of treatment.

4.5.2 Physical Methods Applied for PDT

Topical application of PSs in combination of physical methods is able to improve the PS transport across the skin [174, 175]. Among the physical enhancement techniques, one of the most promising is iontophoresis, in which the constant application of small electric potential forces substance transports across the skin by electromigration or electroosmosis mechanisms. First, there is the occurrence of electrorepulsion between the electrode and drug, with the same charge and, second, there is the convective solvent flow toward anode to cathode transporting cations, neutral and polar compounds. The drug amount transported is directly proportional to the quantity of charge applied and it is advantageous because drug transport occurs when the current is switched on, influencing the delivery kinetics [173, 175].

The association of PSs and iontophoresis has been previously reported. For instance, 5–ALA delivery has been studied in anodal iontophoresis at physiological pH applied on skin, with protoporphyrin IX fluorescence and phototoxicity evaluations [184]. Moreover, iontophoretic delivery *in vitro* of 5–ALA was carried out by electroosmosis [176], while 5–ALA esters with net positive charge were transported by electromigration with great enhancement of transportation in *in vitro* evaluations [177]. Others porphyrins such as *meso*-tetra-(*N*-methylpyridinium-4-yl)-porphyrin (TMPyP), cationic water soluble PS, and *meso*-tetra-(4-sulfonatophenyl)-porphyrin ($TPPS_4$), negatively charged water soluble PS, were studied by iontophoresis at physiological pH to evaluate the influence of electrical charges at skin transportation and homogeneous distribution in *in vitro/in vivo* studies. *In vitro* studies showed high TMPyP transdermal permeation by iontophoresis and passive permeation compared to $TPPS_4$, which only had superficial accumulation in the skin *in vivo*. Unlike, through *in vivo* study, iontophoresis allowed more TMPyP transportation, evidenced by high fluorescence in epidermis/dermis of rats. In addition, TMPyP had superior accumulation around nuclei of skin cells [174].

Sonophoresis is the use of ultrasound for drug delivery. High (≥0.7 MHz) or low (20–100 kHz) frequencies of ultrasound are able to increase the drug transportation through the skin by cavitation effects. Noteworthy, low frequencies of ultrasound are more effective in enhancing skin permeability of drugs; however, both techniques can improve local, regional and systemic drug delivery [178]. For example, low frequency of ultrasound can be applied before or simultaneously to drug delivery, enhancing drug molecules permeation by structural alterations or induced convection flow across the skin. 5–ALA had its permeation enhanced across human skin *ex vivo*, using low frequency sonophoresis before the 5–ALA diffusion assay. The results indicated that sonophoresis exerted a significant accumulation of 5–ALA into full thickness skin compared to passive permeation during 8 h of permeation, also, the same occurred for 5–ALA skin retention, which overcomes the hydrophilicity of 5–ALA, contributing to tumor delivery and increase of 5–ALA clinical response [179].

Considering that PSs have to overcome a thick overlying keratotic layer or disrupted layer in areas of skin pathologies, microneedles are considered a viable method for PS enhanced delivery [180]. Microneedles have needles with length around 25 to 2000 μm and are arranged up to hundreds per square centimeter. They produce channels in microscale, do not provoke pain, and are applied on the skin overpassing stratum corneum to deliver drugs into epidermis/dermis [180, 181]. Microneedles could be solid microneedles made of silicon, metal, or polymer, and they are used in the technique "poke and patch," first, the skin is punctured, second, the patch or formulation containing PS is applied, which has already been described as a successful therapy for PDT, as well as several clinical studies have demonstrated enhancement of PDT efficacy [180, 182]. Because solid microneedles can be broken inside the skin, dissolvable microneedles can be an option for the same purpose for topical application, where the polymeric content starts to dissolve releasing the PS into the skin [180]. Successful results were obtained with hydrogel microneedles composed by poly(methylvinylether/maleic acid) loaded with the PSs: 5–ALA or *meso*-tetra (N-methyl-4-pyridyl) porphine tetra tosylate, showing enhanced *in vitro* effect of released PSs [182].

Finally, a pilot split-face study was carried in patients with multiple actinic keratosis and photodamage. They received treatment with conventional formulation containing methyl aminolevulinate with previous gentle curettage in one side of the face and, on the other side, they received the microneedle device (1500 μm in length) loaded with methyl aminolevulinate, and after 90 min PDT was applied. This device was able to ameliorate photodamage aspects such as mottled pigmentation, roughness and shallowness and a more prominent effect was observed in fine lines and coarse wrinkles compared to conventional formulation. Nonetheless, crusting and pain were more common on the microneedles side [19].

4.6 Potentials and Limitations of Topical PDT

Topical photodynamic therapy by ALA, MAL or phthalocyanines and chlorins is generally well tolerated but contraindications and main side effects after PDT include moderate to severe pain and burning during administration of phototherapy, epithelial exfoliation, erythema, edema, post-inflammatory hyperpigmentation, besides rare complication, such as systemic lupus erythematosus, photosensitive dermatoses, and nonresponsive tumor [2, 21]. Moreover, other limitations of topical PDT make it not currently recommended as a first line treatment. These limitations include (i) the light needed to activate the PSs cannot pass through more than 1 centimeter of tissue thus clinicall limiting PDT to treating superficial lesions; (ii) PDT cannot be used to treatment of metastasized cancer due the light cannot pass far into the tumor; (iii) there is significant variability among PDT practice protocols making it difficult to compare its efficacy to other treatment modalities and (iv) PDT is not yet as effective as surgery.

Topical PDT still appears to be limited to deep skin diseases making necessary the use of drug delivery systems using expensive carriers, thus increasing the therapy cost. Preparations such as nanoparticles, liposomes, nanoemulsions, liquid crystalline phases, and the use of microneedles as a physical method for penetration enhancement, have been receiving a lot of attention as strategies to deliver PSs for topical photodynamic therapy and photodiagnostics. These preparations are able to increase the accumulation of photosensitizer in the deep skin layers with high treatment efficacy and safety. In addition, advances in topical PDT have been a result of combined photosensitizers and the synthesis of new photosensitizers. Perhaps, when other photosensitizers and/or light sources and appropriated skin delivery systems are developed, PDT will become a more reliable and effective treatment.

References

1. Yasmeen K. T., Marjorie F. Y., Elma D. B. Role of photodynamic therapy in psoriasis: A brief review. *Photodermatol. Photoimmunol. Photomed.*, 24 (2008), pp. 222–230.
2. Marilyn T. W., Jennifer Y. L. Current evidence and applications of photodynamic therapy in dermatology. *Clin. Cosmet. Invest. Dermatol.*, 7 (2014), pp. 145–163.
3. Fernández-Guarino M., García-Morales I., Harto A., Montull C., Pérez-García B., Jaén P. Photodynamic therapy: New indications. *Actas Dermosifiliogr.*, 98 (2007), pp. 377–395.
4. Allen C. M., Sharman W. M., Van Lier J. E. Current status of phthalocyanines in the photodynamic therapy of cancer. *J. Porphyrins Phthalocyanines*, 5 (2001), pp. 161–169.
5. McCaughan JS Jr. Photodynamic therapy: A review. *Drugs Aging*, 15 (1999), pp. 49–68.
6. Nowis D., Makowshi M., Stoklosa T., Legat M., Issat T., Golab J. Direct tumor damage mechanism of photodynamic therapy. *Acta Biochim. Pol.*, 52 (2005), pp. 339–352.
7. Juarranz A., Jaen P., Sanz R. F., Cuevas J., Gonzales S. Photodynamic therapy of câncer. Basic principles and applications. *Clin. Transl. Oncol.*, 10 (2008), pp. 148–154.
8. Dennis E. J., Dolmans G. J. Dai Fukumura, Rakesh K. J. Photodynamic therapy for cancer. *Nat. Rev.*, 3 (2003), pp. 380–387.

9. Medina W. S. G., Praça F. S. G., Carollo A. R. H., Bentley M. V. L. B. Nanocarriers to deliver photosensitizers in topical photodynamic therapy and photodiagnostics. In (Beck R., Guterres S., Pohlmann A., eds.), *Nanocosmetics and Nanomedicines*, Springer Berlin Heidelberg, (2011), pp. 285–310.
10. Miller J. D., Baron E. D., Scull H., *et al.* Photodynamic therapy with the phthalocyanine photosensitizer Pc 4: The case experience with preclinical mechanistic and early clinical-translational studies. *Toxicol. Appl. Pharmacol.*, 224 (2007), pp. 290–299.
11. Luo Y., Kessel D. Initiation of apoptosis versus necrosis by photodynamic therapy with chloroaluminum phthalocyanine. *Photochem. Photobiol.*, 66 (1997), pp. 479–483.
12. Annelies, B., Peter A. M., de Witte, R. R. Topical treatment of disseminated superficial actinic porokeratosis with hypericin—photodynamic therapy: A case report. *Photodiagnosis Photodyn. Ther.*, 7 (2010), pp. 123–125.
13. Tardivo, J. P., Del Giglio, A., de Oliveira, C. S., Gabrielli, D. S., Junqueira, H. C., Tada, D. B., Severino, D., de Fátima Turchiello, R., Baptista, M. S. Methylene blue in photodynamic therapy: From basic mechanisms to clinical applications. *Photodiagnosis Photodyn. Ther.*, 2 (2005), pp. 175—191.
14. Jonathan L. S., Richard A. M. Texaphryns: New drugs with diverse clinical applications in radiation and photodynamic therapy. *Biochem. Pharmacol.*, 59 (2000), pp. 733–739.
15. Peng Q., Soler A. M., Warloe T., Nesland J. M., Giercksky K. E. Selective distribution of porphyrins in skin thick basal cell carcinoma after topical application of methyl 5-aminolevulinate. *J. Photochem. Photobiol. B*, 62 (2001), pp. 140–145.
16. Attili S. K., Lesar A., McNeill A., Camacho-Lopez M., Moseley H., Ibbotson S., Samuel I. D., Ferguson J. An open pilot study of ambulatory photodynamic therapy using a wearable low-irradiance organic light-emitting diode light source in the treatment of nonmelonoma skin câncer. *Br J Dermatol.*, 1 (2009), pp. 170–173.
17. Umegaki N., Moritsugu R., Katoh S., *et al.* Photodynamic therapy may be useful in debulking cutaneous lymphoma prior to radiotherapy. *Clin. Exp. Dermatol.*, 29 (2004), pp. 42–45.
18. Dragieva G., Hafner J., Dummer R., Schmid-Grendelmeier P., Ross M., Prinz B. M., *et al.* Topical photodynamic therapy in the treatment of actinic keratoses and Bowen's disease in transplant recipients. *Transplantation*, 77 (2004), pp. 115–121.

19. Torezan L., Chaves Y., Niwa A., Sanches J. A. Jr., Festa-Neto C., Szeimies R. M. A pilot split-face study comparing conventional methyl aminolevulinatephotodynamic therapy (PDT) with microneedling-assisted PDT on actinically damaged skin. *Dermatol. Surg.*, 39 (2013), pp. 1197–1201.

20. Wiegell S. R., Skødt V., Wulf H. C. Daylight-mediated photodynamic therapy of basal cell carcinomas–an explorative study. *J. Eur. Acad. Dermatol. Venereol.*, 28 (2014), pp. 169–175.

21. Wiegell S. R., Wulf H. C. Photodynamic therapy of acne vulgaris using methyl amonilevulinate: A blinded, randomized, controlled trial. *Br. J. Dermatol.*, 154 (2006), pp. 969–976.

22. Moloney F. J., Collins P. Randomized, double-blind, prospective study to compare topical 5-aminolaevulinic acid methylester with topical 5-aminolaevulinic acid photodynamic therapy for extensive scalp actinic keratosis. *Br. J. Dermatol.*, 157 (2007), pp. 87–91.

23. Pinto C., Schafer F., Orellana J. J., Gonzalez S., Hasson A. Efficacy of red light alone and methyl-aminolaevulinate-photodynamic therapy for the treatment of mild and moderate facial acne. *Indian J. Dermatol. Venereol. Leprol.*, 79 (2013), pp. 77–82.

24. Hong J. S., Jung J. Y., Yoon J. Y., Suh D. H. Acne treatment by methyl aminolevulinate photodynamic therapy with red light vs intense pulsed light. *Int. J. Dermatol.*, 52 (2013), pp. 614–619.

25. Akaraphanth R., Kanjanawanitchkul W., Gritiyarangsan P. Efficacy of ALA-PDT vs blue light in the treatment of acne. *Photodermatol. Photoimmunol. Photomed.*, 23 (2007), pp. 186–190.

26. Calzavara-Pinton P. G., Venturini M., Sala R. A comprehensive overview of photodynamic therapy in the treatment of superficial fungal infections of the skin. *J. Photochem. Photobiol. B: Biol.*, 78 (2005), pp. 1–6.

27. Sadick N. An open-label, split-face study comparing the safety and efficacy of levulan kerastick (aminolevulonic acid) plus a 532 nm KTP laser to a 532 nm KTP laser alone for the treatment of moderate facial acne. *J. Drugs Dermatol.*, 9 (2010), pp. 229–233.

28. Szeimies R. M., Radny P., Sebastian M., Borrosch F., Dirschka T., Krähn-Senftleben G., Reich K., Pabst G., Voss D., Foguet M., Gahlmann R., Lübert H., Reinhold U. Photodynamic therapy with BF-200 ALA for the treatment of actinic keratosis: Resultas of a prospective, randomized, double-blind, placebo-controlled phase III study. *Br. J. Dermatol.*, 163 (2010), pp. 386–394.

29. Hauschild A., Stockfleth E., Popp G., Borrosch F., Bruning H., Dominicius R., Mensing H., Reinhold U., Reich K., Moor A. C., Stocker M., Ortland C., Brunnert M. Szeimies R. M. Optimization of photodynamic therapy with a novel self-adhesive 5-aminolaevulinic acid patch: Results of two randomized controlled phase III studies. *Br. J. Dermatol.*, 160 (2009), pp. 1066–1074.

30. Siqueira-Moura M. P., Primo F. L., Espreafico E. M., Tedesco A. C. Development, characterization, and photocytotoxicity assessment on human melanoma of chloroaluminum phthalocyanine nanocapsules. *Mater. Sci. Eng. C*, 33 (2013), pp. 1744–175.

31. Bertoloni G. F., Rossi G., Valduga G., Jori H., Ali J. E. van Lier. Photosensitizing activity of water- and lipid-soluble Phthalocyanines on prokaryotic and eukaryotic microbial cells, *Microbios*, 71 (1992), pp. 33–46.

32. Rhoedes L. E., De Rie M. A., Leifsdottir R., *et al.* Five years follow-up a randomized, prospective Trial of topical methyl aminolevulinate photodynamic therapy VS surgery for nodular basal cell carcinoma. *Arch. Dermatol.*, 143 (2007), pp. 1131–1136.

33. Baron E. D., Malbasa C. L., Santo-Domingo D., Fu P., Miller J. D., Hanneman K. K., Hsia A. H., Oleinick N. L., Colussi V. C., Cooper K. D. Silicon phthalocyanine (Pc 4) photodynamic therapy is a safe modality for cutaneous neoplasms: Results of a phase 1 clinical trial. *Laser Surg. Med.*, 10 (2010), pp. 728–735.

34. Dolmans D. E., Fukumura D., Jain R. K. Photodynamic therapy for cancer. *Nat. Rev. Cancer*, 5 (2003), pp. 380–387.

35. Medina W. S. G., Neife A. G., dos Santos, Carlos C., Antonio Cláudio T., Antonio Cardozo dos Santos. Effects of zinc phthalocyanine tetrasulfonate-based photodynamic therapy on rat brain isolated mitochondria. *Chem. Biol. Interact.*, 179 (2009), pp. 402–406.

36. Colasanti A., Kisslinger A., Kusch D., Liuzzi R., Mastrocinque M., Montforts F. P., Quarto M., Riccio P., Roberti G., Villani F. *In vitro* photo-activation of newly synthesized chlorin derivatives with red light-emitting diodes. *J. Photochem. Photobiol. B Biol.*, 38 (1997), pp. 54–60.

37. Chin W. W. L., Lau W. K. O., Heng P. W. S., Bhuvaneswari R., Olivo M. Fluorescence imaging and phototoxicity effects of new formulation of chlorin e6–polyvinylpyrrolidone, *J. Photochem. Photobiol. B Biol.*, 84 (2006), pp. 103–110.

38. Bose B., Dube A. Photodynamic efficacy of chlorin p6: A pH dependent study in aqueous and lipid environment. *J. Photochem. Photobiol. B Biol.*, 93 (2008), pp. 32–35.

39. Cunderlíková B., Gangeskar L., Moan J. Acid–base properties of chlorin e6: Relation to cellular uptake. *J. Photochem. Photobiol. B Biol.*, 53 (1999), pp. 81–90.

40. De Oliveira K. T., Momo P. B., De Assis F. F., Ferreira M. A. B., Brocksom T. J. Chlorins: Natural sources, synthetic developments and main applications. *Curr. Org. Synth.*, 11 (2014), pp. 42–58.

41. Spikes J. D. New trends in photobiology (invited review) chlorins as photosensitizers in biology and medicine. *J. Photochem. Photobiol. B Biol.*, 6 (1990), pp. 259–274.

42. Juzeniene A. Chlorin e6-based photosensitizers for photodynamic therapy and photodiagnosis. *Photodiagnosis Photodyn. Ther.*, 6 (2009), pp. 94–96.

43. Banfi S., Caruso E., Caprioli S., Mazzagatti L., Canti G., Ravizza R., Gariboldi M., Monti E. Photodynamic effects of porphyrin and chlorin photosensitizers in human colon adenocarcinoma cells. *Bioorg. Med. Chem.*, 12 (2004), pp. 4853–4860.

44. Gupta G., Morton C. A., Whitehurst C., Moore J. V., Mackei R. M. Photodynamic therapy with meso-tetra(hydroxyphenyl) chlorin in the topical treatment of Bowen's disease and basal cell carcinoma. *Br. J. Dermatol.*, 141 (1999), pp. 385–386.

45. Johansson A., Svensson J. Fluorescence and absorption assessment of a lipid mTHPC formulation following topical application in a non-melanotic skin tumor model. *J. Biomed. Opt.*, 12 (2007), pp. 034026/1–034026/9.

46. Park H., Na K. Conjugation of the photosensitizer Chlorin e6 to pluronic F127 for enhanced cellular internalization for photodynamic therapy. *Biomaterials*, 34 (2013), pp. 6992–7000.

47. Lee C. H., Lai P. S., Lu Y. P., Chen H. Y., Chai C. Y., Tsai R. K., FangK. T., Tsai M. H., Hsu C. Y., Hung C. C., Wu D. C., Yu H. S., Chang C. H., Tsai D. P. Real-time vascular imaging and photodynamic therapy efficacy with micelle-nanocarrier delivery of chlorin e6 to the microenvironment of melanoma. *J. Dermatol. Sci.*, 80 (2015), pp. 124–132.

48. Beack S., Won H. K., Ho S. J., In H. D., Seulgi H., Hyemin K., Ki S. K., Seok H. Y., Sei K. H. Photodynamic therapy of melanoma skin cancer using carbon dot – chlorin e6 – hyaluronate conjugate. *Acta Biomater.*, 26 (2015), pp. 295–305.

49. Jeon Y. M., Lee H. S., Jeong D., Oh H. K., Ra K. H., Lee M. Y. Antimicrobial photodynamic therapy using chlorin e6 with halogen light for acne bacteria-induced inflammation. *Life Sci.*, 124 (2015), pp. 56–63.

50. Dube A., Sharma S., Gupta P. K. Evaluation of chlorin p6 for photodynamic treatment of squamous cell carcinoma in the hamster cheek pouch model. *Oral Oncol.*, 42 (2006), pp. 77–82.

51. Wong T. W., Aizawa K., Sheyhedin I., Wushur C., Kato H. Pilot study of topical delivery of mono-L-aspartyl chlorin e6 (NPe6): Implication of topical NPe6-photodynamic therapy. *J. Pharmacol. Sci.*, 93 (2003), pp. 136–142.

52. Kochneva E. V., Filonenko E. V., Vakulovskaya E. G., Scherbakova E. G., Seliverstov O. V., Markichev N. A., Reshetnickov A. V. Photosensitizer Radachlorin®: Skin cancer PDT phase II clinical trials. *Photodiagnosis Photodyn. Ther.* 7 (2010), pp. 258–267.

53. Kennedy J., Pottier R. H. Photodynamic therapy with endogenous protoporphyrin IX: Basic principles and present clinical experience. *J. Photochem. Photobiol. B Biol.*, 6 (1990), pp. 143–148.

54. Gitika B., Kharkwal S. K., Sharma Y. Y., Tianhong D., Michael R. H. Photodynamic therapy for infections: Clinical applications. *Lasers Surg Med.*, 43 (2011), pp. 755–767.

55. Avram D. K., Goldman M. P. Effectiveness and safety of ALA-IPL in treating actinic keratoses and photodamage. *J. Drugs Dermatol.*, 3 (2004), pp. 36–39.

56. Coors E. A., Von den D. P. Topical photodynamic therapy for patients with therapy-resistant lesions of cutaneous T-cell lymphoma. *J. Am. Acad. Dermatol.*, 50 (2004), pp. 363–367.

57. Szeimies R. M., Karrer S., Radakovic-Fijan S., *et al.* Photodynamic therapy using topical methyl 5-aminolevulinate compared with cryotherapy for actinic keratosis: A prospective, randomized study. *J. Am. Acad. Dermatol.*, 47 (2002), pp. 258–262.

58. Jeffes E. W., McCullough J. L., Weinstein G. D., *et al.* Photodynamic therapy of actinic keratoses with topical aminolevulinic acid hydrochloride and fluorescent blue light. *J. Am. Acad. Dermatol.*, 45 (2001), pp. 96–104.

59. Piacquadio D. J., Chen D. M., Farber H. F. Photodynamic therapy with aminolevulinic acid topical solution and visible blue light in the treatment of multiple actinic keratoses of the face and scalp: Investigator-blinded, phase 3, multicenter trials. *Arch. Dermatol.*, 140 (2004), pp. 41–46.

60. Bissonette R., Tremblay J. F., Jucenas P., Boushira M., Lui H. Systemic photodynamic therapy with aminolevulinic acid induces apoptosis in lesional T lymphocytes of psoriatic plaques. *J. Invest. Dermatol.*, 119 (2002), pp. 77–83.

61. Yim Y. C., Lee E. S., Chung P. S., *et al.* Recalcitrant palmoplantar pustular psoriasis successfully treated with topical 5-aminolaevulinic acid photodynamic therapy. *Clin. Exp. Dermatol.*, 30 (2005), pp. 723–724.

62. Radakovic-Fijan S., Blecha-Thalhammer U., Schleyer V., *et al.* Topical aminolaevulinic acid-based photodynamic therapy as a treatment option for psoriasis? Results of a randomized, observerblinded study. *Br. J. Dermatol.*, 152 (2005), pp. 279–283.

63. Enk C. D., Fritsch C., Jonas F., *et al.* Treatment of cutaneous leishmaniasis with photodynamic therapy. *Arch. Dermatol.*, 139 (2003), pp. 432–434.

64. Shieh S., Dee A. S., Cheney R. T., *et al.* Photodynamic therapy for the treatment of extramammary Paget's disease. *Br. J. Dermatol.*, 146 (2002), pp. 1000–1005.

65. Mikasa K., Watanabe D., Kondo C., *et al.* 5-Aminolevulinic acid based photodynamic therapy for the treatment of two patients with extramammary Paget's disease. *J. Dermatol.*, 32 (2005), pp. 97–101.

66. Alexa A. N., Tania S., Devidas M. Effectiveness of photodynamic therapy for mammary and extramammary paget´s disease: A state of the science review. *BMC Dermatol.*, 11 (2011), p. 13.

67. Edstrom D. W., Portwit A., Ros A. M. Photodynamic therapy with topical 5-aminolvulinic acid for mycosis fungoides: Clinical and histological response. *Acta Derm. Venereol.*, 81 (2001), pp. 184–188.

68. Markham T., Sheahan K., Collins P. Topical 5-aminolaevulinic acid photodynamic therapy for tumour-stage mycosis fungoides. *Br. J. Dermatol.*, 144 (2001), pp. 1262–1263.

69. Gold M. H., Boring M. M., Bridges T. M., Bradshaw V. L. The successful use of ALA-PDT in the treatment of recalcitrant molluscum contagiosum. *J. Drugs Dermatol.*, 3 (2004), pp. 187–190.

70. Smucler R., Jatsova E. Comparative study of aminolevulic acid photodynamic therapy plus pulsed dye laser versus pulsed dye laser alone in treatment of viral warts. *Photomed. Laser Surg.*, 23 (2005), pp. 202–205.

71. Goldman M. P., Boyce S. M. A single-center study of aminolevulinic acid and 417 NM photodynamic therapy in the treatment of moderate to severe acne vulgaris. *J. Drugs Dermatol.*, 2 (2003), pp. 393–396.

72. Shaaban D., Abdel-Samad Z., El-Khalawany M. Photodynamic therapy with intralesional 5-aminolevulinic acid and intense pulsed light versus intense pulsed light alone in the treatment of acne vulgaris: A comparative study. *Dermatol. Ther.*, 25 (2012), pp. 86–91.

73. Pollock B., Turner D., Stringer M. R., Bojar R. A., Goulden V., Stables G. I., *et al.* Topical aminolaevulinic acid photodynamic therapy for the treatment of acne vulgaris: A study of clinical efficacy and mechanism of action. *Br. J. Dermatol.*, 151 (2004), pp. 616–622.

74. Horio T., Horio O., Miyauchi-Hashimoto H., *et al.* Photodynamic therapy of sebaceous hyperplasia with topical 5-aminolaevulinic acid and slide projector. *Br. J. Dermatol.*, 148 (2003), pp. 1274–1276.

75. Richey D. F. Aminolevulinic acid photodynamic therapy for sebaceous hyperplasia. *Dermatol. Clin.*, 25 (2007), pp. 59–65.

76. Gold M. H. Intense pulsed light therapy for photorejuvenation enhanced with 20% aminolevulinic acid photodynamic therapy. *J. Lasers Med. Surg.*, 15 (2003), p. 47.

77. Gold M. H. A single-center, open-label investigatory study of photodynamic therapy n the treatment of photoaging with aminolevulinic acid topical solution 20% and intense pulsed light. Poster presented at: 61st Annual Meeting of the American Academy of Dermatology, San Francisco, (2003), http://www.houseofvision.com.br/Medical/Artigos/ala_POSTER_2%20ala.pdf, accessed in october 16, 2015.

78. Gardlo K., Horska Z., Enk C. D., Rauch L., Megahed M., Ruzicka T., *et al.* Treatment of cutaneous leishmaniasis by photodynamic therapy. *J. Am. Acad. Dermatol.*, 48 (2003), pp. 893–896.

79. Rhodes L. E., de Rie M., Enstrom Y., *et al.* Photodynamic therapy using topical methyl aminolevulinate vs surgery for nodular basal cell carcinoma: Results of a multicenter randomized prospective trial. *Arch. Dermatol.*, 140 (2004), pp. 17–23.

80. Vinciullo C., Elliot T., Francis D., Gebauer K., Spelman L., Nguyen R., *et al.* Photodynamic Therapy with topical methyl aminolaevulinate for difficult-to-treat basal cell carcinoma. *Br. J. Dermatol.*, 152 (2005), pp. 765–772.

81. Lee M. R., Ryman W. Erytroplasia of queyrat treated with topical methyl aminolevuliniate photodynamic therapy. *Australas. J. Dermatol.*, 46 (2005), pp. 196–198.

82. Cooper K., Baron E. Photodynamic Therapy Using Silicon Phthalocyanine 4 in Treating Patients With Actinic Keratosis, Bowen's Disease, Skin Cancer, or Stage I or Stage II Mycosis Fungoides. Clinical trial identifier number NCT 00103246 visualized on https://clinicaltrials.gov/ct2/show/NCT00103246.

83. Rossetti F. C., Lopes L. B., Carollo A. R. H., Thomazini J. A., Tedesco A. C., Bentley M. V. L. B. A delivery system to avoid self-aggregation

and to improve *in vitro* and *in vivo* skin delivery of a phthalocyanine derivative used in the photodynamic therapy *J. Control. Release*, 155 (2011), pp. 400–408.

84. Barbugli P. A., Siqueira-Moura M. P., Espreafico E. M., Tedesco A. C. *In vitro* phototoxicity of liposomes and nanocapsules containing chloroaluminum phthalocyanine on human melanoma cell line. *J. Nanosci. Nanotechnol.*, 1 (2010), pp. 569–573.

85. Leman J. Á., Dick D. C., Morton A. C. Topical 5-ALA photodynamic therapy for the treatment of cutaneous T-cell lymphoma. *Clin. Exp. Dermatol.*, 27 (2002), pp. 516–518.

86. Jin Y., Zhang X., Zhang B., Kang H., Du L., Li M. Nanostructures of an amphiphilic zinc phtalocyanines polymer conjugate for photodynamic therapy of psoriasis. *Colloids Surf. B Biointerfaces*, 128 (2015), pp. 405–409.

87. Scot C. R. Photodynamic Therapy in Treating Patients With Skin Cancer or Solid Tumors Metastatic to the Skin. Clinical trial identifier number NCT 00023790 visualized on https://clinicaltrials.gov/ct2/show/NCT00023790. Phase I.

88. Goldman M. P. Using 5-aminolevulinic acid to treat acne and sebaceous hyperplasia. *Cosmetic Dermatol.*, 16 (2003), pp. 57–58.

89. Mbakidi J. P., Drogat N., Granet R., Ouk T. S., Ratinaud H., Rivière E., Verdier M., Sol V. Hydrophilic chlorin-conjugated magnetic nanoparticles—Potential anticancer agent for the treatment of melanoma by PDT. *Bioorg. Med. Chem. Lett.*, 23 (2013), pp. 2486–2490.

90. Ericson M. B., Ann-Marie W., Olle L. Review of photodynamic therapy in actinic keratosis and basal cell carcinoma. *Ther. Clin. Risk Manag.*, 4 (2008), pp. 1–9.

91. Wiegell S. R. Update on photodynamic treatment for actinic keratosis. *Curr. Probl. Dermatol.*, 46 (2015), pp. 122–128.

92. Javier R. A., Natalia R. D. Terapia fotodinamica em dermatologia. *Dermatología Rev. Mex*, 53 (2009), pp. 178–186.

93. Maria Claudia A. I., Monica M. A. Photodynamic therapy: A review of the literature and image documentation. *An Bras Dermatol.*, 4 (2010), pp. 501–511.

94. Marmur E. S., Schmults C. D., Goldberg D. J. A review of laser and photodynamic therapy for the treatment of nonmelanoma skin cancer. *Dermatol. Surg.*, 2 (2004), pp. 264–271.

95. Alexiades-Armenakas M. R., Geronemus R. G. Laser-mediated photodynamic therapy of actinic keratoses. *Arch. Dermatol.*, 139 (2003), pp. 1313–1320.

96. Touma D., Yaar M., Whitehead S., Konnikov N., Gilchrest B. A. A trial of short incubation, broad-area photodynamic therapy for facial actinic keratoses and diffuse photodamage. *Arch. Dermatol.*, 140 (2004), pp. 33–40.

97. Dragieva G., Prinz B. M., Hafner J., Dummer R., Burg G., Binswanger U., *et al.* A randomized controlled clinical trial of topical photodynamic therapy with methyl aminolaevulinate in the treatment of actinic keratoses in transplant recipients. *Br. J. Dermatol.*, 151 (2004), pp. 196–200.

98. Orenstein A., Haik J., Tamir J., Winkler E., Trau H., Malik Z., Kostenich G. Photodynamic therapy of cutaneous lymphoma using 5-aminolevulinic acid topical application. *Dermatol. Surg.*, 26 (2000), pp. 769–770.

99. Andrade C. T., Vollet-Filho J. D., Salvio A. G, Bagnato V. S., Kurachi C. Identification of skin lesions through aminolaevulinic acid-mediated photodynamic detection. *Photodiagnosis Photodyn. Ther.*, 11 (2014), pp. 409–415.

100. Braathen L. R., Szeimies R. M., Basset-Seguin N., Bissonnette R., Foley P., Pariser D., *et al.* Guidelines on the use of photodynamic therapy for nonmelanoma skin cancer: An international consensus. International Society for Photodynamic Therapy in Dermatology, 2005. *J. Am. Acad. Dermatol.*, 56 (2007), pp. 125–143.

101. Tianhong D., Ying-Ying H., Michael R. H. Photodynamic therapy for localized infections-state of the art. *Photodiagnosis Photodyn. ther.*, 6 (2009), pp. 170–188.

102. Ferolla A. C. J. Study of photo aged human skin after photodynamic therapy treatment with topical 5 delta aminolaevulinic acid: Immunohistochemical colagenous and elastic fibre analysis (thesis). São Paulo: Faculdade de Medicina, Universidade de São Paulo (2007), pp. 267.

103. Schroeter C. A., Pleunis J., Van N. P. C., Reineke T., Neumann H. A. Photodynamic therapy: New treatment for therapyresistant plantar warts. *Dermatol. Surg.*, 31 (2005), pp. 71–75.

104. Ichimura H., Yamaguchi S., Kojima A., *et al.* Eradication and reinfection of human papillomavirus after photodynamic therapy for cervical intraepithelial neoplasia. *Int. J. Clin. Oncol.*, 8 (2003), pp. 322–325.

105. Wang J., Xu J., Chen J., *et al.* Successful photodynamic therapy with topical 5-aminolevulinic acid for five cases of cervical intraepithelial neoplasia. *Arch. Gynecol. Obstet.*, 282 (2010), pp. 307–312.

106. Henta T., Itoh Y., Kobayashi M., Ninomiya Y., Ishibashi A. Photodynamic therapy for inoperable vulval Paget's disease using delta-aminolaevulinic acid: Successful management of a large skin lesion. *Br. J. Dermatol.*, 141 (1999), pp. 347–349.

107. Scheinfeld N. Treatment of molluscum contagiosum: A brief review and discussion of a case successfully treated with adapelene. *Dermatol. Online J.*, 13 (2007), p. 15.

108. Chen K., Chang B. Z., Ju M., Zhang X. H., Gu H. Comparative study of photodynamic therapy vs CO_2 laser vaporization in treatment of condylomata acuminata: A randomized clinical trial. *Br. J. Dermatol.*, 156 (2007), pp. 516–520.

109. Szeimies R. M., Schleyer V., Moll I., Stocker M., Landthaler M., Karrer S. Adjuvant photodynamic therapy does not prevent recurrence of condylomata acuminata after carbon dioxide laser ablation–a Phase III, prospective, randomized, bicentric, double-blind study. *Dermatol. Surg.*, 35 (2009), pp. 757–764.

110. Herzinger T., Wienecke R., Weisenseel P., Borelli C., Berking C., Degitz K. Photodynamic therapy of genital condylomata in men. *Clin. Exp. Dermatol.*, 31 (2006), pp. 51–53.

111. Nucci V., Torchia D., Cappugi P. Treatment of anogenital condylomata acuminata with topical photodynamic therapy: Report of 14 cases and review. *Int. J. Infect. Dis.*, 14 (2010), pp. 280–282.

112. Mark W. Local treatment of viral disease using photodynamic therapy. *Int. J. Antimicrob. Agents*, 21 (2003), pp. 510–520.

113. Ben-Hur E., Zuk M. M., Kenney M. E., Oleinick N. L., Mulvihill J., Horowitz B. Action spectra (660/700 nm) for virus inactivation and red cell damage photosensitized by the silicon phthalocyanine Pc4. *Laser Med. Sci.*, 11 (1996), pp. 221–225.

114. Allen C. M., Weber J. M., Van Lier J. E. Sulfophthalocyanines for photodynamic therapy inactivation of viruses in blood products: Effect os structural modifications. *Photochem. Photobiol.*, 62 (1995), pp. 184–189.

115. Rywkin S., Ben-Hur E., Malik Z., *et al.* New phthalocyanines for photodynamic virus inactivation in red blood cell concentrates. *Photochem. Photobiol.*, 60 (1994), pp. 165–170.

116. Hongcharu W., Taylor C. R., Chang Y., Aghassi D., Suthamjariya K., Anderson R. R. Topical ALA-photodynamic therapy for the treatment of acne vulgaris. *J. Invest. Dermatol.*, 115 (2000), pp. 183–192.

117. Taylor M. N., Gonzalez M. The practicalities of photodynamic therapy in acne vulgaris. *Br. J. Dermatol.*, 160 (2009), pp. 1140–1148.

118. Ding H.-L., Wang X.-L., Wang H.-W., Huang Z. Successful treatment of refractory facial acne using repeat short-cycle ALA-PDT: Case study. *Photodiagnosis Photodyn. Ther.*, 8 (2011), pp. 343–346.

119. Itoh Y, Ninomiya Y, Tajima S, Ishibashi A. Photodynamic therapy of acne vulgaris with topical delta-aminolaevulinic acid and incoherent light in Japanese patients. *Br. J. Dermatol.*, 3 (2001), pp. 575–579.

120. Rojanamatin J., Choawawanich P. Treatment of inflammatory facial acne vulgaris with intense pulsed light and short contact of topical 5-aminolevulinic acid: A pilot study. *Dermatol. Surg.*, 32 (2006), pp. 991–997.

121. Miller A. V. C. A. Treatment of acne vulgaris with photodynamic therapy: The use of aminolaevulinic acid and green light. *Cosm. Derm.*, 19 (2006), pp. 624–627.

122. Horfelt C. S. B., Halldin C. B., Ericson M. B., Wennberg A. M. Single low-dose red light is as efficacious as methyl aminolevulinate-photodynamic therapy for treatment of acne: Clinical assessment and fluorescence monitoring. *Acta Derm. Venereol.*, 89 (2009), pp. 372–378.

123. Wardlaw J. L., Sullivan T. J., Lux C. N., Austin F. W. Photodynamic therapy against common bacteria causing wound and skin infections. *Vet. J.*, 192 (2012), pp. 374–377.

124. Ryan F. D., Paul A. M., Michael M. T. Antifungal photodynamic therapy. *Microbiol. Res.*, 163 (2008), pp. 1–12.

125. Calzavara-Pinton P. G., Venturini M., Capezzera R., Sala R., Zane C. Photodynamic therapy of interdigital mycoses of the feet with topical application of 5-aminolevulinic acid. *Photodermatol. Photoimmunol. Photomed.*, 20 (2004), pp. 144–147.

126. Gonzales U., Pinart M., Reveiz L., Alvar J. Interventions for old world cutaneous leishmaniasis. *Cochrane Database Syst. Rev.*, 4 (2008), CD005067.

127. Akilov O. E., Kosaka S., Riordan O K., Hasan T. Photodynamic therapy for cutaneous leishmaniasis: The effectiveness of topical phenothiaziniums in parasite eradication and Th1 immune response stimulation. *Photochem. Photobiol. Sci.*, 10 (2007), pp. 1067–1075.

128. Sohl S., Kauer F., Paasch U., Simon J. C. Photodynamic treatment of cutaneous leishmaniasis. *J. Dtsch. Dermatol. Ges.*, 2 (2007), pp. 128–130.

129. Evangelous G., Krasagakis K., Giannikaki E., Kruger-Krasagakis S., Tosca A. Successful treatment of cutaneous leishmaniasis with intralesional aminolevulinic acid photodynamic therapy. *Photodermatol. Photoimmunol. Photomed.,* 5 (2011), pp. 254–256.

130. Schleyer V., Radakovic F. S., Karrer S., *et al.* Disappointing results and low tolerability of photodynamic therapy with topical 5 aminolevulinic acid in psoriasis. A randomized, double-blind phase I/II study. *JEADV,* 20 (2006), pp. 823–828.

131. Fransson J., Ros A. M. Clinical and inmunohistochemical evaluation of psoriatic plaques treated with topical 5-aminolevulinic acid photodynamic therapy. *Photodermatol. Photoimmunol. Photomed.,* 21 (2005), pp. 326–332.

132. Boehncke W. H., Sterry W., Kaufmann R. Treatment of psoriasis by topical photodynamic therapy with polychromatic light (letter). *Lancet,* 343 (1994), p. 801.

133. Bamujally A. J. A., Judodiharjo H., Gonzales M., *et al.,* Treatment of chronic plaque psoriasis with photodynamic therapy. *J. Eur. Acad. Dermatol. Venereol.,* 11 (1998), pp. 11–28.

134. Nybaek H., Jemec G. B. Photodynamic therapy in the treatment of rosacea, *Dermatology,* 211 (2005), pp. 135–138.

135. Bryld L. E., Jemec G. B. Photodynamic therapy in a series of rosacea patients. *J. Eur. Acad. Dermatol. Venereol.,* 21 (2007), pp. 1199–1202.

136. Mark S. N., Michael H. G., Arielle N. B. K., Amy F. T., Roy G. G., Eva C. R., Mitchel P. G., Dore J. G., Donald F. R. Consensus Conference. The use of photodynamic therapy in dermatology. *J. Drugs Dermatol.,* 5 (2006), pp. 140–154.

137. Gold M. H. The evolving role of aminolaevulinic acid hydrochloride with photodynamic therapy in photoaging. *Cutis,* 69 (2002), pp. 8–13.

138. Szeimies R. M., Ibbotson S., Murrell D. F., *et al.* A clinical study comparing methylaminolevulinic photodynamic therapy and surgery in small superficial basal cell carcinoma (8–20 mm) with a 12 month follow-up. *J. Eur. Acad. Dermatol. Venereol.,* 22 (2008), pp. 1302–1311.

139. Jie Ji L. Z., Hui-Lin D., Hong-Wei W., Zheng H., Xiao-Sin W., Pei-Ru W., Xiu-Li W. Comparison of 5 aminolevulinic acid photodynamic therapy and red light for treatment of photoaging. *Photodiagnosis Photodyn. Ther.,* 11 (2014), pp. 118–121.

140. Fox L. T., Gerber M., Du Plessis J., Hamman J. H. Transdermal drug delivery enhancement by compounds of natural origin, *Molecules,* 16 (2011), pp. 10507–10540.

141. Karadzovska D., Brooks J. D., Monteiro-Riviere N. A., Riviere J. E. Predicting skin permeability from complex vehicles. *Adv. Drug Deliv. Rev.*, 65 (2013), pp. 265–277.

142. Naik A., Kalia Y. N., Guy R. H. Transdermal drug delivery: Overcoming the skin's barrier function. *Pharm. Sci. Technol. Today*, 3 (2000), pp. 318–325.

143. Alnasif N., Zoschke C., Fleige E., Brodwolf R., Boreham A., Rühl E., Eckl K. M., Merk H. F., Hennies H. C., Alexiev U., Haag R., Küchler S., Schäfer-Korting M. Penetration of normal, damaged and diseased skin—an *in vitro* study on dendritic core-multishell nanotransporters, *J. Control. Release*, 185 (2014), pp. 45–50.

144. Davies D. J., Heylings J. R., McCarthy T. J., Correa C. M. Development of an *in vitro* model for studying the penetration of chemicals through compromised skin, *Toxicol. in vitro*, 29 (2015), pp. 176–181.

145. Gattu S., Maibach H. I. Modest but increased penetration through damaged skin: An overview of the *in vivo* human model. *Skin Pharmacol. Physiol.*, 24 (2011), pp. 2–9.

146. Jia X., Jia L. Nanoparticles improve biological functions of phthalocyanine photosensitizers used for photodynamic therapy. *Curr. Drug Metab.*, 13 (2012), pp. 1119–1122.

147. Petrilli R., Praça F. S. G., Carollo A. R. H., Medina W. S. G., Oliveira K. T., Fantini M. C. A., Neves M. G. P. M. S., Cavaleiro J. A. S., Serra O. A., Iamamoto Y., Bentley M. V. L. B. Nanoparticles of lyotropic liquid crystals: A novel strategy for the topical delivery of a chlorin derivative for photodynamic therapy of skin cancer. *Curr. Nanosci.*, 9 (2013), pp. 434–441.

148. Praça F. S. G., Medina W. S. G, Petrilli R., Bentley M. V. L. B. Liquid crystal nanodispersions enable the cutaneous delivery of photosensitizer for topical PDT: Fluorescence Microscopy Study of Skin Penetration. *Curr. Nanosci.*, 8 (2012), pp. 535–540.

149. Pierre M. B. R., Tedesco A. C., Marchetti J. M., Bentley M. V. L. B. Stratum corneum lipids liposomes for the topical delivery of 5-aminolevulinic acid in photodynamic therapy of skin cancer: Preparation and *in vitro* study. *BMC Dermatol.*, 1 (2001), pp. 1–6.

150. Bae B. C., Na K. Development of polymeric cargo for delivery of photosensitizer in photodynamic therapy. *Int. J. Photoenergy*, 2012 (2012), pp. 1–14.

151. Battah S., Balaratnam S., Casas A., O'Neill S., Edwards C., Batlle A., Dobbin P., MacRobert A. J. Macromolecular delivery of 5-aminolevulinic acid for photodynamic therapy using dendrimer conjugates. *Mol. Cancer Ther.*, 6 (2007), pp. 876–835.

152. Li L., Huh K. M. Polymeric nanocarrier system for photodynamic therapy. *Biomater. Res.*, 18 (2013), pp. 1–14.

153. Chatterjee D. K., Fong L. S., Zhang Y. Nanoparticles in photodynamic therapy: An emerging paradigm. *Adv. Drug Deliv. Rev.*, 60 (2008), pp. 1627–1637.

154. Bugaj A. M. Targeted photodynamic therapy–a promising strategy of tumor treatment. *Photochem. Photobiol. Sci.*, 10 (2011), pp. 1097–1109.

155. Da Silva C. L., Del Ciampo J. O., Rossetti F. C., Bentley M. V., Pierri M. B. PLGA nanoparticles as delivery systems for protoporphyrin IX in topical PDT: Cutaneous penetration of photosensitizer observed by fluorescence microscopy. *J. Nanosci. Nanotechnol.*, 13 (2013), pp. 6533–6540.

156. Wang X., Shi L., Tu Q., Wang H., Zhang H., Wang P., Zhang L., Huang Z., Zhao F., Luan H., Wang X. Treating cutaneous squamous cell carcinoma using 5-aminolevulinic acid polylactic-co-glycolic acid nanoparticle-mediated photodynamic therapy in a mouse model. *Int. J. Nanomed.*, 10 (2015), pp. 347–355.

157. Chime S. A., Onyishi I. V. Lipid-based drug delivery systems (LDDS): Recent advances and applications of lipids in drug delivery. *Afr. J. Pharm. Pharmacol.*, 7 (2013), pp. 3034–3059.

158. Eloy J. O., Marina C. S., Raquel P., Juliana P., Robert J. L., Juliana M. M. Liposomes as carriers of hydrophilic small molecule drugs: Strategies to enhance encapsulation and delivery. *Colloids Surf. B Biointerfaces*, 123 (2014), pp. 345–363.

159. Ribeiro J. B. P., Ana L. M. V., Daniel G., Misléia R. A. G., Ana A. S. A., Rafaela D. G., José S. F., Antônio C. T., Fernando L. P., Jonathan R. M., Alexandre V. L., Raimunda N. R. S. Evaluation of the efficacy ofsystemic miltefosine associated with photodynamic therapy with liposomal chloroaluminium phthalocyanine in the treatment of cutaneous leishmaniasis caused by Leishmania (L.) amazonensisin C57BL/6 mice. *Photodiagnosis Photodyn. Ther.*, 13 (2016), pp. 282–290.

160. Fang Y., Tsai Y., Wu P., Huang Y. Comparasion of 5-aminolevulinic acid-encapsulated liposome versus ethosome for skin delivery for photodynamic therapy. *Int. J. Pharm.*, 356 (2008), pp. 144–152.

161. Fang Y. P., Huang Y. B., Wu P. C., Tsai Y. H. Topical delivery of 5-aminolevulinic acid-encapsulated ethosomes in a hyperproliferative skin animal model using the CLSM technique to evaluate the penetration behavior. *Eur. J. Pharm. Biopharm.*, 73 (2009), pp. 391–398.

162. Dragicevic Curic N., Scheglmann D., Albrecht V., Fahr A. Temoporfin—loaded invasomes: Development, characterization and *in vitro* skin penetration studies. *J. Control. Release*, 127 (2008), pp. 59–69.

163. Dragicevic-Curic N., Grafe S., Albrecht V., Fahr A. Topical application of temoporfin-loaded invasomes for photodynamic therapy of subcutaneously implanted tumors in mice: A pilot study. *J. Photochem. Photobiol. B Biol.*, 91 (2008), pp. 41–50.

164. Dragicevic-Curic N., Scheglmann D., Albrecht V., Fahr A. Development of liposomes containing ethanol for skin delivery of temoporfin: Characterization and *in vitro* penetration studies. Colloids SurfB: Biointerfaces 74 (2009), pp. 114–122.

165. Dragicevic-Curic N., Grafe S., Gitter B., Fahr A. Efficacy of temoporfin-loaded invasomes in the photodynamic therapy in human epidermoid and colorectal tumour cell lines. *J. Photochem. Photobiol. B Biol.*, 101 (2010), pp. 238–250.

166. Pardeike J., Hommoss A., Müller R. H. Lipid nanoparticles (SLN, NLC) in cosmetic and pharmaceutical dermal products. *Int. J. Pharm.*, 366 (2009), pp. 170–184.

167. Lima A. M., Dal Pizzol C., Monteiro F. B. F., Creczynski-Pasa T. B., Andrade G. P., Ribeiro A. O., Perussi J. R. Hypericin encapsulated in solid lipid nanoparticles: Phototoxicity and photodynamic efficiency. *J. Photochem. Photobiol. B Biol.*, 125 (2013), pp. 146–154.

168. Garcia F. S. Obtenção e Caracterização Tópica de Geis de Fase Cúbica de Monoleína/Água como Sistema de Liberação para Zinco Ftalocianina Tetrassulfonada na Terapia Fotodinâmica do Câncer de pele: Estudos *In vitro* E *In vivo* PhD Thesis. Faculdade de Ciências Farmacêuticas de Ribeirão Preto, Universidade de São paulo, Ribeirão Preto, August (2004).

169. Monge-Fuentes V., Muehlmann L. A., Azevedo R. B. Perspectives on the application of nanotechnology in photodynamic therapy for the treatment of melanoma. *Nano Rev.*, 5 (2014), pp. 1–14.

170. Navarro J. R. G., Lerouge F., Cepraga C., Micouin G., Favier A., Chateau D., Charreyre M. T., Lanoë P. H., Monnereau C., Chaput F., Marotte S., Leverrier Y., Marvel J., Kamada K., Andraud C., Baldeck P. L., Parola S. Nanocarriers with ultrahigh chromophore loading for fluorescence bio-imaging and photodynamic therapy. *Biomaterials*, 34 (2013), pp. 8344–8351.

171. Zhao B., Yin J. J., Bilski P. J., Chignell C. F., Roberts J. E., He Y. Y. Enhanced photodynamic efficacy towards melanoma cells by

encapsulation of Pc4 in silica nanoparticles, *Toxicol. Appl. Pharmacol.*, 241 (2009), pp. 163–172.

172. Primo F. L., Michieleto L., Rodrigues M. A. M., Macaroff P., Morais P. C., Lacava Z. G. M., Bentley M. V. L. B., Tedesco A. C. Magnetic nanoemulsions as drug delivery system for Foscans: Skin permeation and retention *in vitro* assays for topical application in photodynamic therapy (PDT) of skin cancer. *J. Magn. Magn. Mater.*, 311 (2007), pp. 354–357.

173. Kalia Y. N., Naik A., Garrison J., Guy R. H. Iontophoretic drug delivery, *Adv. Drug Deliv. Rev.*, 56 (2004), pp. 619–658.

174. Gelfuso G. M., Gratieri T., Souza J. G., Thomazine J. A., Lopez R. F. V. The influence of positive or negative charges in the passive and iontophoretic skin penetration of porphyrins used in photodynamic therapy. *Eur J Pharm Biopharm.*, 77 (2011), pp. 249–256.

175. Lopez R. F. V., Lange N., Guy R., Bentley M. V. L. B. Photodynamic therapy of skin cancer: Controlled drug delivery of 5-ALA and its esters. *Adv. Drug Deliv. Rev.*, 56 (2004), pp. 77–94.

176. Lopez R. F. V., Bentley M. V. L. B., Delgado-Charro M. B., Guy R. H. Iontophoretic delivery of 5-aminolevulinic acid (ALA): Effect of pH. *Pharm. Res.*, 18 (2001), pp. 311–315.

177. Lopez R. F. V., Bentley M. V. L. B., Delgado-Charro M. B., Salomon D., Van den Bergh H., Lange N., Guy R. H. Enhanced delivery of 5-aminolevulinic acid esters by iontophoresis *in vitro*. *Photochem. Photobiol.*, 77 (2003), pp. 304–308.

178. Polat B. E., Hart D., Langer R., Blankschtein D. Ultrasound-mediated transdermal drug delivery: Mechanisms, scope, and emerging trends. *J. Control. Release*, 152 (2011), pp. 330–348.

179. Krishnan G., Grice J. E., Roberts M. S., Benson H. A. E., Prow T. W. Enhanced sonophoretic delivery of 5-aminolevulinic acid: Preliminary human *ex vivo* permeation data. *Skin Res. Technol.*, 19 (2013), pp. 283–289.

180. Kearney M. C., Brown S., McCrudden M. T. C., Brady A. J., Donnelly R. F. Potential of microneedles in enhancing delivery of photosensitizing agents for photodynamic therapy. *Photodiagnosis Photodyn. Ther.*, 11 (2014), pp. 459–466.

181. Tuan-Mahmood T. M., McCrudden M. T. C., Torrisi B. M., Mc Alister E., Garland M. J., Singh T. R. R. Microneedles for intradermal and transdermal drug delivery. *Eur. J. Pharm. Sci.*, 50 (2013), pp. 623–637.

182. Donnelly R. F., Morrow D. I. J., McCarron P. A., Woolfson A. D., Morrissey A., Juzenas P., Juzeniene A., Iani V., McCarthy H. O., Moan J. Microneedle-mediated intradermal delivery of 5-aminolevulinic acid: Potential for enhanced topical photodynamic therapy. *J. Control. Release*, 129 (2008), pp. 154–162.

183. Ferolla A. C., Bertha M. T., Luis C. C. Skin rejuvenation with photodynamic therapy: Collagen and elastic fiber analysis, and clinical improvement. *Surg. Cosmet. Dermatol.*, 2 (2010), pp. 87–92.

184. Rhodes L. E., Tsoukas M. M., Rox Anderson R., Kollias N. Iontophoretic delivery of ALA provides a quantitative model for ALA pharmacokinetics and PpIX phototoxicity in human skin. *J. Invest. Dermatol.*, 108 (1997), pp. 87–91.

Part 2

Delivery Systems and Nanocarriers for Topical Route: Examples and Illustration of Some Therapeutic and Cosmetic Applications

Chapter 5

Novel Starch-Derived Topical Delivery Systems

Joana Marto, Inês Jorge, António José de Almeida, and Helena Ribeiro

Research Institute for Medicines (iMed.ULisboa), Faculty of Pharmacy, Universidade de Lisboa, 1649-003 Lisboa, Portugal

helena.ribeiro@ff.ul.pt

5.1 Introduction

5.1.1 Starch: Functional Characteristics and Relevance

In the past decade, the demand for natural and eco-friendly products has been growing. Natural polymers, with emphasis on starch, are a valuable option to comply with this green request, whilst creating products with the desired sensorial attributes.

Starch occurs as an odorless, tasteless, fine, white powder and the chemical formula of its molecule is $(C_6H_{10}O_5)_n$, where n = 300–1000 [1]. It is composed of two polymers of D-glucose (Fig. 5.1): amylose (AM), an essentially unbranched $\alpha[1 \rightarrow 4]$ linked glucan, and amylopectin (AP), which has chains of $\alpha[1 \rightarrow 4]$ linked glucoses arranged in a highly branched structure with $\alpha[1 \rightarrow 6]$ branching links [2].

Carrier-Mediated Dermal Delivery: Applications in the Prevention and Treatment of Skin Disorders
Edited by Andreia Ascenso, Sandra Simões, and Helena Ribeiro

ISBN 978-981-4745-58-1 (Hardcover), 978-1-315-36447-6 (eBook)
www.panstanford.com

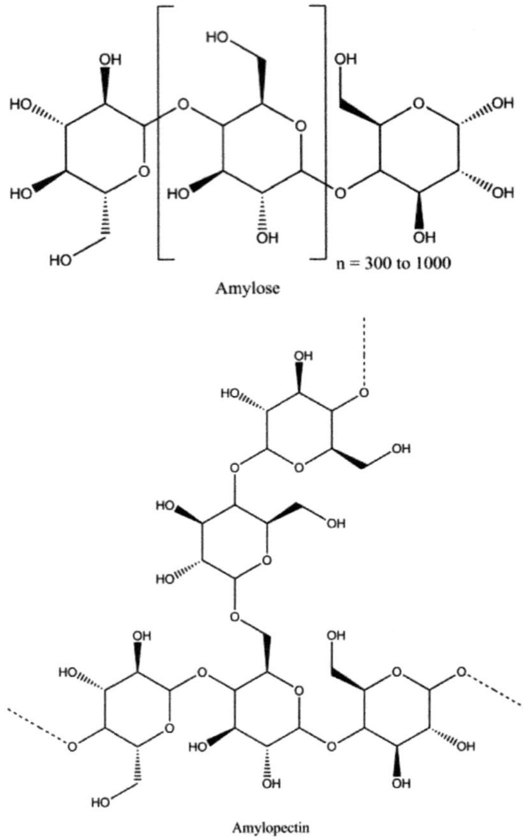

Figure 5.1 Structural formula of starch.

AM has a molecular weight of approximately 10^5–10^6 Da and a degree of polymerization (DP) of 324–10000 glucose units [2–4]. Unlike AP, AM has a low degree of branching, which gives AM the tendency to form insoluble semi-crystalline aggregates [2]. It forms helical complexes with iodine, fatty acids, and monoglycerides [4, 5]. AP, the major component of all starches, is a much larger molecule with a molecular weight of approximately 10^7–10^9 Da and with a heavily branched structure [4]. The basic organization of the chains is described in terms of the A, B and C chains: the outer A chains are unsubstituted and glycosidically linked at their potential reducing group through C_6 of a glucose residue to an inner B chain; these B chains are in turn defined as chains bearing other chains as branches; the single C chain per molecule likewise carries other chains as branches, but contains the sole reducing terminal residue of glucose [2, 5]. In general, the ratio of AM to AP and their structural variability strongly depend on the botanical origin. Normal starches contain approximately 70% to 80% of AP and 20% to 30% of AM (Fig. 5.2) [5, 6].

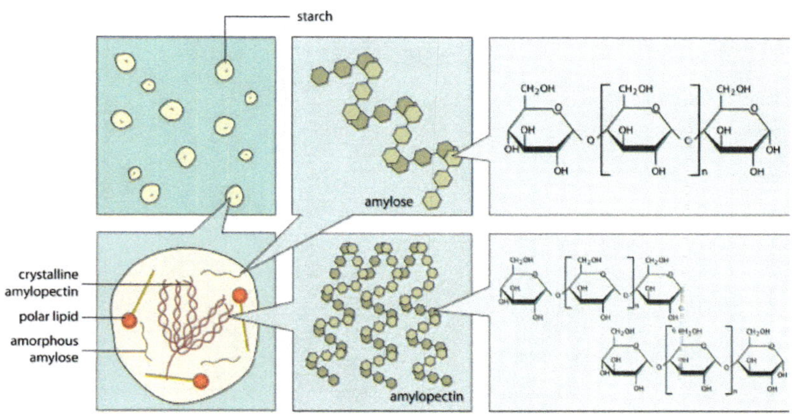

Figure 5.2 Starch granules composition [5, 9].

Minor components of starch include lipids, proteins, and minerals. Starches, in particular cereal starches, contain inner lipids in the form of free fatty acids and lysophospholipids. The minerals found in starch are calcium, magnesium, phosphorus, potassium, and sodium [3, 5].

Phosphorus in starch is mainly present in two forms: phosphate monoesters and phospholipids. Phosphate monoesters are covalently bound to the AM fraction of the starch, increasing its paste clarity and viscosity, and phospholipids gives opaque and lower-viscosity pastes [7].

Starch granules range in size (from 1 to 100 μm diameter), shape (polygonal, spherical, ovoid and lenticular) and in content, structure and organization of the AM and AP molecule, accordingly to each botanical variety and growth conditions (Table 5.1) [1–3]. In addition, starch granules usually contain a central line known as the "Maltese cross" or hilum, and this characteristic reduces the birefringence of the starch granules [8].

Table 5.1 Characteristics of starch granules from different botanical sources [3–6, 8, 10–13]

Botanical origin	Maize	Wheat	Potato	Cassava	Barley	Pea	Oat	Rice	Corn
Type	Cereal	Cereal	Tuber	Root	Cereal	Legume	Cereal	Cereal	Cereal
Granule shape	Round, polygonal	Round, lenticular	Oval, spherical	Oval, truncated	Lenticular, spherical	Oval, rentiform	Polyhedral	Angular, polygonal	Angular
Granule size (μm)	2–30	1–45	5–100	4–35	15–25	5–30	3–10	2–8	11.5 ± 0.3
Phosphate (%, w/w)	0.02	0.06	0.08	0.01	—	—	—	—	—
Protein (%, w/w)	0.35	0.40	0.06	0.1	0.32	—	0.46	0.1	0.4
Lipid (%, w/w)	0.70	0.80	0.05	0.10	0.18	0.30–0.40	0.66	0.6–1.4	0.6–0.8
Amylose content (% total starch)	25–28	25–29	18–21	18–24	21–24	24	22.12	21–25	28.5
Amylopectin content (% total starch)	—	80–90	80–90	—	—	—	—	—	—
Ash (%, w/w)	0.07–0.10	0.16–0.27	0.10–0.33	0.30	0.18	—	0.40	—	—

This variety in form and function gives starch not only advantages but also disadvantages as the variableness of the raw material can affect the proper pharmaceutical manufacturing. Modified starches were developed to allow a wider range of processing conditions and nowadays are extensively used to overcome the variability of the native starches.

5.1.1.1 Modified Starch: A Strategy to Prepare High Performance Starch

Modified starch has been developed to fulfill pharmaceutical industry needs. Modified starch is an excipient that offers functional advantages such as gelling and thickening.

In general, AM synthesis requires just a single gene, while AP modification involves the combined action of several enzymes. Modification of starch can be achieved by using a derivatization technique, such as etherification, esterification, cross-linking and dual modification of starch; by decomposition or conversion using acid or enzymatic hydrolysis and oxidation of starch; or by physical treatment, using heat or moisture. Chemical modification requires the introduction of functional groups into the starch molecule, resulting in markedly altered physicochemical characteristics, such as gelatinization, pasting and retrogradation behavior [14].

Three different enzymatic strategies are available to modify starch biosynthesis in plants, in order to obtain starch polymers with innovative functional properties: knocking out biosynthetic enzymes through selection of mutations; pulling down biosynthetic enzyme expression by mutations or by directing antisense RNA against them; or expressing heterologous genes related or unrelated to the biosynthetic pathway [5]. Mutation of the locus that encodes the granule-bound starch synthase (GBSS[1]) creates a starch with no AM. These waxy starches, high in AP gelatinizes easily, yielding clear pastes. On the other hand, high-AM starches are also of great interest as they have a high gelling strength. This phenotype is caused by a mutation in the gene that encodes the starch-branching enzyme (SBE) IIb, which is also known as "amylose extender" [15].

The antisense inhibition of the gene that encodes the α-glucan water dikinase (GWD) (the enzyme which causes the incorporation of phosphate groups into starch) results in a starch that has low phosphate content and viscosity [5, 10].

5.1.1.2 Starches: From Granules to Novel Applications

Starch is the most abundant storage reserve of carbohydrates in plants. It is found in many different plant organs, including seed, fruits, tubers and roots. This natural material is an edible food substance and is considered as non-toxic and non-irritant. Moreover, starch is also biocompatible, biodegradable, and inexpensive [10, 16–18].

Due to these advantages, starch is widely used: in the food industry as an ingredient, emulsifier, gelling, thickener and encapsulating agent; as a source of energy (after its conversion to ethanol); in the paper industry, as adhesive and coating agent; in the textile industry, as printing thickener and a warp sizing agent; and in the pharmaceuticals industry, where is used as a binder, diluent, and disintegrant [1, 2, 17–19]. However, one potential highly attractive application remains to be explored: its use in topical drug delivery systems. In the dermatological area, generic raw materials requirements include skin compatibility and skin protection, efficacy, sensory properties and environmental compatibility [20].

Skin: The epidermal barrier

The skin is an important part of the integumentary system, as it protects the body from external damage. Skin performs several functions, such as acting as a sensorial organ, promoting the body thermoregulation, allowing the excretion of substances, producing vitamin D (endocrine functions), promoting physical, chemical and immunological defense and protecting against dehydration, UV radiation and external pathogens [21–24].

The skin can be divided in three main layers (Fig. 5.3): (i) the more external being the epidermis constituted by keratinocytes, melanocytes, Langerhans cells, and Merkel cells, among others; (ii) the middle layer, the dermis, which is mainly composed of the fibrillary structural protein collagen and contains capillaries, sebaceous and sweat glands, hair follicles and nerves; and (iii) finally, the hypodermis, the most internal layer and constituted by adipose tissue [22, 24–28]. The epidermis has a multilayer structure consisting of basal, spinous, and granular cell layers. Each layer is defined by position, shape, morphology, composition and state of differentiation of keratinocytes. The last stage of the keratinocyte differentiation is associated with deep changes in their structure, resulting in their transformation into physically and chemically resistant squamous cells, called corneocytes—these cells make the uppermost layer of the epidermis, called stratum corneum (SC). The SC is the layer that controls absorption, constituting a barrier to the delivery of many molecules at therapeutic level [23, 25, 29, 30].

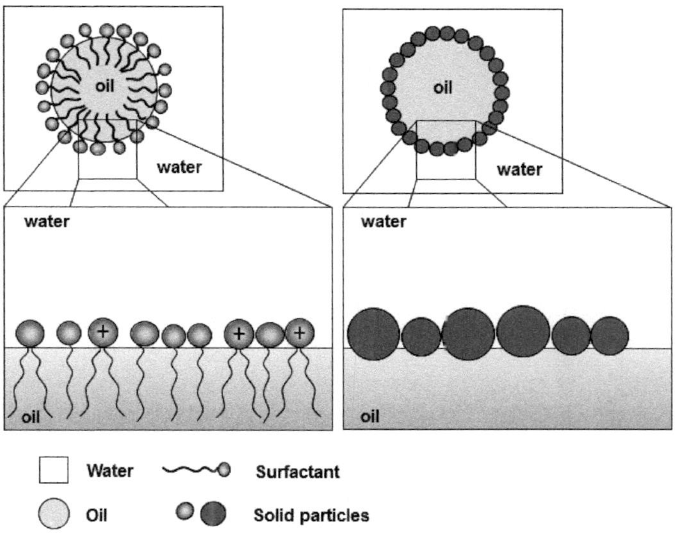

Figure 5.3 Skin structure (adapted from [22]).

Stratum corneum properties are based on the special content and composition of its lipids, the structural arrangement of the intercellular lipid matrix and the lipid envelope surrounding the cells in the SC. The lipids form bi-layers surrounding the corneocytes, producing a "brick-and-mortar" model, the bricks, being the corneocytes packed with keratin, and the intercellular lipids acting as the mortar. These intercellular lipids are mainly ceramides and free fatty acids, but also cholesterol, cholesteryl ester, and a small fraction of cholesterol sulfate [25, 29, 31, 32].

Molecules can penetrate into the skin using one of the following ways: intercellular route, transcellular route or appendageal route (Fig. 5.4). The latter is the least significant mechanism since it takes place through the eccrine glands or hair follicles, and since both represent a low surface area (only 0.1% of the total surface area of human skin). Nonetheless, drug delivery by this route may be important for permeation of slowly diffusing compounds and very high molecular weight substances (nanoparticles, for instance). In the transcellular route, the molecule crosses through the cells. Finally, the intercellular route is the longest pathway, but also the most predominant, in which the molecules passes by the intercellular spaces [21, 25, 27, 30, 32, 34].

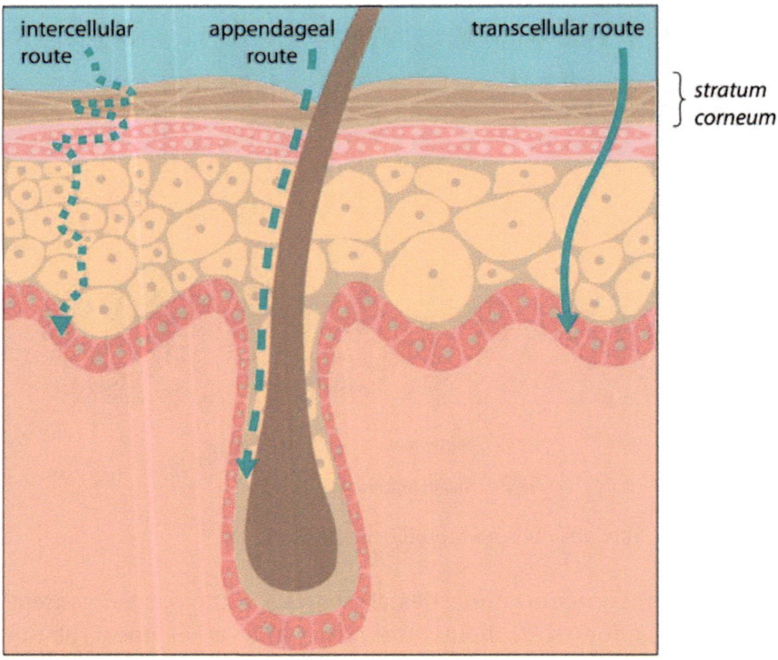

Figure 5.4 Possible transport pathways through the stratum corneum (adapted from [33]).

As such, drug delivery through skin may pose some difficulties due to its tortuous route, problematic sequential diffusion and partitioning between the polar head groups and the alkyl chains of the intercellular bilayers of lipids [27, 30]. Figure 5.5 summarizes some strategies to overcome the SC barrier [32, 35–37]. In most cases, an association of multiple strategies is used in order to assure the dermal and/or transdermal drug delivery.

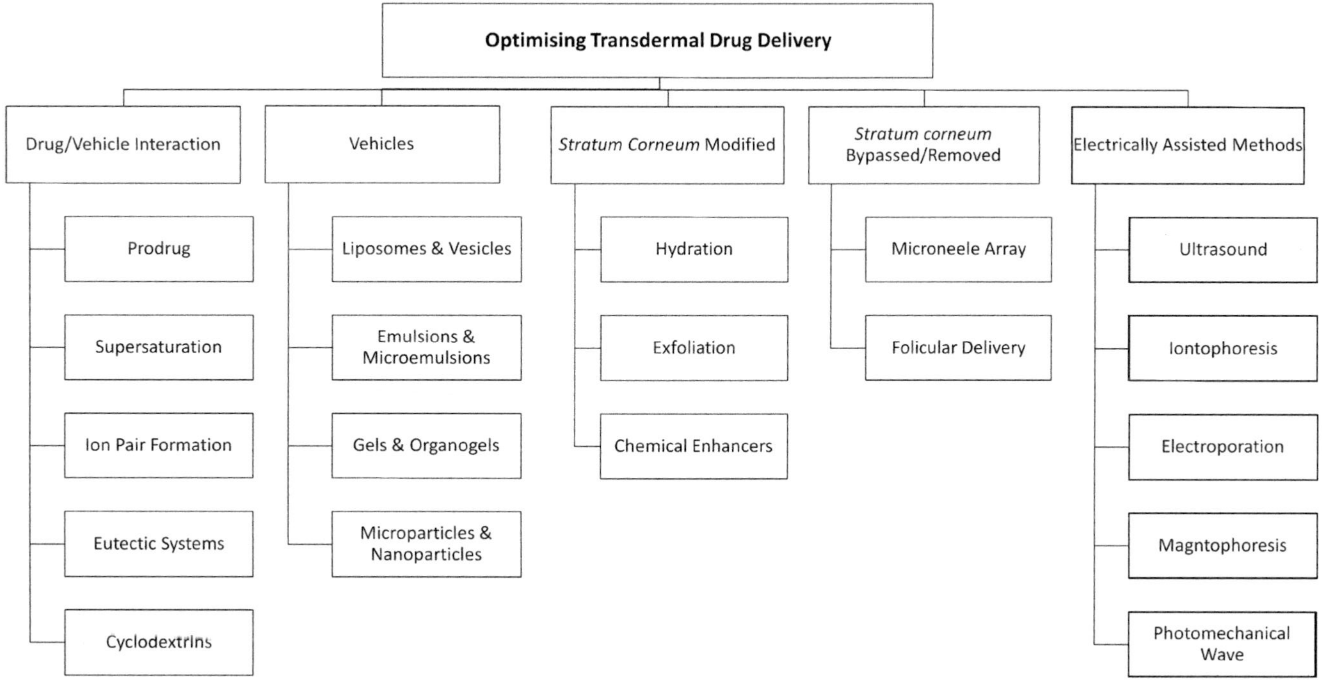

Figure 5.5 Some methods for optimizing transdermal drug delivery (adapted from [32, 35–37]).

Other factors, besides penetration mechanisms, can influence the molecular permeation, such as: skin temperature and peripheral circulation (higher temperature increases penetration due to vasodilation), condition of the skin (normal, abraded or diseased), area of application, contact time and frequency of re-application, moisturizing level of the skin, pre-treatment of the skin, physical properties of the penetrating substance (solubility, molecular size, particle size, crystalline form, volatility, polarity, ionization, partition), vehicle and the penetrating substance-vehicle relationship [21, 27, 38].

The permeation flux (J) of a substance through the SC can be described by Fick's first law of diffusion:

$$J = \frac{D_{\mathrm{m}} c_{\mathrm{s,m}}}{L} \times \frac{c_{\mathrm{v}}}{c_{\mathrm{s,v}}} \tag{5.1}$$

In Eq. 5.1, D_{m} represents the diffusion coefficient of the substance in the skin membrane, $c_{\mathrm{s,m}}$ its solubility in the membrane, L the diffusion path length across the skin, c_{v} the concentration of the substance dissolved in the vehicle and $c_{\mathrm{s,v}}$ the solubility of the substance in the vehicle. Therefore, at least three permeation strategies can be postulated based on Fick's first law of diffusion: (i) to increase the diffusion coefficient of the drug; (ii) to increase the substance partitioning into the skin; and (iii) to increase the saturation level of the substance in the vehicle. The first two strategies require an effect of the vehicle on the barrier function of the SC, whereas the last one is based on vehicle-drug interaction [21, 25].

These permeation strategies can be achieved with chemical penetration enhancers acting within the skin. The ideal enhancer should be non-toxic, non-irritating and non-allergic; appropriate for formulation into diverse topical forms and compatible with other excipients and drugs; and cosmetically acceptable [36, 39, 40].

The diffusion coefficient can be increased by disordering the lipids of the SC. Among the chemical compounds commonly used to achieve this goal are fatty acids, namely oleic acid, azone, a cyclic amide, sulfoxides (e.g., DMSO), terpenes, surfactants, and alcohols. These compounds have shown to induce phase separation in the SC lipid domains, creating a lipid disorder and, reducing barrier function, resulting in a reduced diffusional resistance of the skin [25, 30, 35, 40, 41].

Other chemical enhancers act by increasing the substance solubility in the skin, being of special interest in the delivery of hydrophobic drugs. Examples of this type of agents are propylene glycol, ethanol, diethylene glycol monoethyl ether, and *N*-methyl pyrrolidone. These solvents penetrate the SC and change its properties by altering the chemical environment, thus increasing the partitioning of a second substance into the skin [25, 30, 32, 35, 36].

Finally, the saturation level can be enhanced by increasing the substance concentration in the vehicle or by decreasing the solubility of the substance in the vehicle, both resulting in an enhanced thermodynamic activity and in an increased skin permeation [40]. One way to achieve this, can be through supersaturation, where a high chemical potential, however thermodynamically unstable solution is produced [21, 27].

The perfect chemical enhancer has yet to be discovered and the existing ones are often associated with skin irritation, representing a disadvantage [32]. Additionally, synergy and combinations of chemical enhancers may offer some opportunities in transdermal formulations [39].

Another approach to drug dermal delivery of drugs is the use of prodrugs. In this case, physical modification of the drug facilitates drug permeation, i.e. by addition of a cleavable chemical group for example [39, 42]. From Fick's first law of diffusion, one can deduce the ideal properties of a penetrating skin molecule: low molecular weight, adequate solubility in oil and water (optimal partition coefficient), and high concentration of the substance in the vehicle [35].

General ointments, creams, lotions, gels, and nanoparticulate carriers are the preferred vehicles for dermatological therapy, since they remain *in situ* and deliver the drug over extended periods of time [43].

In addition, to the chemical enhancers and vehicles mentioned above, there are also physical methods for enhancing skin permeation, such as iontophoresis, non-cavitational ultrasound, electroporation, cavitation ultrasound, microneedles, thermal ablation, and microdermabrasion [43].

The main advantages of topical drug delivery systems are the extended duration of action and, consequently the reduction of dosing frequency. Furthermore, topical drug delivery is useful protection tool for potent drugs as it provides a reduction of systemic adverse effects [44]. Finally, it also allows a reduced drug dose due to the shortened metabolization pathway of the transdermal route versus the gastrointestinal pathway, which increases the overall bioavailability of the drug [45].

5.2 Topical Delivery Systems

Delivery systems for topical drug therapy aim to produce the desired therapeutic effect at specific sites in the epidermal tissue. Many studies have been performed to investigate the effect of topical dosage forms on dermal and transdermal drug delivery. Semisolid formulations, such as creams, gels and lotions, are the preferred pharmaceutical vehicles for topical therapy due to the fact that they remain *in situ* and, in general, provide an extended release of the drug. More recently, nanoparticulate carriers have been recognized as alternative topical delivery systems, due to their small size and larger superficial area which increases skin drug uptake [43].

Starch can be a valuable ingredient when formulating semisolid vehicles. Waxy starches, which are poor in AM, gelatinize easily, producing clear pastes that will not be a gel. Hydrophobic modified starches can be used as a stabilizer and thickener. On the contrary, starches with high level of phosphate, for instance potato starch, have higher swelling power and stable-paste properties [10].

5.2.1 Conventional Topical Delivery Systems

5.2.1.1 Emulsions

Emulsions are heterogeneous systems composed of two immiscible liquids, one of which is uniformly dispersed as fine droplets throughout the other. These systems are thermodynamically unstable and their study and development is one of the most difficult and complex subjects in the pharmaceutical field [46]. Nevertheless, emulsions, in the form of creams, lotions, or foams, are extensively used due to their therapeutic properties and as vehicles to deliver drugs and cosmetic agents through the skin. Furthermore, these dosage forms facilitate drug permeation into and through the skin by their occlusive effects and/or by the incorporation of penetration-enhancing molecules.

Current pharmaceutical emulsions are most commonly stabilized by synthetic surfactants, which can be toxic or may alter the pharmacokinetics of co-administered drugs [47]. Consequently, solid-stabilized emulsions, also called Pickering emulsions,[1] may be regarded as an interesting strategy to encapsulate drugs in pharmaceutical formulations [48, 49]. Pickering emulsions are emulsions stabilized by solid particles instead of classic emulsifiers (Fig. 5.6) [50]. Many solid particles can be used in Pickering emulsions and they can be organic, such as polymer latex or starch, or inorganic, such as silica and clay particles [18, 51]. The stabilization of emulsion droplets by solid particles is possible due to their dual wettability. This phenomenon enables the spontaneous accumulation of particles at the oil–water interface and stabilizes it against coalescence by volume exclusion and steric hindrances [17, 52].

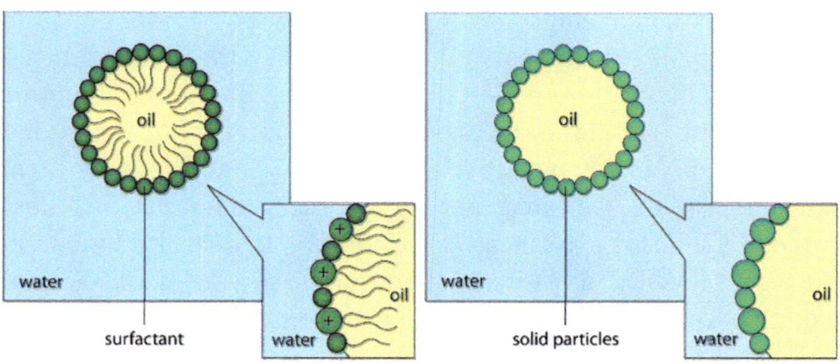

Figure 5.6 Surfactant-based emulsion (left) and a Pickering emulsion (right).

In this type of emulsions, one of the liquids will wet the solid particle more than the other, as the poorest wetting liquid becoming the disperse phase. The importance of the wettability

[1]This type of emulsion was named after S.U. Pickering, who described the phenomenon in 1907 [50].

of the particles at the oil–water interface is quantified by the contact angle (θ) that the particle makes with it, which in turn will determine the type of emulsion. If θ, measured through the aqueous phase, is smaller than 90°, the emulsion will be o/w (oil in water) and, by contrast, if θ is greater than 90°, the emulsion will be w/o (water in oil) [49, 53, 54].

The effectiveness of the solid in stabilizing emulsions will depend on the particle size, particle shape, particle concentration, particle wettability, and the interactions between particles [17, 53].

Since this type of emulsions is free of emulsifiers, it can avoid some adverse effects often linked to surfactants, like skin irritation, making Pickering emulsions very attractive for many application fields, in particular cosmetic and pharmaceutical [50, 53].

This promising approach has already been studied by some authors (Table 5.2). Wille *et al.* [55] developed an o/w Pickering microemulsion using a vegetable oil mix (hydrogenated cotton seed oil, soybean oil and coconut oil), corn starch, glycerol and cationic surfactants, one of which is benzalkonium chloride (BKC), an antimicrobial agent. The lotion proved itself to form an occlusive barrier to water loss and have a skin moisturizing effect, thus enhancing skin hydration. When submitted to a human repeat insult patch test (HRIPT), the lotion, being hypoallergenic, had no potential for inducing dermal irritation or sensitization. This lotion achieved 6 months stability with good antimicrobial action, where the BKC was slowly release upon skin application. This longer stability when compared to others authors' studies is explained by the addition of distearyldimonium chloride (DMDC), a cationic surfactant used as a rheological modifier. This study reinforces the skin compatibility of starch in topical emulsions.

Avoiding chemical surfactants, Marku *et al.* [53] developed an o/w Pickering emulsion, using an octenyl succinic anhydride modified quinoa starch (OSA) and obtained an emulsion with suitable properties for topical application: homogenous white-creamy appearance, high permeability, and stability (8 weeks). During the phase studies, it was observed that no stable emulsion was possible to form at oil concentrations superior to 70% and that, creaming systems were obtained when the oil concentration was inferior to 41%. The increase in starch to oil ratio gave a lower average droplet size and a narrower droplet size distribution. Regarding the three types of oils studied, caprylic/capric triglycerides (Miglyol® 812) and liquid paraffin (LP) provided creams with similar properties (rheological properties and droplet size), while the sheanut oil (a solid fat) showed a higher mean droplet size and higher viscosity and yield stress. Texture and cosmetic properties of the creams were evaluated by a small panel of volunteers and, as result, the difference between the liquid oils and the solid fat was seen: Miglyol® 812 and LP creams were assessed as watery and slippery, while sheanut oil cream showed better permeability, but it was thicker, stickier, and glossier and left residues on the skin. The *in vitro* skin permeation tests showed a steady-state flux of approximately 8.0 μg/(cm^2/h) for all emulsions. When compared to the buffer experiments, where the flux was 4.2 μg/(cm^2/h), it was possible to conclude that the emulsion system increased the permeation into the skin.

The versatility of starch for stabilizing emulsion can be seen in the study performed by Matos *et al.* [17], where OSA was also used in the development of a double w/o/w (water-in-oil-in-water) Pickering emulsion. To the inner water phase was added sodium chloride (NaCl) and the continuous oil phase consisted of Miglyol® 812 and of a lipophilic surfactant: polyglycerol polyricinoleate 90 (PGPR 90). In the outer aqueous phase was added a sodium phosphate buffer, since it has showed to enhance the oil droplets separation. With the addition of PGPR 90, a decreased mean droplet size was possible to obtain and a higher viscosity of the oil phase, which resulted in a less pronounced sedimentation of the inner aqueous phase. This double emulsion showed high encapsulation efficiency (around 98.6%) and high encapsulation stability (91.1–95.2% after 3 weeks of study), confirming the high chemical stability of this double Pickering emulsion.

Concerning emulsions, it is possible to conclude that starch is a very versatile excipient. Its use in pharmaceutics can produce nature-friendly products with suitable and interesting cosmetic properties.

Table 5.2 Pickering emulsions for topical drug delivery

	Qualitative and quantitative composition (%, w/w)						Physical characteristics			
Type of emulsion	Type of starch	Type of lipid	Drug	Other Excipients	Starch quantity	Lipid quantity	Droplet size distribution (μm)	Viscosity (at 25 °C)	Stability	Ref.
o/w	OSA	Miglyol® 812	—	—	214 mg/ml of oil	56%	43 ± 3	7.2×10^6 mPa·s	8 weeks	[53]
		LP					48 ± 2	3.0×10^6 mPa·s	8 weeks	
		Sheanut oil					73 ± 17	1.7×10^7 mPa·s	8 weeks	
o/w	Corn	Hydrogenated vegetables oils	BKC	Glycerol, cyclopentasiloxane, DMDC, dimethicone and Citricidal®	3.3%	8.8%	0.25–5.0	$8–10 \times 10^4$ mPa·s	6 months	[55]
w/o/w	OSA	Miglyol® 812	—	Phosphate buffer, NaCl, PGPR 90	214 mg/ml of oil	33%	40 ± 5	—	3 weeks	[17]

o/w—oil in water emulsions; w/o—water in oil emulsions; w/o/w—water-in-oil-in-water (w/o/w) multiple emulsions; OSA—octenyl succinic anhydride modified quinoa starch; Miglyol® 812—caprylic/capric triglycerides; LP—liquid paraffin; BKC—benzalkonium chloride; DMDC—disteryldimonium chloride; Citricidal®—grape fruit seed extract; PGPR 90—polyglycerol polyricinoleate 90; NaCl—sodium chloride.

5.2.1.2 Gels

Gels can be defined as a semisolid system consisting of a dispersion made up of either small inorganic particle or large organic molecule enclosing and interpenetrated by liquid, forming a three-dimensional interlaced network structure that provides solid-like properties [43, 56]. The European Pharmacopeia (Ph. Eur.) discriminates two types of gels: lipophilic gels (oleogels) and hydrophilic gels (hydrogels) [57].

Gels, as a topical dosage form, have the main advantage of allowing the drug to be absorbed directly at a specific site and avoiding the first-pass metabolism at the liver, offering a great advantage when compared to other delivery systems. Therefore, the therapeutic effects of the drugs are achieved effectively and the systemic side effects are minimized or avoided. Examples of drugs commonly delivered in gels include non-steroidal anti-inflammatory drugs (NSAIDs), antibacterial, antifungal and antihistaminic agents. The drug release from a gel preparation occurs by diffusion of the drug molecules through the gel network or by erosion or dissolution of the gel texture at the interface [58].

The formulation of an effective gel requires the use of an appropriate gelling agent, usually a polymer which can be synthetic (e.g., carbomers or carboxyvinyl polymer), semisynthetic (e.g., methylcellulose, carboxymethylcellulose (CMC) or hydroxyethylcellulose) or natural (e.g., gums or starches) [43]. The main characteristics of such polymer include the inertness, safety, biocompatibility with other ingredients, appropriate adhesion to the skin, allowance of drug permeation, irritation-free and biodegradability. When in formulation, the polymer presents swelling and rheological properties suitable for solidifying or stiffening the system [58]. Lately, natural polymers for topical gels have been reported, such as carrageenan, xanthan gum, chitosan, and several starches.

Starch gelatinization is a process that by dissociation the double-helices and loss of the "Maltese cross" transforms starch from an ordered semicrystalline state to an amorphous one. Starch gelatinization is normally achieved by heating starch above a certain temperature, the "gelatinization temperature" (Table 5.3), in the presence of water or other plasticizers (e.g., glycerol, ethylene glycol, and 1,4-butanediol) (Fig. 5.7), or using alkaline solutions (e.g., sodium hydroxide and potassium hydroxide), neutral salt solutions (e.g., calcium chloride and lithium chloride), and solvents as dimethyl sulfoxide (DMSO) [8, 19, 59]. Wheat starch has the lowest gelatinization temperature, followed by potato, cassava, and maize starches. Regular rice starches show a high variation in gelatinization temperature, which can in part be attributed to their high variation of AM content [59]. After gelatinization, the amorphous starch readily absorbs water and develops viscosity to form a paste. Upon cooling, some starch pastes can create gels. The difference between starch paste and gel lies in that starch paste possesses a certain fluidity, whereas starch gel does not, owing to its defined shape [60].

Table 5.3 Gelatinization properties of native starches (adapted from [60])

Type	T_o(°C)	T_p(°C)	T_c(°C)	Range (°C)	ΔH (J/g)
Normal maize	64.1 ± 0.2	69.4 ± 0.1	74.9 ± 0.6	10.8	12.3 ± 0.0
Waxy maize	64.2 ± 0.2	69.2 ± 0.0	74.2 ± 0.4	10.4	15.4 ± 0.2
Normal rice	70.3 ± 0.2	76.2 ± 0.0	80.2 ± 0.0	9.9	13.2 ± 0.6
Waxy rice	56.9 ± 0.3	63.2 ± 0.3	70.3 ± 0.7	13.4	15.4 ± 0.2
Sweet rice	58.6 ± 0.2	64.7 ± 0.0	71.4 ± 0.5	12.8	13.4 ± 0.6
Wheat	57.1 ± 0.3	61.6 ± 0.2	66.2 ± 0.3	9.1	10.7 ± 0.2
Barley	56.3 ± 0.0	59.5 ± 0.0	62.9 ± 0.1	6.6	10.0 ± 0.3
Mung bean	60.0 ± 0.4	65.3 ± 0.4	71.5 ± 0.4	11.5	11.4 ± 0.5
Chinese taro	67.3 ± 0.1	72.9 ± 0.1	79.8 ± 0.2	12.5	15.0 ± 0.5
Tapioca	64.3 ± 0.1	68.3 ± 0.2	74.4 ± 0.1	10.1	14.7 ± 0.7
Potato	58.2 ± 0.1	62.6 ± 0.1	67.7 ± 0.1	9.5	15.8 ± 0.4
Green leaf canna	59.3 ± 0.3	65.4 ± 0.4	80.3 ± 0.3	21.0	15.5 ± 0.4
Lotus root	60.6 ± 0.0	66.2 ± 0.0	71.1 ± 0.2	10.5	13.5 ± 0.1
Green banana	68.6 ± 0.2	72.0 ± 0.2	76.1 ± 0.4	7.5	17.2 ± 0.1
Water chestnut	58.7 ± 05	70.1 ± 0.1	82.2 ± 0.2	24.1	13.6 ± 0.5

T_o: onset temperature; T_p: peak temperature; T_c: conclusion temperature; range of gelatinization: T_c–T_o; ΔH: enthalpy change.

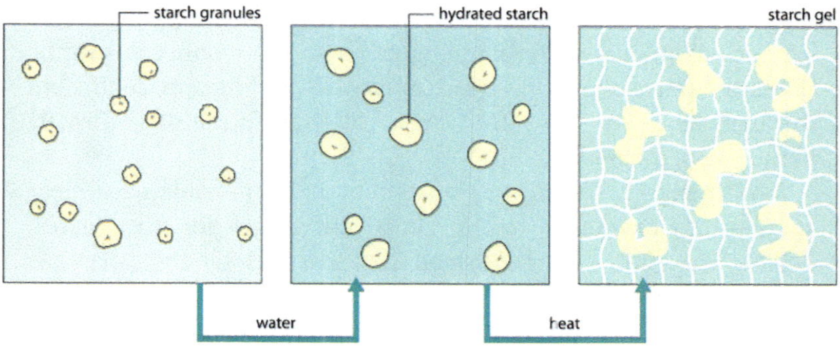

Figure 5.7 Mechanism of starch-based gel formation.

Amylose content, relative crystallinity, AP chain length distribution, and phosphorus content affect the gelatinization behavior of starches [59, 61, 62]. Short AP chains (DP < 14) prevent crystalline order, whereas longer chains (DP > 18) produce more stable crystals. Consequently, increased levels of longer chains shift gelatinization to higher temperatures, whereas short chains decrease the gelatinization temperatures [59, 61]. In potato and cassava starches, it was showed that increased crystallinity and increased AM content resulted in higher ΔH values [61]. Other studies on potato starch showed that at high level of AM content, higher phosphorus content, enhanced re-crystallization of AP during retrogradation of starch gel, associated with a well-formed gel structure and more ordered molecules in the gel, producing a more structured gel [62].

The use of starch gels for topical delivery of pharmaceutical drugs reports back to the 1900s when El-Khordagui *et al.* [63] studied the physicochemical properties of a maize starch gel, using riboflavin as a model drug. A clear gel was obtained with a suitable swelling ability. Riboflavin showed diffusion-controlled release kinetics, influenced mainly by initial loading levels and starch concentration. This was one of the first reported evidence that starch has the ability to produce gels with adequate properties for topical vehiculation of drugs.

More recently, Pal *et al.* [64] formulated a corn starch hydrogel with salicylic acid to protect injured skin and keep the wound surface appropriately moist to increase the healing process. The resulting hydrogel membrane had a tensile strength of 35.92 ± 1.87 mPa, a value similar to the failure strength of skin, meaning that this hydrogel membrane can have a potential application as artificial skin, giving a cushioning effect to the wound. Regarding the drug release, salicylic acid presented a diffusion coefficient through the gel of 4.1×10^{-6} cm^2s^{-1}, delivering the drug directly to the site of action.

Kittipongtana *et al.* [65] on their studies formulated a clear sodium carboxymethyl mung bean starch gel for the topical delivery of ibuprofen, confirmed its clearness by the very low absorption at UV 700 nm ($A < 0.10$). The starch gel presented high viscosity and acceptable spreadability, sticking well onto the skin. The gel was subject to satisfaction evaluation in human volunteers. Statistical analysis showed that the factors significantly affecting satisfaction were the spreading of gel base and the ease of rubbing, as such these properties are the ones to improve, in order to maximize end user satisfaction. The gel also received high score on grittiness and disappearance after applying. Texture and skin irritation were also evaluated through volunteer's opinion. Despite the lack of safety and effectiveness tests, the subjective test performed on volunteers testified for the starch gel satisfactory macroscopic and organoleptic properties.

In another study, Kittipongtana *et al.* [58] prepared a carboxymethyl mung bean starch gel that showed a shear thinning flow with thixotropy behavior, suggesting adequate spreadability. Concerning the organoleptic characteristics, the gel had a good macroscopic appearance, tackles, greaseless and easily washable. Piroxicam was used as a model drug and a controlled release was observed, however, the amount of piroxicam released was low (<5%). In addition, they also tested the stability of the gel production and concluded that it remained stable after 2 months of storage at room temperature.

Nazim *et al.* [66] investigated the use of a potato starch gel as a topical vehicle for rofecoxib, a NSAID, with the purpose to avoid the gastrointestinal disorders often associated with oral administration of these drugs. The physical appearance of the gels was found to be white opaque to white translucent with good homogeneity. By testing different concentrations of potato starch (5–15%), they concluded that higher concentrations of starch give slightly higher pHs and reduce gel spreadability by increasing viscosity. The rofecoxib release was 16.65% and 16.39% for 5% (w/w) and 10% (w/w) potato starch, respectively.

This difference can be attributed to the higher viscosity of gels containing more starch. These formulations were stable up to 24 weeks.

Gabriel *et al.* [67] studied the use of modified *Enset*[2] and cassava starches as gelling agents. Translucent gels with good and smooth homogeneous appearance were obtained, with pH values ranging from 6.8–7.2, within physiologically accepted pH. The formulations with lower starch concentration presented better extrudability and higher spreadability. The release studies conducted using ibuprofen demonstrated that the cumulative percentage of drug released over 12 h ranged from 43.8% to 84.5%. The release profiles exhibited a burst effect in the first hour followed by a sustained and controlled release profile. The authors concluded that the nature of the modified starches influenced the rheological properties (spreadability and extrudability) and the release properties (cumulative release and diffusion coefficient). The modified release of the drug can be of valuable advantage, when delivering drugs such as, analgesics or anti-inflammatories, providing extended drug action.

Nazim *et al.* [56] conducted another work where a clotrimazole starch gel was developed. This antifungal gel presented good mechanical and macroscopic properties. It was easily washable and had no greasiness. The rheological parameters with a viscosity of 6.9×10^3 mPa·s and a spreadability of 2.6 cm, were suitable for topical application and the pH of 6.96 assured skin compatibility. *In vitro* drug release was 85% during 6 h. Stability studies were carried for a period of 3 months according to ICH guidelines and all parameters maintained within the specifications during the period of storage.

As summarized in Table 5.4, it is possible to conclude that a lot of investigation on starch gels has been done and continuous advances on the subject have been achieved. The starch on these gels helps forming a strong and stable matrix due to the AP and AM presence, providing a controlled and sustaining drug release. This slower release, in addition to the starch biocompatibility, is what makes starch gels an attractive delivery system for the pharmaceuticals industry.

[2]*Ensete ventricosum*, Musaceae is starch-rich staple food widely used in the southern and south-western regions of Ethiopia and closely related to the banana tree, but does not produce banana. Enset is often referred as "false banana" [67].

Table 5.4 Pharmaceutical starch gels for topical delivery

Starch	Drug	Other excipients	Concentration of starch	pH	Viscosity	Spreadability	Drug release	Type of membrane	Kinetic model	Stability	Macroscopic appearance	*In vivo* studies	Ref.
Maize starch	Riboflavin 0.044% (w/v)	Sodium salicylate	10% (w/v)	—	—	—	28.5 ± 1.7%	—	Higuchi kinetic *D*: 2.6 ± 0.1 1 $\times 10^{-6}$ cm^2s^{-1}	—	Optically clear gel	—	[63]
Corn starch	Salicylic acid 0.025% (w/w)	Ethanol, glutaraldehyde	2.26% (w/w)	6.37–7.57	Tensile strength of 35.92 ± 1.87 mPa	—	—	Corn starch hydrogel membrane	*D*: 4.1 × 10^{-6} cm^2s^{-1}	—	—	—	[64]
Sodium carboxymethyl mungbean starch	Ibuprofen 0.5% (w/w)	Propylene glycol, denatured alcohol, triethanolamine	2% (w/w)	9.0–9.22	High viscosity; 5.1–15.3 mPa·s	Good spreadability on the skin	—	—	—	—	Clear gel	The satisfaction was influenced by the spreadability and the ease of rubbing.	[65]
Carboxymethyl mungbean starch	Piroxicam 0.5% (w/w)	Propylene glycol, denatured alcohol, triethanolamine	3% (w/w)	9.0–9.2	Shear-thinning fluid with thixotropy behavior; 4.8 × 10^4 mPa·s	Good spreadability	5%	Spectra/Por®7 regenerated cellulose membrane	—	2 months	Clear gel; Tackles, greaseless and easily washable	—	[58]

Starch	Drug	Other excipients	Concentration of starch	pH	Viscosity	Spreadability	Drug release	Type of membrane	Kinetic model	Stability	Macroscopic appearance	In vivo studies	Ref.
Potato starch	Rofecoxib 1% (w/w)	Sodium salicylate 15% (w/w)	5–10% (w/w)	6.37–6.38	2.0×10^3–11.9×10^4 mPa·s	47.14–110 $g.cm{\cdot}s^{-1}$	16.4 – 6.6%	Cellophane membrane	—	6 months	White opaque and good homogeneity	—	[66]
		Sodium benzoate 15% (w/w)		7.54–7.57	3.2×10^3 -22.4×10^4 mPa.s	5.36–82.5 $g.cm{\cdot}s^{-1}$							
Carboxymethyl cassava starch	Ibuprofen 5% (w/w)	—	8–12% (w/w)	6.80–6.86	Shear-thinning fluid; 2.9–3.0×10^5 mPa·s	5.4 ± 1.2 –8.8 ± 1.7 cm	50.6–77.6%	Cellulose acetate membrane		Higuchi kinetic *D*: 4.6±0.04 $\times 10^{-6}$ cm^2s^{-1} to 14.8 ± 0.12 $\times 10^{-6}$ cm^2s^{-1}	3 months	Translucent gels with good homogeneity	[67]
Carboxymethyl *Enset* starch		—	6–10% (w/w)	7.17–7.22			43.8–84.5%						
Corn starch	Clotrimazole 1% (w/w)	Sodium salicylate 15% (w/w)	10% (w/w)	6.0–7.5	6960 mPa.s	2.6 cm	85%	Cellophane membrane		—	3 months	Opaque in appearance; good homogeneity, extrudability, and gel strength	[56]

D: Drug diffusion coefficient.

5.2.1.3 Starch in Personal Care: A Multifunctional Ingredient

Not only starch is edible, but it has also a huge potential on the creation of innovative and skin-friendly cosmetics. Starch, due to its high derma-compatibility and unique properties, is used in cosmetic products for all skin types and, specially, for sensitive skin.

Personal care starches range from basic unmodified starches for body powders to very specific and innovative starches for gels, films, and other unique applications.

Peigen *et al.* [68] proposed to explore new starch resource for the cosmetics industry. The plants *Paeonia suffruticosa Andr., Paeonia lactiflora Pall.*, and *Curcuma phaeocaulis Val.* were explored and the studies indicated different moisture, fat and protein contents, as well as high water binding capacity and clarity, bringing forward high potential application in the cosmetics industry.

Other researchers extracted an innovative and cost-effective native starch from sago. The sago starch, a fine and white powder, was added in perfumed and cool body powders and demonstrated to provide the required physicochemical properties and acceptable levels of specific properties according to Thai Industrial Standards (TIS 443–2525—Skin Powders properties—i.e., slip, covering power, adhesiveness, absorbency, bloom, and spreading power), without causing skin irritation. Sago starch-containing body powders showed good results of users' satisfaction, evidencing to be an excellent ingredient for skin care and one of the best applications to exploit endemic plants [69].

Also, an interesting work of extraction of starch from *Okenia hypogaea*, which belongs to the family of the Nyctaginaceae, was performed by Solorza-Feriaa *et al.*, and the results obtained suggested that *Okenia* starch could be used as a cosmetic raw material, due to the fact that starches with small size granules have a high adsorbent capacity, making it suitable to regulate the oily and shiny appearance of skin, or could also be used as carriers, because it can adsorb substances such as colorants and perfumes [70].

Modified starches, namely, thermally inhibited starches are also used in cosmetic compositions, such as skin and hair care products as emulsifier, thickener, and aesthetic control agent [71].

In the cosmetics industry, starch is used in skin and hair care products where it can substitute the silicone oils. Concerning hair care products, the advantage of starch in this type of product consists in simpler formulation technology and improved care performance with regard to volume of the hair, due to the absence of depot-forming effects, as typical of silicone oils. Albrecht *et al.* [72] formulated a hair shampoo with pregelatinized cross-linked starch derivatives in a concentration of 0.2% to 10% (w/w), and concluded that shampoos with starch derivatives can improve the volume, shine, and after-feel of hair.

McCuag [73] also used the unique properties of starch to develop a long life deodorant, using starch treated with *N*-alkanoyl amino acid, as a substitute for talc in a composition, comprising also citric acid. The starch improved the spreadability and the organoleptic characteristics of the product, once it gives a smooth and silky after-feel.

Starch can also function as an oil control agent, as described by McCook *et al.* [74] when developing formulations for topical treatment for oily skin. Biedermann *et al.* [75] have also referred to this property of starch in their work on skin care products for regulating the oily and shiny appearance of skin.

Formulations containing natural polymer hydrophobically modified starch-based have been studied for their sensory modifier quality. Polonka *et al.* [76] studied a method for making a sensory modifier comprising a polysaccharide carbohydrate rich in AM, such as waxy corn starch or tapioca starch. After treatment with an anhydrous solvent (for example, dipropylene glycol, polyethylene glycol and/or diglycerine), the combined starch and solvent mixture was heated to a temperature from 70 to 80°C for 1.5 to 4.5 h. The sensory modifiers prepared by this method yielded no gelation in a 70% water formulation and present very desirable sensory benefits. Chorilli *et al.* [77] evaluated the volunteers' acceptance of a sunscreen formulation containing aluminum starch octenylsuccinate, compared with a control formulation (without polymer), and determined that the sensory modifier starch added to the formulation was able to promote softness and velvet feel to the sunscreen and it was able to mitigate and noticeably reduce the oiliness of the skin. In addition, the starch showed a soft and dry after-feel, while also improving the spreadability of the product.

Further to these applications of modified starch in cosmetic, Guth *et al.* [78] reported that the addition of 5% aluminum starch octenylsuccinate can enhance sun protection factor (SPF) of a titanium dioxide formulation by 40% (w/w) [79]. Martino *et al.* [80] described the use of 0.5–30% (w/w) starch as a sunscreen agent, providing protection against ultraviolet radiation (UV) A and B. The starch can be used alone in these formulations, in mixtures or in combination with other known UVA or UVB filters to provide several SPF values (Fig. 5.8).

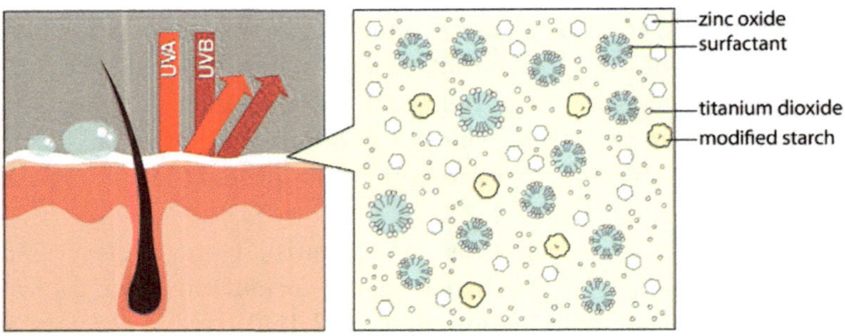

Figure 5.8 Mechanism of action of sunscreens/solid particles with UV radiation.

Personal care products have traditionally been sold as semi-solid or liquid dosage forms. These cosmetics normally include a considerable amount of water in the formula. However, due to the water content in cosmetics, the inclusion of lipophilic active ingredients is limited to either the solubilization or emulsification within aqueous phase. As a result, Gott *et al.* [81] described a successful application of a modified starch in another category of cosmetic products. The concept was to develop a multifaceted starch-based solid cosmetic product, especially foamed solids. When immersed in water, the foamed starch rapidly dissolves, releasing the active cosmetic agent or the perfume, which can then deposit or treat human skin or hair.

The same authors proved that a perfume is perceived with greater intensity when added to a modified starch rather than one incorporated into an unmodified type of starch. The selected starch was a destructurized starch, which is generated under certain conditions of temperature, pressure, shear, limited water, and sufficient time. Other authors [82] developed a starch-based porous dissolvable solid substrate in the form of a unit dose personal care product. This cosmetic can be conveniently and quickly dissolved in the palm of the consumer's hand to reconstitute a semi-solid or a liquid cosmetic, for a suitable application on the skin or the hair, respectively, while efficiently deliver lipophilic active ingredients, during consumer usage.

Thus, depending on the application, the level of starch added can range from 1–100%. These results demonstrated that starch is a multifaceted, alternative, and eco-friendly ingredient, compatible with a wide variety of other personal care ingredients.

5.2.2 Non-Conventional Topical Delivery Systems

5.2.2.1 Polymeric Nanoparticles

Polymeric nanoparticles (particle size between 1 and 1000 nm) can be classified as nanospheres or nanocapsules. Nanospheres are solid-core structures and nanocapsules are hollow-core structures, as schematized in Fig. 5.9 [44, 83, 84]. Thus, nanocapsules differ from nanospheres for being a reservoir type of system, while nanospheres are a polymeric matrix system. The core acts as a reservoir for drugs or active substances, or several, which can be dissolved or dispersed. In addition, the core itself can have biological activity or effects (octylmethocinnamate is a UV chemical sunscreen; turmeric oil has antibacterial, antifungal, and antioxidant properties, among others). It is usually composed of triglycerides, an active ingredient, and other chemical substances, such as capric/caprylic triglycerides and sorbitan monostearate or vegetable oils and sorbitan monostearate [85]. On the other hand, surface is directly related to the nature of the polymer, natural or synthetic (ideally non-toxic and biodegradable) and the surfactants used. Usually, synthetic polymers make more reproducible and purer formulations [85–87].

Polymeric nanoparticles can be classified as ionic (anionic or cationic) or non-ionic (presence or absence of charge on their surfaces). The presence of charge can increase the hydration forces at the surface of the nanocapsules, supporting repulsion among the particles in suspension and reducing aggregation. On the other hand, non-ionic nanocapsules have a great physicochemical compatibility with other ingredients of dosage forms [85].

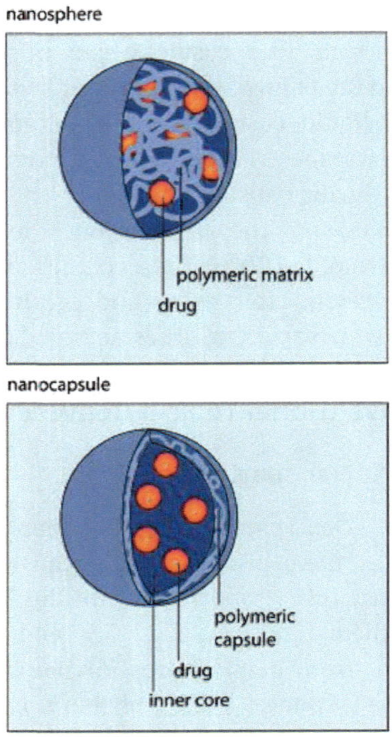

Figure 5.9 Schematic representation of nanospheres and nanocapsules (adapted from [88]).

These nanocarriers can be useful as reservoirs or matrixes for lipophilic or hydrophilic drugs, delivering them in the SC. The use of polymeric materials for encapsulating drugs or other active substances is an important approach to cover the physical and chemical intrinsic properties of substances. The skin penetration and transport seem to be largely dependent on the chemical composition of ingredients, encapsulation mechanism (that will influence the drug release mechanism), the nanoparticles' size and formulations' viscosity [85, 89].

The active ingredient in nanocapsules and nanospheres can be incorporated in different patterns: dissolved in the nanosphere matrix, adsorbed at the nanosphere surface, dissolved in the liquid-phase nanocapsules and adsorbed at the nanocapsules surface (Fig. 5.10) [90, 91].

Nanoparticles preparation techniques are based on their physicochemical properties. They are made by emulsification-diffusion by solvent displacement, emulsification-polymerization, *in situ*-polymerization (interfacial polymerization), gelation (cross-linking), nanoprecipitation (solvent displacement), solvent evaporation/extraction, inverse salting out, dispersion polymerization and other techniques derived from these ones [44, 83]. Drug loading into the nanoparticles can be achieved essentially by two methods: either by incorporating the drug at the time of the nanoparticle production, or by adsorbing the drug after nanoparticle formation, upon incubation in a drug solution. The capacity of adsorption is related to the hydrophobicity of the polymer and the superficial area of the nanoparticles, when comparing the two methods, which it can be possible to conclude that a larger amount of drug can be entrapped using the incorporation method rather than the adsorption one [92].

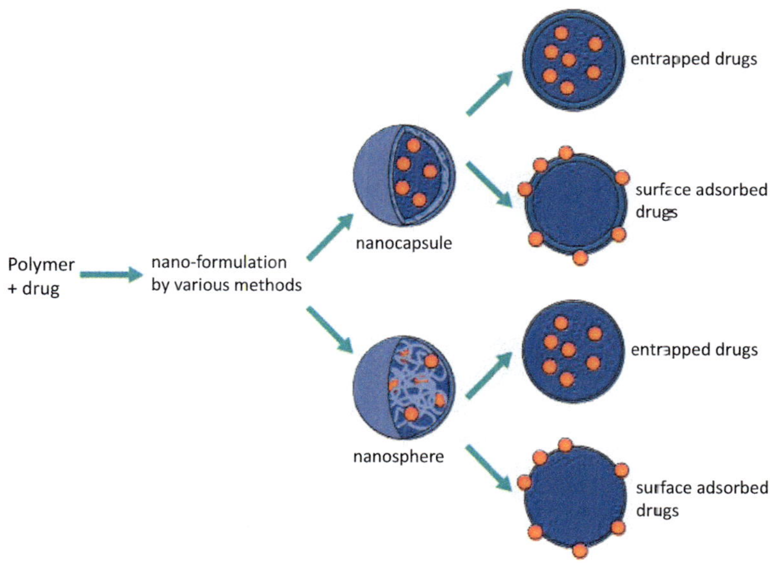

Figure 5.10 Different mechanisms of drug incorporation into polymeric nanoparticles (adapted from [84, 88]).

The advantages and limitations of using nanocarriers arise from their peculiar features: small size, high surface energy, composition, and architecture, among others. However, comprehensive characterization, analytical evaluation, toxicological and pharmacological assessment are necessary to determine the efficacy of using these nanostructures as drug delivery systems [83]. Nonetheless, the use of well-known polysaccharides (for example, chitosan, alginate, starch, and so on) as drug vehicles has the additional benefit of the safety, toxicity and availability issues being substantially simplified [93]. Polymeric nanoparticles, for instance, starch nanoparticles, are an excellent option for topical delivery, because they can be tailor-made in different sizes and it is possible to modify surface polarity in order to improve skin penetration [44, 83]. Moreover, nanoparticle carriers made of bioadhesive materials, such as starch, can form an occlusive film and prolong the residence time, increasing the absorbance of the loaded drugs (Fig. 5.11) [93].

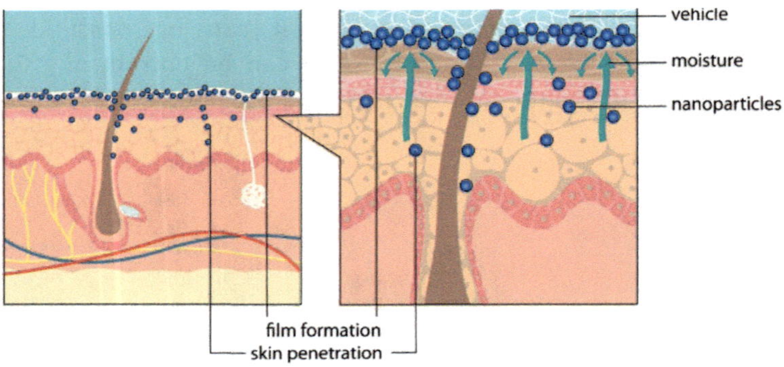

Figure 5.11 Film formation associated with the occlusion effect of nanoparticles (adapted from [85]).

Santander-Ortega *et al.* [16] formulated nanocapsules using propyl-maize starch with high degree of substitution. A simple o/w emulsion diffusion technique was used to formulate the nanocapsules, therefore avoiding the use of hazardous solvents. The incorporation of polyvinyl alcohol (PVA) improved the nanocapsule hydrodynamic size distribution and a mono-modal size distribution was obtained. The drug encapsulation did not produce any intelligible change in the spherical shape or soft surface of the nanoparticles and did not alter their original mean size. The encapsulation efficiency was higher than 95% for flufenamic acid (FFA) and for testosterone and, higher than 80% for caffeine. Nanocapsules loaded with FFA and with testosterone showed a sustained release without any burst effect and a nearly linear profile in the *in vitro* release tests, while the nanocapsules loaded with caffeine showed a much faster release within the first 10 h before reaching a *plateau* phase (close to zero release kinetic). This difference in the release patterns might be explained by the more hydrophilic character of the caffeine. The analysis of the drug permeation profiles, using Franz diffusion cells, suggested that for FFA the release from the starch nanoparticles was the rate-limiting factor, whereas for caffeine the skin barrier was the rate-limiting factor. Regarding testosterone, there was no effect of the nanoencapsulation on skin permeation since very similar permeation profiles were obtained with both, free and encapsulated drug. The high encapsulation efficiency of the nanocapsules in this study is of great importance, when demonstrating the advantages of polymeric starch nanoparticles for topical delivery of drugs.

In different studies, Saboktakin *et al.* [45] used modified carboxymethyl corn starch and hyperbranched 1,4-*cis* polybutadiene (1,4-PBD) as novel polymer matrix for transdermal drug delivery systems. The nanospheres showed a solid and consistent structure, with good spherical geometry. They presented a relatively broad size distribution, ranging from 40 to 100 nm. The incorporation of the drug clonidine into the nanospheres produced a smooth surface and compact structure, being the average drug entrapment 92.2 ± 0.1%. The primary mechanism for release of clonidine from the matrix system *in vitro* is swelling, diffusion and disintegration. In this research, clonidine showed a delayed release, only starting after 2 h. This delayed release profile reinforces the importance of starch nanoparticles for the drug retention onto the skin, providing a long lasting drug action.

More recently, El-Feky *et al.* [94] investigated the use of native maize starch in the production of nanospheres for the transdermal delivery of indomethacin (IND) and acyclovir (ACV), two poorly soluble drugs. The nanospheres were produced using the nanoprecipitation method and the loaded nanospheres were analyzed and characterized in what concerns mean particle size, potential zeta, entrapment efficiency and *in vitro* release. For the nanospheres loaded with 20 mg of IND, the mean particle diameter was 32.67 nm and showed a polydispersity index (PDI), an indicator of particle size distribution, 0.519, suggesting some heterogeneity. The zeta potential was –29.1 mV, which is sufficiently high to ensure physical stability, due to the electrostatic repulsion forces, avoiding aggregation. The nanospheres loaded with 50 mg of ACV showed a mean particle diameter of 15.69 nm with a PDI of 0.423. In what concerns to the surface charge, the ACV nanoparticles showed a zeta potential of –35.9 mV, indicating that the prepared nanospheres are stable for a long time. The entrapment efficiency was determined by a spectrophotometric method and was 63.72% for IND and 74.24% for ACV, where the smallest spheres are capable of entrapping more drug inside and in the outer sphere, due to the larger surface area. The *in vitro* release of IND showed an initial burst in the first 60 min, releasing 4.65% of the drug, followed by a more sustained release. On the other hand, in the ACV nanospheres, no initial burst was observed, suggesting that the ACV drug was not weakly adsorbed onto the nanosphere surface.

Finally, an interesting work of encapsulation of a NSAID resulted in suitable and cost-effective cross-linked starch nanospheres, for the transdermal delivery, using diclofenac sodium (DS) as a model drug. The nanocarrier is based on cross-linked starch nanospheres, which were synthesized using native starch. Cross-linking was achieved by reacting sodium tripolyphosphate (STPP) at different concentrations and DS-loaded cross-linked starch nanospheres were synthesized according to the nanoprecipitation method in the presence of Tween® 80, as a surfactant. A two-level factorial design was selected for the prediction of optimized formulation for DS-loaded cross-linked starch nanospheres. The resultant nanospheres were characterized using world-class facilities such as TEM, DLS, FTIR, XRD and DSC. The efficiency of DS loading was also evaluated by entrapment efficiency, as well as *in vitro* release and histopathological study on rat skin. The optimum nanoparticles formulation selected by the JMP® software was the formula that composed of 5% (w/w) maize starch, 0.058% (w/w) DS, 0.5% (w/w) STPP and 0.4% (w/w) Tween® 80, with particle diameter of about 21.04 nm, PDI of 0.2 and zeta potential of −35.3 mV. It is also worth noting that this selected formula shows an average entrapment efficiency of 95.01% and sustained DS release up to 6 h. Histophathological studies using the best formula on rat skin advocate the use of designed transdermal DS-loaded cross-linked starch nanospheres as it is safe and non-irritant to rat skin. The authors concluded that the starch nanospheres could be considered a good carrier for DS drug, regarding the enhancement in its controlled release and successful permeation, thus offering a promising nanoparticulate system for the transdermal delivery of NSAID [95].

The previously referred works are summarized in Table 5.5. Despite the fact that there are numerous proofs of nanoparticle use for topical drug delivery, starch nanoparticles remain little explored.

Table 5.5 Nanoparticles for topical drug delivery

Type of Nanoparticles	Starch	Drug(s)	Other excipient(s)	Encapsulation efficiency (%)	Size (nm)	PDI	Zeta Potential (mV)	Drug Release (%)	Permeation (cm s^{-1})	*In vivo* studies	Ref.
Nanocapsules	Maize starch	FFA	Ethyl acetate 0.1% (w/w) and PVA 1% (w/w)	> 95	185.5	0.06	−12.3 ± 1.5	26	3.11×10^{-6}	—	[16]
		Testosterone		> 95	176.6	0.43	−12.7 ± 0.6	56	0.43×10^{-6}	—	
		Caffeine		> 80	183.3	0.11	−10.3 ± 1.2	75	0.11×10^{-6}	—	
Nanospheres	Carboxymethyl corn starch 3% (w/w)	Clonidine 0.04% (w/w)	1,4 - PBD	92.2 ± 0.1	40–100	—	—	100	—	—	[45]
Nanospheres	Native maize starch 5% (w/w)	IND 0.02% (w/w)	Tween® 80 and STPP	63.7	32.67	0.519	-29.1	12	—	—	[94]
		ACV 0.05% (w/w)		74.2	15.69	0.423	-35.9	11	—	—	
Nanospheres	Native maize starch 5% (w/w)	DS 0.058% (w/w)	Tween® 80 0.4% (w/w) and STPP 0.5% (w/w)	95.01	21.04	0.2	-35.3	20 (4 h)	—	The histopathological studies advocate the use of starch nanoparticles as it is safe and non-irritant to rat skin	[95]

FFA—Flufenamic acid; IND—Indomethacin; ACV—Acyclovir; DS—Diclofenac sodium; PVA—Polyvinyl alcohol; PBD—1,4-*cis* Polybutadine; Tween® 80—polysorbate 80; STPP—Sodium tripolyphosphate; PDI—Polydispersity index.

5.3 Conclusions

Properties of starch and its advantages as a pharmaceutical excipient are well studied, and it is widely used in the technology of oral formulations. Its use in topical formulations is emerging only now, mostly due to the consumers' demand for natural products and to the industry quest for low cost excipients.

In what concerns starch emulsions, formulators went back in time and now are reinventing the Pickering emulsions. At present, none of the commercialized product uses this technology, but some patents have been submitted and some work is yet to be done: increasing the stability and reduced the aging problems of flocculation and coalescence. On the other hand, the use of starch in cosmetics is rising faster with its use in skin care and hair care products, due to starch ability to decrease oiliness, and in sunscreen products, due to the SPF-enhancing power of starch.

So far, starch gels have been the most investigated. Several studies using a model drug have proved their success in topical drug delivery, allied with good organoleptic appearance and skin feeling. The elevated gelatinization temperature of starch remains the limitative factor of these topical systems, requiring more energy in its preparation. In that line, modified starches have been intensively studied since they gelatinize at room temperature and they have great potential to solve this drawback.

Although the nanoparticles are the most recent trend in pharmaceutical technology, starch nanoparticles remain roughly unexplored. This represents a valuable opportunity for research and investment, since they present themselves as the future. The advances in modified starches result in tailor-made excipients, where properties can be shaped and enhanced in order to assure the desired ones. Starch nanoparticles have many advantages, starting with the use of natural and environment-friendly materials, passing through low immunogenicity and ending with modified drug release and better drug permeation.

In addition, there is a lack of comparative studies between different formulations and different pharmaceutical dosage forms, which makes it challenging to understand what are the critical physicochemical factors to study when developing topical delivery systems. Furthermore, it is difficult to correlate each topical delivery system with its efficacy, since there is also a lack of quantitative percutaneous data.

In this review, in the field of topical starch-based delivery systems, no study presented *in vivo* skin permeation studies, neither *in vivo* skin safety studies. Moreover, the scale-up to industry has not yet been carried out, which brings along some difficulties to anticipate their cost-benefit, thus still remaining a challenge to be overcome by industry and their researchers.

Considering the consumers' demand for more eco-friendly cosmetics, pharmaceutical industry should take that trend into consideration and develop novel starch-based vehicles. Starch's high skin compatibility and recognized property as sensory modifier, both important features when addressing topical vehicles, make it the perfect candidate for novel topical delivery systems.

References

1. Rowe RC, Sheskey PJ, Quinn ME. *Handbook of Pharmaceutical Excipients*. 6 ed. London: Pharmaceutical Press; 2009.
2. Copeland L, Blazek J, Salman H, Tang MC. Form and functionality of starch. *Food Hydrocolloids*. 2009; 23(6): 1527–1534.
3. Tester RF, Karkalas J, Qi X. Starch—composition, fine structure and architecture. *Journal of Cereal Science*. 2004; 39(2): 151–165.
4. Hoover R, Hughes T, Chung HJ, Liu Q. Composition, molecular structure, properties, and modification of pulse starches: A review. *Food Research International*. 2010; 43(2): 399–413.
5. Buléon A, Colonna P, Planchot V, Ball S. Starch granules: Structure and biosynthesis. *International Journal of Biological Macromolecules*. 1998; 23(2): 85–112.
6. Schirmer M, Höchstötter A, Jekle M, Arendt E, Becker T. Physicochemical and morphological characterization of different starches with variable amylose/amylopectin ratio. *Food Hydrocolloids*. 2013; 32(1): 52–63.
7. Singh N, Singh J, Kaur L, Singh Sodhi N, Singh Gill B. Morphological, thermal and rheological properties of starches from different botanical sources. *Food Chemistry*. 2003; 81(2): 219–231.
8. Alcázar-Alay SC, Meireles MAA. Physicochemical properties, modifications and applications of starches from different botanical sources. *Food Science and Technology (Campinas)*. 2015; 35(2): 215–236.

9. Pérez S, Bertoft E. The molecular structures of starch components and their contribution to the architecture of starch granules: A comprehensive review. *Starch-Stärke*. 2010; 62(8): 389–420.
10. Jobling S. Improving starch for food and industrial applications. *Current Opinion in Plant Biology*. 2004; 7(2): 210–218.
11. Jenkins PJ, Donald AM. The influence of amylose on starch granule structure. *International Journal of Biological Macromolecules.* 1995; 17(6): 315–321.
12. Majzoobi M, Saberi B, Farahnaky A, Mesbahi G. Comparison of physicochemical and gel characteristics of hydroxypropylated oat and wheat starches. *International Journal of Food Engineering*. 2014; 10: 657–667.
13. Hoover R, Ratnayake WS. Starch characteristics of black bean, chick pea, lentil, navy bean and pinto bean cultivars grown in Canada. *Food Chemistry*. 2002; 78(4): 489–498.
14. Singh J, Kaur L, McCarthy OJ. Factors influencing the physico-chemical, morphological, thermal and rheological properties of some chemically modified starches for food applications—A review. *Food Hydrocolloids*. 2007; 21(1): 1–22.
15. Seung D, Soyk S, Coiro M, Maier BA, Eicke S, Zeeman SC. Protein targeting to starch is required for localising granule-bound starch synthase to starch granules and for normal amylose synthesis in arabidopsis. *PLoS Biology*. 2015; 13(2): e1002080–e1002080.
16. Santander-Ortega MJ, Stauner T, Loretz B, Ortega-Vinuesa JL, Bastos-González D, Wenz G, Schaefer UF, Lehr CM. Nanoparticles made from novel starch derivatives for transdermal drug delivery. *Journal of Controlled Release*. 2010; 141(1): 85–92.
17. Matos M, Timgren A, Sjöö M, Dejmek P, Rayner M. Preparation and encapsulation properties of double pickering emulsions stabilized by quinoa starch granules. *Colloids and Surfaces A: Physicochemical and Engineering Aspects*. 2013; 423(0): 147–153.
18. Timgren A, Rayner M, Sjöö M, Dejmek P. Starch particles for food based Pickering emulsions. *Procedia Food Science*. 2011; 1(0): 95–103.
19. Ai Y, Jane J-l. Gelatinization and rheological properties of starch. *Starch-Stärke*. 2015; 67(3–4): 213–224.
20. Röper H. Renewable raw materials in Europe: industrial utilisation of starch and sugar. *Starch-Stärke*. 2002; 54(3–4): 89–99.

21. Hadgraft J. Skin deep. *European Journal of Pharmaceutics and Biopharmaceutics.* 2004; 58(2): 291–299.

22. Menon GK. New insights into skin structure: Scratching the surface. *Advanced Drug Delivery Reviews.* 2002; 54, Supplement: S3–S17.

23. Prow TW, Grice JE, Lin LL, Faye R, Butler M, Becker W, Wurm EM, Yoong C, Robertson TA, Soyer HP, Roberts MS. Nanoparticles and microparticles for skin drug delivery. *Advanced Drug Delivery Reviews.* 2011; 63(6): 470–491.

24. Junqueira LC, Carneiro J. *Histologia Básica* 11 ed. Rio de Janeiro: Guanabara Koogan; 2008.

25. Moser K, Kriwet K, Naik A, Kalia YN, Guy RH. Passive skin penetration enhancement and its quantification *in vitro. European Journal of Pharmaceutics and Biopharmaceutics.* 2001; 52(2): 103–112.

26. Jungersted JM, Hellgren LI, Jemec GBE, Agner T. Lipids and skin barrier function—a clinical perspective. *Contact Dermatitis.* 2008; 58(5): 255–262.

27. Hadgraft J. Skin, the final frontier. *International Journal of Pharmaceutics.* 2001; 224(1–2): 1–18.

28. James WD, Berger T, Elston D. *Andrews' Diseases of the Skin: Clinical Dermatology:* Elsevier Health Sciences; 2015.

29. Bouwstra JA, Honeywell-Nguyen PL, Gooris GS, Ponec M. Structure of the skin barrier and its modulation by vesicular formulations. *Progress in Lipid Research.* 2003; 42(1): 1–36.

30. Lane ME. Skin penetration enhancers. *International Journal of Pharmaceutics.* 2013; 447(1–2): 12–21.

31. Fartasch M. The nature of the epidermal barrier: Structural aspects. *Advanced Drug Delivery Reviews.* 1996; 18(3): 273–282.

32. Prausnitz MR, Mitragotri S, Langer R. Current status and future potencial of transdermal drug delivery. *Nature Reviews Drug Discovery.* 2004; 3(2): 115–124.

33. Williams AC, Barry BW. Penetration enhancers. *Advanced Drug Delivery Reviews.* 2012; 64: 128–137.

34. Barrett C. Skin penetration. *Journal of the Society of Cosmetic Chemists,* 1969; 20(48): 499.

35. Barry BW. Novel mechanisms and devices to enable successful transdermal drug delivery. *European Journal of Pharmaceutical Sciences.* 2001; 14(2): 101–114.

36. Williams AC, Barry BW. Penetration Enhancers. *Advanced Drug Delivery Reviews.* 2012; 64, Supplement(0): 128–137.

37. Neubert RHH. Potentials of new nanocarriers for dermal and transdermal drug delivery. *European Journal of Pharmaceutics and Biopharmaceutics.* 2011; 77(1): 1–2.

38. Barrett CW. Skin penetration. *Journal of Society of Cosmetic Chemists.* 1968; 20: 487–499.

39. Prausnitz MR, Langer R. Transdermal drug delivery. *Nature Biotechnology.* 2008; 26(11): 1261–1268.

40. Hadgraft J. Passive enhancement strategies in topical and transdermal drug delivery. *International Journal of Pharmaceutics.* 1999; 184(1): 1–6.

41. Walker RB, Smith EW. The role of percutaneous penetration enhancers. *Advanced Drug Delivery Reviews.* 1996; 18(3): 295–301.

42. Naik A, Kalia YN, Guy RH. Transdermal drug delivery: Overcoming the skin's barrier function. *Pharmaceutical Science & Technology Today.* 2000; 3(9): 318–326.

43. Raposo SC, Simões SD, Almeida AJ, Ribeiro HM. Advanced systems for glucocorticoids' dermal delivery. *Expert Opinion on Drug Delivery.* 2013; 10(6): 857–877.

44. Escobar-Chávez JJ. Nanocarriers for transdermal drug delivery. *Skin.* 2012; 19: 22.

45. Saboktakin MR, Akhyari S, Nasirov FA. Synthesis and characterization of modified starch/polybutadiene as novel transdermal drug delivery system. *International Journal of Biological Macromolecules.* 2014; 69(0): 442–446.

46. Ribeiro HM, Morais JA, Eccleston GM. Structure and rheology of semisolid o/w creams containing cetyl alcohol/non-ionic surfactant mixed emulsifier and different polymers. *International Journal of Cosmetic Science.* 2004; 26(2): 47–59.

47. Bouyer E, Mekhloufi G, Rosilio V, Grossiord J-L, Agnely F. Proteins, polysaccharides, and their complexes used as stabilizers for emulsions: Alternatives to synthetic surfactants in the pharmaceutical field? *International Journal of Pharmaceutics.* 2012; 436(1–2): 359–378.

48. Chen J, Vogel R, Werner S, Heinrich G, Clausse D, Dutschk V. Influence of the particle type on the rheological behavior of Pickering emulsions. *Colloids and Surfaces A: Physicochemical and Engineering Aspects.* 2011; 382(1–3): 238–245.

49. Binks BP. Particles as surfactants—similarities and differences. *Current Opinion in Colloid & Interface Science.* 2002; 7(1–2): 21–41.

50. Chevalier Y, Bolzinger M-A. Emulsions stabilized with solid nanoparticles: Pickering emulsions. *Colloids and Surfaces A: Physicochemical and Engineering Aspects*. 2013; 439(0): 23–34.

51. Kaewsaneha C, Tangboriboonrat P, Polpanich D, Eissa M, Elaissari A. Preparation of Janus colloidal particles via Pickering emulsion: An overview. *Colloids and Surfaces A: Physicochemical and Engineering Aspects*. 2013; 439(0): 35–42.

52. Fan H, Striolo A. Mechanistic study of droplets coalescence in Pickering emulsions. *Soft Matter*. 2012; 8(37): 9533–9538.

53. Marku D, Wahlgren M, Rayner M, Sjöö M, Timgren A. Characterization of starch Pickering emulsions for potential applications in topical formulations. *International Journal of Pharmaceutics*. 2012; 428(1–2): 1–7.

54. Schmitt V, Ravaine V. Surface compaction versus stretching in Pickering emulsions stabilised by microgels. *Current Opinion in Colloid & Interface Science*. 2013; 18(6): 532–541.

55. Wille JJ. Encapsulation of a biocide in a starch-oil microemulsion lotion: Antimicrobial activity and clinical safety of benzalkonium chloride. In (Méndez-Vilas A, eds), *Science Against Microbial Pathogens: Communicating Current Research and Technological Advances:* Formatex; 2011.

56. Nazim S, Shaikh S. Formulation and evaluation of clotrimazole hydrotropic starch gel. *Indo American Journal of Pharmaceutical Research*. 2014; 4(2): 1181–1186.

57. EP. European Pharmacopoeia 8. Strasbourg: Council of Europe: European Directorate for the Quality of Medicines and Healthcare; 2014.

58. Kittipongpatana OS, Burapadaja S, Kittipongpatana N. Carboxymethyl mungbean starch as a new pharmaceutical gelling agent for topical preparation. *Drug Development and Industrial Pharmacy*. 2009; 35(1): 34–42.

59. Waterschoot J, Gomand SV, Fierens E, Delcour JA. Production, structure, physicochemical and functional properties of maize, cassava, wheat, potato and rice starches. *Starch-Stärke*. 2015; 67(1–2): 14–29.

60. Ai Y, Jane J-I. Gelatinization and rheological properties of starch. *Starch-Stärke*. 2015; 67(3–4): 213–224.

61. Gomand SV, Lamberts L, Derde LJ, Goesaert H, Vandeputte GE, Goderis B, Visser RGF, Delcour JA. Structural properties and gelatinisation characteristics of potato and cassava starches and mutants thereof. *Food Hydrocolloids*. 2010; 24(4): 307–317.

62. Lu Z-H, Donner E, Yada RY, Liu Q. The synergistic effects of amylose and phosphorus on rheological, thermal and nutritional properties of potato starch and gel. *Food Chemistry*. 2012; 133(4): 1214–1221.
63. El-Khordagui LK. Hydrotrope-gelled starch: Study of some physicochemical properties. *International Journal of Pharmaceutics*. 1991; 74(1): 25–32.
64. Pal K, Banthia A, Majumdar D. Starch based hydrogel with potential biomedical application as artificial skin. *African Journal of Biomedical Research*. 2006; 9(1): 23–29.
65. Kittipongpatana OS, Burapadaja S, Kittipongpatana N. Development of pharmaceutical gel base containing sodium carboxymethyl mungbean starch. *CMU Journal of Natural Sciences*. 2008; 7(1): 23–32.
66. Nazim S, Dehghan M, Shaikh S, Shaikh A. Studies on hydrotrope potato starch gel as topical carrier for rofecoxib. *Pelagia Research Library*. 2011: 227–235.
67. Gabriel T, Belete A, Gebre-Mariam T. Preparation and Evaluation of Carboxymethyl Enset and Cassava Starches as Pharmaceutical Gelling Agents. *Journal of Drug Delivery and Therapeutics*. 2013; 3(5): 1–10.
68. Xia Y, Gao W, Wang H, Jiang Q, Li X, Huang L, Xiao P. Characterization of tradition Chinese medicine (TCM) starch for potential cosmetics industry application. *Starch–Stärke*. 2013; 65(5-6): 367–373.
69. Boonme P, Pichayakorn W, Prapruit P, Boromthanarat S, eds. Application of sago starch in cosmetic formulations. *Advances in Sago Research and Development*; 2012; Kota Samarahan.
70. Solorza-Feriaa J, Paredes-Lópezb O, Bello-Péreza LA. Isolation and partial characterization of Okenia (Okenia hypogaea) starch. *Starch-Stärke*. 2002; 54: 193–197.
71. Jeffcoat RP, J., Ronco, D. L., Solarek, D. B., Hanchett, D. J., inventor National Starch and Chemical Investment Holding Corporation, assignee. Cosmetics Containing Thermally-Inhibited Starches. United States. 1999.
72. Albrecht H, Heitmann B, Ruppert S, inventors, Beiersdorf Ag, assignee. Hair shampoo containing pregelatinized, cross-linked starch derivatives. United States. 2007.
73. McCuaig D, inventor Dorothy McCuaig, assignee. Of zinc oxide, citric acid and a starch spreading agent. United States. 1997.
74. McCook J, Stephens T, inventors, Mccook John P, Stephens Thomas J, assignee. Topical treatment of acne, seborrheic dermatitis, and oily

skin with formulations containing histamine antagonists. United States. 2004.

75. Biedermann KA, Schubert HL, Parran Jr JJ, inventors, The Procter & Gamble Company, assignee. Topical compositions for regulating the oily/shiny appearance of skin. United States. 2000.

76. Polonka J, inventor Conopco, Inc., assignee. Sensory modifier United States. 2013.

77. Rigon RB, Piffer AR, Lima AAS, Bighetti AE, Chorilli M. Influence of natural polymer derived from starch as a sensory modifier in sunscreen formulations. *International Journal of Pharmacy and Pharmaceutical Sciences.* 2013; 5(1): 306–309.

78. Guth J, Martino G, Pasapane J, RONC D. Polymeric approaches to skin protection. *Cosmetics and Toiletries.* 1991; 106(12): 71–74.

79. Nair B, Yamarik TA. Final report on the safety assessment of aluminum starch octenylsuccinate. *International Journal of Toxicology.* 2002; 21 Suppl 1: 1–7.

80. Martino GT, Pasapane J, Nowak FA, inventors, National Starch And Chemical Investment Holding Corporation, assignee. Granular starch as sunscreen agent in aqueous compositions. United States. 1993.

81. Gott R, Schmitt W, Sabin R, Londin J, Dobkowski B, Cheney M, Vinski P, Slavtcheff C, Paredes R, inventors, Unilever Home & Personal Care Usa, Division Of Conocpo, Inc., assignee. Fragranced solid cosmetic compositions based on a starch delivery system. United States. 2004.

82. Glenn RW, Kaufman KM, Hutchins VTH, Dubois ZG, inventors; The Procter & Gamble Company, assignee. Porous, dissolvable solid substrates and surface resident starch perfume complexes. United States. 2013.

83. Escobar-Chávez JJ. *Current Technologies to Increase the Transdermal Delivery of Drugs:* Bentham Science Publishers; 2010.

84. Kumari A, Yadav SK, Yadav SC. Biodegradable polymeric nanoparticles based drug delivery systems. *Colloids and Surfaces B: Biointerfaces.* 2010; 75(1): 1–18.

85. Poletto FS, *et al.* Polymeric nanocapsules: concepts and applications. In *Nanocosmetics and Nanomedicines: New Approaches for Skin Care* (Beck R, Guterres S, Pohlmann A, eds.), Springer, 2011, pp. 49–68.

86. Patravale V, Mandawgade S. Novel cosmetic delivery systems: An application update. *International Journal of Cosmetic Science.* 2008; 30(1): 19–33.

87. Arora N, Agarwal S, Murthy R. Latest technology advances in cosmaceuticals. *International Journal of Pharmaceutical Sciences & Drug Research.* 2012; 4(3): 168–182.

88. Guterres SS, Alves MP, Pohlmann AR. Polymeric nanoparticles, nanospheres and nanocapsules, for cutaneous applications. *Drug Target Insights.* 2007; 2: 147–157.

89. Guterres SS, Alves MP, Pohlmann AR. Polymeric nanoparticles, nanospheres and nanocapsules, for cutaneous applications. *Drug Target Insights.* 2007; 2: 147.

90. Patravale VB, Mandawgade SD. Novel cosmetic delivery systems: An application update. *International Journal of Cosmetic Science.* 2008; 30(1): 19–33.

91. Arora N, Agarwal S, Murthy R. Latest technology advances in cosmaceuticals. *International Journal of Pharmaceutical Sciences and Drug Research.* 2012; 4(3): 168–182.

92. Soppimath KS, Aminabhavi TM, Kulkarni AR, Rudzinski WE. Biodegradable polymeric nanoparticles as drug delivery devices. *Journal of Controlled Release.* 2001; 70(1–2): 1–20.

93. Liu Z, Jiao Y, Wang Y, Zhou C, Zhang Z. Polysaccharides-based nanoparticles as drug delivery systems. *Advanced Drug Delivery Reviews.* 2008; 60(15): 1650–1662.

94. El-Feky GS, El-Rafie MH, El-Sheikh MA, El-Naggar ME, Hebeish A. Utilization of crosslinked starch nanoparticles as a carrier for indomethacin and acyclovir drugs. *Nanomedicine & Nanotechnology.* 2015; 6(1): 254.

95. El-Naggar ME, El-Rafie M, El-Sheikh M, El-Feky GS, Hebeish A. Synthesis, characterization, release kinetics and toxicity profile of drug-loaded starch nanoparticles. *International Journal of Biological Macromolecules.* 2015; 81: 718–729.

Chapter 6

Solid Lipid Nanoparticles and Nanostructured Lipid Carriers as Topical Delivery Systems for Antioxidants

Carla Vitorino[a,b] and António J. Almeida[c]

[a]*Pharmacometrics Group of the Centre for Neurosciences and Cell Biology (CNC), University of Coimbra, Coimbra, Portugal*
[b]*Faculty of Pharmacy, University of Coimbra, Pólo das Ciências da Saúde, Azinhaga de Santa Comba, Coimbra, Portugal*
[c]*Research Institute for Medicines and Pharmaceutical Sciences (iMed.UL), Faculty of Pharmacy, University of Lisbon, Lisbon, Portugal*

aalmeida@ff.ulisboa.pt

6.1 Introduction

The stratum corneum (SC) is the main barrier to the percutaneous absorption of topically applied drugs. It is a multilayer matrix of hydrophobic and hydrophilic components whose structural integrity is maintained by the presence of modified desmosomes, called corneodesmosomes, which lock the corneocytes together and provide tensile strength for the SC to resist to shearing forces [1]. The barrier nature of the SC depends critically on its unique constituents, such as the ceramides that account for

Carrier-Mediated Dermal Delivery: Applications in the Prevention and Treatment of Skin Disorders
Edited by Andreia Ascenso, Sandra Simões, and Helena Ribeiro

ISBN 978-981-4745-58-1 (Hardcover), 978-1-315-36447-6 (eBook)
www.panstanford.com

45–50% of the hydrophobic lipids present in the intercellular spaces, cholesterol (25%), long-chain free fatty acids mostly with chain lengths C22 and C24 (15%), and 5% other lipids (e.g., cholesterol sulfate, cholesterol esters, and glucosylceramides) [2]. These lipids, which are organized in multilamellar bilayers, regulate the passive flux of water through the SC and are considered to be very important for skin barrier function [3].

The absorption of drugs through the skin can occur through intact epidermis (transepidermal route) and/or skin appendages (transappendageal route) [4]. Since skin appendages occupy <0.1% of the total human skin surface, the transappendageal route has generally been considered to contribute minimally to the overall permeation. However, recent advances in this area have demonstrated the important role of hair follicles as penetration pathways and reservoir structures for topically applied compounds [3]. This is particularly important when topically applied drugs are formulated in nanoparticulate systems, because it has been demonstrated that the penetration depth of the particles can be influenced by their size, resulting in the possibility of a differentiated targeting of specific follicular structures [5]. Nevertheless, under normal circumstances, the predominant route is the transepidermal, with a diffusion path length larger than the thickness of the SC (20 μm) and has been estimated as large as 500 μm. Importantly, the intercellular spaces contain structured lipids and a diffusing molecule has to cross a variety of lipophilic and hydrophilic domains before it reaches the junction between the SC and the viable epidermis [3].

To overcome low absorption rates, lipid-based nanoparticulate systems and vesicular colloidal carriers have been investigated not only to enhance percutaneous absorption but also for drug targeting to the skin or even to its substructure. Thus, they might have the potential for an improved benefit/risk ratio of topical drug therapy [6]. The literature is rich in publications on the ability of lipid-based particles and vesicular colloidal carriers to penetrate the SC and the possibility of using such particles for topical drug delivery has been widely discussed. Actually, topical application, either for therapeutic or cosmetic purposes is a research area in which lipid nanoparticles, including solid lipid nanoparticles (SLNs) and later nanostructured lipid carriers (NLCs), have been used since the 1990s with promising results

[7]. The cosmetic field was the first to benefit from the positive features of lipid nanoparticles, having marketed more than 40 products [6, 8], including Nanobase® (Yamanouchi) and a series of Q10-containing NLC-based antiaging treatment products from Cutanova (Dr. Rimpler GmbH). Besides that, the large number of research works focused on dermopharmaceutical products shows the interest about these nanosystems, being expected their entrance in the pipeline of clinical programs of pharma companies in a near future.

SLNs are composed of solid lipids, that is, lipids that are solid at both room and body temperatures, whereas NLCs are mixtures of solid and liquid lipids, both of them stabilized by an aqueous solution of surfactant(s). In practice, SLNs are considered "solidified" o/w emulsions, in which the oil droplets have been replaced by fat droplets, with a typical solid lipid content of 0.1–30% w/w and a surfactant concentration ranging from 0.5 to 5.0% w/V. On the other hand, in NLCs, the lipidic phase is a blend of solid and liquid lipids (oils) in a solid to liquid ratio, typically, between 70/30 and 99.9/0.1 [6]. However, it should be noted that higher oil loadings have been employed so far to prepare NLCs [9, 10]. Their fat content may be as high as 95%. As a consequence, NLCs have a significantly higher loading capacity in comparison to SLNs.

Lipid nanoparticles protect incorporated active molecules against chemical degradation and allow modulating drug release, thus leading to the use of labile active ingredients, which cannot be utilized in traditional formulations [11]. In addition, they are reported to enhance the bioavailability of topically applied drugs, increasing skin penetration. Their small particle size ensures close contact to the SC and thus, the amount of encapsulated drug reaching the site of action will be increased. To explain this phenomenon, Müller *et al.* proposed a model of film formation on the skin dependent on the particle size, thus causing an occlusion effect after application of SLNs and NLCs onto the skin [7]. Therefore, topical application of aqueous SLNs or NLC dispersions creates a mono-layered lipid film onto the skin, which prevents transepidermal water loss, thus increasing skin's moisture and hydration [12]. Due to their rigidity and the surfactants in their composition, the thermodynamic stability of SLNs and NLCs is enhanced. The presence of surfactants plays also a role on skin's permeability because they may cause skin structure disruption. This mechanism has been pointed out as allowing a physical UV blockage, thus explaining the putative sunscreen effect of lipid nanoparticles [13] and attempts have also been made to explore the potential of NLCs in sunscreen-loaded formulations with promising results in terms of drug loading and stability [14].

Undoubtedly, lipid nanoparticles present occlusive properties, which facilitate drug permeation through SC. However, the drug penetration may also be affected by parameters, such as the carriers themselves, the type and concentration of the lipid, and the drug localization in the nanoparticle structure [15, 16]. Research on the mechanism of interaction between NLCs and the skin showed a clear change in intercellular packing after topical application of an NLC dispersion, with reduced corneocyte packing and wider inter-corneocytes gaps. It was hypothesized that NLCs had an effect on the skin barrier, thus promoting drug permeation [16]. Studies performed with Nile red-labeled NLCs showed not only that the nanoparticles promoted distribution and penetration of the dye into the skin *ex vivo*, with epidermal targeting [17], but also that they may constitute an adequate reservoir system for transdermal delivery with a good *in vivo*/*in vitro* correlation [18].

Suitable formulation in NLCs may therefore control drug penetration in the skin. Studies on the cutaneous absorption of betamethasone valerate into excised human skin showed the importance of a close association of drug molecule and carrier, while emphasizing the role of drug localization within the lipid matrix [15]. The latter may be influenced by drug physicochemical properties, surfactant type and concentration, lipid type, and production method. Therefore, besides the good tolerability, simple and cost-effective large-scale production, stability, targeting, controlled drug release, and protection of liable drugs from degradation, it is very important to ensure the solubility of the drug in the lipid matrix. In NLCs, this is enabled by the oil (liquid lipid) that solubilizes the drugs to a much higher extent than solid lipids [11].

Concerning the transappendageal route, it is well known that sebaceous glands are of particular interest for topical delivery of corticoids for treating diseases like seborrheic dermatitis, although scientific publications demonstrating skin permeation through follicular orifices are scarce [3, 19]. However, follicular penetration has also been studied for NLCs, since hair follicles represent interesting target sites for topically applied substances, such as diphencyprone, minoxidil [20, 21] and spironolactone [22] for topical treatment of alopecia. Once they penetrate into a hair follicle, particles can follow different routes, according to their size thus providing some sort of selective targeting. Smaller particles can penetrate through the follicular epithelium into blood circulation [5].

Many studies have demonstrated the great potential of NLCs, either for local drug delivery or to improve drug absorption through the skin, with a wide variety of drug molecules intended for topical treatment of multiple diseases, making the dermal route perhaps the most studied for the application of NLCs (Table 6.1). It should be noticed that the number of reports in the literature involving SLNs is much larger.

Table 6.1 Dermal applications of NLC formulations

Encapsulated drug	NLC composition	Outcome	Ref.
Acitretin	Precirol ATO 5/oleic acid/Tween 80	Clinical studies with psoriasis patients showed significant improvement in therapeutic response and reduction in local side effects with NLCs	[23]
All-*trans* retinoic acids	Oleic acid/cetyl palmitate/cineole/limonene/Transcutol/butylated hydroxytoluene/Tween 20/Tween 80	NLCs showed higher epidermal permeation across the skin, suggesting their potential use as dermal drug delivery carriers for all-*trans* retinoic acids	[24]
Artemether	Gelucire 43/01/Transcutol/Phospholipon 85G	Artemether permeates excised human epidermis, where the formulation served as a reservoir to gradually control drug release over an extended period of time	[25]

(Conitnued)

Table 6.1 (*Continued*)

Encapsulated drug	NLC composition	Outcome	Ref.
Benzocaine	Compritol 888 ATO/ Miglyol 812/Lutrol F68	Radiant heat tail-flick test was carried out in mice to determine the antinociceptive effect of benzocaine from NLCs; the results showed NLCs act as an effective drug reservoir, prolonging the anesthetic effect	[26]
Celecoxib	Compritol 888 ATO/Miglyol/1,2-dioleoyl-sn-glycero-3-[(*N*-(5-amino-1-carboxypentyl) imidodiacetic acid) succinyl nickel sal/ Tween 80; NLCs modified with cell penetrating peptides	Cell penetrating peptides increased the permeability of celecoxib encapsulated in NLCs; *in vivo* pre-treatment with the formulation inhibited PGE2 and IL-6 expression compared to controls	[27]
	Glyceryl dilaurate/ Capmul MCM/ Cremophor RH 40/ Transcutol	The NLC-based gel showed faster onset and elicited prolonged activity until 24 h	[28]
Clotrimazole	Glyceryl tripalmitate/ Miglyol 812/Tyloxapol	Only *in vitro* characterization was performed; NLCs presented prolonged release properties	[29]
Cyproterone	Precirol ATO 5/Miglyol 812/Poloxamer 188	Drug encapsulation into NLCs resulted in a 2- to 3-fold increase in absorption	[30]
Diphenciprone	Precirol/Squalene/ Pluronic F68/Lecithin	Follicular uptake by NLCs was 2-fold higher for DPCP compared to the free control; great accumulation of NLCs in the follicles and the deeper skin strata	[20]

Encapsulated drug	NLC composition	Outcome	Ref.
Econazole	Precirol ATO 5/Oleic acid/Poloxamer 407	Drug-loaded NLCs showed better permeability and thermodynamic stability as effective topical delivery system for deep-seated fungal infection	[31]
Epidermal growth factor (rhEGF)	Precirol ATO 5/Miglyol 812/Poloxamer F68/ Tween 80	The bioactivity of NLC formulations was even higher than that of free rhEGF; topical administration of NLC-rhEGF improved wound healing in terms of wound closure, restoration of the inflammatory process, and re-epithelization grade	[32, 33]
Finasteride	Precirol ATO 5/Mygliol 812/Tween 60	Loading efficiency was 70–90% and formulations showed low penetration levels in pig ear skin	[34]
Fluconazole	Compritol 888 ATO/ oleic acid/ Pluronic F68/lecithin	*In vivo* skin-retention studies showed drug a 5-fold higher accumulation in the case of NLC formulation, which also revealed maximum antifungal efficacy	[35]
Flufenamic acid	Precirol ATO 5/Mygliol 812/Plantacare 810	Good *in vitro* skin permeation and penetration properties for flufenamic acid formulated in NLCs	[36]
Fluticasone propionate	Precirol ATO 5/ Labrasol/Tween 80/ lecithin Precirol ATO 5/Softigen 767/ Tween 80/lecithin	Reduction of adverse effects/not demonstrated	[37]

(*Continued*)

Table 6.1 (*Continued*)

Encapsulated drug	NLC composition	Outcome	Ref.
Indometacin	Compritol 888 ATO/ Miglyol 812/Lutrol F68	The anti-inflammatory effect, following topical application, was more prolonged with the NLC gel formulation; topical bioavailability in the SC depended from the formulations	[38]
Ketoconazole	Compritol 888 ATO/ α-tocopherol/ Poloxamer 188/ sodium deoxycholate	Only *in vitro* characterization was performed; in contrast to SLNs, the NLCs were able to stabilize the drug	[39]
Lidocaine	Compritol/Miglyol/ Lutrol F68	Results from the radiant heat tail-flick test showed NLCs can act as an effective drug reservoir, prolonging the anesthetic effect	[26]
Loratadine	1-Hexadecanol/ Mygliol 812/ TegoCare 450	Penetration profiles of drug encapsulated in NLCs was lower when compared to a nanoemulsion	[40]
Metotrexate	Witepsol S51/oleic acid/Tween 60 or Tween 80	NLCs promote drug skin penetration *in vitro*	[41]
Minoxidil	Stearic acid/oleic acid/ Poloxamer 188	Only physicochemical characterization was performed	[42]
	Precirol ATO 5/ squalene/ Pluronic F68/lecithin	NLCs reduced minoxidil penetration through the skin; this may indicate a minimized absorption into systemic circulation; great accumulation of NLCs in the follicles and the deeper skin strata	[20]

Encapsulated drug	NLC composition	Outcome	Ref.
	Tristearin/oleic acid/ cholesterol/Lecithin/ Tween 80	Only physicochemical characterization was performed. *In vitro* release suggests a faster onset and prolonged activity up to 16 h	[21]
	Cetyl palmitate/ oleic acid/ Tween 60	Loading efficiency was <30% and formulations showed low penetration levels in pig ear skin.	[34]
Oxybenzone	Glyceryl monostearate/ Mygliol 812/ polyvinyl alcohol Glyceryl monostearate/oleic acid/polyvinyl alcohol	The incorporation of oxybenzone into NLCs greatly increased the *in vitro* sun protection factor and erythemal UVA protection factor of oxybenzone more, while providing the advantage of overcoming side effects of free oxybenzone	[43]
Spironolactone	Compritol 888 ATO/ olive oil/Transcutol/ Tween 80	Confocal laser scanning microscopy confirmed the potential of delivering NLCs within the hair follicles, suggesting the possibility of localized delivery into the scalp hair follicles	[22]
Valdecoxib	Glyceryl dilaurate/ Capmul MCM/ Cremophor RH 40/ Transcutol	The NLC-based gel showed faster onset and elicited prolonged activity until 24 h	[44]

Importantly, NLCs must be either applied as an aqueous dispersion or incorporated in suitable liquid or semi-solid preparations for cutaneous use that provide an appropriate formulation consistency for application upon the skin. The development of such pharmaceutical dosage forms containing SLNs or NLCs usually involves (a) the incorporation of SLNs/NLCs in preformed topical products (e.g., lotions, gels or creams); (b) addition of viscosity enhancers to the aqueous phase of SLNs/NLCs to obtain a gel (e.g., xanthan gum, hydroxyethylcellulose, hydroxypropyl methylcellulose, Carbopol® and chitosan); or (c) the direct production of a final product containing only nanoparticles in a one-step process using high lipid concentrations [6]. Attention should be paid to avoid excessive dilution of the lipid nanoparticles and therefore of the encapsulated drug, as well as instability phenomena such as particle aggregation or dissolution.

The literature clearly indicates that skin penetration depends not only on nanoparticle formulation, but also on vehicle in which NLCs are included. Vitorino *et al.* [10, 45] reported a Carbopol® hydrogel as a vehicle for NLCs, combined with classical skin enhancers (ethanol and limonene), in proportions optimized in what pertains the permeation rate. The combination of the occlusive effect promoted by NLCs, with the SC lipid disturbance attributed to the chemical enhancers, arises as an appealing strategy to overcome stratum corneum, resulting in transdermal delivery [10, 18].

6.2 Antioxidants

Nowadays, the use of antioxidants (AOs) has been boosted, from either a prophylactic or a therapeutic point of view. Such trend has been prompt by the discovery of the involvement of free radicals and other reactive oxygen species (ROS) in physiopathological mechanisms of diseases, including infection, inflammation and cancer, either systemically or at a topical level [46, 47]. Therefore, the use of AOs as a therapeutic approach to overcome the occurrence of these pathologies appears promising. Besides that, AOs are molecules that exert multifunctional activities not only applied in a specific condition, but also common to several diseases. Free radicals are molecules or chemical species with unpaired electrons, which makes them unstable and highly reactive, able to cause oxidative stress (OS). OS is a condition defined as

Table 6.2 Principal studies involving the incorporation of antioxidants into lipid nanoparticles and respective outcomes

Encapsulated drug	LN composition	Outcome	Ref.
Vitamin A and derivatives	Compritol® 888 ATO (glyceryl behenate)/Miglyol® 812 (capric caprylic triglycerides)	Controlled release for retinol and retinyl palmitate-loaded SLNs over 24 h, with a release rate higher than the corresponding nanoemulsions; permeation studies suggested a drug localizing effect, with high retinol concentrations found in the upper skin; prolonged targeting was obtained with the incorporation of SLNs into the o/w cream vs. hydrogel	[53, 54]
	Palm oil/egg phosphatidylcholine/Tween® 80// butylate hydroxyanisol (BHA)/butylate hydroxytoluene (BHT)	SLN formulations decelerated the degradation of retinol, compared with the respective solution in methanol; co-loading of antioxidants greatly enhanced the stability of retinol-SLNs in a antioxidant-dependent manner	[56, 57]
	Precirol ATO 5®/dicetyl phosphate (DCP)/ Gelucire 50/13®	Surface-modification of the SLNs with DCP enhanced the skin distribution of retinyl palmitate 4.8-fold and delivered it to a greater depth than did neutral SLNs	[55]
Ascorbyl-palmitate	Imwitor® 900 (glyceryl monostearate)/Labrafil® M1944 (apricot kernel oil polyethylene glycol-6 ester)/Hydrine® (PEG-2 stearate)/Apifil® (non-ionic hydrophilic white beeswax)/glyceryl monostearate/Miranol Ultra® C32 (sodium cocamphoacetate)/Lorol C16® (cetyl alcohol)/ Tego® Care 450 (polyglyceryl-3 methylglucose distearate)/Tween® 80	Chemical stability of AP-loaded NLCs can be improved by selecting suitable composition and proper storage conditions; NLC gel can be used as a colloidal carrier for topical application of AP	[63, 143]

(*Continued*)

Table 6.2 *(Continued)*

Encapsulated drug	LN composition	Outcome	Ref.
	Compritol® 888 ATO/ Miglyol® 812/Phospholipon 80/Emulmetic 320/ Cholesterol/Labrasol/Plurol oleique/Pluronic F68	AP was most resistant against oxidation in non-hydrogenated soybean lecithin liposomes, followed by SLNs, w/o and o/w ME, and, finally, in hydrogenated soybean lecithin liposomes	[64]
	Witepsol® E85 (hard fat)/Miglyol1® 812/TegoCare® 450	SLNs and NLCs ensured higher chemical stability, degree of skin moisturizing and penetration of AP as compared to corresponding NEs	[40, 65]
Vitamin E derivatives	Cetyl palmitate 30/ Tego Care® 450	*In vitro* assays revealed a UV-blocking effect of sole SLNs and a synergistic effect of tocopherol acetate and SLNs	[67]
β-Carotene	Stearyl ferulate/stearic acid/sodium taurocholate/ butanol Tween® 20	The antioxidants entrapped (β-carotene and α-tocopherol) in SF-SLNs were extremely stable after being submitted to treatment with a pro-oxidant and/or exposition to sunlight	[75]
	Hydrogenated palm kernel and palm glycerides/isopropyl palmitate/ Span 40/Tween 80/Phenonip	Enhancement of the chemical stability of heat-sensitive bioactive compounds and the physical stability of NLCs was attained through the development of two new methods of production	[76]
Lutein	Cetyl palmitate/Glyceryl tripalmitate/ carnauba wax/Miglyol® 812/Plantacare® 810	Remarkable photoprotection of the labile lutein was achieved by the nanocarriers, being highest for SLNs (0.06%), followed by NLCs (6–8%), as compared to 50% of lutein powder in oil	[81]

Encapsulated drug	LN composition	Outcome	Ref.
Co-enzyme Q10	Precifac® ATO (cetyl palmitate)/Labrasol® (PEG-8 caprylic/capric triacylglycerols)/ Miglyol® 812/TegoCare® 450 (polyglyceryl-3 methylglucose distearate)	NLCs showed a biphasic release pattern, i.e., in comparison to NEs, NLCs provided an initial fast release ideal for skin saturation, followed by a slow and prolonged release profile to maintain the skin concentration of CoQ10, which was dependent on the oil content	[86]
	Precifac® ATO/Labrasol®/Miglyol® 812/ TegoCare® 450	*In vitro* permeation studies revealed that the amount of CoQ10 in the skin and acceptor medium was affected by the amount of oil content in NLCs and the occlusive effect	[89]
	Soybean lecithin/octyl decyl acid triglyceride (GTCC)/glycerol	CoQ10-NLCs significantly enhanced the antioxidative capacity of fibroblasts submitted to oxidative stress induced by UVA-irradiation; *in vitro* and *in vivo* CoQ10 skin penetration was also improved by NLCs when compared to a non-particulate formulation	[82]
	Precifac® ATO/Labrasol®/Miglyol® 812/ TegoCare® 450	In a tape-stripping test it was found that penetration of coenzyme Q10 into the skin was enhanced by using NLCs as compared with emulsion and liquid paraffin coenzyme Q10	[90]
	Cetylpalmitate/Miglyol® 812/TegoCare® 450	A preservative classification system to help suggesting a mechanistic model of the key parameters affecting the physical stability of NLCs was developed	[88]
	Precirol® ATO 5/Miglyol® 812/Lutrol® F68 //Tween® 80	Reduced CoQ10-NLCs, differently from oxidized CoQ10-NLCs, were able to efficiently counteract UVA-associated mitochondrial depolarization, thus suggesting a potential role of this molecule in antiaging cosmetological formulations	[87]

(*Continued*)

Table 6.2 *(Continued)*

Encapsulated drug	LN composition	Outcome	Ref.
	Cetyl palmitate/Miglyol® 812/TegoCare® 450/ Cetiol® OE (dioctylether)/Span® 20/Tween® 80	The reduced particle size (80 nm) obtained with ultra-small NLCs resulted in improved dermal delivery of CoQ10 when compared to larger, conventional NLCs (about 230 nm)	[144]
	Cetyl palmitate/Miglyol® 812/TegoCare® 450/ Cetiol® OE (dioctylether)/Span® 20/Tween® 80	CoQ10-loaded usNLCs showed a higher release, a higher antioxidant capacity, and a better skin penetration for Q10, due to a flip–flop core–shell structure of the lipid matrix	[9]
	Cetyl palmitate/Miglyol® 812/TegoCare® 450/ Cetiol® OE/Span® 20/Tween® 80	CoQ10-loaded usNLCs were taken up into HaCaT keratinocytes and conferred them an increased antioxidant potential compared to classical nanocarrier systems (NLCs and NEs)	[91]
Idebenone	Ceteth-20 (polyoxyethylene-20-cetyl ether)/ Isoceteth-20 (polyoxyethylene-20-isohexadecyl ether)/ Oleth-20 (polyoxyethylene-20-oleyl ether)/ Tegin O® (Glyceryl oleate)/Cutina CP® (cetyl palmitate)	SLNs were prepared by the PIT method; IDE penetration into the different skin layers was dependent on the type of SLNs used; no IDE permeation occurred from all the SLNs studied, suggesting that these carriers could be interesting for IDE targeting to the upper skin layers	[95, 145]
	Precirol ATO 5®/medium chain fatty acid triglycerides (MCT)/Tween® 80/ Soybean phosphatidylcholine	NLCs were found to achieve a significant improvement with respect to the chemical stability of IDB, skin permeation, and formulation stability when compared to NEs and oil solution	[93]
	Compritol 888®/Captex 500 P (triacetin)/ethanol/ Labrasol/Transcutol P	IBD-NLC high-skin deposition and SPF led to superior antioxidant effect of formulations (vs. plain IDB gel); this indicated that NLCs have potential as an antioxidant and sun protection for topical drug delivery	[97]
Alpha-lipoic acid	Softisan 601/Miranol® Ultra C32	SLNs with sufficiently higher loading capacity of LA were prepared	[100]

Encapsulated drug	LN composition	Outcome	Ref.
	Miglyol® 812/Apifil/Pluronic® F68	SLNs and NLCs ensured a sustained release of LA and similar antioxidant activity as compared with pure compound	[99]
Quercetin	Glyceryl monostearate/Stearic acid/Medium chain triglycerides/d-α-Tocopheryl polyethylene glycol 1000 succinate	NLCs promoted drug permeation, increased drug retention in epidermis and dermis, and enhanced the antioxidant and anti-inflammatory effects	[116]
	Compritol® 888 ATO/Oleic acid/Polysorbate 20(Tween® 20)/dioctyl sodium sulfosuccinate	SLNs and NLCs promoted a biphasic release profile, with an initial burst release followed by controlled release for up to 30 h; both systems exhibited superior topical delivery of quercetin compared to the control formulation (particles in the micrometer range), with NLCs presenting the highest improvement and stability	[109]
Genistein	Compritol® 888 ATO/Miglyol® 812/ Polysorbate 80 (Tween® 80)/sorbitan trioleate/ cetylpyridinium chloride	GN was released more slowly from NLCs than from SLNs, although more fluidity has been observed for NLCs; LN increased GN skin retention, with NLCs favoring drug penetration into deeper skin layers	[125]
Resveratrol	Cetylpalmitate/Tricaprin/polyglyceryl-3-methyl glucose distearate (Tego Care 450)/potassium cetyl phosphate (Amphisol K)/1-tetradecanol/ lecithin (Epikuron 200)/Tween® 60/tetradecyl-γ-cyclodextrin	Varying surfactant mixtures yield different SLN properties (particle size and zeta potential); SLNs improved RV stability, reducing its photodegradation rate, when compared with RV freely dispersed in the same medium; the presence of tetradecyl-γ-cyclodextrin improved the physicochemical parameters of SLN formulation and enhanced the photostability of RV entrapped in SLNs, also increasing its skin uptake and antilipoperoxidative activity	[133]

(Continued)

Table 6.2 *(Continued)*

Encapsulated drug	LN composition	Outcome	Ref.
	Compritol® 888 ATO/Myglyol® 812, Poloxamer® 188/Tween® 80	RV–loaded SLNs and NLCs revealed antioxidant properties at a concentration of 50 μM; when the two systems were compared, NLCs, which presented smaller particle size and higher drug loading penetrated deeper into the skin	[135]
Gamma-oryzanol	Cetyl palmitate/Dynasan114 (short chain triglycerides)/Witepsole E85 (semi-synthetic glycerides or hard fat: a mixture of mono-, di- and triglyceride)/Miglyol® 812/Poloxame®r 188	Solid lipid type in SLNs played an important role only on EE and RI, not affecting chemical stability; a significant increase in size and decreased EE was found only with NEs after 1 month of storage at 45°C; biphasic release patterns were observed only for SLNs, whereas a fast and much higher release from NEs was observed; cetyl palmitate and Witepsole E85-based SLNs demonstrated enhanced antioxidant activity	[128]
Curcuminoids	Stearic acid/glyceryl monostearate (GMS)/Poloxamer® 188/dioctyl sodium sulfosuccinate (AOT)/Ethanol	Stability of CM in a cream containing CM-SLNs was significantly improved as compared to free CM in the cream formulation; The light and oxygen sensitivity of CM was strongly reduced by incorporating them into SLNs	[141]
	Beeswax/Tween® 80/Lecithin	Beeswax showed to be an excellent natural carrier for CM nanoencapsulation, resulting in nanocapsules with adequate size and zeta potential; hydrophilic gel containing solid lipid nanocapsules showed to be stable	[142]

an impairment of the balance between antioxidants and pro-oxidants/oxidants in favor of the latter, which leads to a potential damage [48]. It affects lipids, carbohydrates, proteins, enzymes and DNA in cells, causing membrane damage, fragmentation or random cross linking of molecules such as DNA, enzymes and structural proteins or even lead to cell death induced by DNA fragmentation and lipid peroxidation [49, 50]. To protect against potential damages, human body ensures physiological antioxidant mechanisms, which counteract free radicals and prevent the damage induced by them by crumbling them before they reach the biological targets, preventing chain reactions or the activation of oxygen to highly reactive products, the so-called ROS. The superoxide radical ($O_2^{\bullet -}$), hydroxyl radical ($^{\bullet}OH$), nitric oxide radical ($NO^{\bullet}$), peroxyl radical ($ROO^{\bullet}$), and hydrogen peroxide are the most common ROS that occur in the human body. These can be divided into two major groups, i.e., enzymatic and non-enzymatic. The first one includes enzymatic scavengers, e.g., superoxide dismutase (SOD), catalase (CAT) and glutathione peroxidase and other repair enzymes involved in the reduction of oxidized forms of molecular antioxidants, such as glutathione reductase and dehydroascorbate reductase. Among non-enzymatic AOs are lipophilic radical scavengers like tocopherols, carotenoids, and coenzyme Q10 (ubiquinone) and hydrophilic scavengers such as ascorbate, glutathione, and flavonoids [49].

Some of these agents are endogenously synthesized, which includes enzymes, low molecular weight molecules and enzyme cofactors; however, the majority are provided by the diet. Dietary AOs can be classified into various classes, such as vitamins, polyphenols, carotenoids, cofactors, among others (Fig. 6.1). The absence of these nutrients compromise physiological defense mechanisms.

The role of these agents in the skin is crucial to cope with the protection functions ensured by this vital organ. Human skin is very susceptible to OS, since it is continuously exposed to exogenous and endogenous factors that contribute to ROS formation and, consequently, to cell/structural modifications that results in skin damage. Some of the environmental causes that makes skin a major target of ROS attack include the exposure to ultraviolet (UV) light, cigarette smoke, xenobiotics, poor diet, harsh soaps or aggressive detergent-based moisturizers. Aging and many biological disorders are additional factors that contribute to a decrease in the antioxidant levels.

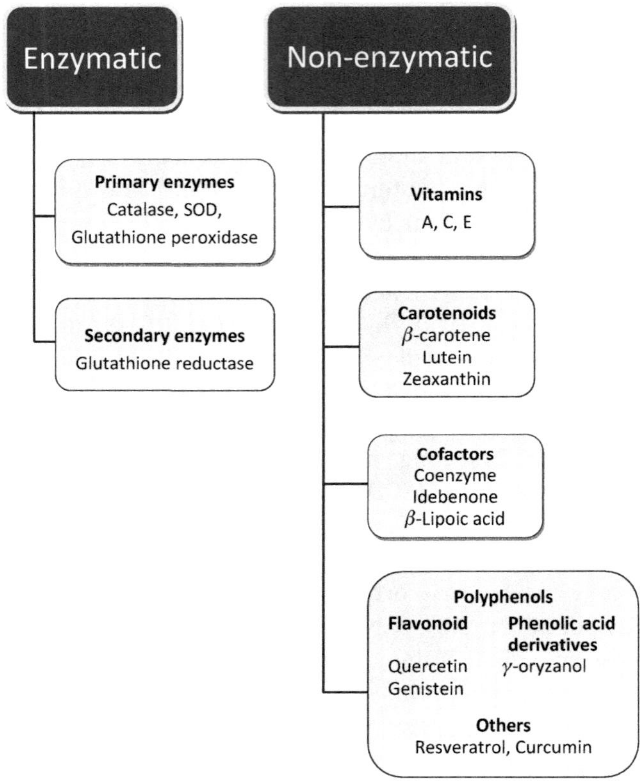

Figure 6.1 Classification of antioxidants. Some AOs are simultaneously obtained by synthesis in the human body and from diet. Within the later, polyphenols represent the major class.

Under these conditions, wherein the protection against ROS is weakened along with the increase in oxidants species, the antioxidants produced by the body are inadequate. Thus, the external supply of AOs is crucial to counteract the deleterious effects of oxidative stress. Antioxidant therapeutic strategies encompass drug administration via the skin for either a local (topical delivery) or a systemic effect (transdermal delivery). The delivery of AOs in conventional dosage forms (e.g., gels, creams, ointments) is challenged due to several reasons like poor solubility along with instability issues. In this context, lipid nanoparticles arise as an appealing and versatile approach to overcome the skin permeability barrier, and facilitate the transport to deeper layers, especially for large and hydrophilic molecules, and the stability issues associated to compounds. This type of carriers has been employed to a wide range of AO classes, either on diseased or healthy skin. An overview of the studies carried out hitherto is presented in the following sections and resumed in Table 6.2.

6.2.1 Vitamins

6.2.1.1 Vitamin A and derivatives

Retinoids show promise in the treatment of skin aging. However, irritant reactions, such as burning, scaling or dermatitis associated with retinoid therapy have limited their acceptance by patients. This problem is more critical with tretinoin and tazarotene, whereas other retinoids mainly represented by retinaldehyde and retinol are considerably less irritating [51].

Vitamin A (Vit A) is a 20-carbon molecule composed of a cyclohexenyl ring, a side chain with four double bonds in *trans* configuration, and an alcohol end group (hence the name *all-trans-retinol*). Vit A is not synthesized by the body, so that it needs to be externally supplied [51]. It is widely used as a major component in functional cosmetics. However, it presents limited skin penetration ability [52]. Thus, there is a strong need for the development of new delivery systems capable of enhancing the skin permeation of this lipophilic vitamin. Jenning *et al.* studied the topical drug release properties of SLNs encapsulating vitamin A, employing both the free alcohol retinol and the respective ester retinyl palmitate [53]. SLNs loaded with Vit A either in suspension or incorporated in a hydrogel and o/w-cream were subsequently tested with respect to their influence on drug penetration into porcine skin [54]. A controlled release for both forms was observed over 24 h, with a release rate even exceeding the one obtained from the corresponding nanoemulsions [53]. Vitamin A concentrations in the skin suggested, however, a certain drug localizing effect, with high retinol concentrations found in the upper skin layers, whereas the deeper regions showed only very low vitamin A levels. A prolonged targeting was obtained when SLN-retinol were incorporated into the o/w cream versus hydrogel, which slowed down the polymorphic transition and thus drug expulsion. In turn, the penetration of retinyl palmitate, an occlusion sensitive drug, was even more influenced by SLN incorporation [54]. In another study, retinyl palmitate was loaded in a surface-modified SLNs with dicetyl phosphate (DCP), a safe cosmetic excipient that conferred them a negative charge. The SLN system was characterized and its skin permeation *in vitro* and antiwrinkle effect *in vivo* further evaluated. DCP-modified negative SLNs allowed an enhancement of the retinyl palmitate skin distribution of 4.8-fold and delivered it to a greater depth than did neutral SLNs. In addition, the *in vivo* studies revealed that the antiwrinkle effects promoted by DCP-SLN formulations were significantly different from that of the negative control and effectively prevented the reduction of elastin and superoxide dismutase activity by UV irradiation [55].

Jee and coworkers also investigated whether stabilization of retinol could be achieved through SLNs [56, 57]. It was observed that despite the incorporation of retinol into SLNs could not perfectly stabilize it the instability of the active could be overcome by the co-loading of antioxidants, such as butylate hydroxyanisol (BHA) and butylate hydroxytoluene (BHT) in SLNs. Moreover, the presence of antioxidants greatly increased the loading efficiency of retinol in SLNs [56].

6.2.1.2 Ascorbic acid derivatives

Vitamin C or L-ascorbic acid (AA) and its derivatives have been widely used in cosmetic and pharmaceutical products to protect and reduce the signs of aging, mainly due to their antioxidant properties. Due to its excellent reducing efficiency, AA is extremely unstable especially under aerobic conditions, in the presence of metal ions and at light exposure, being irreversible and quickly degraded to the biologically inactive form (2,3-diketo-l gulonic acid) by oxidation [40]. Because of its low stability in aqueous solutions, its use in topical formulations is difficult. The structural modification of the ascorbic acid ring at positions 2, 3, 5, or 6 contributes not only to its stabilization as an antioxidant, but also to the formulation of a variety of products with antioxidant activity with a broaden use in nonaqueous media [58]. One of its derivatives is ascorbyl-palmitate (AP), an amphiphilic molecule, which is chemically more stable and penetrates into the skin more effectively as compared to its acidic form. In addition, AP moisturizing activity in various topical formulations was found to be considerably high [59, 60]. However, chemical stability issues have still been associated to many AP preparations [61]. The use of SLNs and NLCs to enhance AP stability has been employed by several research groups. Teeranachaideekul *et al.* investigated the formulation parameters affecting the stability of AP physicochemical characterization after incorporating it into NLCs [62]. It was observed that the chemical stability of AP-loaded NLCs can be improved by selecting adequate types of lipid, surfactant, and proper storage conditions, i.e., cold temperature and flushing with nitrogen gas. Moreover, an improvement of long-term (90 days storage) chemical stability of A-P was achieved by the addition of extra antioxidants (butylate hydroxyanisol (BHA), butylate hydroxytoluene (BHT) and Vitamin E) [62]. The further inclusion of AP-NLCs into a gel supported the suitability of these nanocarriers for topical application [63].

Kristl *et al.* studied the stabilizing effect of different nanocarrier systems for AP: microemulsions (ME), liposomes, and SLNs [64]. It was also observed that AP was resistant against oxidation in the following order: nonhydrogenated soybean lecithin liposomes, SLNs, w/o and o/w ME, and hydrogenated soybean lecithin liposomes. The location of the nitroxide group of AP in the carrier system revealed being crucial to its stability.

In turn, Üner *et al.* examined the stability of AP incorporated into SLNs, NLCs, and nanoemulsions (NEs) [40]. According to this study, nondegraded AP content in NLCs, SLNs, and NEs was found to be 71.1 ± 1.4%, 67.6 ± 2.9% and 55.2 ± 0.3% after 3 months at 4°C, respectively, thus demonstrating that the carrier structure is important to ensure the AP stability. The highest degradation was observed with NEs, which was attributed to the fact that oxidation rate increased with mobility of the AP in the oil droplets of NEs, as a result of the liquid state of the internal phase. Another reason was the higher solubility of oxygen in oil than in water, which led to a preferential distribution into the oil phase of the NEs. In the case of SLNs and NLCs, their solid lipid content did not allow oxygen diffusion into the carrier easily, thus keeping the entrapped AP more protected. Apart the stability, the degree of skin moisturizing and penetration of AP entrapped in

SLNs, NLCs, and NEs incorporated into hydrogel was significantly higher as compared to that of NEs [65].

6.2.1.3 Vitamin E derivatives

Alpha-Tocopherol, the major biologically active form of vitamin E, is a frequently added lipophilic compound of cosmetic and dermopharmaceutical products as a potent antioxidant, hence able to reduce skin damage caused by UV radiation. The UV-blocking capacity of vitamin E mainly in the UV-B region is already known [66]. However, owing to stability problems, tocopherol acetate (TA) is usually used instead. In turn, it has been reported that SLNs exhibit a pigment-like whitening effect, as is given by titanium dioxide, thus providing characteristics of physical UV-blockers on their own. This opens the possibility of developing a more effective sunscreen system with reduced side effects. Such an approach was investigated by Wissing *et al.* [67]. According to this study, the incorporation of the chemical sunscreen TA into SLNs prevented chemical degradation and increased the UV-blocking capacity at least twice in comparison to the respective reference emulsions (conventional emulsions with identical lipid content). Placebo SLNs even exhibited greater UV-blocking efficacy than did emulsions containing tocopherol acetate as the molecular sunscreen. The subsequent inclusion of TA into SLNs led to a synergistic UV-blocking effect. These data highlight the potential of SLNs as basis of an improved carrier system for potentially harmful sunscreens [67].

6.2.2 Carotenoids

Carotenoids are common natural lipophilic pigments, derived from a 40-carbon basal structure, which includes a system of conjugated double bonds. This pattern of conjugated double bonds in the polyene backbone allows them to extract or donate electrons to or from suitable coreactants, leading to free radicals anions and cations, which in turn can react with oxygen and other molecules [68]. Thus, they behave as efficient antioxidants scavenging singlet molecular oxygen and peroxyl radicals, also interacting in a synergistic manner with other antioxidants [69]. Carotenoids can be grouped into provitamin A carotenoids (those that can be converted to vitamin A in the human body) and the non-provitamin A carotenoids (those that cannot be converted to vitamin A in the human body) [70].

6.2.2.1 Beta-carotene

Beta-carotene is the most prominent provitamin A carotenoid. The protective effects of this carotenoid against UV and infrared radiation have been reported in several *ex vivo* and *in vitro* studies, thus preventing skin photooxidative damage [71, 72]. There is evidence that β-carotene regenerates tocopherol from the tocopheroxyl radical, being the resulting carotenoid radical cation further repaired by vitamin C. Thus, vitamins E, C and β-carotene exhibit antioxidant cooperative effects [73]. Despite the wide range of pharmacological effects, the dermatological application of β-carotene is still limited due to its photochemical instability. In fact, its chemical structure based on a conjugated double-bond system prompt it to oxidative cleavage and isomerization by oxygen, light and heat [74].

Hence, the development of formulations that improve the chemical stability of β-carotene and present an increased penetration potential through the stratum corneum is deemed necessary. Trombino *et al.* employed stearyl ferulate-SLNs (SF-SLNs) as vehicle for both β-carotene and α-tocopherol, as a strategy to increase the photochemical stability of these compounds [75]. In this study, ferulic acid, a potent antioxidant on its own, was coupled to stearyl alcohol to synthesize stearyl ferulate, since it exhibits synergistic effects with other antioxidant systems, being also able to protect and stabilize them from degradation. The SF-SLNs obtained were characterized for entrapment efficiency, size, shape, stability, antioxidant and cytotoxicity activity, and compared to traditional stearic acid-SLNs (SA-SLNs). The pro-oxidant and/or exposition to sunlight indicated that SF-SLNs are a good vehicle for β-carotene and α-tocopherol, as they were able to stabilize them, thus preventing the oxidation and the degradation of both compounds. SF-SLNs displayed better entrapment efficiency than the corresponding SA-SLNs, which enabled them to ensure a higher driving force for drug release and penetration. Although cytotoxicity studies have suggested that SF-SLNs tolerability is lower than SA-SLNs, even at a low micromolar concentration they are still considered suitable as a dermatological formulation [75].

Hung *et al.* employed NLCs to incorporate palm oil, which is composed by carotenoids, including 13 different types of carotenes, among others, α- and β-carotenes, being also a rich source of tocols (vitamin E) [76]. In this study, two modified hot-high pressure homogenization NLC production methods were developed by minimizing the working high production temperatures usually employed: 60°C was used instead the previous 85°C, for method A and 60°C with admixture of surfactants in the lipid phase, for method B. Reducing the working temperature minimizes not only the degradation of labile actives, but also improve the emulsifying capability of surfactants exhibiting cloud points lower than 85°C. Thus, these two methods allowed enhancing the chemical stability of these heat-sensitive bioactive compounds and the physical stability of NLCs [76].

6.2.2.2 Lutein

Lutein and its isomer zeaxanthin are non-provitamin A carotenoids, also known as xanthophylls (contain oxygenated substituents in their chemical structure). Since they are not endogenously synthesized, they must be provided by the diet. In nature, they are found in high concentrations in green leafy vegetables (e.g., spinach and kale) and egg yolk. In human body, lutein and zeaxanthin are highly concentrated in the macula lutea of human eye and also in the skin as components of the antioxidant system. They have been shown to provide protection against tissue damage (skin swelling and hyperplasia) caused by UV light, due to their blue light filter properties, known to induce photo-oxidative damage by ROS generating [70, 77, 78]. Given the relationship established between UV exposure and skin cancer, a protective role against skin cancer, either as a filter of blue light or as a ROS scavenger seems also to be called to these carotenoids [78]. The efficacy of lutein and zeaxanthin topical application was demonstrated in a clinical trial [79]. However, as many other carotenoids their conjugated double bond isoprenoid structure dictates a poor chemical stability [80]. To overcome this issue, SLNs and NLCs encapsulating lutein formulations were prepared and characterized in what concerns particle size, zeta potential, performance (release and penetration behavior), physical and photo-stability [81]. The corresponding NEs and free drug formulations were used as references. The nanocarriers were able to protect lutein against UV degradation. Only 0.06% and 6–8% degradation was observed after irradiation with 10 Minimal Erythema Dose (MED) in SLNs and NLCs, respectively, as compared to 14% in the NEs, and to 50% as lutein powder suspended in corn oil. Moreover, SLNs and NLCs exhibited suitable physicochemical properties, with release and permeation patterns compatible with a topical and prolonged administration [81].

6.2.3 Co-factors

6.2.3.1 Coenzyme Q10

Coenzyme Q10 (CoQ10), also called ubiquinone-10, due to its ubiquitous presence in nature, is a biological cofactor in the mitochondrial respiratory chain. Much attention has been devoted to the potential beneficial effects of CoQ10. As an essential electron carrier in cellular respiration during oxidative phosphorylation and ATP synthesis, it acts as a chain breaking antioxidant, clearing ROS and protecting cells from oxidative stress [82, 83]. When topically applied, CoQ10 has also been shown to prevent many of the detrimental effects of photoaging and, therefore, causing a decrease in wrinkle depth [84]. These effects could be related to the ability of CoQ10 to induce production of basal membrane components, fibroblast proliferation and to decrease oxidative damage caused by lipid peroxidation [85]. Considering its chemical structure (a tyrosine-derived quinone ring linked to a long side chain containing ten isoprenoid units), CoQ10 has been barely satisfactory in topical drug delivery because of its lipid solubility. It means that systems that can deliver the antioxidant more efficiently into the skin are required. In this context, many have been the literature reports evaluating the potential of lipid nanoparticles as topical delivery carriers to improve its antioxidative efficiency [82, 86–90]. According to these studies, NLCs loaded with CoQ10 with a size ranging from 100 nm to 250 nm improved its skin penetration *in vitro* and *in vivo* compared to non-particulate formulations as reference [82, 86, 89]. The amount of CoQ10 released and the occlusiveness ensured by NLCs were considered major key factors to promote deep penetration into the skin [89]. In addition, when incorporated into NLCs, it significantly enhanced the antioxidative capacity of fibroblasts in which oxidative stress was induced by UVA irradiation. Chemical stability of Q10 entrapped into NLCs was also reported as higher than that in equivalent NE systems and dependent on the amount of oil content in NLCs, i.e., the higher oil content, the lower percent Q10 remaining. Rheological properties of a commercial cream based on CoQ10-NLCs (Cutanova Nanorepair Q10) indicated a better subjective impression in terms of consistency and spreadability on the skin than that found for conventional o/w emulsion as control cream, which makes NLC-based formulations more attractive from a user point of view. All these features showed how NLCs and the latest ultra-small lipid nanoparticles (usNLCs) are beneficial for the dermal delivery of CoQ10.

Recently, a further development of the previously analyzed NLC systems yielded a new (third) generation of nanocarrier systems, the usNLCs [9, 91, 92]. These nanosystems take advantage from the decrease in particle size (around 80 nm), as a result of an oil-enrichment of the lipid phase. In practice, this relies on the use of liquid/solid lipid ratios of 75/25 and the determination of the required hydrophilic lipophilic balance (HLB). Therefore, an improvement in the dermal delivery of CoQ10 by usNLCs over conventional, larger NLCs was achieved. Higher amounts of CoQ10 in deeper skin layers were detected, as well as higher total amount penetrated according to *in vitro* diffusion studies. The penetration/permeation from the usNLCs was even superior to identically sized NE systems, demonstrating that rather than particle size, the influence of the physical state of the particle matrix (solid vs. liquid) and the associated properties (film formation properties, occlusive effect, hydration state, chemical composition of lipid and drug release profile from matrix) is crucial for their performance [92]. Moreover, HaCaT uptake studies indicated that usNLC-CoQ10 presented the strongest reduction of the radical formation, compared to classical nanocarriers (NLCs and NEs), reaching up to 23% (taken the control cells without nanocarrier treatment as reference) [91]. A flip–flop core–shell structure of the lipid matrix, i.e., the oil with dissolved active surrounding the solid lipid based core, thus creating a higher gradient concentration of CoQ10, was suggested as possible reason for their efficacy [9].

6.2.3.2 Idebenone

Idebenone (IDE), a synthetic analogue of coenzyme Q10 with a shorter carbon side chain, is a free radical quencher, able to counteract skin damage and aging due to ROS overproduction. As an antioxidant, the maintenance of its chemical stability is a requirement to warrant proper antioxidant action. However, IDB presents poor stability, being described its degradation by 60% and lose of antioxidant activity by 30% when exposed to a relative humidity of 75% and at 40°C during 45 days [93, 94]. Therefore, a formulation that can increase the chemical stability of IDE needs to be developed. A strategy to decrease this undesired effect could be incorporation of the active into lipid nanoparticles. Li *et al.* developed a carrier system for IDB based on NLCs and compared the chemical stability of IDB and the respective skin delivery against other two formulations, an IDB-encapsulated nanoemulsion and an IDB-dissolved oil solution. NLCs were found to achieve a significant improvement concerning the chemical stability of IDB, skin permeation, and formulation stability when compared to NEs and oil solution [93]. Montenegro *et al.* assessed the feasibility of targeting IDE into the upper layers of the skin by topical application of IDE-loaded SLNs prepared by the phase inversion temperature (PIT) method. Irrespectively of the loading capacity and stabilizer used, no IDE permeation occurred, conversely to penetration behavior. IDE loaded into SLNs mainly accumulated in the upper skin layers instead, i.e., stratum corneum and epidermis [95, 96].

A third study carried out by Salunkhe *et al.* aimed at developing an IDB-NLC formulation and assess its sun protection efficacy. An improvement of the antioxidant effect promoted by the entrapment of IDB into NLCs was observed, along with a high significance ($p < 0.001$) of drug deposition in skin and enhanced sun protection factor (SPF), in comparison to plain IDB gel. Such results support their use as anti-skin aging topical system [97].

6.2.3.3 Alpha-lipoic acid

Alpha-lipoic acid (LA, 5-(1,2-dithiolan-3-yl)pentanoic acid) is a co-factor in the citric acid cycle and a well-known potent natural antioxidant found in certain food products (e.g., red meat, spinach, broccoli, among others). It can be also produced in the human body, but only in very small amounts. Therefore, it needs to be absorbed from exogenous sources. LA has been linked to skin care as a primary antioxidant that prevents aging and inflammation, since it is able to directly scavenge ROS and possesses metal chelating activity [98]. Despite its promising properties, problems related to its poor water solubility, stability (LA polymerizes when heated above its melting point, 47.5°C, or under the influence of light) and consequent strong sulfur smell remains a major challenge [98, 99]. Souto *et al.* investigated the incorporation of LA into protective lipid nanoparticles [100]. According to this study, a topical SLNs (20% lipid content) formulation containing 1% of LA, a sufficiently high concentration to ensure a possible therapeutic effect, was achieved. More recently, Ruktanonchai *et al.* developed water-soluble formulations also with 1% of LA loading using SLNs, NLCs, and NEs (9% total lipid content) and characterized them in terms of physical and biological properties. A sustained release of LA was observed only for SLNs and NLCs. The less-ordered structure of the lipid matrix of NLC formulations resulted in unchanged EE apart from a delayed percentage recrystallization index over 90 days of storage. In what concerns the biological performance, the antioxidant activity of LA-loaded formulations was similar to magnitude of pure LA. Both SLNs and NLCs were considered adequate alternative carriers to administer LA topically [101].

6.2.4 Polyphenols

6.2.4.1 Flavonoids

Flavonoids are natural antioxidants derived from the plant kingdom, being widely present in the human diet in the form of numerous edible fruits and vegetables, such as apples, berries, red grapes, and onions. This opens novel possibilities for the treatment and prevention of oxidative stress-mediated skin diseases [102]. The mechanism of antioxidant action of flavonoids has been attributed to their ability to scavenge reactive oxygen species, to chelate transition metal ions involved in oxygen activation via redox reactions and to the synergistic effects with other antioxidants [103].

6.2.4.1.1 Quercetin

Quercetin (QC), chemically described as 3,3′,4′,5,7-pentahydroxyflavone, is a natural flavonoid, belonging to the subclass of flavonols. A wide range of pharmacological activities related to the antioxidant systems of the skin have been addressed to QC. It encompasses the inhibition of lipid peroxidation, complexation of redox-active transition metal ions to form inert chelate complexes and the scavenging of oxygen radicals, also ensuring protection against the UVB-induced oxidative skin damage [104–107]. Such a high antiradical/antioxidant activity exhibited by QC has been ascribed to the presence of three active functional groups in its chemical structure: the ortho-dihydroxy moiety in the B ring, the C_2–C_3 double bond in conjugation with a 4-oxo function and the hydroxyl substitution at positions 3, 5 and 7 [108].

Thus, QC gathers a set of beneficial therapeutic effects along with a safety profile, which make it a very promising candidate for incorporation into topical formulations [109]. Additionally, it is also extensively metabolized by the gut microorganisms when administered orally, which reduces its absorption from the gastrointestinal tract and bioavailability [110, 111]. However, despite it presents a suitable log P (log P = 1.82 ± 0.32 [112]) for cutaneous administration, its low aqueous solubility (0.55 μM [113]) limits its penetration to deeper skin layers [114]. One of the possibilities to enhance QC transport to/through the skin relies on the use of a carrier system, such as LN. This strategy was employed in a research work carried out by Bose *et al.* [109, 115]. These authors have previously optimized the composition and production variables of a SLN formulation based on Compritol® 888 and a mixture of Tween 20/dioctyl sodium sulfosuccinate, manufactured via probe sonication. The *in vitro* release studies showed an initial burst release followed by prolonged release for up to 24 h. In turn, higher amounts of QC were localized within the skin, when compared to a control formulation containing particles in the micrometer range, thus demonstrating the feasibility of topical delivery of QC via a solvent-free solid lipid based nanosystem [115]. In a second study carried out by the same authors, the impact of other formulation factors, such as the inclusion of a liquid lipid (i.e., NLCs versus SLNs), and drug loading in the formulation previously optimized were investigated.

No statistically significant differences ($p > 0.05$) were observed between the SLN and NLC formulations loaded with 0.05% of QC for particle size (D50 and D90 values) at the initial time point and after 8 weeks at 2–8°C, also presenting similar zeta potential values. Release studies corroborated the previous studies indicating a biphasic release profile, characterized by an initial burst release followed by a more controlled release pattern from both SLN and NLC systems. In terms of *in vitro* permeation studies, QC was not detected in the receptor media, which discards a transdermal delivery. This was not considered a critical aspect, since the efficacy of QC in delaying UV radiation-mediated damage occurs at a topical level. NLCs exhibited the highest improvement in topical delivery of QC manifested by the amount of quercetin retained in full thickness human skin, compared to the control formulation (QC non-encapsulated). This second study demonstrated that NLCs could be considered a better carrier system for improving topical delivery of QC, being able to deliver an adequate amount to achieve the desired pharmacological action [109]. These results were corroborated by Guo *et al.* [116].

6.2.4.1.2 Genistein

Genistein (GN) is a soy-derived isoflavone and phytoestrogen with antineoplastic, antioxidant, and antiproliferative activity [117, 118]. GN substantially inhibits skin carcinogenesis and cutaneous aging induced by UV light in mice, and photodamage in humans. The mechanisms of action involve protection of oxidative and photodynamically damaged DNA, downregulation of UVB-activated signal transduction cascades, and antioxidant activities [119, 120]. Therefore, the topical administration of GN may be a suitable route for the prevention and treatment of pre-malignant skin lesions. In addition to this, topical administration of GN may also be an alternative route to attain systemic bioavailability, as a result of its rapid clearance from plasma [121]. Previous studies have reported the feasibility of administration of GN to/through the skin [122–124]. However, some difficulty in permeating the skin when administered in conventional dosage forms has been denoted. This called for the need to improve GN skin penetration employing, for example, lipid nanoparticles. Andrade *et al.* studied the effect of the incorporation of GN into SLNs and NLCs and their ability to deliver it to deeper skin layers [125]. Both nanocarrier systems allowed a sustained release of GN, with NLCs, although more flexible (as suggested by the electron paramagnetic resonance measurements), has led to a slower profile. On the other hand, permeation studies showed that the encapsulation of GN in lipid nanoparticles significantly increased drug retention in different skin layers, when compared to unloaded GN, which was not quantifiable. Again, similar amounts of GN were retained in the SC when SLNs and NLCs were topically applied. However, NLCs increased the retention of GN 1.7-fold in the remaining skin layers, which is highly desirable to ensure photoprotection and the antioxidant activity. Thus, the targeting effect promoted by NLCs by delivering GN into deeper skin layers highlights its role as an enhancement drug penetration strategy [125].

6.2.4.2 Phenolic acid derivatives

6.2.4.2.1 Gamma-oryzanol

Gamma-oryzanol (γ-O), a phytosteryl ferulate mixture extracted from rice bran oil, has a wide spectrum of biological activities, among others, its antioxidant properties either with a therapeutic or a technological function purpose. It is often used in cosmetic formulations not only as a natural antioxidant to improve the stability, but also as a natural UV-A filter in sunscreens [126, 127]. However, the utility of most of these preparations has been compromised by the poor water solubility of γ-O. Ruktanonchai *et al.* studied the effects of the incorporation of γ-O into three different types of solid lipid: wax, triglycerides and a mixture of glycerides as a SLN system, and compared against the a liquid lipid (Miglyol® 812) as a nanoemulsion [128]. Comparisons were established in terms of mean particle diameter, zeta potential, particle morphology, entrapment efficiency (EE), *in vitro* release profile and physicochemical stability during long-term storage. Moreover, their cytotoxicity and biological activity in terms of antioxidation were still investigated. SLNs demonstrated good physical stability presenting an EE comprised between 40 and 60%, whereas a significant increase in size and decreased EE was found with NEs after 1 month of storage at 45°C. The solid lipid type chosen to produce SLNs played an important role in terms of EE and recrystallization index (RI), not impacting though their chemical stability. In particular, a decrease in the crystallinity of SLNs was observed after incorporating the γ-O and low RI were found with two glycerides-based SLNs. *In vitro* release studies revealed biphasic release patterns in the case of SLNs, whereas a fast and much higher release from NEs was observed. This supports the ability of this type of carriers to promote a sustained drug release. The lowest cytotoxicity was observed for the wax-based SLNs. From an efficacy point of view, NEs, wax-based SLNs, and a mixture of glycerides-based SLNs were considered to enhance the antioxidant activity of γ-O [128]. An overall analysis highlights the better performance obtained from SLNs.

6.2.4.3 Other polyphenols

6.2.4.3.1 Resveratrol

Resveratrol (RV), chemically known as 3,5,4′-trihydroxystilbene, is a naturally occurring polyphenol and a phytoalexin found in grapes, nuts, fruits and red wine. Numerous pleiotropic health beneficial effects, including antioxidant, anti-inflammatory, cardioprotective and antitumor activities have been ascribed to RV [129]. Nevertheless, it must been taken into consideration that it exists as both *cis* and *trans* isomers, although the most abundant and biologically active form is attributed to the *trans*-isomer [130]. Besides that, it is a highly photosensitive compound susceptible to UV-induced isomerization, being the *cis*-form easily obtained after 1 h exposure to light of a *trans*-form solution, which clearly reduces its efficacy and limits its topical administration [131, 132]. One of the strategies to increase the stability of the drug consists in the use of solid lipid nanoparticles. Such an approach was addressed in a study carried out by Carlotti *et al.* [133], which assessed the UVA-induced photodegradation of both RV free and loaded formulations, in the absence or presence of tetradecyl-γ-cyclodextrin. It was found that the loading of RV in SLNs significantly reduced the RV photodegradation, which became even more pronounced when the tetradecyl-γ-cyclodextrin was included. In addition to this, the presence of the cyclodextrin improved the nanoparticle properties in terms of reducing particle size and respective distribution, turning zeta potential values more negative and slightly increasing the entrapment efficiency. On the other hand, a notable *in vitro* porcine skin accumulation and an increased antioxidative efficacy were observed by entrapping RV in SLNs, as elucidated by skin uptake and antilipoperoxidative activity studies [133]. The evidence for SLN-mediated cell uptake of RV and with potential benefits for prevention of skin cancer has also been reported in a cell culture study [134].

In turn, the performance of the nanocarrier type, i.e., SLNs and NLCs, for dermal application of RV was investigated by Gokce *et al.* [135]. According to this study, several nanocarrier systems were developed and the optimal formulations evaluated both in terms of cytotoxicity and antioxidant capacity (by determining the amount of ROS produced in fibroblasts) and the accumulation of RV in the different layers of skin. The results indicated a higher particle size for SLNs (287.2 nm ± 5.1) in comparison with NLCs (110.5 nm ± 1.3), along with a higher RV entrapment efficiency for the latter and similar zeta potential values. Both RV-loaded SLN and NLC systems were characterized by high biocompatibility, showing normal human dermal fibroblasts at concentrations below 50 μM of RV, considered effective in decreasing ROS accumulation. However, NLC was characterized by a better performance, indicating again that NLCs had a noticeably higher tendency to penetrate into deeper layers of skin with respect to SLNs. Such a behavior was ascribed to the smaller size, which possibly favoring particle and cell endocytosis [135].

6.2.4.3.2 Curcuminoids

Curcuminoids (CM), comprising curcumin, desmethoxycurcumin, and bisdesmethoxycurcumin as major phenolic compounds, are powerful bioactive agents and natural antioxidants obtained from the rhizome of *Curcuma longa* [136]. They are able to prevent against oxidative damage mediated by hydrogen peroxide [137], playing therefore important functions in retarding skin aging. However, despite their several utilities and low toxicity, these molecules present poor bioactivity and instability under neutral and alkaline aqueous solutions, being also susceptible to degradation by oxidation and light exposition [138–140]. A possible solution to overcome this obstacle would be the development of lipid nanoparticle based formulations to improve their stability. CM-loaded SLN included in a conventional cream was successfully prepared through a microemulsion technique at moderate temperature by Tiyaboonchai *et al.* [141]. In this study, the primary goal was to characterize the processing factors affecting the characteristics of the CM-SLNs, such as the optimal conditions for their preparation, as well as their stability during storage. It was found that after 6-month storage at room temperature in the absence of sunlight, the physicochemical stability of the lyophilized CM-loaded SLNs was maintained, since no significant changes ($p > 0.05$) in the mean particle size were observed, when compared to the freshly prepared formulation. Moreover, the chemical stability of CM incorporated into SLNs, inspected by dispersing them into a model cream base, revealed that after storage in the absence of sunlight for 6 months, the percentages of the remaining curcumin, bisdemethoxycurcumin and desmethoxycurcumin were 91, 96, and 88, respectively. In addition, the light and oxygen sensitivity of CM was strongly reduced by incorporating them into SLNs [141]. These results were corroborated by a recent study wherein SLNs, with a different composition and obtained from the hot melting homogenization, were used as a carrier system, thus demonstrating that the stability of CM can be improved when incorporated into SLNs [142].

6.3 Conclusions

Cutaneous anti-oxidants play a pivotal role in the defense mechanisms against ROS induced skin disorders. Topical delivery of different antioxidants is deemed necessary, especially when external supplementation by oral administration is compromised, either because physicochemical or pharmacokinetic reasons. Due to their positive feature, i.e., increased AO penetration and stability, re-enforcement of the lipid barrier, increase in skin hydration, and controlled release and biocompatibility properties, lipid nanoparticles, in particular, NLCs come across as appealing carrier systems. At present, more than 40 NLC products are available on the cosmetic market, and NLCs for pharmaceutical use are already in development, which clearly highlights the potential of these nanosystems for AO topical delivery.

A refinement to further improve and optimize cell/tissue penetration properties of AOs spurred the development of a recent generation of carriers, the usNLCs, with very promising results hitherto. The use of the later could be challenging to mediate the transport through the skin of antioxidant enzymes, including SOD and CAT. These special active proteins have already been employed in many studies as antioxidant therapeutic agents, but with restricted success related to its limited skin penetration. Thus, this class of AOs could be the next one to benefit from the advantages of lipid nanoparticles in a near future.

Acknowledgements

The authors acknowledge the Fundação para a Ciência e a Tecnologia (FCT), for the financial support through projects Pest-UID/DTP/04138/2013 to iMed.ULisboa, PEst-UID/NEU/04539/2013 to CNC, COMPETE (POCI-01-0145-FEDER-007440) and research project PTDC/CTM-NAN/2658/2014.

References

1. Cork MJ, Danby SG, Vasilopoulos Y, Hadgraft J, Lane ME, Moustafa M, *et al.* Epidermal barrier dysfunction in atopic dermatitis. *J Invest Dermatol.* 2009; 129: 1892–1908.
2. Madison KC. Barrier function of the skin: "la raison d'etre" of the epidermis. *J Invest Dermatol.* 2003; 121: 231–241.
3. Raposo SC, Simoes SD, Almeida AJ, Ribeiro HM. Advanced systems for glucocorticoids' dermal delivery. *Expert Opin Drug Deliv.* 2013; 10: 857–877.
4. Trommer H, Neubert RHH. Overcoming the stratum corneum: the modulation of skin penetration. *Skin Pharmacol Physiol.* 2006; 19(2): 106–121.
5. Patzelt A, Richter H, Knorr F, Schafer U, Lehr CM, Dahne L, *et al.* Selective follicular targeting by modification of the particle sizes. *J Control Release.* 2011; 150: 45–48.
6. Pardeike J, Hommoss A, Müller RH. Lipid nanoparticles (SLN, NLC) in cosmetic and pharmaceutical dermal products. *Int J Pharm.* 2009; 366: 170–184.
7. Müller RH, Radtke M, Wissing SA. Solid lipid nanoparticles (SLN) and nanostructured lipid carriers (NLC) in cosmetic and dermatological preparations. *Adv Drug Deliv Rev.* 2002; 54: S131–S55.

8. Müller, RH, Rimpler, C, Petersen, R, Hommoss, A, Schwabe, K. A new dimension in cosmetic products by nanostructured lipid carriers (NLC) technology. *Eurocosmetics*. 2007; 3: 2–7.

9. Keck CM, Baisaeng N, Durand P, Prost M, Meinke MC, Müller RH. Oil-enriched, ultra-small nanostructured lipid carriers (usNLC): A novel delivery system based on flip–flop structure. *Int J Pharm*. 2014; 477: 227–235.

10. Vitorino C, Almeida J, Gonçalves LM, Almeida AJ, Sousa JJ, Pais AACC. Co-encapsulating nanostructured lipid carriers for transdermal application: From experimental design to the molecular detail. *J Control Release*. 2013; 167: 301–314.

11. Souto EB, Müller RH, Alemieda AJ. Topical delivery of oily actives by means of solid lipid particles. *Pharm Tech Eur*. 2007; 19: 28–32.

12. Souto EB, Almeida AJ, Müller RH. Lipid nanoparticles (SLN®, NLC®) for cutaneous drug delivery: structure, protection and skin effects. *J Biomed Nanotechnol*. 2007; 3: 317–331.

13. Wissing SA, Muller RH. Solid lipid nanoparticles (SLN)—a novel carrier for UV blockers. *Pharmazie*. 2001; 56: 783–786.

14. Xia Q, Saupe A, Muller RH, Souto EB. Nanostructured lipid carriers as novel carrier for sunscreen formulations. *Int J Cosmet Sci*. 2007; 29: 473–482.

15. Sivaramakrishnan R, Nakamura C, Mehnert W, Korting HC, Kramer KD, Schafer-Korting M. Glucocorticoid entrapment into lipid carriers—characterisation by parelectric spectroscopy and influence on dermal uptake. *J Control Release*. 2004; 97: 493–502.

16. Zhai Y, Zhai G. Advances in lipid-based colloid systems as drug carrier for topic delivery. *J Control Release*. 2014; 193: 90–99.

17. Teeranachaideekul V, Boonme P, Souto EB, Müller RH, Junyaprasert VB. Influence of oil content on physicochemical properties and skin distribution of Nile red-loaded NLC. *J Control Release*. 2008; 128: 134–141.

18. Vitorino C, Almeida A, Sousa J, Lamarche I, Gobin P, Marchand S, *et al.* Passive and active strategies for transdermal delivery using co-encapsulating nanostructured lipid carriers: *In vitro* vs. *in vivo* studies. *Eur J Pharm Biopharm*. 2014; 86: 133–144.

19. Frum Y, Eccleston GM, Meidan VM. Factors influencing hydrocortisone permeation into human hair follicles: Use of the skin sandwich system. *Int J Pharm*. 2008; 358: 144–150.

20. Aljuffali IA, Sung CT, Shen FM, Huang CT, Fang JY. Squarticles as a lipid nanocarrier for delivering diphencyprone and minoxidil to hair follicles and human dermal papilla cells. *Aaps J*. 2014; 16: 140–150.

21. Uprit S, Kumar Sahu R, Roy A, Pare A. Preparation and characterization of minoxidil loaded nanostructured lipid carrier gel for effective treatment of alopecia. *Saudi Pharm J*. 2013; 21: 379–385.

22. Shamma RN, Aburahma MH. Follicular delivery of spironolactone via nanostructured lipid carriers for management of alopecia. *Int J Nanomed*. 2014; 9: 5449–5460.

23. Agrawal Y, Petkar KC, Sawant KK. Development, evaluation and clinical studies of Acitretin loaded nanostructured lipid carriers for topical treatment of psoriasis. *Int J Pharm*. 2010; 401: 93–102.

24. Charoenputtakun P, Pamornpathomkul B, Opanasopit P, Rojanarata T, Ngawhirunpat T. Terpene composited lipid nanoparticles for enhanced dermal delivery of all-trans-retinoic acids. *Biol Pharm Bull*. 2014; 37: 1139–1148.

25. Nnamani PO, Hansen S, Windbergs M, Lehr CM. Development of artemether-loaded nanostructured lipid carrier (NLC) formulation for topical application. *Int J Pharm*. 2014; 477: 208–217.

26. Puglia C, Sarpietro MG, Bonina F, Castelli F, Zammataro M, Chiechio S. Development, characterization, and *in vitro* and *in vivo* evaluation of benzocaine- and lidocaine-loaded nanostructrured lipid carriers. *J Pharm Sci*. 2011; 100: 1892–1899.

27. Desai PR, Shah PP, Patlolla RR, Singh M. Dermal microdialysis technique to evaluate the trafficking of surface modified lipid nanoparticles upon topical application. *Pharm Res*. 2012; 29: 2587–2600.

28. Joshi M, Patravale V. Nanostructured lipid carrier (NLC) based gel of celecoxib. *Int J Pharm*. 2008; 346: 124–132.

29. Souto EB, Wissing SA, Barbosa CM, Müller RH. Development of a controlled release formulation based on SLN and NLC for topical clotrimazole delivery. *Int J Pharm*. 2004; 278: 71–77.

30. Stecova J, Mehnert W, Blaschke T, Kleuser B, Sivaramakrishnan R, Zouboulis CC, *et al.* Cyproterone acetate loading to lipid nanoparticles for topical acne treatment: Particle characterisation and skin uptake. *Pharm Res*. 2007; 24: 991–1000.

31. Keshri L, Pathak K. Development of thermodynamically stable nanostructured lipid carrier system using central composite design for zero order permeation of econazole nitrate through epidermis. *Pharm Dev Technol*. 2013; 18: 634–644.

32. Gainza G, Pastor M, Aguirre JJ, Villullas S, Pedraz JL, Hernandez RM, *et al.* A novel strategy for the treatment of chronic wounds based on the topical administration of rhEGF-loaded lipid nanoparticles: *In vitro* bioactivity and *in vivo* effectiveness in healing-impaired db/db mice. *J Control Release.* 2014; 185: 51–61.
33. Gainza G, Bonafonte DC, Moreno B, Aguirre JJ, Gutierrez FB, Villullas S, *et al.* The topical administration of rhEGF-loaded nanostructured lipid carriers (rhEGF-NLC) improves healing in a porcine full-thickness excisional wound model. *J Control Release.* 2015; 197: 41–47.
34. Gomes MJ, Martins S, Ferreira D, Segundo MA, Reis S. Lipid nanoparticles for topical and transdermal application for alopecia treatment: Development, physicochemical characterization, and *in vitro* release and penetration studies. *Int J Nanomed.* 2014; 9: 1231–1242.
35. Gupta M, Vyas SP. Development, characterization and *in vivo* assessment of effective lipidic nanoparticles for dermal delivery of fluconazole against cutaneous candidiasis. *Chem Phys Lipids.* 2012; 165: 454–461.
36. Schwarz JC, Weixelbaum A, Pagitsch E, Low M, Resch GP, Valenta C. Nanocarriers for dermal drug delivery: Influence of preparation method, carrier type and rheological properties. *Int J Pharm.* 2012; 437: 83–88.
37. Doktorovova S, Araujo J, Garcia ML, Rakovsky E, Souto EB. Formulating fluticasone propionate in novel PEG-containing nanostructured lipid carriers (PEG-NLC). *Colloids Surf B Biointerfaces.* 2010; 75: 538–542.
38. Ricci M, Puglia C, Bonina F, Di Giovanni C, Giovagnoli S, Rossi C. Evaluation of indomethacin percutaneous absorption from nanostructured lipid carriers (NLC): *In vitro* and *in vivo* studies. *J Pharm Sci.* 2005; 94: 1149–1159.
39. Souto EB, Muller RH. SLN and NLC for topical delivery of ketoconazole. *J Microencapsul.* 2005; 22: 501–510.
40. Üner M, Wissing SA, Yener G, Müller RH. Solid lipid nanoparticles (SLN) and nanostructured lipid carriers (NLC) for application of ascorbyl palmitate. *Die Pharmazie—An Int J Pharm Sci.* 2005; 60: 577–582.
41. Pinto MF, Moura CC, Nunes C, Segundo MA, Costa Lima SA, Reis S. A new topical formulation for psoriasis: Development of methotrexate-loaded nanostructured lipid carriers. *Int J Pharm.* 2014; 477: 519–526.

42. Silva AC, Santos D, Ferreira DC, Souto EB. Minoxidil-loaded nanostructured lipid carriers (NLC): Characterization and rheological behaviour of topical formulations. *Pharmazie*. 2009; 64: 177–182.

43. Sanad RA, AbdelMalak NS, elBayoomy TS, Badawi AA. Formulation of a novel oxybenzone-loaded nanostructured lipid carriers (NLCs). *AAPS PharmSciTech*. 2010; 11: 1684–1694.

44. Joshi M, Patravale V. Formulation and evaluation of nanostructured lipid carrier (NLC)–based gel of valdecoxib. *Drug Dev Ind Pharm*. 2006; 32: 911–918.

45. Vitorino C, Alves L, Antunes FE, Sousa JJ, Pais AACC. Design of a dual nanostructured lipid carrier formulation based on physicochemical, rheological, and mechanical properties. *J Nanopart Res*. 2013; 15.

46. Harman D. Aging: A theory based on free radical and radiation chemistry. *J Gerontol*. 1956; 11: 298–300.

47. Beckman KB, Ames BN. The free radical theory of aging matures. *Physiol Rev*. 1998; 78: 547–581.

48. Sies H. Oxidative stress: Oxidants and antioxidants. *Exp Physiol*. 1997; 82: 291–295.

49. Ratnam DV, Ankola DD, Bhardwaj V, Sahana DK, Kumar MN. Role of antioxidants in prophylaxis and therapy: A pharmaceutical perspective. *J Control Release*. 2006; 113: 189–207.

50. Beckman KB, Ames BN. The free radical theory of aging matures. *Physiol Rev*. 1998; 78: 547–581.

51. Mukherjee S, Date A, Patravale V, Korting HC, Roeder A, Weindl G. Retinoids in the treatment of skin aging: An overview of clinical efficacy and safety. *Clin Int Aging*. 2006; 1: 327–348.

52. Oh YK, Kim MY, Shin JY, Kim TW, Yun MO, Yang SJ, *et al.* Skin permeation of retinol in Tween 20-based deformable liposomes: In-vitro evaluation in human skin and keratinocyte models. *J Pharm Pharmacol*. 2006; 58: 161–166.

53. Jenning V, Schäfer-Korting M, Gohla S. Vitamin A-loaded solid lipid nanoparticles for topical use: Drug release properties. *J Control Release*. 2000; 66: 115–126.

54. Jenning V, Gysler A, Schäfer-Korting M, Gohla SH. Vitamin A loaded solid lipid nanoparticles for topical use: Occlusive properties and drug targeting to the upper skin. *Eur J Pharm Biopharm*. 2000; 49: 211–218.

55. Jeon HS, Seo JE, Kim MS, Kang MH, Oh DH, Jeon SO, *et al.* A retinyl palmitate-loaded solid lipid nanoparticle system: Effect of surface modification with dicetyl phosphate on skin permeation *in vitro* and anti-wrinkle effect *in vivo*. *Int J Pharm.* 2013; 452: 311–320.

56. Jee J-P, Lim S-J, Park J-S, Kim C-K. Stabilization of all-trans retinol by loading lipophilic antioxidants in solid lipid nanoparticles. *Eur J Pharm Biopharm.* 2006; 63: 134–139.

57. Lim S-J, Kim C-K. Formulation parameters determining the physicochemical characteristics of solid lipid nanoparticles loaded with all-trans retinoic acid. *Eur J Pharm Biopharm.* 2002; 243: 135–146.

58. Moribe K, Limwikrant W, Higashi K, Yamamoto K. Drug nanoparticle formulation using ascorbic acid derivatives. *J Drug Deliv.* 2011; 2011: 138929.

59. Gonullu U, Yener G, Uner M, Incegul T. Moisturizing potentials of ascorbyl palmitate and calcium ascorbate in various topical formulations. *Int J Cosmet Sci.* 2004; 26: 31–36.

60. Kristl J, Volk B, Gašperlin M, Šentjurc M, Jurkovič P. Effect of colloidal carriers on ascorbyl palmitate stability. *Eur J Pharm Sci.* 2003; 19: 181–189.

61. Špiclin P, Gašperlin M, Kmetec V. Stability of ascorbyl palmitate in topical microemulsions. *Eur J Pharm Biopharm.* 2001; 222: 271–279.

62. Teeranachaideekul V, Muller RH, Junyaprasert VB. Encapsulation of ascorbyl palmitate in nanostructured lipid carriers (NLC)—effects of formulation parameters on physicochemical stability. *Int J Pharm.* 2007; 340: 198–206.

63. Teeranachaideekul V, Souto EB, Muller RH, Junyaprasert VB. Physicochemical characterization and *in vitro* release studies of ascorbyl palmitate-loaded semi-solid nanostructured lipid carriers (NLC gels). *J Microencapsul.* 2008; 25: 111–120.

64. Kristl J, Volk B, Gasperlin M, Sentjurc M, Jurkovic P. Effect of colloidal carriers on ascorbyl palmitate stability. *Eur J Pharm Biopharm Sci.* 2003; 19: 181–189.

65. Uner M, Wissing SA, Yener G, Muller RH. Skin moisturizing effect and skin penetration of ascorbyl palmitate entrapped in solid lipid nanoparticles (SLN) and nanostructured lipid carriers (NLC) incorporated into hydrogel. *Pharmazie.* 2005; 60: 751–755.

66. McVean M, Liebler DC. Inhibition of UVB induced DNA photodamage in mouse epidermis by topically applied alpha-tocopherol. *Carcinogenesis.* 1997; 18: 1617–1622.

67. Wissing SA, Muller RH. A novel sunscreen system based on tocopherol acetate incorporated into solid lipid nanoparticles. *Int J Cosmet Sci.* 2001; 23: 233–243.

68. Olson JA, Krinsky NI. Introduction: The colorful, fascinating world of the carotenoids: Important physiologic modulators. *FASEB J.* 1995; 9: 1547–1550.

69. Stahl W, Sies H. Antioxidant activity of carotenoids. *Mol Aspects Med.* 2003; 24: 345–351.

70. Roberts RL, Green J, Lewis B. Lutein and zeaxanthin in eye and skin health. *Clin Dermatol.* 2009; 27: 195–201.

71. Darvin ME, Fluhr JW, Meinke MC, Zastrow L, Sterry W, Lademann J. Topical beta-carotene protects against infra-red-light–induced free radicals. *Exp Dermatol.* 2011; 20: 125–129.

72. Stahl W, Sies H. beta-Carotene and other carotenoids in protection from sunlight. *Am J Clin Nutr.* 2012; 96: 10.

73. Bohm F, Edge R, McGarvey DJ, Truscott TG. Beta-carotene with vitamins E and C offers synergistic cell protection against NOx. *FEBS Lett.* 1998; 436: 387–389.

74. Boon CS, McClements DJ, Weiss J, Decker EA. Factors influencing the chemical stability of carotenoids in foods. *Crit Rev Food Sci Nutr.* 2010; 50: 515–532.

75. Trombino S, Cassano R, Muzzalupo R, Pingitore A, Cione E, Picci N. Stearyl ferulate-based solid lipid nanoparticles for the encapsulation and stabilization of beta-carotene and alpha-tocopherol. *Colloids Surf B Biointerfaces.* 2009; 72: 181–187.

76. Hung LC, Basri M, Tejo BA, Ismail R, Nang HLL, Hassan HA, *et al.* An improved method for the preparations of nanostructured lipid carriers containing heat-sensitive bioactives. *Colloids Surf B Biointerfaces.* 2011; 87: 180–186.

77. Gonzalez S, Astner S, An W, Goukassian D, Pathak MA. Dietary lutein/zeaxanthin decreases ultraviolet B-induced epidermal hyperproliferation and acute inflammation in hairless mice. *J Invest Dermatol.* 2003; 121: 399–405.

78. Lee EH, Faulhaber D, Hanson KM, Ding W, Peters S, Kodali S, *et al.* Dietary lutein reduces ultraviolet radiation-induced inflammation and immunosuppression. *J Invest Dermatol.* 2004; 122: 510–517.

79. Palombo P, Fabrizi G, Ruocco V, Ruocco E, Fluhr J, Roberts R, *et al.* Beneficial long-term effects of combined oral/topical antioxidant treatment with the carotenoids lutein and zeaxanthin on human

skin: A double-blind, placebo-controlled study. *Skin Pharmacol Physiol*. 2007; 20: 199–210.

80. Shi XM, Chen F. Stability of lutein under various storage conditions. *Food/Nahrung*. 1997; 41: 38–41.
81. Mitri K, Shegokar R, Gohla S, Anselmi C, Müller RH. Lipid nanocarriers for dermal delivery of lutein: Preparation, characterization, stability and performance. *Int J Pharm*. 2011; 414: 267–275.
82. Yue Y, Zhou H, Liu G, Li Y, Yan Z, Duan M. The advantages of a novel CoQ10 delivery system in skin photo-protection. *Int J Pharm*. 2010; 392: 57–63.
83. James AM, Smith RAJ, Murphy MP. Antioxidant and prooxidant properties of mitochondrial Coenzyme Q. *Arch Biochem Biophys*. 2004; 423: 47–56.
84. Hoppe U, Bergemann J, Diembeck W, Ennen J, Gohla S, Harris I, *et al.* Coenzyme Q10, a cutaneous antioxidant and energizer. *Biofactors*. 1999; 9: 371–378.
85. Muta-Takada K, Terada T, Yamanishi H, Ashida Y, Inomata S, Nishiyama T, *et al.* Coenzyme Q10 protects against oxidative stress-induced cell death and enhances the synthesis of basement membrane components in dermal and epidermal cells. *Biofactors*. 2009; 35: 435–441.
86. Teeranachaideekul V, Souto EB, Junyaprasert VB, Müller RH. Cetyl palmitate-based NLC for topical delivery of Coenzyme Q10—Development, physicochemical characterization and *in vitro* release studies. *Eur J Pharm Biopharm*. 2007; 67: 141–148.
87. Brugè F, Damiani E, Puglia C, Offerta A, Armeni T, Littarru GP, *et al.* Nanostructured lipid carriers loaded with CoQ10: Effect on human dermal fibroblasts under normal and UVA-mediated oxidative conditions. *Int J Pharm*. 2013; 455: 348–356.
88. Pardeike J, Schwabe K, Müller RH. Influence of nanostructured lipid carriers (NLC) on the physical properties of the Cutanova Nanorepair Q10 cream and the *in vivo* skin hydration effect. *Int J Pharm*. 2010; 396: 166–173.
89. Junyaprasert VB, Teeranachaideekul V, Souto EB, Boonme P, Müller RH. Q10-loaded NLC versus nanoemulsions: Stability, rheology and *in vitro* skin permeation. *Int J Pharm*. 2009; 377: 207–214.
90. Obeidat WM, Schwabe K, Muller RH, Keck CM. Preservation of nanostructured lipid carriers (NLC). *Eur J Pharm Biopharm*. 2010; 76: 56–67.

91. Lohan SB, Bauersachs S, Ahlberg S, Baisaeng N, Keck CM, Muller RH, *et al.* Ultra-small lipid nanoparticles promote the penetration of coenzyme Q10 in skin cells and counteract oxidative stress. *Eur J Pharm Biopharm.* 2015; 89: 201–207.

92. Schwarz JC, Baisaeng N, Hoppel M, Löw M, Keck CM, Valenta C. Ultra-small NLC for improved dermal delivery of coenyzme Q10. *Int J Pharm.* 2013; 447: 213–217.

93. Li B, Ge ZQ. Nanostructured lipid carriers improve skin permeation and chemical stability of idebenone. *AAPS PharmSciTech.* 2012; 13: 276–283.

94. Amorim CdM, Couto AG, Netz DJA, de Freitas RA, Bresolin TMB. Antioxidant idebenone-loaded nanoparticles based on chitosan and N-carboxymethylchitosan. *Nanomed Nanotechnol Biol Med.* 6: 745–752.

95. Montenegro L, Sinico C, Castangia I, Carbone C, Puglisi G. Idebenone-loaded solid lipid nanoparticles for drug delivery to the skin: *In vitro* evaluation. *Int J Pharm.* 2012; 434: 169–174.

96. Montenegro L, Sarpietro MG, Ottimo S, Puglisi G, Castelli F. Differential scanning calorimetry studies on sunscreen loaded solid lipid nanoparticles prepared by the phase inversion temperature method. *Int J Pharm.* 2011; 415: 301–306.

97. Salunkhe S, Bhatia N, Pokharkar V, Thorat J, Bhatia M. Topical delivery of Idebenone using nanostructured lipid carriers: Evaluations of sun-protection and anti-oxidant effects. *J Pharm Invest.* 2013; 43: 287–303.

98. Goraca A, Huk-Kolega H, Piechota A, Kleniewska P, Ciejka E, Skibska B. Lipoic acid—biological activity and therapeutic potential. *Pharmacol Rep.* 2011; 63: 849–858.

99. Ruktanonchai U, Bejrapha P, Sakulkhu U, Opanasopit P, Bunyapraphatsara N, Junyaprasert V, *et al.* Physicochemical characteristics, cytotoxicity, and antioxidant activity of three lipid nanoparticulate formulations of alpha-lipoic acid. *AAPS Pharm Sci Tech.* 2009; 10: 227–234.

100. Souto EB, Muller RH, Gohla S. A novel approach based on lipid nanoparticles (SLN) for topical delivery of alpha-lipoic acid. *J Microencapsul.* 2005; 22: 581–592.

101. Iqbal MA, Md S, Sahni JK, Baboota S, Dang S, Ali J. Nanostructured lipid carriers system: Recent advances in drug delivery. *J Drug Target.* 2012; 20: 813–830.

102. Aquino R, Morelli S, Tomaino A, Pellegrino M, Saija A, Grumetto L, *et al.* Antioxidant and photoprotective activity of a crude extract of Culcitium reflexum H.B.K. leaves and their major flavonoids. *J Ethnopharmacol.* 2002; 79: 183–191.

103. Filipe P, Silva JN, Haigle J, Freitas JP, Fernandes A, Santus R, *et al.* Contrasting action of flavonoids on phototoxic effects induced in human skin fibroblasts by UVA alone or UVA plus cyamemazine, a phototoxic neuroleptic. *Photochem Photobiol Sci.* 2005; 4: 420–428.

104. Bors W, Heller W, Michel C, Saran M. Flavonoids as antioxidants: Determination of radical-scavenging efficiencies. *Methods Enzymol.* 1990; 186: 343–355.

105. Laughton MJ, Evans PJ, Moroney MA, Hoult JR, Halliwell B. Inhibition of mammalian 5-lipoxygenase and cyclo-oxygenase by flavonoids and phenolic dietary additives. Relationship to antioxidant activity and to iron ion-reducing ability. *Biochem Pharmacol.* 1991; 42: 1673–1681.

106. Cao G, Sofic E, Prior RL. Antioxidant and prooxidant behavior of flavonoids: Structure-activity relationships. *Free Radic Biol Med.* 1997; 22: 749–760.

107. Casagrande R, Georgetti SR, Verri Jr WA, Dorta DJ, dos Santos AC, Fonseca MJV. Protective effect of topical formulations containing quercetin against UVB-induced oxidative stress in hairless mice. *J Photochem Photobiol B Biol.* 2006; 84: 21–27.

108. Saija A, Scalese M, Lanza M, Marzullo D, Bonina F, Castelli F. Flavonoids as antioxidant agents: Importance of their interaction with biomembranes. *Free Radic Biol Med.* 1995; 19: 481–486.

109. Bose S, Michniak-Kohn B. Preparation and characterization of lipid based nanosystems for topical delivery of quercetin. *Eur J Pharm Biopharm.* 2013; 48: 442–452.

110. Graefe EU, Derendorf H, Veit M. Pharmacokinetics and bioavailability of the flavonol quercetin in humans. *Int J Clin Pharmacol Ther.* 1999; 37: 219–233.

111. Walle T. Absorption and metabolism of flavonoids. *Free Radic Biol Med.* 2004; 36: 829–837.

112. Rothwell JA, Day AJ, Morgan MR. Experimental determination of octanol-water partition coefficients of quercetin and related flavonoids. *J Agric Food Chem.* 2005; 53: 4355–4360.

113. Azuma K, Ippoushi K, Ito H, Higashio H, Terao J. Combination of lipids and emulsifiers enhances the absorption of orally administered quercetin in rats. *J Agric Food Chem.* 2002; 50: 1706–1712.

114. Tan Q, Liu W, Guo C, Zhai G. Preparation and evaluation of quercetin-loaded lecithin-chitosan nanoparticles for topical delivery. *Int J Nanomed.* 2011; 6: 1621–1630.

115. Bose S, Du Y, Takhistov P, Michniak-Kohn B. Formulation optimization and topical delivery of quercetin from solid lipid based nanosystems. *Int J Pharm.* 2013; 441: 56–66.

116. Chen-yu G, Chun-fen Y, Qi-lu L, Qi T, Yan-wei X, Wei-na L, *et al.* Development of a quercetin-loaded nanostructured lipid carrier formulation for topical delivery. *Int J Pharm.* 2012; 430: 292–298.

117. Banerjee S, Li Y, Wang Z, Sarkar FH. Multi-targeted therapy of cancer by genistein. *Cancer Lett.* 2008; 269: 226–242.

118. Borrás C, Gambini J, López-Grueso R, Pallardó FV, Viña J. Direct antioxidant and protective effect of estradiol on isolated mitochondria. *Biochim Biophys Acta (BBA)—Mol BasisDis.* 2010; 1802: 205–211.

119. Wei H, Saladi R, Lu Y, Wang Y, Palep SR, Moore J, *et al.* Isoflavone genistein: Photoprotection and clinical implications in dermatology. *J Nutr.* 2003; 133: 3811S–3819S.

120. Moore JO, Wang Y, Stebbins WG, Gao D, Zhou X, Phelps R, *et al.* Photoprotective effect of isoflavone genistein on ultraviolet B-induced pyrimidine dimer formation and PCNA expression in human reconstituted skin and its implications in dermatology and prevention of cutaneous carcinogenesis. *Carcinogenesis.* 2006; 27: 1627–1635.

121. Busby MG, Jeffcoat AR, Bloedon LT, Koch MA, Black T, Dix KJ, *et al.* Clinical characteristics and pharmacokinetics of purified soy isoflavones: Single-dose administration to healthy men. *Am J Clin Nutr.* 2002; 75: 126–136.

122. Huang Z-R, Hung C-F, Lin Y-K, Fang J-Y. *In vitro* and *in vivo* evaluation of topical delivery and potential dermal use of soy isoflavones genistein and daidzein. *Int J Pharm.* 2008; 364: 36–44.

123. Minghetti P, Cilurzo F, Casiraghi A, Montanari L. Evaluation of *ex vivo* human skin permeation of genistein and daidzein. *Drug Deliv.* 2006; 13: 411–415.

124. Georgetti SR, Casagrande R, Verri WA, Jr., Lopez RF, Fonseca MJ. Evaluation of *in vivo* efficacy of topical formulations containing soybean extract. *Int J Pharm.* 2008; 352: 189–196.

125. Andrade LM, de Fatima Reis C, Maione-Silva L, Anjos JL, Alonso A, Serpa RC, *et al.* Impact of lipid dynamic behavior on physical stability, *in vitro* release and skin permeation of genistein-loaded lipid nanoparticles. *Eur J Pharm Biopharm.* 2014; 88: 40–47.

126. Juliano C, Cossu M, Alamanni MC, Piu L. Antioxidant activity of gamma-oryzanol: Mechanism of action and its effect on oxidative stability of pharmaceutical oils. *Int J Pharm.* 2005; 299: 146–154.

127. (DE) KB. Cosmetic sunscreen compsns. contg. ferulic acid and gamma-oryzanol. A61K8/27 A61K8/29 A61K8/365 A61K8/37 A61K8/63 A61Q17/04 (IPC1-7): A61K7/44 ed: Goldwell Gmbh, 64297 Darmstadt, DE 12/21/1995.

128. Ruktanonchai U, Sakulkhu U, Bejrapha P, Opanasopit P, Bunyapraphatsara N, Junyaprasert V, *et al.* Effect of lipid types on physicochemical characteristics, stability and antioxidant activity of gamma-oryzanol-loaded lipid nanoparticles. *J Microencapsul.* 2009; 26: 614–626.

129. Kundu JK, Surh Y-J. Cancer chemopreventive and therapeutic potential of resveratrol: Mechanistic perspectives. *Cancer Lett.* 2008; 269: 243–261.

130. Stervbo U, Vang O, Bonnesen C. A review of the content of the putative chemopreventive phytoalexin resveratrol in red wine. *Food Chem.* 2007; 101: 449–457.

131. Vian MA, Tomao V, Gallet S, Coulomb PO, Lacombe JM. Simple and rapid method for *cis*- and *trans*-resveratrol and piceid isomers determination in wine by high-performance liquid chromatography using chromolith columns. *J Chromatogr A.* 2005; 1085: 224–229.

132. Neves AR, Lucio M, Lima JL, Reis S. Resveratrol in medicinal chemistry: A critical review of its pharmacokinetics, drug-delivery, and membrane interactions. *Curr Med Chem.* 2012; 19: 1663–1681.

133. Carlotti ME, Sapino S, Ugazio E, Gallarate M, Morel S. Resveratrol in solid lipid nanoparticles. *J Dispersion SciTechnol.* 2012; 33: 465–671.

134. Teskač K, Kristl J. The evidence for solid lipid nanoparticles mediated cell uptake of resveratrol. *Int J Pharm.* 2010; 390: 61–69.

135. Gokce EH, Korkmaz E, Dellera E, Sandri G, Bonferoni MC, Ozer O. Resveratrol-loaded solid lipid nanoparticles versus nanostructured lipid carriers: Evaluation of antioxidant potential for dermal applications. *Int J Nanomed.* 2012; 7: 1841–1850.

136. Zhou H, Beevers CS, Huang S. Targets of curcumin. *Curr Drug Target.* 2011; 12: 332–347.

137. Phan TT, See P, Lee ST, Chan SY. Protective effects of curcumin against oxidative damage on skin cells *in vitro*: Its implication for wound healing. *J Trauma.* 2001; 51: 927–931.

138. Bernabe-Pineda M, Ramirez-Silva MT, Romero-Romo M, Gonzalez-Vergara E, Rojas-Hernandez A. Determination of acidity constants

of curcumin in aqueous solution and apparent rate constant of its decomposition. *Spectrochim Acta A Mol Biomol Spectrosc.* 2004; 60: 1091–1097.

139. Tonnesen HH, Karlsen J. Studies on curcumin and curcuminoids. VI. Kinetics of curcumin degradation in aqueous solution. *Z Lebensm Unters Forsch.* 1985; 180: 402–404.

140. Tonnesen HH, Karlsen J, van Henegouwen GB. Studies on curcumin and curcuminoids. VIII. Photochemical stability of curcumin. *Z Lebensm Unters Forsch.* 1986; 183: 116–122.

141. Tiyaboonchai W, Tungpradit W, Plianbangchang P. Formulation and characterization of curcuminoids loaded solid lipid nanoparticles. *Int J Pharm.* 2007; 337: 299–306.

142. Zamarioli CM, Martins RM, Carvalho EC, Freitas LAP. Nanoparticles containing curcuminoids (Curcuma longa): Development of topical delivery formulation. *Revista Brasileira de Farmacognosia.* 2015; 25: 53–60.

143. Teeranachaideekul V, Müller RH, Junyaprasert VB. Encapsulation of ascorbyl palmitate in nanostructured lipid carriers (NLC)—Effects of formulation parameters on physicochemical stability. *Int J Pharm.* 2007; 340: 198–206.

144. Schwarz JC, Baisaeng N, Hoppel M, Low M, Keck CM, Valenta C. Ultra-small NLC for improved dermal delivery of coenyzme Q10. *Int J Pharm.* 2013; 447: 213–217.

145. Sarpietro MG, Accolla ML, Puglisi G, Castelli F, Montenegro L. Idebenone loaded solid lipid nanoparticles: Calorimetric studies on surfactant and drug loading effects. *Int J Pharm.* 2014; 471: 69–74.

Chapter 7

Manufacture and Applications of Gelatin Nanoparticles: A Practical Approach

Diogo Pineda Rivelli and Silvia Berlanga de Moraes Barros

Department of Clinical Chemistry and Toxicology,
School of Pharmaceutical Sciences, University of São Paulo, São Paulo, Brazil

silviaberlanga@gmail.com

7.1 Introduction

Nanostructured materials are usually classified as a product with dimensions ranging from 1 nm to 100 nm (although the nanoscale goes up to 1000 nm). At this size, the particles may develop some unique properties and functions and may easily interact with membranes and protein complexes since the usual biological target is also at nanometric scale [1].

Among the several possibilities of drug formulation are the nanostructured systems. A great number of such systems are described in the literature, being the polymeric ones holding a special ranking. They are colloidal particles (such as liposomes, niosomes, and microemulsions), but may be constituted of natural or artificial polymers. Nanoparticles offer many advantages over the regular colloidal carriers, including higher stability/

Carrier-Mediated Dermal Delivery: Applications in the Prevention and Treatment of Skin Disorders
Edited by Andreia Ascenso, Sandra Simões, and Helena Ribeiro

ISBN 978-981-4745-58-1 (Hardcover), 978-1-315-36447-6 (eBook)
www.panstanford.com

compatibility with biological fluids, enhanced incorporation capacity, and protection of the active compound against degradation, besides the possibility of controlled releasing. According to the preparation method, the active molecules may be dissolved on the liquid core or adsorbed in the dense polymeric matrix [2].

Over the past few years, there has been a growing interest on gelatin nanoparticles as a drug delivery system. Among the reasons for that, gelatin is classified as Generally Recognized as Safe (GRAS), since it is a traditional ingredient used for food purposes. Gelatin is a protein polymer easily obtained by acid (type A) or alkali (type B) hydrolysis of collagen [4], and characterized by its Bloom number. This number represents the pressure needed to depress the surface of a gel by 4 mm (higher numbers indicate higher gel strength) [5]. Acid is usually employed to prepare gelatin from pig or fish skin, while alkali hydrolyzes more complex matrices, with heavily cross-linked bounds, such as bovine collagen [4].

Other important points to be considered are its remarkable features, such as biodegradability, biocompatibility, non-antigenicity, high nutritional value, abundant renewable sources, among others [6].

7.2 Preparation methods

7.2.1 Desolvation

In this process, a gelatin aqueous solution is added to a desolvating agent, usually an organic solvent (such as ethanol or acetone), leading to a phase separation and a conformational change in protein structure. Then a cross-link agent (usually glutaraldehyde or glyoxal) is combined with the formed nanoparticles to promote hardening. Other variables of the system can be altered, mainly temperature, ionic strength and pH. One of the drawbacks of this technique is the high heterogeneity of gelatin weight distribution, which results on broad size distribution. In order to solve this problem, a second desolvation step may be added, removing the low molecular weight gelatin from the reaction medium. The drug loading can be obtained either by adsorption on the surface after the nanoparticle formation or by entrapment during the preparation [2, 6].

Kaur and Jain [7, 8] have obtained type A gelatin nanoparticles through double desolvation method using acetone as a desolvating agent and glutaraldehyde as a cross-link agent. The formed nanoparticles presented a small particle size (120 nm and 173 nm, respectively) and positive zeta potential (18.6 mV and 19.83 mV, respectively). In a similar work, other authors entrapped rifampicin, but used cystein to neutralize the excess of glutaraldehyde. These authors obtained bigger particles (250 nm) with a lower zeta potential (15.3 mV) [9].

Others [10] used type B gelatin to produce nanoparticles loaded with cysplatin by double desolvation method. The authors neutralized the excess of glutaraldehyde with glycine and were able to freeze-dry the structures using trehalose as a cryoprotectant. The particles presented an adequate size (140 nm) and were considered quite stable due to the high zeta potential value (–85.8 mV).

Azarmi *et al.* [11] compared the influence of several factors on the formation of gelatin nanoparticles. These authors concluded that the desolvating agent, the type of gelatin and temperature were the most important factors. Smaller nanoparticles were obtained using type B gelatin and acetone as the desolvating agent at 40°C. Apparently, the pH was strictly correlated with zeta potential, being the optimal conditions at the extremes of acidic or alkali ranges. The amount of cross-link agent did not produce any effect.

Likewise, Kuntworbe *et al.* [12] prepared type A gelatin nanoparticles, obtaining similar results. In addition, these authors evaluated the stability of the nanoparticles. Since the nanoparticles were unstable at room temperature, they freeze-dried the structures with several cryoprotectants. The results showed that the best cryoprotectant was sucrose, which was able to stabilize the particles up to 52 weeks.

Nahar *et al.* [13] compared type A and B gelatins regarding entrapment efficiency of amphotericin B. Both systems were made using double desolvation technique. The results indicated that gelatin type A was more suitable for this purpose and the particle size was inversely proportional to the amount of glutaraldehyde.

The gelatin nanoparticles are usually cross-linked with an aldehyde, such as glutaraldehyde or formaldehyde. Although those agents help to improve the stability of the nanosystem, they possess an intrinsic toxicity. In order to overcome this problem, Lam *et al.* [14] proposed the use of D-glucose as cross-link agent. Nanoparticles with *Calendula officinalis* extract presented good entrapment efficiency (74%), satisfactory release rate, and particles size around 500 nm.

Won *et al.* [15] used human recombinant gelatin as raw material for the nanoparticle formation and genipin as the cross-link agent, less toxic than glutaraldehyde. The obtained nanoparticles were considered stable (after 6 months storage) and with 250 nm diameter. Fuchs *et al.* [16] evaluated the effect of another possible cross-link agent. These authors successfully used transglutaminase, an enzyme isolated from *Escherichia coli*. However, the reaction conditions were slightly different from the conventional desolvation protocol. In this method, the final nanoparticles suspension was diluted in purified water in order to obtain an adequate medium for the enzyme. Qazvini *et al.* [17] also assessed another non-toxic cross-link agent for gelatin. A mixture of carbodiimide and *N*-hydroxysuccinimide (CDI/NHS) in water was used, and its cross-link effect was compared with glutaraldehyde (GA). CDI/NHS nanoparticles were much smaller than the GA ones. Paracetamol encapsulation efficiency was found to be 10% and 27% in GA and CDI/NHS, respectively.

Azimi *et al.* [18] compared the desolvation process between one and two steps on the production of bovine gelatin nanoparticles concluding that two steps were more suitable due to the smaller particle size and narrower polydispersity index obtained. The authors demonstrated that the first desolvation step eliminated the low molecular weight gelatin, leading to a more uniform matrix. This fact has already been reported before by Coester *et al.* [19], who firstly described the two steps desolvation method.

Ofokansi *et al.* [20] introduced a single-step desolvation protocol using a hydroethanolic alcalinated solution instead of acetone. The final nanoparticles presented a proper size (around 300 nm) and moderate zeta potential (–20 mV).

Another interesting feature of the gelatin nanoparticles is the possibility of conjugation with other chemical moieties. For instance, Jain *et al.* [8] produced the nanostruture by double desolvation method followed by conjugation with mannose. The mannosylation process allows the nanosystem to be recognized by several cell receptors in order to drive a molecule to a specific target.

7.2.2 Emulsification-Solvent Evaporation

In this method, nanoparticles are formed by adding an aqueous solution containing both gelatin and the drug to an oil phase (paraffin oil, castor oil, olive oil, polymethylmethacrylate (PMMA), etc.) followed by cross-linking. This process is usually performed in ultrasonic bath or in high-speed homogenizer. At the end, the oil phase is removed by extraction with organic solvents. Protein concentration, water/oil ratio and emulsification time dramatically influence the particle size [2, 6].

Several authors [21–23] have used gelatin dissolved in phosphate buffer solution as the aqueous phase and PMMA dissolved in chloroform/toluene as the oil phase. Then, the aqueous phase was mixed with the oil phase by high-speed homogenizer and the final solution was cross-linked with glutaraldehyde. The nanoparticles were extracted with toluene and acetone, and their particle size was on 100–200 nm range.

Regarding another similar process [24], paraffin was used as the oil phase with comparable results. These authors also impregnate the nanoparticles with iron oxide in order to obtain a magnetized system focusing on drug targeting.

7.2.3 Reverse-Phase Microemulsion

This approach uses an aqueous solution of gelatin combined with an emulsifier dissolved in a non-polar solvent. This surfactant forms reverse micelles where its hydrophobic portion is driven by the non-polar solvent and the hydrophilic moiety encloses the aqueous solution. After cross-linking the solvent is evaporated, recovering the nanoparticles [25].

Gupta *et al.* [25] used sodium bis(2-ethylhexyl sulfosuccinate) as a surfactant diluted in hexane, followed by the addition of gelatin and glutaraldehyde as the cross-linker. The mixture was stirred overnight and the nanoparticles were recovered after evaporating the hexane and precipitation with acetone-methanol. Following this procedure, the nanoparticle sizing was around 37 nm.

7.2.4 Nanoprecipitation

Nanoprecipitation, known as the solvent displacement or interfacial deposition, is a simple and fast technique which uses a low organic solvent amount and produces relatively small nanoparticles with a narrow polydispersity index [26, 27].

In this method, both gelatin and drug are solubilized in water (solvent phase) and this solution is slowly added under stirring over a miscible non-solvent containing a stabilizer agent (e.g., polyvinyl alcohol or poloxamer 188). It is very important that the polymer (gelatin) is not soluble in the non-solvent, leading to the precipitation of the nanoparticles on the interface [26, 27].

Khan *et al.* [28] optimized the nanoprecipitation technique for gelatin nanoparticle preparation varying several parameters of this process. They verified that polysorbate 20 and 80 were unsuitable stabilizers, whereas poloxamer 188 and 407 could be used at a minimum concentration of 7%. Regarding the nanoparticle formation, methanol and ethanol could be successfully used as non-solvents, while isopropyl alcohol, acetone, and acetonitrile caused a great aggregation. The average size at optimal conditions was 250 nm.

These results are similar to those found by Lee *et al.* [29] who worked in analogous conditions, using ethanol and Poloxamer 407. The authors were also able to successfully encapsulate antibiotic agents (gatifloxacin) and a muscle relaxant (tizanidine). Both articles used glutaraldehyde as the cross-link agent.

7.2.5 Self-Assembly

The hydrophilic structure of gelatin can be conjugated with hydrophobic molecules, forming an amphiphilic polymer. This polymer is able to self-assemble in an aqueous environment, forming a micelle-like structure entrapping hydrophobic compounds into the core [30].

Chen *et al.* [31] evaluated the influence of gelatin and catechin concentration on the formation of self-assembled nanoparticles. The authors concluded that higher catechin concentrations contributed for a less negative zeta potential (i.e., less stability) because the proton dissociation of this flavonoid counterbalanced the negative zeta potential of gelatin. In contrast, lower catechin concentrations increased the mean particle size, because the excess of gelatin molecules entangles the surface of the nanoparticles.

Li and Gu [32] performed a similar experiment using purified pomegranate ellagitannins (PPE), and assessing other experimental variables such as temperature, pH and reaction time. According to the results obtained, they established that the ideal PPE: gelatin mass ratio was around 6:5. Regarding the pH, no nanoparticle formation was obtained on the pH extremes, being the ideal pH values between 4.0 and 5.3 (correspondent to small particle size and high zeta potential). Another important issue was the temperature influence since below 20°C or above 30°C might result in large or unstable nanoparticles (depending on PPE:gelatin mass ratio).

7.2.6 Layer-by-Layer Coating

This is not a nanostructure building technique, but rather a procedure to modify the surface of already formed nanoparticles. In this system, the surface charge of the particles are characterized and then incubated in a solution containing the opposite charge (i.e., if the particle has positive surface charge, it will be incubated in a polyanion solution). Due to the adsorption of these electrolytes, the surface charge is reversed. The particles are then rinsed with water and incubated again with other electrolyte solution. This process aims to anchor the drug between the layers of electrolytes [33, 34].

Shutava *et al.* [34] produced gelatin nanoparticles by two-step desolvation method. After that, the authors performed several layer-by-layer coatings on the formed structures using polyanions/polycations combinations. They verified that the coating increased the particles size only by 20 nm. This represents a viable approach to encapsulate substances with the most diverse physicochemical features and to control the drug release profile.

7.3 Uses

Lai *et al.* [35] developed a collagen skin equivalent embedded in gelatin nanoparticles containing several angiogenic growth factors (vascular endothelial growth factor (VEGF), platelet-derived growth factor (PDGF), basic fibroblast growth factor (bFGF) and endothelial growth factor (EGF)). The aim of this work was to produce an artificial system to supply those factors for the treatment of chronic wounds. The results indicated a satisfactory releasing rate, with an initial delivery of EGF and bFGF, accelerating the epithelialization, followed by the release of PDGF and VEGF in a later stage, inducing the blood vessels growth. Authors concluded that this could be a powerful tool to restore the properties of the damaged skin.

This same wound healing potential was evaluated by others [36] who used a bi-layer nanoparticles. This structure had an external core constituted of gelatin nanoparticle with an inner core of poly vinyl alcohol hydrogel combined with epigallocatechin gallate. The system demonstrated a suitable releasing profile and was able to induce angiogenesis and re-epithelialization, leading to a faster healing rate.

Gelatin nanostructured system was used to encapsulate the aqueous extract of *Centella asiatica* with good efficiency (67%). The crude extract was able to reduce the expression of metalloproteinase type I (MMP-1) (enzyme related with the extracellular matrix degradation) on UV irradiated fibroblasts. However, when the extract was incorporated into the gelatin nanoparticles, the activity was enhanced. Another interesting feature was the penetration across the skin. The encapsulated extract showed accumulation in the dermis and a doubled flux compared to the crude extract. These results demonstrated the possible application of this complex system for cosmetic and dermatological purposes [37].

7.3.1 Ilex paraguariensis Extract Gelatin Encapsulation

Ilex paraguariensis St. Hilaire (Aquifoliaceae), popularly known as yerba mate, is a South American plant, broadly harvested in northeastern of Argentina, south of Brazil and east of Paraguay. Its aerial parts are used to prepare a tea recognized by its stimulant properties and particular flavor [38].

Some authors [38] confirmed the presence of high amounts of chlorogenic acid derivatives, such as 5-caffeoylquinic acid, 3,5-dicaffeoylquinic acid and 4,5-dicaffeoylquinic acid in the aqueous extract of mate leaves. These compounds have received particular attention due to the antioxidant properties and its role on preventing several conditions, including UVB induced erythema [39, 40], and especially atherosclerosis and cancer. Interestingly, chlorogenic acid has a chemical structure similar to classic sunscreens, such as octyl methoxycinnamate (Parsol MCX). In fact, its sun protection factor (SPF) was considered the highest among several natural compounds [41].

Considering the promising results of sunscreen activity previously described for *Ilex paraguariensis* extract, a formulation was developed with this extract. This formulation should enhance the skin permeation of some extract components, enabling the extract to act as sunscreens (on skin surface) as well as photochemoprotectors (in the deeper layers) by preventing, slowing, or reversing UV radiation induced lesions [42].

The concentration gradient drives the passive permeation of the molecules through the skin, while the physicochemical properties of the compound, such as the solubility on the dosage form, partition coefficient between vehicle and skin and molecular weight influence the rate and extent of permeation [43]. Many photoprotective compounds of the hydroethanolic extract are hydrosoluble and with impaired cutaneous permeation due to the low affinity of these compounds with the skin [44]. To solve this problem, the design of a gelatin based nanostructured system was selected.

Regarding the methodology, gelatin (0.5 g of type B bovine gelatin) was solubilized in 10 mL of milliQ water (water bath at 40°C) followed by slowly addition of 2.5 mL of acetone under stirring. The mixture was allowed to rest during 10 min and the supernatant was discarded.

The precipitate was resuspended with 25 mL of milliQ water under stirring (40°C) and pH adjusted to 2.5. After that, 100 mg of hydroethanolic extract of *Ilex paraguariensis* (previously solubilized in water) and acetone were slowly incorporated (40°C under stirring) until the remaining solution turned bluish grey. Finally, 200 μL of glutaraldehyde was added and the acetone was evaporated under stirring at 40°C.

The nanoparticles were characterized by size determination, zeta potential, and encapsulation efficiency. For size and zeta potential determination the sample was properly diluted in milliQ water and the reading was performed twice at 90° angle (stabilization time: 5 min) on Beckman Coulter Counter (Brea, CA, EUA) and Malvern zetasizer (Worcestershire, UK), respectively.

For the encapsulation efficiency, chlorogenic acid and caffeine were used as extract biomarkers [45]. Nanoparticles suspension (100 μL) was added to 900 μL of metaphosphoric acid 10%, and the mixture was centrifuged at 10,000 rpm during 4 min and the supernatant analyzed by HPLC. Briefly, the system was constituted of a Constametric pump model 3500 (Fremont, CA, EUA), a Rheodyne 7125 manual injector (Contati, CA, EUA) with 20 μL loop and a UV-Vis Lab Alliance detector model 525 (San Jose, CA, EUA). Separation was carried out using 0.2% metaphosphoric acid:acetonitrile (80:20 v/v) as the mobile phase, flow rate at 1 mL/min and wavelength detection at 330 nm and 275 nm for chlorogenic acid and caffeine, respectively [45]. For the quantification of free material (non-encapsulated) a new fraction of nanoparticles suspension was filtrated through a Millipore Amicon 100 K membrane and the solution was analyzed as described above.

The results of particle size, zeta potential and encapsulation efficiency are presented in Table 7.1.

Table 7.1 Particle size, encapsulation efficiency and zeta potential of gelatin nanoparticles containing *Ilex paraguariensis* hydroethanolic extract

	Size (nm)	Polydispersity index	Zeta potential (mV)	Encapsulation efficiency (%)
Blank	153	0.166	+16.0	—
Encapsulated	263	−0.182	+13.4	Chlorogenic acid: 9.4 Caffeine: 7.4

In order to evaluate the formulation stability after dryness, an aliquot of the nanosuspension was dried at room temperature in an Eppendorf 5301 concentrator (Hamburg, Germany). After this procedure, the material clotted becoming insoluble. To solve this problem, glucose and sucrose were chosen as stabilizers at different concentrations [46] and added to 100 μL of the suspension. After dryness, the material was resuspended and analyzed regarding its size and polydispersity index.

The results are presented in Table 7.2.

Table 7.2 Particle size after resuspension of nanoparticles stabilized by sucrose or glucose

	Glucose		Sucrose	
Sugar amount (mg)	Size (nm)	Polydispersity index	Size (nm)	Polydispersity index
10	614	0.385	350	0.112
20	829	−0.896	274	−0.057
40	304	−0.153	258	0.094
100	274	−0.063	267	0.021
200	261	−0.040	259	0.338

Based on the Table 7.2, 40 mg of sucrose was the minimum amount of sugar necessary to keep the nanoparticles size stable, and therefore it was chosen as the best dryness condition. The nanoparticle suspension was also considered stable when stored at 4°C during 60 days regarding the particle size (data not shown).

In summary, these preliminary studies will continue in order to improve the formulation development and characterization of gelatin nanoparticles containing *Ilex paraguariensis* hydroethanolic extract, including topical delivery studies.

In conclusion, gelatin nanoparticles can be obtained by different procedures and can be generally considered stable systems which may be a good option to incorporate a whole sort of bioactive compounds. This delivery system has opened new perspectives on reaching and targeting hydrophilic molecules to the living entities.

Acknowledgments

Thanks to the National Research Council Brasil (CNPq) for the postdoctoral fellowship to Diogo Pineda Rivelli at the Department of Clinical Chemistry and Toxicology, School of Pharmaceutical Sciences, University of São Paulo.

References

1. A. Mahapatro, D. K. Singh, Biodegradable nanoparticles are excellent vehicle for site directed *in vivo* delivery of drugs and vaccines, *J. Nanobiotechnol.* 9 (2011) 55.

2. S. R. D'Mello, S. K. Das, N. G. Das, Polymeric nanoparticles for small-molecule drugs: biodegradation of polymers and fabrication of nanoparticles, in: Y. Pathak, D. Thassu (eds.), *Drug Delivery Nanoparticles Formulation and Characterization*, Informa Healthcare, New York, 2009, pp. 16–34.
3. FDA—Food and Drug Administration. Select Committee on GRAS Substances (SCOGS) Opinion: Gelatin. Online. 2015 March 30. Available from URL: http://www.fda.gov/Food/Ingredients Packaging Labeling/GRAS/SCOGS/ucm261307.htm.
4. D. Hudson, A. Margaritis, Biopolymer nanoparticle production for controlled release of biopharmaceuticals, *Crit. Rev. Biotechnol.* 34 (2014) 161–179.
5. A. Saxena, K. Sachin, H. B. Bohidar, A. K. Verma, Effect of molecular weight heterogeneity on drug encapsulation efficiency of gelatin nano-particles, *Colloids Surf. B: Biointerfaces* 45 (2005) 42–48.
6. A. O. Elzoghby, Gelatin-based nanoparticles as drug and gene delivery systems: reviewing three decades of research, *J. Control. Release* 172 (2013) 1075–1091.
7. A. Kaur, S. Jain, A. K. Tiwary, Mannan-coated gelatin nanoparticles for sustained and targeted delivery of didanosine: *in vitro* and in vivo evaluation, *Acta Pharm.* 58 (2008) 61–74.
8. S. K. Jain, Y. Gupta, A. Jain, A. R. Saxena, P. Khare, A. Jain, Mannosylated gelatin nanoparticles bearing an anti-HIV drug didanosine for site-specific delivery, *Nanomed.-Nanotechnol. Biol. Med.* 4 (2008) 41–48.
9. G. K. Saraogi, P. Gupta, U. D. Gupta, N. K. Jain, G. P. Agrawal, Gelatin nanocarriers as potential vectors for effective management of tuberculosis, *Int. J. Pharm.* 385 (2010) 143–149.
10. N. Dixit, K. Vaibhav, R. S. Pandey, U. K. Jain, O. P. Katare, A. Katyal, J. Madan, Improved cisplatin delivery in cervical cancer cells by utilizing folate-grafted non-aggregated gelatin nanoparticles, *Biomed. Pharmacother.* 69 (2015) 1–10.
11. S. Azarmi, Y. Huang, H. Chen, S. McQuarrie, D. Abrams, W. Roa, W. H. Finlay, G. G. Miller, R. Löbenberg, Optimization of a two-step desolvation method for preparing gelatin nanoparticles and cell uptake studies in 143B osteosarcoma cancer cells, *J. Pharm. Pharmaceut. Sci.* 9 (2006) 124–132.
12. N. Kuntworbe, R. AL-Kassas, Design and *in vitro* haemolytic evaluation of cryptolepine hydrochloride-loaded gelatine nano-particles as a novel approach for the treatment of malaria, *AAPS PharmSciTech.* 13 (2012) 568–581.

13. M. Nahar, D. Mishra, V. Dubey, N. K. Jain, Development, characterization, and toxicity evaluation of amphotericin B-loaded gelatin nanoparticles, *Nanomed.-Nanotechnol. Biol. Med.* 4 (2008) 252–261.

14. P. L. Lam, S. H. Kok, Z. X. Bian, K. H. Lam, J. C. Tang, K. K. Lee, R. Gambari, C. H. Chui, D-Glucose as a modifying agent in gelatin/collagen matrix and reservoir nanoparticles for *Calendula officinalis* delivery, *Colloids Surf. B Biointerfaces.* 117 (2014) 277–283.

15. Y. W. Won, Y. H. Kim, Recombinant human gelatin nanoparticles as a protein drug carrier, *J. Control. Release.* 127 (2008) 154–161.

16. S. Fuchs, M. Kutscher, T. Hertel, G. Winter, M. Pietzsch, C. Coester, Transglutaminase: new insights into gelatin nanoparticle cross-linking, *J. Microencapsul.* 27 (2010) 747–754.

17. N. T. Qazvini, S. Zinatloo, Synthesis and characterization of gelatin nanoparticles using CDI/NHS as a non-toxic cross-linking system, *J. Mater. Sci.: Mater. Med.* 22 (2011) 63–69.

18. B. Azimi, P. Nourpanah, M. Rabiee, S. Arbab, Producing gelatin nanoparticles as delivery system for bovine serum albumin, *Iran. Biomed. J.* 18 (2014) 34–40.

19. C. J. Coester, K. Langer, H. Briesen, J. Kreuter, Gelatin nanoparticles by two step desolvation: a new preparation method, surface modifications and cell uptake, *J. Microencapsul.* 17 (2000) 187–193.

20. K. Ofokansi, G. Winter, G. Fricker, C. Coester, Matrix-loaded biodegradable gelatin nanoparticles as new approach to improve drug loading and delivery, *Eur. J. Pharm. Biopharm.* 76 (2010) 1–9.

21. M. G. Cascone, L. Lazzeri, Z. Zho, Gelatin nanoparticles prepared by a simple W/O emulsion as delivery system for methotrexate, *J. Mater. Sci. Mater. Med.* 13 (2002) 523–526.

22. A. K. Bajpai, J. Choubey, Design of gelatin nanoparticles as swelling controlled delivery system for chloroquine phosphate, *J. Mater. Sci. Mater. Med.* 17 (2006) 345–358.

23. A. K. Bajpai, J. Choubey, *In vitro* release dynamics of an anticancer drug from swellable gelatin nanoparticles, *J. Appl. Polym. Sci.* 101 (2006) 2320–2332.

24. J. Choubey, A. K. Bajpai, Investigation on magnetically controlled delivery of doxorubicin from superparamagnetic nanocarriers of gelatin crosslinked with genipin, *J. Mater. Sci. Mater. Med.* 21 (2010) 1573–1586.

25. A. K. Gupta, M. Gupta, S. J. Yarwood, A. S. Curtis, Effect of cellular uptake of gelatin nanoparticles on adhesion, morphology and cytoskeleton organisation of human fibroblasts, *J. Control. Release* 95 (2004) 197–207.

26. M. R. Kulterer, M. Reischl, V. E. Reichel, S. Hribernik, M. Wu, S. Köstler, R. Kargl, V. Ribitsch, Nanoprecipitation of cellulose acetate using solvent/nonsolvent mixtures as dispersive media, *Colloids Surf. A: Physicochem. Eng. Aspects* 375 (2011) 23–29.

27. S. Schubert, T. Delaney, L. Schubert, Nanoprecipitation and nano formulation of polymers: from history to powerful possibilities beyond poly(lactic acid), *Soft Matter* 7 (2010) 1581–1588.

28. S. A. Khan, M. Schneider, Improvement of nanoprecipitation technique for preparation of gelatin nanoparticles and potential macromolecular drug loading, *Macromol. Biosci.* 13 (2013) 455–463.

29. E. J. Lee, S. A. Khan, J. K. Park, K. H. Lim, Studies on the characteristics of drug-loaded gelatin nanoparticles prepared by nanoprecipitation, *Bioprocess Biosyst. Eng.* 35 (2012) 297–307.

30. Y. W. Won, S. M. Yoon, C. H. Sonn, K. M. Lee, Y. H. Kim, Nano self-assembly of recombinant human gelatin conjugated with α-tocopheryl succinate for Hsp90 inhibitor, 17-AAG, delivery, *ACS Nano.* 5 (2011) 3839–3848.

31. Y. C. Chen, S. H. Yu, G. J. Tsai, D. W. Tang, F. L. Mi, Y. P. Peng, Novel technology for the preparation of self-assembled catechin/gelatin nanoparticles and their characterization, *J. Agric. Food Chem.* 58 (2010) 6728–6734.

32. Z. Li, L. Gu, Effects of Mass Ratio, pH, Temperature, and reaction time on fabrication of partially purified pomegranate ellagitannin–gelatin nanoparticles, *J. Agric. Food Chem.* 59 (2011) 4225–4231.

33. H. Ai, S. A. Jones, Y. M. Lvov, Biomedical applications of electrostatic layer-by-layer nano-assembly of polymers, enzymes, and nanoparticles, *Cell Biochem. Biophys.* 39 (2003) 23–43.

34. T. G. Shutava, S. S. Balkundi, P. Vangala, J. J. Steffan, R. L. Bigelow, J. A. Cardelli, D. P. O'Neal, Y. M. Lvov, Layer-by-layer-coated gelatin nanoparticles as a vehicle for delivery of natural polyphenols, *ACS Nano.* 3 (2009) 1877–1885.

35. H. J. Lai, C. H. Kuan, H. C. Wu, J. C. Tsai, T. M. Chen, D. J. Hsieh, T. W. Wang, Tailored design of electrospun composite nanofibers with staged release of multiple angiogenic growth factors for chronic wound healing, *Acta Biomater.* 10 (2014) 4156–4166.

36. M. Jaiswal, A. Gupta, A. K. Agrawal, M. Jassal, A. K. Dinda, V. Koul, Bi-layer composite dressing of gelatin nanofibrous mat and poly vinyl alcohol hydrogel for drug delivery and wound healing application: *in vitro* and in vivo studies, *J. Biomed. Nanotechnol.* 9 (2013) 1495–1508.

37. M. C. Kwon, W. Y. Choi, Y. C. Seo, J. S. Kim, C. S. Yoon, H. W. Lim, H. S. Kim, J. H. Ahn, H. Y. Lee, Enhancement of the skin-protective activities of *Centella asiatica* L. Urban by a nano-encapsulation process, *J. Biotechnol.* 157 (2012) 100–106.

38. R. Filip, P. López, G. Giberti, J. Coussio, G. Ferraro, Phenolic compounds in seven South American *Ilex* species, *Fitoterapia.* 72 (2001) 774–778.

39. P. Mattila, J. Kumpulainen, Determination of free and total phenolic acids in plant-derived foods by HPLC with diode-array detection, *J. Agric. Food Chem.* 50 (2002) 3660–3667.

40. Y. Yamada, H. Yasui, H. Sakurai, Suppressive effect of caffeic acid and its derivatives on the generation of UVA-induced reactive oxygen species in the skin of hairless mice and pharmacokinetic analysis on organ distribution of caffeic acid in ddY mice, *Photochem. Photobiol.* 82 (2006) 1668–1676.

41. B. Choquenet, C. Couteau, E. Paparis, L. J. Coiffard, Flavonoids and polyphenols, molecular families with sunscreen potential: determining effectiveness with an *in vitro* method, *Nat. Prod. Commun.* 4 (2009) 227–230.

42. F. Afaq, V. M. Adhami, H. Mukhtar, Photochemoprevention of ultraviolet B signaling and photocarcinogenesis, *Mutat. Res.* 57 (2005) 153–173.

43. V. V. Venuganti, O. P. Perumal, Nanosystems for dermal and transdermal drug delivery, in: Y. Pathak, D. Thassu (eds.), *Drug Delivery Nanoparticles Formulation and Characterization*, Informa Healthcare, New York, 2009, pp. 126–155.

44. C. D. Ropke, T. M. Kaneko, R. M. Rodrigues, V. V. Silva, S. Barros, T. C. H. Sawada, M. J. Kato, S. B. Barros, Evaluation of percutaneous absorption of 4-nerolidylcathecol from four topical formulations, *Int. J. Pharm.* 249 (2002) 109–116.

45. D. P. Rivelli, R. L. Almeida, C. D. Ropke, S. B. M. Barros, Hydrolysis influence on phytochemical composition, antioxidant activity, plasma concentration and tissue distribution of hydroethanolic extract of *Ilex paraguariensis* extract components, *J. Agric. Food Chem.* 59 (2011) 8901–8907.

46. J. C. Zillies, K. Zwiorek, F. Hoffmann, A. Vollmar, T. J. Anchordoquy, G. Winter, C. Coester, Formulation development of freeze-dried oligonucleotide-loaded gelatin nanoparticles, *Eur. J. Pharm. Biopharm.* 70 (2008) 514–521.

Chapter 8

Lipid Vesicles for Skin Delivery: Evolution from First Generation

Tiago Mendes,[a] **Maria Manuela Gaspar,**[b] **Sandra Simões,**[b] **and Andreia Ascenso**[b]

[a]*Faculdade de Farmácia, Universidade de Lisboa, Av. Prof. Gama Pinto, 1649-003 Lisboa, Portugal*
[b]*Instituto de Investigação do Medicamento (iMed.ULisboa), Faculdade de Farmácia, Universidade de Lisboa, Av. Prof. Gama Pinto, 1649-003 Lisboa, Portugal*

tiagormendes@hotmail.com

8.1 Introduction

Skin is the largest human organ, and although topical administration is a promising route, it acts as an extraordinary main barrier for drug delivery, diffusion, and bioavailability [1]. The horny outermost skin layer, *stratum corneum*, is the major permeation obstacle for most molecules, especially hydrophilic ones, mainly due to corneocytes cells with less than 10–30% water content and intercellular lipid matrix [1, 2]. The *stratum corneum* structure has been described as "brick and mortar model" that represents the corneocytes and the surrounding intercellular lipids [3].

Carrier-Mediated Dermal Delivery: Applications in the Prevention and Treatment of Skin Disorders
Edited by Andreia Ascenso, Sandra Simões, and Helena Ribeiro

ISBN 978-981-4745-58-1 (Hardcover), 978-1-315-36447-6 (eBook)
www.panstanford.com

Topical formulations can be advantageous for being easily administrated with a minimal invasion leading to an increased patient compliance, and avoiding many variables like stomach content, intestinal motility, and first pass metabolism associated to oral and rectal administration routes.

Liposomes were the first lipid vesicles introduced by Bangham and Horne in 1964 [4]. They found that lipids tend to form vesicles in aqueous environment by establishing different interactions between them. Van der Waals forces of the apolar chains tend to keep the tails together and the hydrogen bonds between the water molecules and polar heads lead to the liposomes architecture and stabilization [5]. These vesicles have been used as models of biological membranes and as carriers of bioactive agents in different areas (cosmetics, therapeutics, vaccines, tissue engineering, imaging, food technology, etc.) [6]. Several vesicle generations have been developed, including vectorization strategies to escape from capture by the mononuclear phagocytic system (MPS) upon parenteral administration. Among them, the use of polyethylene glycol (PEG)-coated liposomes ("stealth liposomes") is one of the most studied approaches. Other targeting strategies consist of the inclusion of immunoglobulins, ligands, peptides, glycoproteins, growth factors, transferrin, and folate, among others, at liposome surface providing an additional targeting mechanism to receptors overexpressed in solid tumors [7, 8]. In general, liposomes are suitable for intravenous administration, but their rigid structure prevents the dermal and transdermal drug delivery [9]. The impaired skin drug delivery by this system has led to a demanded progress on their design [10, 11]. Therefore, a new type of vesicles called elastic or (ultra) deformable liposomes/vesicles (UDV) was developed. Lipid vesicles can have unilamellar or multilamellar concentric arranged bilayers. Their structure is composed by a hydrophilic core surrounded by a lipid bilayer that can contain a surfactant or an edge activator [12]. Hydrophilic drugs can be encapsulated in the aqueous core while lipophilic, amphipathic, and charged drugs can be incorporated in the lipid bilayer by hydrophobic and electrostatic interactions, respectively [12, 13].

UDV have numerous advantages to promote a non-invasive skin dermal and/ or transdermal delivery; to present high thermodynamic stability and elasticity; to be able to carry proteins, peptides among other larger molecules; to be easily produced in large scale; and to improve patients compliance, etc. [2, 12, 14]. However, some of these systems present limited physical stability hampering the existence of these systems in the market [10, 15].

Transfersomes, the first described UDV, were introduced by Cevc and Blume in 1992 [16]. These vesicles are characterized by having an edge activator (surfactant) responsible for their deformability allowing a deeply skin penetration [16]. In 2000, Touitou *et al.* presented another deformable lipid system, Ethosomes, with high ethanol content [9], and more recently in 2012, Song *et al.* introduced Transethosomes as a promising combination of Transfersomes and Ethosomes [17]. In addition, many other UDV have been investigated as well [18].

In this chapter, the evolution of lipid vesicles as drug delivery systems is described and summarized. Their potential as nanocarrier systems will be highlighted as well as their actual and future applications for the development of new cosmetic, therapeutic or diagnostic strategies for several skin disorders and/ or diseases.

8.2 Vesicles Composition

Lipid molecules such as phospholipids and cholesterol are the major components of the vesicular carriers [12]. Several factors related to the lipid molecules have a great influence on the drug incorporation and stability. Among them, the hydrophobic carbon tail, concentration and acidity of phospholipids are the most important factors [19]. The nature and type of lipid vesicle bilayers influence the incorporation efficiency of lipophilic drugs by affecting their partitioning coefficient and solubility. On the other hand, the encapsulation of hydrophilic compounds depends on the aqueous volume core that is closely related to the preparation method and also to the thickness and polarity of the vesicles membrane [20]. Although cholesterol endorses rigidity in the bilayer structure of liposomes prepared with fluid phospholipids, its presence may result in high encapsulation levels of hydrophilic drugs in the aqueous compartment by reducing the permeability of lipid bilayers. [21]. In addition, the environmental pH can also modify the drug-vesicle interactions [20].

The phospholipids mainly used for the lipid vesicles preparation are phosphatidylcholine from egg yolk or soybean, hydrogenated phosphatidylcholine, phosphatidylethanolamine, phosphatidyl glycerol, phosphatidyl inositol, phosphatidylserine, phosphatidic acid which are constituted by a complex mixture of fatty acids (C14–C20) [10, 22]. These phospholipids are constituted by two hydrophobic fatty acid chains or tails, a glycerol backbone and a polar head group. The purity, chain length and saturation degree affect the phase transition temperature (T_m) of the phospholipids [6, 23]. T_m corresponds to the temperature at which a particular phospholipid changes from a condensed gel-state (compressed and ordered structure) to a liquid-crystalline and less ordered condition. Shorter and unsaturated hydrocarbon chains present lower T_m. This can be justified considering that in unsaturated fatty acids, C=C bonds have less free energy than C–H bonds and the interactions among shorter acyl chains are also weaker when compared to longer chains. Liposomes prepared only phospholipids and maintained at a temperature lower than T_m are in a gel-state being lipid bilayers in a rigid, compressed and ordered structure. An increase of the temperature for higher values than T_m will enable movements of the lipid bilayers and the disordered state may consequently influence the stability of liposomes in terms of release of incorporated material. The modification of the apolar and polar regions leads to the synthesis of more stable phospholipids than the respective natural forms. Examples of synthetic lipids are dipalmitoyl phosphatidylcholine, dimyristoyl phosphatidylcholine, distearoyl phosphatidylcholine, hydrogenated soy phosphatidylcholine, 1-palmitoyl-2-oleoyl-snglycero-3-phosphocholine,1-palmitoyl-2-stearoyl(5-doxyl)-sn-glycero-3-phosphocholine,1,2-dioleoyl-snglycero-3-phosphocholine, 1,2-distearoyl-sn-glycero-3-phospho-(10-rac-glycerol), 1,2-dipalmitoyl-sn-glycero-3-phosphoglycero and 1,2-distearoyl-sn-glycero-3-phosphate [6].

Concerning the UDV preparation, unsaturated phosphatidylcholine from soybean or egg yolk are usually used due to its lower T_m (<0°C), contributing to the lipid bilayers fluidity and vesicles elasticity [23]. In addition, this property is also enhanced by the inclusion of edge activators in lipid composition which significantly reduces the T_m as well. In fact, the modification of this parameter could be the key factor to obtain both vesicular flexibility and stability [23]. The chemical and physical characteristics of the edge activator also influence the penetration and permeation through the skin layers besides affecting the drug interaction with the vesicle bilayer [24]. The main edge activators are alcohols (e.g., ethanol and isopropyl), amides (e.g., azone), several esters, glycols (e.g., propylene glycol), several fatty acids (e.g., oleic acid), sulfoxides (e.g., dimethyl sulfoxide (DMSO)), anionic (e.g., sodium lauryl sulfate, sodium cholate and sodium deoxycholate), cationic (e.g., ammonium halides) and nonionic surfactants (e.g., Brij®, Tweens® and Spans®) [25, 26]. The ingredients of vesicular formulations should be chosen according to their characteristics and optimized on a case-by-case manner (e.g., amount and ratio of PC: surfactant; surfactant type, preparation method, etc.). A multifactorial and space design analysis should be used to further study the variables of the formulation development. For example, the concentration of the edge activator should be high enough to guarantee the vesicles flexibility but below the critical micellar concentration (CMC) [23]. Phospholipids with compact and short chains tend to form micelles [6]. Mixed micelles have usually low drug entrapment efficiency due to their low elasticity and small size. The hydrophilic/lipophilic balance (HLB) value of edge activator should also be considered during the formulation studies [23]. A lower HLB may contribute to decrease the vesicles size [23]. Finally, nonionic surfactants would be preferred in order to prevent toxic effects even though a lower electrostatic repulsion between vesicles will be obtained.

The composition of the aqueous phase is also another important subject. Actually, the osmotic gradient may influence the permeability and elasticity of UDV probably by modifying the area per lipid on the membrane surface. The pH value of the vesicular formulation also optimizes the UDV properties. It was reported a higher flux for acidic formulations (pH 5) [27].

Other components could also be added in lipid vesicles formulation, such as PEG (for steric stabilization, higher skin permeation and active targeting by coupling ligands to the terminus of PEG molecules); cyclodextrins (for hydrophobic drugs incorporation in the aqueous vesicle core); DMSO (as permeation enhancer), and polymers like Carbopol® (for rheology improvement especially for topical administration) [6, 23]. Poly (aminoacid)s, heparin, dextran and chitosan were also suggested to replace PEG. Antioxidants, complexing agents (EDTA) and inert atmosphere (N_2) should also be used to prevent the lipids hydrolysis and oxidation [6].

8.3 Preparation Methods

Liposomes can be classified according to the preparation method, lamellarity and size: unilamellar vesicles (ULVs) (25 nm to 1 μm), multilamellar vesicles (MLVs) (0.1–15 μm), oligolamellar vesicles (OLV) (0.1–1 μm), multi-vesicular vesicles (MVV) (1.6–10.5 μm) large unilamellar vesicles (LUV) (100 nm to 1 μm) and small unilamellar vesicles (SUV) (25–50 nm) [6].

The preparation method influences size, lipid vesicles structure and stability of incorporated molecules. Several preparation methods have been applied to produce liposomes with different arrangements (Fig. 8.1). The classic, thin-film hydration method includes the formation of a lipid film after evaporation of organic solvents followed by the addition of an aqueous solution (hydration step) [28]. The final steps include the homogenization and stabilization processes. Sterilization and separation of the non-incorporated drug fraction can also be done. The thin film hydration method produces heterogeneous MLVs and allows the use of several types of lipids at high concentrations. It is the simplest method to encapsulate hydrophilic or hydrophobic compounds [6]. Nevertheless, it appears to have some disadvantages including the heterogeneous vesicular size and shape obtained as well as the requirement of size reduction techniques and the use of organic solvents [11].

Different mechanical procedures may be performed in order to obtain homogenous mean size lipid vesicles. The French press technique is based on the application of a very high pressure on multilamellar or LUV that have to pass through a small orifice. It is possible to obtain SUV, being this method suitable for either sensitive lipids or drug molecules. Some drawbacks of this technique are the use of very high working pressures and achievement of larger vesicles than those produced by sonication. In addition, it is not indicated for large-scale production [29]. Although vesicles sonication by ultrasonic waves is a simple process and can be used to obtain small size vesicles, it is not suitable for unstable molecules due to a high risk of lipid or drug degradation. In addition, a low entrapment efficiency is usually obtained [30]. Freeze-thawed method leads to the formation of unilamellar vesicles, being the lipid concentration and ionic strength the major factors that influence this method [7]. The most widely performed downsizing method is the Extrusion through polycarbonate membranes mainly because of its high reproducibility. The size of the track-etched pores, the number of extrusion cycles and the applied pressure will influence the homogeneity of the vesicles mean size [31]. The pressure applied to liposomal suspension can be created using an inert gas such as nitrogen [11]. Different extrusion systems are marketed enabling the production of liposomes from 1 mL scale to several liters either at discontinuous or continuous flows [11, 32].

Some methods of lipid vesicles production have in common the removal of organic solvents. The ethanol or ether injection method is one of them, and it is based on the ability of forming lipid vesicles when an organic solvent solubilizing the phospholipid mixture is further injected into an aqueous environment. Although the final step includes the organic solvent evaporation or tangential flow filtration, the final product may contain residues of these organic solvents. Nevertheless, this rapid, simple and reproducible method allows sterilization and avoids the chemical or physical treatment of lipids [6, 32]. The major advantage of this method is the possibility of incorporating large proteins or antigens in the lipid bilayer [33]. Hydration of dried and lyophilized, free-flowing granular particles containing the mixture of phospholipid and the drug (*proliposomes*) with water or organic solvent that can be further removed by rotatory vacuum evaporation, fluidization or spray dryer [34–36]. Although the time procedure can be extended, these dried vesicles have an enriched stability [36].

Reversed phase evaporation is another method which consists of the initial dissolution of phospholipids in organic solvents followed by water addition under stirring to obtain a water-in-oil emulsion followed by evaporation of the organic phase [37]. The major advantage of this technique is the commonly high drug incorporation efficiency obtained despite drug exposure to organic solvents [6, 38].

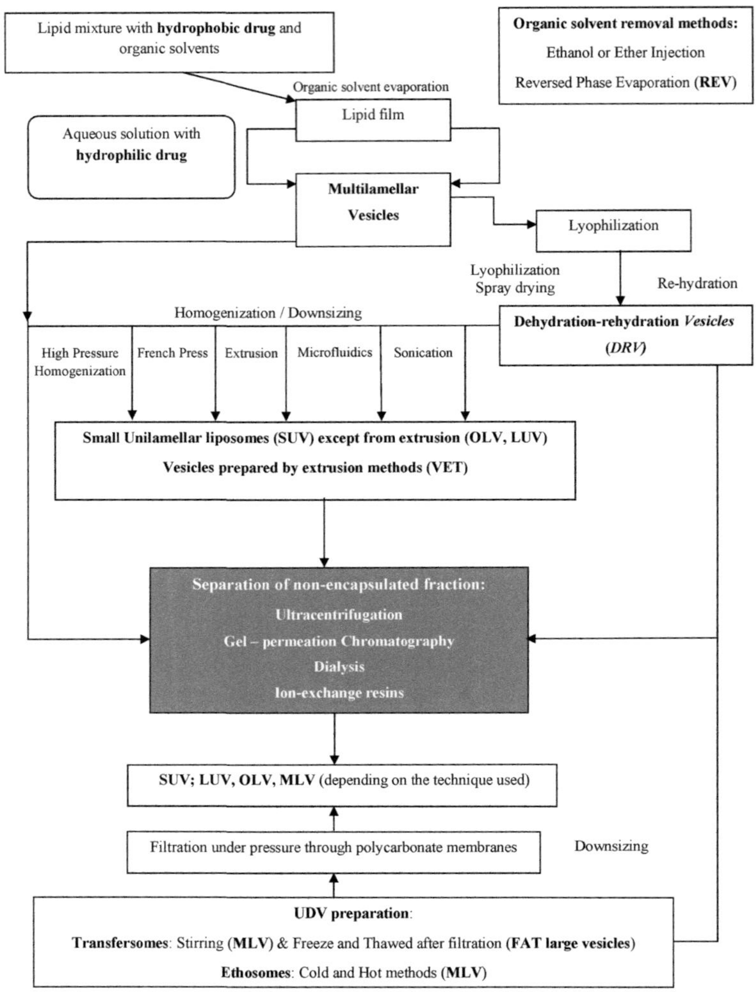

Figure 8.1 Lipid vesicles preparation methods (adapted from [47]).

Detergent removal method using bile salts and alkyl glycosides for lipid solubilization is also used to produce micellar systems incorporating proteins in the lipid bilayer. The detergent can be removed by dialysis or column chromatography with possible separation of low molecular weight molecules [39].

Microfluidic methods, involve fluid flow through channels in the range of 5–500 μm using flow focusing, pulsed jet, thin film hydration and microfluidic droplets techniques. Altering the flow rate, size, polydispersity index and encapsulation efficiency can be modified [40]. The major advantage of this method is dispensing a rigorous volume of liposomal formulation and achieving a precise control of size distribution. Although monodisperse vesicles are obtained in a continuous operating mode, more extensive studies are still needed to solve certain scale-up issues [41].

The use of dense gas or supercritical fluids like carbon dioxide is another method for lipid vesicles formation with the possibility of replacing organic solvents leading to more effective purification steps besides the ability of solubilizing a wide range of drugs in a continuous operating mode. Some drawbacks include the use of a high pressure and the initial costly investment [42].

Drug loading can be obtained either passively or actively if the drug is encapsulated during or after liposome formation, respectively. The active loading is used for encapsulating molecules in preformed liposomes by creating a pH and/or salts gradient between the intra and extra-liposomal medium. This technique has been applied for encapsulating bases or weak acids. It presents several advantages, over passive methods, such as increased entrapment efficiency, drug stabilization in the aqueous space of liposomes, reduced and controlled drug release, and possibility of extemporaneous preparation [43].

Regarding UDV preparation, above-mentioned techniques may be applied. However, for Transfersomes some details are specific for these UDV. After mixing all the components under stirring for several days, the application of several freeze and thawing cycles is commonly used [44]. In the case of Ethosomes, they can be prepared by the addition of the aqueous phase drop by drop to the mixture of lipid and ethanol solubilized under stirring at 30°C (cold method), or the addition of ethanol drop by drop to the lipid mixture and aqueous phase at 40°C (hot method) [2, 9]. Filtration under nitrogen pressure through polycarbonate membranes for size homogenization (150 ± 50 nm), and eventually sterilization through 0.2 μm filter under aseptic conditions can also be performed [45]. However, this sterilization process is not effective in removing virus [6]. Concerning Transethosomes, the production method is quite similar to Ethosomes preparation; the difference lies in the addition of a surfactant to the lipid mixture [17]. UDV may suffer irreversible aggregation when rehydrated after freeze-drying even in the presence of a high concentration of cryoprotectants (sugars such as trehalose) [46]. Therefore, these vesicles should be stored in its liquid/ gel state preferentially at 4°C.

Nowadays, the transfer of technology to industry is primarily focused on large-scale production of liposomes. A strict control of the final products is required to ensure the predictable therapeutic effect of liposomal formulations. For this achievement several quality criteria have to be defined during the all process, such as unwanted by-products like residues of organic solvents, degradation products, drug release, instability, pyrogens, endotoxins, and sterility conditions. Alternative strategies, such as lyophilization and production processes in closed containments equipped with sterile filter barriers, are currently used by industry to solve some of the above-described problems [11].

Real-time, high sensitive and low-costing monitoring devices are useful for the quality control. Nowadays, quality cannot any longer be tested simply on the final medicinal product, but it should be performed during the all process to improve the product quality and manufacturing efficiency [48].

8.4 Vesicles Characterization

The characterization of lipid vesicles often involves the use of several methodologies, including physical, chemical, and biological techniques [12].

Physical characterization corresponds to the analysis of the vesicles structure, several parameters, such as shape, lamellarity, size, zeta potential, elasticity, fluidity, viscosity, crystalline or amorphous state, etc. Morphological examinations are frequently performed by optical microscopy for large MLV, and transmission and scanning electron microscopy for smaller vesicles [48]. Laser diffraction using dynamic light scattering or photon correlation spectroscopy allows obtaining the vesicular size and polydispersity index (PdI) by measuring the volume of equivalent spheres [49]. PdI should present a value below 0.2 corresponding to a homogeneous population. Nanoparticle tracking analysis is a recently developed method that simultaneously records size, aggregation, zeta potential, and enables the use of florescent probes

[50]. Zeta potential determined by laser Doppler anemometry or electrophoresis allows the study of the superficial charge that influences both stability and adhesive properties to the skin of the lipid vesicles included in topical formulations. High zeta potential values (negative or positive) stimulate the vesicles repulsion avoiding their aggregation. Consequently, it will lead to constant mean size and PdI [51]. The deformability or elasticity is a crucial topic to evaluate the ability of the vesicles to overpass pores with much lower diameter than their mean size. Extrusion method and filtration under pressure are the usual techniques used to evaluate the deformability index. These methods allow an initial and indirect *in vitro* prevision of the skin permeation and penetration of the tested formulations [12, 52]. The surface tension influences the vesicles elasticity, and can be determined by capillarity methods [49]. Fluorescence anisotropy using polarized excitation light is inversely correlated with lipid bilayer fluidity of the vesicles which can also affect drug release [53]. Viscosity evaluated through rheological analysis is an additional parameter to be considered regarding its influence on the formulation stability, spreadability, and drug release profile. [54–56]. Differential scanning calorimetry can also be employed for thermal studies of both interactions between drug and excipients as well as lipid vesicles and skin, according to the alteration of the T_m as well as endothermic and/or exothermic events [49, 57]. X-ray diffraction can be performed for analyzing the crystalline or amorphous state of the carrier [58].

Concerning the chemical characterization, the entrapment efficiency is the most important parameter [59]. Several factors may influence the loading capacity of the vesicles, namely the pH, drug polarity, type, and concentration of both lipid and/or edge activator selected. Regarding the drug polarity, hydrophilic (Log Poct < 1.7) and hydrophobic (Log Poct > 5) molecules will be preferably entrapped in the aqueous core and in the lipid membrane, respectively [6]. Consequently, higher entrapment efficiencies are obtained for all hydrophobic drugs. This drawback could be minimized by proliposomal formulations. Gupta *et al.* [60] have proposed a liquid protransfersomal formulation which is only converted to Transfersomes *in situ* by absorbing water from skin. The determination of entrapment efficiency is based on the drug and lipid quantification being HPLC or spectrophotometry the most commonly used techniques [61]. A crucial step is the separation of the non-incorporated drug by ultracentrifugation, gel filtration or dialysis methods [49]. The pH evaluated by potentiometry affects the molecules ionization, and consequently its interaction and entrapment efficiency. The pH level should be suitable for achieving a balance between formulation properties and biological applications, including the administration route [25, 62]. Besides the differential scanning calorimetry, Fourier transform infrared spectroscopy is also used to study the chemical interactions between molecules and the influence of the lipid vesicles in the *stratum corneum* destabilization [56].

Both physical and chemical characterizations can be carried out during real-time or in accelerated stability conditions at different temperatures, relative humidity percentage, and exposure to light and air (e.g., evaluation of peroxidation) according to the guidelines Q1A ICH. According to the results from these studies, it is possible to estimate the optimal storage conditions [63].

Biological characterization includes the microbiological stability, safety, and biological activity. The microbiological stability is usually performed in aerobic and anaerobic conditions using suspension or adherent cell cultures. Intravenous formulations can be sterilized by filtration through 0.2 μm membranes or prepared under aseptic conditions. These formulations should also be free from any endotoxins/pyrogens.

Flow cytometry could be used for the assessment of apoptosis and necrosis (cytotoxicity assays) in order to ensure the safety and biocompatibility of the drug-loaded lipid vesicles [64]. For cell internalization studies, bio-imaging assays using bioluminescence compounds and fluorescence confocal microscopy are also used [65]. These methods are quite rigorous and allow a more complex study compared to the *in vitro* classic methods.

The biological activity and safety of the vesicular formulations are tested through *in vitro* and *in vivo* studies. Different cell lines can be used including keratinocytes, melanocytes, or fibroblasts (for topical use), and for the treatment of certain parasitic diseases, *Leishmania* and fungal dermatophytes like *Trichophyton rubrum* and *T. Mentagrophytes* depending on the proposed study [53, 66–68]. Different animal models can also be used to evaluate the pharmacological effect [69].

Topical delivery of vesicular formulations can be characterized by *in vitro*, *ex vivo*, and *in vivo* experiments [49]. *In vitro* and *ex vivo* experiments allow the determination of the delivery/release and bioavailability/permeation of topical formulations. To perform these studies Franz type diffusion cells are employed using synthetic membranes such as cellulose, silicone or polysulfone with different pore sizes as well as human or animal skin. Fluorescent probes (e.g., rhodamine) can be incorporated in lipid vesicles to assess carrier skin retention. An important aspect is the selection of the appropriated receptor phase, a buffer that ensures both physiological and sink conditions according to the regulatory guidelines [61, 70, 71]. The permeation profile will depend on the amount of drug present in the receptor phase or blood through *in vitro* and *in vivo* experiments, respectively. Tape striping studies are recommended to evaluate the penetration profile of the lipid vesicles in the different skin layers [70, 72]. However, this technique has some limitations, as it is an invasive method (i.e., the tissue must be destroyed) and with a difficult and variable step of calibrating the amount of tissue removed. Thus, this procedure should be rigorously validated. Other methodologies like laser scanning confocal microscopy and multiphoton excitation fluorescence microscopy with fluorescent probes have been used as alternatives to the tape stripping technique. These two methods are noninvasive and provide information about the skin depth [73]. In addition, confocal Raman microscopy and enzyme treatment with trypsin are also other possible techniques for the evaluation of skin penetration.

8.5 Pharmacokinetics

The release kinetic is dependent on the entrapped active molecule, lipid composition, vesicles preparation method and administration route. On the other hand, it should be suitable for the intended application, i.e., the entrapped drug should not leak out of the vesicles too fast (before reaching the target) or too slow (not reaching the therapeutic level) [5].

Intravenous administration is the most common administration route for conventional liposomes. Lipid vesicles interact with plasma proteins including opsonins leading to their capture by the MPS. High- and low-density lipid proteins may also lead to the vesicles disruption and consequently only a small fraction is able to reach the site of action. Only in case of a high solubility in the body fluids, an increased half-life with a low plasma clearance and a selective delivery with low distribution volume, the liposomal formulation will have reduced immunogenicity and adverse effects [74]. These characteristics can be achieved with PEGylated liposomes, also called stealth liposomes, or with ligands at liposome surface, targeted liposomes. PEG decreases the phagocytosis process and reduces the opsonin adsorption at the liposome surface due to the steric stabilization and protection of liposomal surface by the chains of this polymer [75, 76]. This will lead to an increased circulating time of these vesicles, and consequently, there will be a higher possibility of PEG liposomes to accumulate at the site of action. In addition, other variables such as the pH, temperature, ultrasonic waves, magnetic field and light have been investigated in order to improve the target release of the therapeutic agent [6]. The major sites of accumulation of lipid vesicles are the liver, spleen and solid tumors. This can be advantageous if these organs are the intended target. For topical administration, rigid liposomes do not penetrate in healthy skin in its intact form. However, according to literature, the liposomal content has been found in skin layers based on some penetration mechanisms (Fig. 8.2). The free drug delivery mechanism is based on the ability of free drug molecules (non-incorporated drug fraction and remained fraction delivered after the liposomes interaction with the skin cells) permeate the skin. Another theory is related to the possibility of liposomes to penetrate through the hair follicles (transfollicular route) [3]. Adsorption to the *stratum corneum* surface was also demonstrated even though collapsed vesicles could be an obstacle by forming a rigid film at the skin surface [77]. The penetration enhancing mechanism may explain the penetration of both entrapped and non-entrapped drug fractions (especially hydrophilic compounds) by the formation of new pathways enhanced by osmotic gradient [78]. Although the intercellular route is the main one, the transcellular route can also occur following the lipid vesicles fusion with the skin cells and further drug diffusion [13].

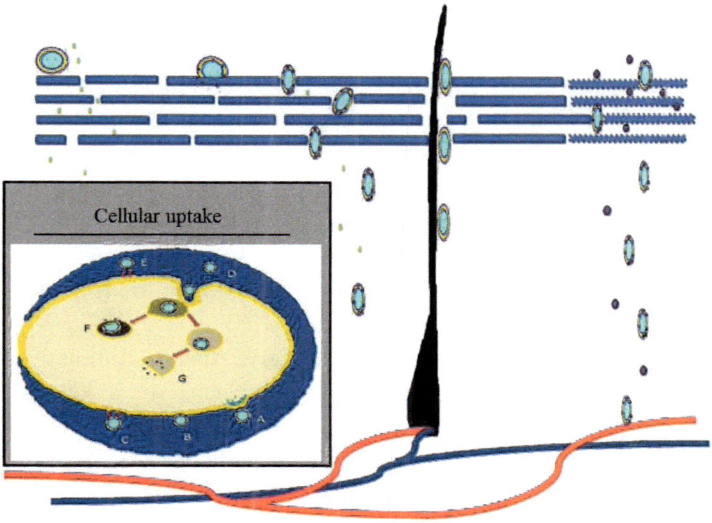

Figure 8.2 Skin penetration and cellular uptake of lipid vesicles (adapted from Bozzuto and Molinari and Ascenso *et al.* [75, 84]). Skin penetration: (I) Free drug delivery (non-incorporated fraction); (II) Drug delivery after the vesicles fusion with the skin surface; (III) Drug delivery through UDV penetration into the skin layers; (IV) Drug delivery through UDV penetration into transappendageal route; (V) Drug delivery through lipids disruption and UDV penetration with a skin enhancer (e.g., ethanol). Cellular uptake: (A) Vesicle fusion with cell membrane with content release; (B) Adsorption of vesicle on the cell membrane by electrostatic interactions (especially with specific ligands); (C) Vesicle biding by specific receptors; (D) Endocytosis or phagocytosis of lipid vesicles; (E) Lipid exchange between vesicle bilayer and cell membrane; (F) Vesicle fusion with lysosome with possible drug release; (G) Vesicle fusion with endosome and destabilization with possible drug release.

In the case of Transfersomes, the presence of an edge activator contributes to the vesicles elasticity, and therefore, to a lower energy required for curved structures formation. The combination of elasticity properties of these vesicles and the hydration driven force (hydrotaxis) which occurs with the aqueous phase evaporation from these vesicles applied on skin surface (under non occlusive conditions) will contribute for Transfersomes to squeeze through intracellular spaces between corneocytes in the *stratum corneum* in its intact form and to achieve the deeper skin layers [77]. The ethanol incorporation in both Ethosomes and Transethosomes causes reversible perturbations, including the lipids melting and disruption, on the cells membrane of the *stratum corneum*, increasing the skin fluidity and permeability. In fact, alcohols are well-known skin enhancers which increase the transepidermal water loss [23, 25]. Ethosomes, unlike Transfersomes, can be applied under occlusive conditions, as it is possible to apply them in transdermal patches [10].

Regarding the transdermal delivery, the maintenance of the incorporated drug fraction in the vesicles is a key factor that affects the penetration and permeation profiles of UDV. General requirements to obtain an ideal transdermal delivery are compounds with an aqueous solubility higher than 1 mg/mL, oil–water partition coefficient between 10 and 1000, molecular weight lower than 500 Da, melting point lower than 200°C, and formulation pH between 5 and 9 (nonionized form) [79]. Besides physical approaches like electrically assisted methods (ultrasounds, iontophoresis, electroporation, and magnetophoresis) and horny layer removal procedures (microneedles and ballistic droplets), lipid vesicles play an important role in the transdermal route [1, 80]. The synergistic effect of permeation enhancer presence in vesicles formulation has been used to optimize the transdermal delivery. The quantification by a mathematical equation of such synergy or the extent of interaction is an important parameter in pharmaceutical development [81] and for transdermal application the synergy should be higher than 1. In addition, it was reported that cationic surfactants contributed for a higher permeation rate due to the interaction with the anionic charge on skin surface [23]. UDV containing 1,2-dioleoyl-3-trimethylammonium-propane (DOTAP) have been shown to be a potential formulation for transcutaneous gene delivery [23]. Depending on the lipid vesicles properties, carriers may reach the systemic circulation via the lymphatic vessels and Langerhans cells uptake, and especially when the blood vascularization is altered like in tissue inflammation and tumors [82].

Regarding the cellular uptake (Fig. 8.2), lipid vesicles can interact with the cells surface by a variety of ways, such as enzymatic reactions, interaction with membrane receptors, lipids exchange between cellular and vesicular membranes, absorption due to electrostatic interactions (especially on the presence of specific ligands on vesicles surface), fusion with the cellular membrane, and internalization of vesicles by endocytosis and phagocytosis [75]. The vesicles internalization may lead to endosomes formation and/or fusion with lysosomes, and consequently, to the release of the drug content due to the destabilization or degradation of the vesicles membrane [83]. The cellular uptake by this last mechanism is not favorable due to the degradative effect of lysosomal enzymes and low pH. However, this step enhances the release content from pH sensitive lipid vesicles to the cytoplasm [59].

8.6 Toxicology

The most used lipid component, i.e., phosphatidylcholine used in lipid vesicles formulations is considered to have a GRAS (generally regarded as safe) status. However, the chemical modification of lipid compounds and the incorporation of other components might increase the toxicity of the final formulation [35]. Skin penetration promotion by chemical enhancers proportionally increases with the possibility of causing lipid disturbance which may lead to the keratinocytes cytotoxicity and skin irritation. In fact, it is quite challenging to achieve an ideal balance between safety and skin penetration enhancer activity [86]. Several *in vitro* and *in vivo* tests should be performed for each particular drug, route of administration and lipid composition to access the safety of lipid vesicles formulations [88].

Among *in vitro* tests, the evaluation of erythrocyte hemolysis, platelet aggregation, cytotoxicity (MTT, MTS, or other similar assays) and formation of reactive oxygen species after cells exposure to vesicular formulation are the most recognized [88]. *In vivo* studies should also be included as the phagocytic index determination; pyrogenicity determination by rabbit body temperature control after the formulation administration or by LAL (limulus amebocyte lysate) test; systemic and topical tolerability (e.g., acute dermal toxicity), and monitoring survival tests [43,

85, 87]. The choice of these tests will depend on the therapeutic activity and administration route of the vesicular formulation. For example, a recent study evaluated the genotoxicity and gene expression in rats after intravenous administration of cationic empty liposomes. Even though any hepatic toxicity was observed, it was found DNA damage in lung cells followed by inflammatory cytokines release (IL6 and CCL2) [88]. These results might be associated to the superficial charge of liposomes that could affect the intracellular components [89].

In recent years, a new application of lipid vesicles has been used as a model for toxicological evaluation of xenobiotics [92]. These studies tend to biomimetic the fluid mosaic model of cytoplasmatic membrane by replacing the traditional *in vitro* assays by lipid vesicles. This replacement can be justified considering that a large number of xenobiotics influence the membrane permeability and porosity. In particular, these studies based on toxic effects in the lipid bilayer may be applied to evaluate acrylates used in dental materials, antibiotics like aminoglycosides, detergents and bacterial peptides. The results assessment can be based on different techniques mainly the fluorescence confocal microscopy, with fluorescent dyes-loaded vesicles, and differential scanning calorimetry for the evaluation of the T_m, lipid melting or temperature destabilization [90].

8.7 Evolution from First Generation

Liposomes are the oldest lipid vesicles that were designed for the first time in 1960s. Liposomes ability to incorporate both hydrophobic and hydrophilic drugs has led to an intense investigation and several optimizations in the last 50 years, becoming the most widely accepted carrier system (Fig. 8.3).

In order to become undetectable by the host immune system, PEG was added at the lipid vesicle surface in 1990, as already mentioned. These new liposomes with a longer half-life after intravenous administration were called stealth liposomes.

Ligand target liposomes are capable of recognizing and binding to receptors at the cells of interest with increased drug accumulation [91]. Peptides, glycoproteins, specific cell receptors (e.g., epidermal growth factor receptor) and monoclonal antibodies are the most used targeting ligands [92, 93]. They can be attached at the lipid bilayer or bound via a spacer arm like PEG for an increased time circulation [94]. Folates, transferring, and hyaluronic acid moieties can also be anchored at liposome surface to specifically targeting receptors that are overexpressed at tumor sites [95].

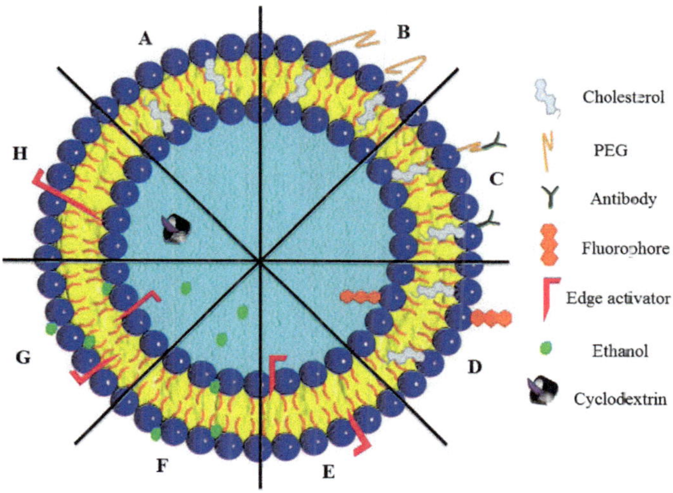

Figure 8.3 Cross section of different types of lipid vesicles. (A) Conventional liposome; (B) Stealth liposome; (C) Immunoliposome; (D) Fluorescent liposome; (E) Transfersome; (F) Ethosome; (G) Transethosome; (H) Cyclodextrin-loaded liposome. Adapted from Monteiro *et al.* [6].

Charged liposomes are another type of innovative vesicles. They can be anionic or cationic depending on the type of lipids used in their composition. Negative charged liposomes have been extensively used due to their enhanced interaction with MPS and particularly used with high success in pathologies localized in liver and spleen [96]. Unlike anionic charged liposomes, cationic ones were firstly developed in 1987 for gene delivery. Negative charged nucleic acids bind to positive charged vesicles through electrostatic interactions enabling cell transfection. More recently, chemotherapeutic drug–loaded cationic liposomes have shown an increased efficacy in the tumor blood vessels [97]. In fact, the formed tumor blood vessels exhibit anionic sites of angiogenesis which establish electrostatic interactions with the cationic liposomes [97]. Another advantage of this type of liposomes is their potential to pass through the blood–brain barrier enhancing the drug delivery in the central nervous system [98]. Concerning the transdermal drug delivery, charged liposomes have been studied as an interesting tool due to their enhanced skin penetration.

Stimuli-responsive liposomes have been developed as another targeting strategy and especially to overcome the problem of a slow drug release. When the stimuli are exclusively related to the disease pathology, it is possible to allow the liposomes to release the incorporated drugs in the presence of the pathological trigger. The stimuli can be generated from the tissues like pH or redox potential or from external sources like hyperthermia, ultrasound or electromagnetic field [99]. pH responsive liposomes tend to destabilize and release the incorporated drugs under acidic conditions such as those in tumors, inflammation and infections sites (pH < 6.5). The major lipid components used for pH responsive liposomes formulation are phosphatidylethanolamine, terminally alkylated copolymer of poly(N-isopropylacrylamide) and dioleoyl phosphatidyl ethanolamine combined with cholesteryl hemisuccinate and polyorganophosphazenes [100]. Redox responsive liposomes use the redox potential of the intracellular environment for the content release. Additionally, lipids with disulfide bridges can be included in e lipid vesicles composition in order to influence their membrane stability. Upon reaching the intracellular spaces, redox responsive liposomes destabilize in response to disulfide reduction by glutathione [100]. Temperature sensitive liposomes, studied since 1978, respond to temperatures above the physiological one releasing their content when heating is applied [75]. The T_m of the phospholipids may cause changes in membrane permeability of the lipid bilayers, and consequently, to the content release (e.g., DPPC has a T_m of +41°C). The hyperthermia applicators can be based on radio frequency, ultrasound, microwaves and magnetic nanoparticles (iron oxide like magnetite). The main lipid used in temperature responsive liposomes is DPPC which can be combined with polymers like poly(N-isopropylacrylamide-co-acrylamide) and PEG [100].

Concerning the topical and transdermal administrations, a new type of elastic vesicles already described as UDV has appeared for an enhanced skin drug delivery. The first UDV to appear were Transfersomes in 1992 [15]. The name means "carrying body" and derives from the Latin and Greek words "transferre" and "soma," respectively [48]. Transfersomes composition enables the deformability of the lipid membrane being maximal when the optimized local composition is achieved in response to an anisotropic stress. This characteristic allows Transfersomes to squeeze through hydrophilic skin pathways five to ten times narrower than their own diameter without loss of components [12]. The first-generation of Transfersomes includes the vesicles composed by phospholipids with edge activators [10]. The second-generation of Transfersomes is composed by a combination of a bilayer component (e.g., phosphatidylcholine) and more than one amphipathic membrane destabilizing components [101]. The third-generation of Transfersomes is constituted by amphipathic non-phospholipidic bilayers, but unlike the first and second generations, the surfactant is replaced by a water-soluble modulator compound (e.g., organic ions) that have the same effectivity [102]. Ethosomes were the second UDV to be introduced in 2000 [9]. Besides lipid molecules, these vesicles are composed by a high ethanol content (≥ 30%) [23] which is a skin enhancer according to the mechanism already described [49]. *Transethosomes* were one of the most recently described UDV, being introduced in 2012. Surfactant and ethanol conjugation synergically contributes for a high elasticity of the vesicles and skin permeability and penetration [17]. In the recent years, PEG liposomes have also been introduced for topical application resulting in a highly stable formulation with increased entrapment efficiencies and skin deposition when compared to classical liposomes, Transfersomes (first-generation) and Ethosomes [103].

Table 8.1 summarizes the numerous generations of lipid vesicles according to their structure and/or preparation method that have been developed for specific applications or purposes.

Table 8.1 Description of lipid vesicles according to their structure and/or method of preparation

Lipid vesicle	Definition
Aquosomes	Composed by a tree layer structure and a central solid nanocrystalline core coated with polyhydroxylated compounds in which active molecules are adsorbed; the core provides stability and the coating protects against dehydration; these vesicles are applied for peptide and proteins delivery (e.g., insulin) [104]
Archaeosomes	Composed by Archaeal bacterial-type lipids exhibiting high stability to temperature, pH, oxidative environment, high pressure, action of phospholipases, bile salts, and serum proteins; these vesicles are used for immunization [105, 106] among other applications
Dendrosomes	Vesicles with spherical dendritic structure for direct gene delivery; the major advantages are: inexpensive production, high stability and easy manipulation [107]
Discosomes	Derived from niosomes by the addition of lanolin derivate; used for ophthalmic administration; the major advantages are the suitable size and shape which prevents the passage to the systemic circulation [108]
Emulsomes	Colloidal system with an internal core of lipids and triglycerides with a high concentration of lecithin, for stabilization, forming an o/w emulsion; used for antifungal and antileishmanial drugs [109]
Enzymosomes	Liposomal vesicles with surface ligand enzymes, e.g., superoxide dismutase; application in inflammation [110]
Erythrosomes	Large lipid vesicles with cytoskeletons from red blood cells; applied for the membrane protein reconstitution [111]
Genosomes	Cationic lipid vesicles used for gene transfection into eukaryotic cells; these vesicles are less immunogenic, can be targetable and are easy for large-scale production than the viral vehicles [112]
Immuno-liposomes	Lipid vesicles containing ligands at their surface; one type of specific target liposome with application in drug delivery to cancer cells and central nervous system (due to the ability to overpass the brain blood barrier), gene therapy and molecular imaging [113]
Immunosomes	Lipid vesicles coated with glycoprotein (e.g., T cell receptor and envelope of Influenza virus glycoproteins) used for immunity stimulation [114, 115]

Invasomes	Lipid vesicles incorporating terpenes for enhanced skin penetration; they may be used for the treatment of cutaneous malignant or non-malignant diseases [116, 117]
Marinosomes	Lipid vesicles based on a natural marine lipid extract containing a high concentration of a polyunsaturated fatty acids which metabolites have an anti-inflammatory action [118]
Menthosomes	Lipid vesicles with cetylpyridinium chloride and L-menthol as penetration enhancers for transdermal delivery [119]
Niosomes	Lipid vesicles constituted by nonionic surfactants which are usually biodegradable and nontoxic; applied for cosmetic and/ or medical purposes [120]
Novasomes	Non-phospholipid liposomes constituted by several bilayers and an amorphous core composed of dioxyethylene cetyl ether, cholesterol and oleic acid; applied in vaccines [121]
Pharmacosomes	Amphiphilic phospholipid complexes of drugs bearing active hydrogen that bind to phospholipids; applied in a variety of inflammatory and heart diseases, cancer, for protein and herbal drug delivery [122, 123]
Photosomes	Lipid vesicles constituted by a bacterial enzyme (photolyase) that can repair ultraviolet B radiation induced by cyclobutane pyrimidine dimers; these vesicles are applied in cosmetic (e.g., sunscreens) and in photodynamic therapy due to their potential as a DNA repair system [124]
Phytosomes	Lipid vesicles developed to incorporate plant extracts (e.g., ginsenosides, flavonoids, epigallocatechin, procyanidins, quercetin, curcumin and naringenin) in order to increase their bioavailability [125]
Sphingosomes	Lipid vesicles mainly constituted by sphingolipids which increase the formulation stability [126]
Vesosomes	Structures with interdigitated bilayer compartments constituted by ethanol and saturated phospholipids; the numerous compartments can encapsulate other vesicles, biological macromolecules and colloidal particles; the major advantage is the incorporation of different compounds with modified release [127]
Virosomes	Vesicles including viral surface proteins without the genetic material (no replication ability); the major advantage is its ability to introduce the encapsulated macromolecule into the cytoplasm of cells by the viral proteins; used as drug delivery systems and adjuvants for cancer therapy [128]

8.8 Therapeutic, Diagnostic, and Cosmetic Applications

Lipid vesicular systems are often used for a variety of cosmetic, medical, and diagnostic purposes. Nevertheless, this chapter will be mainly focused on topical applications of these systems.

Topical lipid vesicles are commonly used considering the absence of skin irritation besides the exceptional low acute and chronic toxicity which leads to a very high tolerability [129]. At the moment, cancer treatment constitutes a very important application of liposomes with improved effectiveness and decreased side effects of new and/or existing drugs. Globally 2–3 million non-melanoma and 132,000 melanoma skin cancers occur each year [130]. In fact, UV radiation is a well-known factor responsible for inducing both photoaging and photocarcinogenesis processes. The incorporation of a UV absorber as octyl methoxycinnamatein in multilamellar liposomes showed an increased skin protection factor compared to conventional emulsions [131]. The incorporation of antioxidant compounds (e.g., α-tocopherol, resveratrol, lycopene, flavanols from green tea, Coenzyme Q10, melatonin, among others) in liposomes has been also used to treat skin aging, inflammatory skin diseases and cancer [66, 129, 132, 133]. In particular, melatonin-loaded ethosomal formulation showed ability for transdermal delivery of this endogenous antioxidant [134]. The co-incorporation of resveratrol and 5-fluorouracil in ultradeformable liposomes improved the anticancer activity as compared to both free drugs and single entrapped drug, revealing a potential application in squamous cell carcinoma context [135]. The 5-aminolevulinic acid–loaded liposomal formulations have been used in photodynamic therapy for skin cancer with increased local drug concentration [18]. Cisplatin and imiquimod have been also incorporated in a transfersomal gel formulation revealing an optimal antitumor activity against a cutaneous squamous cell carcinoma cell line [136]. More recently, Transethosomes loading imiquimod have provided an effective dermal delivery compared to Ethosomes in the treatment of actinic keratose, a pre-stage of squamous cell carcinoma [137].

Bacterial enzyme T4 endonuclease V incorporated in pH-sensitive liposomes has been studied for the skin cancer treatment. The results of this clinical trial showed an annual rate of new actinic keratosis of 8.2 among the patients assigned for the liposome formulation and 25.9 among those assigned placebo. In addition, the annual rates of new lesions of basal cell carcinoma were 3.8 in the treatment group and 5.4 in the placebo group [138].

Treatment of acne is another medical application of these carriers. Benzoyl peroxide, lauryl, salicylic and retinoic acids formulated in UDV have demonstrated a much better risk–benefit relationship compared to non-lipid vesicles pharmaceutical formulations [139, 140].

Psoriasis management has also been investigated using lipid vesicles formulations as drug carriers. In particular, methotrexate-loaded Ethosomes showed a high *in vitro* dermal delivery [141]. Atopic dermatitis and psoriasis are usually treated with steroidal anti-inflammatory drugs. Although betamethasone dipropionate–loaded liposomes revealed an increased anti-inflammatory activity, an antiproliferative effect was not observed in the atopic eczema [142]. A phase II clinical trial with HL-009 liposomal gel is currently being developed for atopic dermatitis [143]. Topical liposomal recombinant human Cu/Zn-superoxide dismutase is also being tested for prophylactic treatment of dermatitis induced by breast cancer radiotherapy [144].

Non-steroidal anti-inflammatory drugs (e.g., diclofenac, ketoprofen, and meloxicam) have been also incorporated in lipid vesicles for clinical application of several inflammatory skin diseases [63, 145, 146].

Considering that the first-line treatment of cutaneous leishmaniasis is based on pentavalent antimonials administered through multiple painful injections poorly tolerated by most of the patients, a new topical formulation of meglumine antimoniate–loaded liposomes has been recently tested [147].

Anti-viral (e.g., indinavir, terbinafine) and anti-fungal (e.g., itraconazol, clotrimazol) drugs have been formulated with UDV in order to increase the respective therapeutic index as well [148, 149].

Peptide (e.g., insulin)-loaded Transfersomes epicutaneously applied proved to be as efficient as injected subcutaneously [44]. Topical immunization with UDV has been used for transdermal delivery of plasmid encoding for hepatitis B surface antigen.

However, this research area still needs much more investigation and optimized studies [150, 151].

Liposomes have also been applied in the wound treatment. For example, the incorporation of povidone iodine presented a better antiseptic efficacy and burn wound-healing quality compared to silver-sulfadiazine cream [152].

Diagnostic application using topical lipid vesicles is still a largely unexplored field. However, recent studies have reported the application of Ethosomes loaded with contrast agents for computed tomography imaging applications showed *in vitro* a good macrophage cellular uptake and no cytotoxicity [153].

Regarding cosmetic applications, hyperpigmentation and melasma have been successfully treated with 4-n-butylresorcinol and linoleic acid incorporated in a liposomal formulation [154, 155]. Minoxidil-loaded lipid vesicles have shown interesting results for alopecia treatment. Neutral liposomal formulations were the most retained in skin and pilosebaceous unit [156]. In addition, finasteride- and cyclosporine A-loaded lipid vesicles have also been applied for androgenic alopecia and alopecia areata, respectively [157, 158].

Cyproterone-loaded liposomes have been studied for hirsutism treatment. According to the results obtained, the permeation rate of this drug was mainly controlled by the type of excipients and size of the vesicles [159]. Regarding cellulite management, caffeine-loaded liposomes and niosomes have been successfully formulated for an enhanced stimulation of lipolysis in epidermal adipocytes [160].

8.9 Regulatory Considerations

In the recent years, regulatory authorities have made an effort to create guidelines dealing with liposomal related products. However, any final legislation has not yet been approved. In 2001, Food and Drug Administration (FDA) published the first draft guidance concerning this subject. According to this guidance, liposomes manufacturers should submit to the regulatory authorities the description of the lipid vesicles formulation, manufacturing process and characterization. It is also recommended the submission of bioavailability and pharmacokinetics studies including *in vitro* and *in vivo* assays and the information regarding the product labeling [161]. In 2006, the European Medicines Agency (EMA) reviewed nanotechnology-based medicinal products, emphasizing the need of appropriated technologies for risk assessment and safety purposes [162]. More recently, the EMA has published another reflection paper in 2013 about the special requirements of intravenous liposomal products. The quality control should include physical, chemical, and kinetics parameters. In addition, clinical studies for efficacy and toxicological assessments are also needed as well as comparative studies between liposomes already commercialized and new liposomal products. All processes must follow the International Conference on Harmonization (ICH) guidelines and more control and information are needed regarding the evaluation of quality, safety, and efficacy of lipid vesicles products. Therefore, the Innovation Task Force EMA multidisciplinary group continues to discuss the inherent problems linked to nanomaterials [163].

8.10 Conclusion and Future Perspectives

The lack of efficiency of classical topical formulations led to the design of new vesicles for skin application. UDV became a promising delivery system especially for the treatment of several skin diseases even though the number of commercialized lipid vesicle–based formulations is still quite low. This may be associated to high production costs and problems related to stability of these vesicles. A large number of UDV compositions have been described and related mechanism of action has been proposed. However, this technological area still constitutes an intense research area [23]. Accordingly, new ingredients will continue to be studied in order to improve stability, elasticity, and high drug entrapment efficiency. Liposome manufacturing optimization including the scale-up production constitutes another area of intense research. New ultradeformable vesicles related formulations are expected to reach the market in the near future.

References

1. Cevc G, Vierl U. Nanotechnology and the transdermal route: A state of the art review and critical appraisal. *Journal of Controlled Release.* 2010; 141: 277–299.
2. Akhtar N. Vesicles: A recently developed novel carrier for enhanced topical drug delivery. *Current Drug Delivery*. 2014; 11: 87–97.
3. El Maghraby GM, Barry BW, Williams AC. Liposomes and skin: From drug delivery to model membranes. *European Journal of Pharmaceutical Sciences*. 2008; 34: 203–222.
4. Bangham AD, Horne RW. Negative staining of phospholipids and their structured modofication by surface active agents as observed in the electron microscope. *Journal of Molecular Biology*. 1964; 8: 660–668.
5. Frolov VA, Shnyrova AV, Zimmerberg J. Lipid polymorphisms and membrane shape. *Cold Spring Harbor Perspectives in Biology*. 2011; 3: a004747.
6. Monteiro N, Martins A, Reis RL, Neves NM. Liposomes in tissue engineering and regenerative medicine. *Interface*. 2014; 11: 2014–2059.
7. Akbarzadeh A, Rezaei-Sadabady R, Davaran S, Joo SW, Zarghami N, Hanifehpour Y, *et al.* Liposome: Classification, preparation, and applications. *Nanoscale Research Letters*. 2013; 8: 102.
8. Duncan R, Gaspar R. Nanomedicine(s) under the microscope. *Molecular Pharmaceutics*. 2011; 8: 2101–2141.
9. Touitou E, Dayan N, Bergelson L, Godin B, Eliaz M. Ethosomes–novel vesicular carriers for enhanced delivery: Characterization and skin penetration properties. *Journal of Controlled Release*. 2000; 65: 403–418.
10. Romero EL, Morilla MJ. Highly deformable and highly fluid vesicles as potential drug delivery systems: Theoretical and practical considerations. *International Journal of Nanomedicine*. 2013; 8: 3171–3186.
11. Wagner A, Vorauer-Uhl K. Liposome technology for industrial purposes. *Journal of Drug Delivery*. 2011; 2011: 5913–5925.
12. Kumar A, Pathak K, Bali V. Ultra-adaptable nanovesicular systems: A carrier for systemic delivery of therapeutic agents. *Drug Discovery Today*. 2012; 17: 1233–1241.

13. Barry BW. Novel mechanisms and devices to enable successful transdermal drug delivery. *European Journal of Pharmaceutical Sciences*. 2001; 14: 101–114.

14. Estanqueiro M, Amaral MH, Conceição J, Sousa Lobo JM. Nanotechnological carriers for cancer chemotherapy: The state of the art. *Colloids and Surfaces B: Biointerfaces*. 2015; 126: 631–648.

15. Fireman S, Toledano O, Neimann K, Loboda N, Dayan N. A look at emerging delivery systems for topical drug products. *Dermatologic Therapy*. 2011; 24: 477–488.

16. Cevc G, Blume G. Lipid vesicles penetrate into intact skin owing to the transdermal osmotic gradients and hydration force. *Biochimica et Biophysica Acta*. 1992; 1104: 226–232.

17. Song CK, Balakrishnan P, Shim CK, Chung SJ, Chong S, Kim DD. A novel vesicular carrier, transethosome, for enhanced skin delivery of voriconazole: Characterization and *in vitro/in vivo* evaluation. *Colloids and Surfaces B: Biointerfaces*. 2012; 92: 299–304.

18. Pierre MB, Dos Santos Miranda Costa I. Liposomal systems as drug delivery vehicles for dermal and transdermal applications. *Archives of Dermatological Research*. 2011; 303: 607–621.

19. Kato N, Ishijima A, Inaba T, Nomura F, Takeda S, Takiguchi K. Effects of lipid composition and solution conditions on the mechanical properties of membrane vesicles. *Membranes*. 2015; 5: 22–47.

20. Eloy JO, Claro de Souza M, Petrilli R, Barcellos JP, Lee RJ, Marchetti JM. Liposomes as carriers of hydrophilic small molecule drugs: Strategies to enhance encapsulation and delivery. *Colloids and Surfaces B: Biointerfaces*. 2014; 123: 345–363.

21. Kępczyński M, Nawalany K, Kumorek M, Kobierska A, Jachimska B, Nowakowska M. Which physical and structural factors of liposome carriers control their drug-loading efficiency? *Chemistry and Physics of Lipids*. 2008; 155: 7–15.

22. van Hoogevest P, Wendel A. The use of natural and synthetic phospholipids as pharmaceutical excipients. *European Journal of Lipid Science and Technology*. 2014; 116: 1088–1107.

23. Chen J, Lu WL, Gu W, Lu SS, Chen ZP, Cai BC. Skin permeation behavior of elastic liposomes: Role of formulation ingredients. *Expert Opinion on Drug Delivery*. 2013; 10: 845–856.

24. El Zaafarany GM, Awad GA, Holayel SM, Mortada ND. Role of edge activators and surface charge in developing ultradeformable vesicles with enhanced skin delivery. *International Journal of Pharmaceutics*. 2010; 397: 164–172.

25. Sinico C, Fadda AM. Vesicular carriers for dermal drug delivery. *Expert Opinion on Drug Delivery*. 2009; 6: 813–825.
26. Lane ME. Skin penetration enhancers. *International Journal of Pharmaceutics*. 2013; 447: 12–21.
27. Gonzalez-Rodriguez ML, Rabasco AM. Charged liposomes as carriers to enhance the permeation through the skin. *Expert Opinion on Drug Delivery*. 2011; 8: 857–871.
28. Reeves JP, Dowben RM. Formation and properties of thin-walled phospholipid vesicles. *Journal of Cellular Physiology*. 1969; 73: 49–60.
29. Barenholzt Y, Amselem S, Lichtenberg D. A new method for preparation of phospholipid vesicles (liposomes)-French press. *FEBS Letters*. 1979; 99: 210–214.
30. Parente RA, Lentz BR. Phase behavior of large unilamellar vesicles composed of synthetic phospholipids. *Biochemistry*. 1984; 23: 2353–2362.
31. Popa R, Vranceanu M, Nikolaus S, Nirschl H, Leneweit G. Entrance effects at nanopores of nanocapsules functionalized with poly (ethylene glycol) and their flow through nanochannels. *Langmuir*. 2008; 24: 13030–13036.
32. Batzri S, Korn ED. Single bilayer liposomes prepared without sonication. *Biochimica et Biophysica Acta*. 1973; 298: 1015–1019.
33. Wagner A, Stiegler G, Vorauer-Uhl K, Katinger H, Quendler H, Hinz A, *et al.* One step membrane incorporation of viral antigens as a vaccine candidate against HIV. *Journal of Liposome Research*. 2007; 17: 139–154.
34. Maitani Y, Soeda H, Junping W, Takayama K. Modified ethanol injection method for liposomes containing beta-sitosterol beta-D-glucoside. *Journal of Liposome Research*. 2001; 11: 115–125.
35. Alves GP, Santana MHA. Phospholipid dry powders produced by spray drying processing: Structural, thermodynamic and physical properties. *Powder Technology*. 2004; 145: 139–148.
36. Chen CM, Alli D. Use of fluidized bed in proliposome manufacturing. *Journal of Pharmaceutical Sciences*. 1987; 76: 419.
37. Deamer D, Bangham AD. Large volume liposomes by an ether vaporization method. *Biochimica et Biophysica Acta-Nucleic Acids and Protein Synthesis*. 1976; 443: 629–634.
38. Szoka F, Jr., Papahadjopoulos D. Procedure for preparation of liposomes with large internal aqueous space and high capture by reverse-phase evaporation. *Proceedings of the National Academy of Sciences of the United States of America*. 1978; 75: 4194–4198.

39. Jackson ML, Litman BJ. Rhodopsin-phospholipid reconstitution by dialysis removal of octyl glucoside. *Biochemistry*. 1982; 21: 5601–5608.

40. Patil YP, Jadhav S. Novel methods for liposome preparation. *Chemistry and Physics of Lipids*. 2014; 177: 8–18.

41. van Swaay D, deMello A. Microfluidic methods for forming liposomes. *Lab on a Chip*. 2013; 13: 752–767.

42. Meure LA, Foster NR, Dehghani F. Conventional and dense gas techniques for the production of liposomes: A review. *PharmSciTech*. 2008; 9: 798–809.

43. Storm G, Crommelin DJA. Liposomes: Quo vadis? *Pharmaceutical Science & Technology Today*. 1998; 1: 19–31.

44. Cevc G, Gebauer D, Stieber J, Schatzlein A, Blume G. Ultraflexible vesicles, Transfersomes, have an extremely low pore penetration resistance and transport therapeutic amounts of insulin across the intact mammalian skin. *Biochimica et Biophysica Acta*. 1998; 1368: 201–215.

45. Cevc G. Transfersomes, liposomes and other lipid suspensions on the skin: Permeation enhancement, vesicle penetration, and transdermal drug delivery. *Critical Reviews in Therapeutic Drug Carrier Systems*. 1996; 13: 257–388.

46. Montanari J, Roncaglia DI, Lado LA, Morilla MJ, Romero EL. Avoiding failed reconstitution of ultradeformable liposomes upon dehydration. *International Journal of Pharmaceutics*. 2009; 372: 184–190.

47. Cruz MEM, Almeida AJ, Corvo ML. Sistemas de libertação controlada. In: LIDEL, editor. *Biotecnologia: Fundamentos e Aplicações*. Portugal, 2003. p. 370.

48. Vinod KR, Kumar MS, Anbazhagan S, Sandhya S, Saikumar P, Rohit RT, *et al.* Critical issues related to transfersomes: Novel vesicular system. *Acta Scientiarum Polonorum Technologia Alimentaria*. 2012; 11: 67–82.

49. Mbah CC, Builders PF, Attama AA. Nanovesicular carriers as alternative drug delivery systems: Ethosomes in focus. *Expert Opinion on Drug Delivery*. 2014; 11: 45–59.

50. NanoSight. Nanoparticle tracking analysis. 2015, 29 march. http://www.nanosight.com/technology/nanoparticle-tracking-analysis-nta.

51. Celia C, Trapasso E, Cosco D, Paolino D, Fresta M. Turbiscan Lab® Expert analysis of the stability of ethosomes® and ultradeformable liposomes containing a bilayer fluidizing agent. *Colloids and Surfaces B: Biointerfaces*. 2009; 72: 155–160.

52. Cevc G, Blume G, Schätzlein A, Gebauer D, Paul A. The skin: A pathway for systemic treatment with patches and lipid-based agent carriers. *Advanced Drug Delivery Reviews*. 1996; 18: 349–378.

53. Marianecci C, Paolino D, Celia C, Fresta M, Carafa M, Alhaique F. Non-ionic surfactant vesicles in pulmonary glucocorticoid delivery: Characterization and interaction with human lung fibroblasts. *Journal of Controlled Release*. 2010; 147: 127–135.

54. Zhong Y, Chen L, Zhang Y, Li W, Sun X, Gong T, *et al.* Vesicular phospholipid gels using low concentrations of phospholipids for the sustained release of thymopentin: Pharmacokinetics and pharmacodynamics. *Die Pharmazie*. 2013; 68: 811–815.

55. Gupta A, Aggarwal G, Singla S, Arora R. Transfersomes: A novel vesicular carrier for enhanced transdermal delivery of sertraline: Development, characterization, and performance evaluation. *Scientia Pharmaceutica*. 2012; 80: 1061–1080.

56. Schwarz JC, Weixelbaum A, Pagitsch E, Löw M, Resch GP, Valenta C. Nanocarriers for dermal drug delivery: Influence of preparation method, carrier type and rheological properties. *International Journal of Pharmaceutics*. 2012; 437: 83–88.

57. Tejera-Garcia R, Parkkila P, Zamotin V, Kinnunen PK. Principles of rational design of thermally targeted liposomes for local drug delivery. *Nanomedicine: Nanotechnology, Biology, and Medicine*. 2014; 10: 1243–1252.

58. Kang M, Huang G, Leal C. Role of lipid polymorphism in acoustically sensitive liposomes. *Soft Matter*. 2014; 10: 8846–8854.

59. Elizondo E, Moreno E, Cabrera I, Cordoba A, Sala S, Veciana J, *et al.* Liposomes and other vesicular systems: Structural characteristics, methods of preparation, and use in nanomedicine. *Progress in Molecular Biology and Translational Science*. 2011; 104: 1–52.

60. Gupta V, Karthikeyan C, Trivedi P. Localized delivery of cisplatin for the effective management of squamous cell carcinoma from protransfersome formulation. *Archives of Pharmacal Research*. 2012; 35: 851–859.

61. Ascenso A, Salgado A, Euletério C, Praça FG, Bentley MVLB, Marques HC, *et al.* *In vitro* and *in vivo* topical delivery studies of tretinoin-loaded ultradeformable vesicles. *European Journal of Pharmaceutics and Biopharmaceutics*. 2014; 88: 48–55.

62. Lambers H, Piessens S, Bloem A, Pronk H, Finkel P. Natural skin surface pH is on average below 5, which is beneficial for its resident flora. *International Journal of Cosmetic Science*. 2006; 28: 359–370.

63. Ghanbarzadeh S, Arami S. Enhanced transdermal delivery of diclofenac sodium via conventional liposomes, ethosomes, and transfersomes. *Bio Med Research International.* 2013; 2013: 616810.

64. Zhang YT, Shen LN, Wu ZH, Zhao JH, Feng NP. Comparison of ethosomes and liposomes for skin delivery of psoralen for psoriasis therapy. *International Journal of Pharmaceutics.* 2014; 471: 449–452.

65. Mishra D, Mishra PK, Dubey V, Nahar M, Dabadghao S, Jain NK. Systemic and mucosal immune response induced by transcutaneous immunization using Hepatitis B surface antigen-loaded modified liposomes. *European Journal of Pharmaceutical Sciences.* 2008; 33: 424–433.

66. Ascenso A, Pinho S, Eleuterio C, Praca FG, Bentley MV, Oliveira H, *et al.* Lycopene from tomatoes: Vesicular nanocarrier formulations for dermal delivery. *Journal of Agricultural and Food Chemistry.* 2013; 61: 7284–7293.

67. Bavarsad N, Fazly Bazzaz BS, Khamesipour A, Jaafari MR. Colloidal, *in vitro* and *in vivo* anti-leishmanial properties of transfersomes containing paromomycin sulfate in susceptible BALB/c mice. *Acta Tropica.* 2012; 124: 33–41.

68. Steinberg G. Cytoplasmic fungal lipases release fungicides from ultra-deformable vesicular drug carriers. *PloS One.* 2012; 7: e38181.

69. Caddeo C, Sales OD, Valenti D, Sauri AR, Fadda AM, Manconi M. Inhibition of skin inflammation in mice by diclofenac in vesicular carriers: Liposomes, ethosomes and PEVs. *International Journal of Pharmaceutics.* 2013; 443: 128–136.

70. Nava G, Pinon E, Mendoza L, Mendoza N, Quintanar D, Ganem A. Formulation and *in vitro, ex vivo* and *in vivo* evaluation of elastic liposomes for transdermal delivery of ketorolac tromethamine. *Pharmaceutics.* 2011; 3: 954–970.

71. Al-Mahallawi AM, Khowessah OM, Shoukri RA. Nano-transfersomal ciprofloxacin loaded vesicles for non-invasive trans-tympanic ototopical delivery: In-vitro optimization, ex-vivo permeation studies, and in-vivo assessment. *International Journal of Pharmaceutics.* 2014; 472: 304–314.

72. Puglia C, Bonina F, Rizza L, Cortesi R, Merlotti E, Drechsler M, *et al.* Evaluation of percutaneous absorption of naproxen from different liposomal formulations. *International Journal of Pharmaceutics.* 2010; 99: 2819–2829.

73. Brewer J, Bloksgaard M, Kubiak J, Sorensen JA, Bagatolli LA. Spatially resolved two-color diffusion measurements in human skin applied

to transdermal liposome penetration. *Journal of Investigative Dermatology*. 2013; 133: 1260–1268.

74. Gabizon A, Shmeeda H, Barenholz Y. Pharmacokinetics of pegylated liposomal Doxorubicin: Review of animal and human studies. *Clinical Pharmacokinetics*. 2003; 42: 419–436.
75. Bozzuto G, Molinari A. Liposomes as nanomedical devices. *International Journal of Nanomedicine*. 2015; 10: 975–999.
76. Allen TM, Cullis PR. Liposomal drug delivery systems: From concept to clinical applications. *Advanced Drug Delivery Reviews*. 2013; 65: 36–48.
77. Elsayed MM, Abdallah OY, Naggar VF, Khalafallah NM. Lipid vesicles for skin delivery of drugs: Reviewing three decades of research. *International Journal of Pharmaceutics*. 2007; 332: 1–16.
78. Elsayed MM, Cevc G. The vesicle-to-micelle transformation of phospholipid-cholate mixed aggregates: A state of the art analysis including membrane curvature effects. *Biochimica et Biophysica Acta*. 2011; 1808: 140–153.
79. Naik A, Kalia YN, Guy RH. Transdermal drug delivery: Overcoming the skin's barrier function. *Pharmaceutical Science & Technology Today*. 2000; 3: 318–326.
80. Barry BW. Is transdermal drug delivery research still important today? *Drug Discovery Today*. 2001; 6: 967–971.
81. Karande P, Jain A, Mitragotri S. Discovery of transdermal penetration enhancers by high-throughput screening. *Nature Biotechnology*. 2004; 22: 192–197.
82. Mishra D, Dubey V, Asthana A, Saraf DK, Jain NK. Elastic liposomes mediated transcutaneous immunization against Hepatitis B. *Vaccine*. 2006; 24: 4847–4855.
83. Fanciullino R, Ciccolini J. Liposome-encapsulated anticancer drugs: Still waiting for the magic bullet? *Current Medicinal Chemistry*. 2009; 16: 4361–4371.
84. Ascenso A, Marques HC, Ribeiro HM, Simões S. Topical delivery of antioxidants. *Current Drug Delivery*. 2011; 8, 640–660.
85. Parnham MJ, Wetzig H. Toxicity screening of liposomes. *Chemistry and Physics of Lipids*. 1993; 64: 263–274.
86. Karande P, Mitragotri S. Enhancement of transdermal drug delivery via synergistic action of chemicals. *Biochimica et Biophysica Acta*. 2009; 1788: 2362–2373.

87. Chaudhary H, Kohli K, Kumar V. A novel nano-carrier transdermal gel against inflammation. *International Journal of Pharmaceutics*. 2014; 465: 175–186.

88. Knudsen KB, Northeved H, Kumar PE, Permin A, Gjetting T, Andresen TL, *et al. In vivo* toxicity of cationic micelles and liposomes. *Nanomedicine: Nanotechnology, Biology, and Medicine*. 2015; 11: 467–477.

89. Dokka S, Toledo D, Shi X, Castranova V, Rojanasakul Y. Oxygen radical-mediated pulmonary toxicity induced by some cationic liposomes. *Pharmaceutical Research*. 2000; 17: 521–525.

90. Zepik HH, Walde P, Kostoryz EL, Code J, Yourtee DM. Lipid vesicles as membrane models for toxicological assessment of xenobiotics. *Critical Reviews in Toxicology*. 2008; 38: 1–11.

91. Noble GT, Stefanick JF, Ashley JD, Kiziltepe T, Bilgicer B. Ligand-targeted liposome design: Challenges and fundamental considerations. *Trends in Biotechnology*. 2014; 32: 32–45.

92. Noble CO, Kirpotin DB, Hayes ME, Mamot C, Hong K, Park JW, *et al.* Development of ligand-targeted liposomes for cancer therapy. *Expert Opinion on Therapeutic Targets*. 2004; 8: 335–353.

93. Lorenzer C, Dirin M, Winkler AM, Baumann V, Winkler J. Going beyond the liver: Progress and challenges of targeted delivery of siRNA therapeutics. *Journal of Controlled Release*. 2015; 203: 1–15.

94. Maruyama K. PEG-immunoliposome. *Bioscience Reports*. 2002; 22: 251–266.

95. Maruyama K. Intracellular targeting delivery of liposomal drugs to solid tumors based on EPR effects. *Advanced Drug Delivery Reviews*. 2011; 63: 161–169.

96. Gaspar MM, Calado S, Pereira J, Ferronha H, Correia I, Castro H, *et al.* Targeted delivery of paromomycin in murine infectious diseases through association to nano lipid systems. *Nanomedicine: Nanotechnology, Biology, and Medicine*. 2015.

97. Abu Lila AS, Ishida T, Kiwada H. Targeting anticancer drugs to tumor vasculature using cationic liposomes. *Pharmaceutical Research*. 2010; 27: 1171–1183.

98. Schnyder A, Huwyler J. Drug transport to brain with targeted liposomes. *NeuroRx*. 2005; 2: 99–107.

99. Torchilin V. Multifunctional and stimuli-sensitive pharmaceutical nanocarriers. *European Journal of Pharmaceutics and Biopharmaceutics*. 2009; 71: 431–444.

100. Ganta S, Devalapally H, Shahiwala A, Amiji M. A review of stimuli-responsive nanocarriers for drug and gene delivery. *Journal of Controlled Release*. 2008; 126: 187–204.

101. Cevc G, Vierl U. Aggregates with increased deformability, comprising at least three amphipats, for improved transport through semi-permeable barriers and for the non-invasive drug application *in vivo*, especially through the skin. 2012; https://www.google.com/patents/WO2004032900A1?cl=ja.

102. Cevc G. Optimised preparations of highly adaptable aggregates. 2014.

103. Elsayed MM, Abdallah OY, Naggar VF, Khalafallah NM. PG-liposomes: Novel lipid vesicles for skin delivery of drugs. *Journal of Pharmacy and Pharmacology*. 2007; 59: 1447–1450.

104. Umashankar MS, Sachdeva RK, Gulati M. Aquasomes: A promising carrier for peptides and protein delivery. *Nanomedicine: Nanotechnology, Biology, and Medicine*. 2010; 6: 419–426.

105. Benvegnu T, Lemiegre L, Cammas-Marion S. New generation of liposomes called archaeosomes based on natural or synthetic archaeal lipids as innovative formulations for drug delivery. *Recent Patents on Drug Delivery & Formulation*. 2009; 3: 206–220.

106. Kaur G, Garg T, Rath G, Goyal AK. Archaeosomes: An excellent carrier for drug and cell delivery. *Drug Delivery*. 2015: 1–16.

107. Sarbolouki MN, Sadeghizadeh M, Yaghoobi MM, Karami A, Lohrasbi T. Dendrosomes: A novel family of vehicles for transfection and therapy. *Journal of Chemical Technology & Biotechnology*. 2000; 75: 919–922.

108. Sahoo SK, Dilnawaz F, Krishnakumar S. Nanotechnology in ocular drug delivery. *Drug Discovery Today*. 2008; 13: 144–151.

109. Gupta S, Vyas SP. Development and characterization of amphotericin B bearing emulsomes for passive and active macrophage targeting. *Journal of Drug Targeting*. 2007; 15: 206–217.

110. Gaspar MM, Martins MB, Corvo ML, Cruz ME. Design and characterization of enzymosomes with surface-exposed superoxide dismutase. *Biochimica et Biophysica Acta*. 2003; 1609: 211–217.

111. Cuppoletti J, Mayhew E, Zobel CR, Jung CY. Erythrosomes: Large proteoliposomes derived from crosslinked human erythrocyte cytoskeletons and exogenous lipid. *Proceedings of the National Academy of Sciences of the United States of America*. 1981; 78: 2786–2790.

112. Zhdanov RI, Podobed OV, Vlassov VV. Cationic lipid-DNA complexes-lipoplexes-for gene transfer and therapy. *Bioelectrochemistry*. 2002; 58: 53–64.

113. Paszko E, Senge MO. Immunoliposomes. *Current Medicinal Chemistry*. 2012; 19: 5239–5277.

114. Derdak SV, Kueng HJ, Leb VM, Neunkirchner A, Schmetterer KG, Bielek E, *et al.* Direct stimulation of T lymphocytes by immunosomes: Virus-like particles decorated with T cell receptor/CD3 ligands plus costimulatory molecules. *Proceedings of the National Academy of Sciences of the United States of America*. 2006; 103: 13144–13149.

115. el Guink N, Kris RM, Goodman-Snitkoff G, Small PA, Jr., Mannino RJ. Intranasal immunization with proteoliposomes protects against influenza. *Vaccine*. 1989; 7: 147–151.

116. Aqil M, Ahad A, Sultana Y, Ali A. Status of terpenes as skin penetration enhancers. *Drug Discovery Today*. 2007; 12: 1061–1067.

117. Dragicevic-Curic N, Grafe S, Gitter B, Fahr A. Efficacy of temoporfin-loaded invasomes in the photodynamic therapy in human epidermoid and colorectal tumour cell lines. *Journal of Photochemistry and Photobiology B: Biology*. 2010; 101: 238–250.

118. Moussaoui N, Cansell M, Denizot A. Marinosomes, marine lipid-based liposomes: Physical characterization and potential application in cosmetics. *International Journal of Pharmaceutics*. 2002; 242: 361–365.

119. Duangjit S, Obata Y, Sano H, Kikuchi S, Onuki Y, Opanasopit P, *et al.* Menthosomes, novel ultradeformable vesicles for transdermal drug delivery: Optimization and characterization. *Biological & Pharmaceutical Bulletin*. 2012; 35: 1720–1728.

120. Rajera R, Nagpal K, Singh SK, Mishra DN. Niosomes: A controlled and novel drug delivery system. *Biological & Pharmaceutical Bulletin*. 2011; 34: 945–953.

121. Gupta RK, Varanelli CL, Griffin P, Wallach DF, Siber GR. Adjuvant properties of non-phospholipid liposomes (Novasomes) in experimental animals for human vaccine antigens. *Vaccine*. 1996; 14: 219–225.

122. Semalty A, Semalty M, Rawat BS, Singh D, Rawat MS. Pharmacosomes: The lipid-based new drug delivery system. *Expert Opinion on Drug Delivery*. 2009; 6: 599–612.

123. Pandita A, Sharma P. Pharmacosomes: An emerging novel vesicular drug delivery system for poorly soluble synthetic and herbal drugs. *International Scholarly Research Notices Pharmaceutics*. 2013; 2013: 348186.

124. Decome L, De Méo M, Geffard A, Doucet O, Duménil G, Botta A. Evaluation of photolyase (Photosome®) repair activity in human

keratinocytes after a single dose of ultraviolet B irradiation using the comet assay. *Journal of Photochemistry and Photobiology B: Biology*. 2005; 79: 101–108.

125. Ajazuddin, Saraf S. Applications of novel drug delivery system for herbal formulations. *Fitoterapia*. 2010; 81: 680–689.

126. Kumar D, Sharma D, Singh G, Singh M, Rathore MS. Lipoidal soft hybrid biocarriers of supramolecular construction for drug delivery. *International Scholarly Research Notices Pharmaceutics*. 2012; 2012: 474830.

127. Kisak ET, Coldren B, Evans CA, Boyer C, Zasadzinski JA. The vesosome: a multicompartment drug delivery vehicle. *Current Medicinal Chemistry*. 2004; 11: 199–219.

128. Saga K, Kaneda Y. Virosome presents multimodel cancer therapy without viral replication. *BioMed Research International*. 2013; 2013: 764706.

129. Rahimpour Y, Hamishehkar H. Liposomes in cosmeceutics. *Expert Opinion on Drug Delivery*. 2012; 9: 443–455.

130. WHO. Ultraviolet radiation and the Intersun Program. 2015, 06 June. http://www.who.int/uv/en/.

131. Mota Ade C, de Freitas ZM, Ricci Junior E, Dellamora-Ortiz GM, Santos-Oliveira R, Ozzetti RA, *et al. In vivo* and *in vitro* evaluation of octyl methoxycinnamate liposomes. *International Journal of Nanomedicine*. 2013; 8: 4689–4701.

132. Sharma B, Sharma A. Future prospect of nanotechnology in development of anti-ageing formulations. *International Journal of Pharmacy and Pharmaceutical Sciences*. 2012; 4: 57–66.

133. Scognamiglio I, De Stefano D, Campani V, Mayol L, Carnuccio R, Fabbrocini G, *et al.* Nanocarriers for topical administration of resveratrol: A comparative study. *International Journal of Pharmaceutics*. 2013; 440: 179–187.

134. Dubey V, Mishra D, Jain NK. Melatonin loaded ethanolic liposomes: Physicochemical characterization and enhanced transdermal delivery. *European Journal of Pharmaceutics and Biopharmaceutics*. 2007; 67: 398–405.

135. Cosco D, Paolino D, Maiuolo J, Marzio LD, Carafa M, Ventura CA, *et al.* Ultradeformable liposomes as multidrug carrier of resveratrol and 5-fluorouracil for their topical delivery. *International Journal of Pharmaceutics*. 2015; 489: 1–10.

136. Gupta V, Dhote V, Paul BN, Trivedi P. Development of novel topical drug delivery system containing cisplatin and imiquimod for dual

therapy in cutaneous epithelial malignancy. *Journal of Liposome Research*. 2014; 24: 150–162.

137. Ma M, Wang J, Guo F, Lei M, Tan F, Li N. Development of nano-vesicular systems for dermal imiquimod delivery: Physicochemical characterization and *in vitro/in vivo* evaluation. *Materials in Medicine*. 2015; 26: 5524.

138. Yarosh D, Klein J, O'Connor A, Hawk J, Rafal E, Wolf P. Effect of topically applied T4 endonuclease V in liposomes on skin cancer in xeroderma pigmentosum: A randomised study. Xeroderma Pigmentosum Study Group. *Lancet*. 2001; 357: 926–929.

139. Patel VB, Misra AN, Marfatia YS. Preparation and comparative clinical evaluation of liposomal gel of benzoyl peroxide for acne. *Drug Development and Industrial Pharmacy*. 2001; 27: 863–869.

140. Ascenso A, Salgado A, Euleterio C, Praca FG, Bentley MV, Marques HC, *et al. In vitro* and *in vivo* topical delivery studies of tretinoin-loaded ultradeformable vesicles. *European Journal of Pharmaceutics and Biopharmaceutics*. 2014; 88: 48–55.

141. Dubey V, Mishra D, Dutta T, Nahar M, Saraf DK, Jain NK. Dermal and transdermal delivery of an anti-psoriatic agent via ethanolic liposomes. *Journal of Controlled Release*. 2007; 123: 148–154.

142. Korting HC, Zienicke H, Schafer-Korting M, Braun-Falco O. Liposome encapsulation improves efficacy of betamethasone dipropionate in atopic eczema but not in psoriasis vulgaris. *European Journal of Clinical Pharmacology*. 1990; 39: 349–351.

143. NHI US. Phase 2 Study of HL-009 Liposomal Gel to Treat Mild to Moderate Atopic Dermatitis. June 13, 2015. https://clinicaltrials.gov/ct2/show/NCT01568489?term=HL-009+Gel&rank=1.

144. NHI US. Study of APN201 (Liposomal Recombinant Human Cu/Zn-Superoxide Dismutase) for the Prevention of Radiation-induced Dermatitis in Women With Breast Cancer. June 13, 2015. https://clinicaltrials.gov/ct2/show/NCT01513278?term=APN201&rank=1.

145. Duangjit S, Obata Y, Sano H, Onuki Y, Opanasopit P, Ngawhirunpat T, *et al.* Comparative study of novel ultradeformable liposomes: Menthosomes, transfersomes and liposomes for enhancing skin permeation of meloxicam. *Biological & Pharmaceutical Bulletin*. 2014; 37: 239–247.

146. Chourasia MK, Kang L, Chan SY. Nanosized ethosomes bearing ketoprofen for improved transdermal delivery. *Results in Pharma Sciences*. 2011; 1: 60–67.

147. NHI US. Efficacy of Topical Liposomal Form of Drugs in Cutaneous Leishmaniasis. 2015, 13 june. https://clinicaltrials.gov/ct2/show/NCT01050777?term=glucantime+liposomes&rank=1.

148. Maheshwari RG, Tekade RK, Sharma PA, Darwhekar G, Tyagi A, Patel RP, *et al.* Ethosomes and ultradeformable liposomes for transdermal delivery of clotrimazole: A comparative assessment. *Saudi Pharmaceutical Journal.* 2012; 20: 161–170.

149. Priyanka K, Singh S. A review on skin targeted delivery of bioactives as ultradeformable vesicles: Overcoming the penetration problem. *Current Drug Targets.* 2014; 15: 184–198.

150. Mahor S, Rawat A, Dubey PK, Gupta PN, Khatri K, Goyal AK, *et al.* Cationic transfersomes based topical genetic vaccine against hepatitis B. *International Journal of Pharmaceutics.* 2007; 340: 13–19.

151. Bal SM, Ding Z, van Riet E, Jiskoot W, Bouwstra JA. Advances in transcutaneous vaccine delivery: Do all ways lead to Rome? *Journal of Controlled Release.* 2010; 148: 266–282.

152. Homann HH, Rosbach O, Moll W, Vogt PM, Germann G, Hopp M, *et al.* A liposome hydrogel with polyvinyl-pyrrolidone iodine in the local treatment of partial-thickness burn wounds. *Annals of Plastic Surgery.* 2007; 59: 423–427.

153. Shin H, Cho YM, Lee K, Lee CH, Choi BW, Kim B. Synthesis and characterization of ethosomal contrast agents containing iodine for computed tomography (CT) imaging applications. *Journal of Liposome Research.* 2014; 24: 124–129.

154. Huh SY, Shin JW, Na JI, Huh CH, Youn SW, Park KC. The efficacy and safety of 4-*n*-butylresorcinol 0.1% cream for the treatment of melasma: A randomized controlled split-face trial. *Annals of Dermatology.* 2010; 22: 21–25.

155. Celia C, Cilurzo F, Trapasso E, Cosco D, Fresta M, Paolino D. Ethosomes (R) and transfersomes (R) containing linoleic acid: Physicochemical and technological features of topical drug delivery carriers for the potential treatment of melasma disorders. *Biomedical Microdevices.* 2012; 14: 119–130.

156. Jain B, Singh B, Katare OP, Vyas SP. Development and characterization of minoxidil-loaded liposomal system for delivery to pilosebaceous units. *Journal of Liposome Research.* 2010; 20: 105–114.

157. Kumar R, Singh B, Bakshi G, Katare OP. Development of liposomal systems of finasteride for topical applications: Design, characterization, and *in vitro* evaluation. *Pharmaceutical Development and Technology.* 2007; 12: 591–601.

158. Verma DD, Verma S, McElwee KJ, Freyschmidt-Paul P, Hoffman R, Fahr A. Treatment of alopecia areata in the DEBR model using Cyclosporin A lipid vesicles. *European Journal of Dermatology*. 2004; 14: 332–338.

159. Valenta C, Janisch M. Permeation of cyproterone acetate through pig skin from different vehicles with phospholipids. *International Journal of Pharmaceutics*. 2003; 258: 133–139.

160. Khazaeli P, Pardakhty A, Shoorabi H. Caffeine-loaded niosomes: Characterization and *in vitro* release studies. *Drug Delivery*. 2007; 14: 447–452.

161. FDA. Guidance for Industry: Liposome Drug Products Chemistry, Manufacturing, and Controls; Human Pharmacokinetics and Bioavailability; and Labeling Documentation 2001.

162. EMA. Reflection paper on nanotechnology-based medicinal products for human use. 2006.

163. EMA. Reflection paper on the data requirements for intravenous liposomal products developed with reference to an innovator liposomal product. 2013.

Chapter 9

Archaeosomes for Skin Injuries

Monica Vazzana,[a] Joana F. Fangueiro,[c] Caterina Faggio,[a] Antonello Santini,[b] and Eliana B. Souto[c,d]

[a]*Department of Biological and Environmental Sciences, University of Messina, Viale Fernando Stagno d'Alcontres, 31, 98166 S. Agata—Messina, Italy*
[b]*Department of Pharmacy, Università degli Studi di Napoli Federico II, Napoli, Via D. Montesano 49, 80131 Napoli, Italy*
[c]*Department of Pharmacutical Technology, Faculty of Pharmacy, University of Coimbra (FFUC), Pólo das Ciências da Saúde, Azinhaga de Santa Comba, 3000-548 Coimbra, Portugal*
[d]*REQUIMTE/LAQV, Group of Pharmaceutical Technology, Faculty of Pharmacy, University of Coimbra, Coimbra, Portugal*

ebsouto@ff.uc.pt

9.1 Introduction

Artificially vesicular carriers, such as liposomes, have attracted great interest showing important success in the delivery of drugs, genes, vaccines, and other bioactive agents [1]. In addition, these systems are ideal models for biological membranes, becoming useful tools in diagnostics, targeted cancer, and gene therapy [2].

The word liposome has been generally adopted to describe mesomorphic lipid–water structures, in which chemical components are lipids and/or phospholipids. When phospholipids are present

Carrier-Mediated Dermal Delivery: Applications in the Prevention and Treatment of Skin Disorders
Edited by Andreia Ascenso, Sandra Simões, and Helena Ribeiro

ISBN 978-981-4745-58-1 (Hardcover), 978-1-315-36447-6 (eBook)
www.panstanford.com

in water, by sufficient uptake of energy (sonication, heating, homogenization) [3], form a series of closed bilayers vesicles [2]. Therefore, liposomes result in spherical vesicles with one or several concentric phospholipid bilayers and an aqueous phase inside and between the lipid bilayers [4], which are formed spontaneously in aqueous solution and whose membrane consists of amphiphilic lipids, i.e., lipids that are hydrophilic on one side and lipophilic on the other side, and they enclose an aqueous core, same as the bilayer membranes of living cells [5]. The thickness of the phospholipid bilayer measures approximately 5 to 6 nm [6]. The surface charge of liposomes can be neutral, negative, or positive in physiological pH ranges [2].

Liposomes can encapsulate both hydrophobic and hydrophilic [7]. Hydrophilic drugs may get entrapped in the central aqueous core of the vesicles, while lipophilic drugs are entrapped within the bilayer membrane [8]. Drug-loaded liposomes can specifically or non-specifically adsorb onto the cell surface, or fuse with the cell membrane, and release their contents into the cell cytoplasm, or can be destabilized by certain cell membrane components when adsorbed on the surface so that the released drug can enter cell via micropinocytosis. Liposomes can undergo direct or transfer-protein-mediated exchange of lipid components with the cell membrane or be subjected to a specific or non-specific endocytosis by phagocytic cells of the reticulo-endothelial system, such as macrophages and neutrophils. In the case of endocytosis, a liposome can be delivered by the endosome into the lysosome or, en route to the lysosome, the liposome can provoke endosome destabilization [9].

Several techniques have been presented for liposome preparation giving rise to vesicles of different size diameter (from 20 nm to several microns) and number of bilayers. According to the number of lamellae, size, and preparation method, phospholipid vesicles can be classified into several groups.

The type of liposomes is also dependent on the processing technique, such as, single or oligolamellar vesicles made by reverse-phase evaporation (REV), multilamellar vesicles made by the reverse phase method (MLV REV), stable plurilamellar vesicles (SPLV), frozen and thawed MLV (FAT MLV), vesicles prepared by extrusion technique (VET) and dehydration-rehydration method (DRV). The most relevant techniques for the production of liposomes are described in Chapter 8. However, most of techniques are not suitable for the encapsulation of sensitive substances due to their exposure to the mechanical stresses (e.g., sonication, high pressure, or shear tension), potentially harmful chemicals (e.g., organic solvents and detergents) or change of pH during the preparation. In the reduction of liposomes, it is not only expected the assembling of the membranes, but instead getting the membranes to form vesicles with right size, structure, and efficiency without any leakage. A new method has been introduced for fast production of liposomes without the use of any dangerous chemical or process. This method involves the hydration of the liposome components in an aqueous medium followed by the heating of these components in the presence of glycerol (3% v/v). There are indications that this heating method simulates the formation of the early cell membranes under the conditions of the primordial earth [2].

Triton X-100, sodium chlorate, sodium dodecyl sulfate, chloroform, octyl-β-glucoside, methanol, bile salts, and acidified isopropanol can be used to extract the encapsulated drugs from liposome. It has been shown that cryoprotective agents such as trehalose, sucrose, mannitol, dimethylsulfoxide, and glycerol protect phospholipid bilayers from damage during freeze-drying and freeze-thawing. The protection mechanism by the sugars is to form an amorphous matrix during freezing and exhibiting a low molecular mobility after drying [3].

The problems related to the lipid oxidation and hydrolysis during the shelf-life of the liposomal product can be reduced by the storage of liposomal dispersion in the dry state by freeze-drying (lyophilization), without compromising their physical state or encapsulation capacity [10].

Liposomes are used to administer drugs by several routes such as topical, oral, and parenteral and have many applications in the field of immunology, tumor therapy, vaccine adjuvant, antimicrobial therapy, gene therapy, and delivery of radiopharmaceuticals for diagnostic imaging. Furthermore, liposomes can be used for cosmetic applications to skin, delivery of dyes to textiles, enzymes and nutritional supplements in foods, pesticides on plants and can be used as biosensors [3].

These vesicular systems have several advantages including improvement of drug penetration into tissues, ability to entrap small molecules and macromolecules, reducing the toxicity of incorporated drugs, prolonging release of active pharmaceutical agents, protecting encapsulated molecules from metabolic processes, biodegradability, and biocompatibility [11].

Despite these advantages, only a small number of liposomal formulations have been approved for human use so far, due to their instability, high cost of production especially in scale up, relatively short half-life, toxicity of some liposomal formulations, relative low solubility, low entrapment of molecules and compounds into vesicles and sometimes phospholipid undergoes oxidation and hydrolysis [3, 11].

Liposomes have been used to deliver peptide and protein drugs for oral route of administration, imparting some protection on entrapped drugs against gastrointestinal (GI) enzymes. However, one of the remaining obstacles, especially for most of the liposomes composed of ester-lipids, was their poor stability against low pH, enzymes in the GI tract and bile salts. Therefore, it was desirable to develop more stable liposome formulations to protect peptide and protein drugs [12, 13].

A large number of improvements have been introduced in order to target liposomes and enhance the therapeutic effect of drugs loaded in liposomes. Among of them, passive targeting and active targeting are of major relevance. Active targeting is employed to enhance liposomal drug accumulation in the tissues, by means of targeted liposomes with surface-attached ligands capable of recognizing and binding to cells has been used. Antibodies such as IgG and their fragments are the most widely used targeting ligands for liposomes. Folate receptors are frequently overexpressed in a range of tumor cells, so targeting tumors with folate-modified liposomes is another strategy [3].

In passive targeting, biocompatible polymers, such as poly-ethyleneglycol (PEG), form a protective layer over the liposome surface and decrease liposome recognition by opsonins and cause long circulation of liposomes. Studies have carried out on the conjugation of drug to PEG on PEGylated liposomes to combine longevity and targetability for drug delivery into solid tumors [3].

Additionally, liposomes are frequently used as vehicles in pharmaceuticals and cosmetics for drug delivery in controlled manner to particular areas of skin or its layers. Because liposomes offer an amphiphilic environment, they may encapsulate hydrophilic substances in their aqueous core and lipophilic substances in their lipid bilayer. This unique dual release capability enables the delivery of two types of substances once they are applied on the skin; each differs in its effects on skin permeability which may enhance the desired therapeutic benefit [5].

A new type of liposomes shows important application for skin—the so-called ultradeformable liposomes (UDL), where high deformability is a specific mechanical property. These are liposomes prepared with phospholipids plus a certain proportion of edge activators, usually detergents such as sodium cholate (NaChol), Tween 80 or Spam 80. Applied on the skin surface under non-occlusive conditions, the lipid bilayer of UDL efficiently penetrates the outermost layer of the skin, the stratum corneum (SC), carrying small molecules up to the viable epidermis nearly tens of micrometers depth. Besides of being used as drug delivery systems, there is a growing interest on their application as adjuvants for topical vaccination to elicit antigen-dependent and memory systemic and mucosal responses [14].

When UDL are submitted to a local stress of deformation, such as elongation or lost of spherical shape, the individual components, initially homogeneously mixed, are partly separated in the bilayer (demixing). The most soluble ingredients, (NaChol), are accumulated within the sites of highest curvature, while those less soluble (phospholipids) become more abundant on the zones of lower deformation. The demixing of components alleviates the highly curved zones and allows the reversible local protrusion in these sites [14].

Although there are various liposomal formulations already on the market or in clinical trials, a major limitation to the use of these conventional phospholipid bilayer products is their instability [15], especially during the transit to the site of action. Attempts to improve their circulation time along the extracellular route to the target site, either by incorporation of high quantities of cholesterol or by coating the liposome surface with hydrophilic PEG polymers have led to limited success [7]. The disadvantages of traditional liposomes as drug delivery systems include (i) low stability; (ii) leakage and fusion of encapsulated drug molecules; (iii) sometimes phospholipid undergoes oxidation and hydrolysis; (iv) short half-life; (v) low solubility; and (vi) high production cost [9, 16].

A novel family of liposomes, called archaeosomes, demonstrated higher stability to several conditions in comparison with conventional liposomes, mainly due to the remarkable molecular and physicochemical characteristics of archaeal-type lipids. This presents a significant advancement for the delivery of medicines, allowing for more varied routes of administration. Furthermore, the stability, tissue distribution profiles, and adjuvant activity of archaeosome formulations clearly indicate that they may represent a new solution for improving vaccine therapies. The great ability of Archaea to adapt their membrane lipid compositions to many different extreme environments renders archaeal lipids ideal for developing nanodelivery vesicles capable of overcoming the various biological, biophysical, and biomedical barriers that the body stages against drug/gene or vaccine therapies against infections and cancer [7]. The structural features of the archaeal polar lipids allow the preparation of stable archaeosomal formulations in the absence of the normal precautions required (e.g., protection from air to avoid oxidation) for conventional ester liposome formulations. The knowledge about archaeosome applications in the vaccine field is much more advanced than for drug delivery [17].

9.2 Archaeosomes: Definitions and Properties

Archaeosomes are a new generation of liposomes, made from one or more polar ether lipids extracted from natural archaeal membrane lipids or synthetic archaeal lipids, exhibiting higher stability under different conditions, such as high temperatures, alkaline or acidic pH, presence of bile salts that would be encountered in the GI tract, and they are more resistant to oxidation and chemical hydrolysis than liposomes [2, 10, 11]. Due to their extraordinary stability, which allows sterilization and filtration, archaeosomes have found many applications and can be used as successful drug, gene and vaccine delivery systems [18].

In the late 1970s, the discovery of the Archaea domain as a new taxonomy for living systems revealed the existence of a novel class of microorganisms encountered in unusual but exceptional ecological niches such as high (thermophiles and hyperthermophiles) or low (psychrophiles) temperatures, low pH (acidophiles and thermoacidophiles), anaerobic atmosphere (methanogens) and high salinity (halophiles) [19]. The molecular adaptations responsible for archaeal ability to survive and grow in harsh environments have clearly emphasized the key role of membrane lipid components in overcoming the destabilizing conditions encountered in such extreme environments [20].

In contrast to Bacteria and Eukarya, where the unsaturated acyl chains of variable length of the membrane phospholipids are ester-linked to the sn-1,2 position of glycerol-3-phosphate scaffold, the backbone of archaeal lipids consists of methyl branched isoprenoid chains (5-carbon repeating units), which are fully saturated and connected via ether linkages to sn-2,3 carbon of glycerol-1-phosphate [21, 22]. Archaeal-type lipids consist of archaeol (diether) (2,3-di-O-diphytanil-snglicerol consisting mainly of C20,20 alkyl chains) and/or caldarchaeol (tetraether) (2,2′,3,3′-tetra-O-dibiphytanil-sndiglycerol with C40,40 carbon chains) core structures [20].

The lipid membrane of archaeosomes may be entirely of the bilayer form if made exclusively from monopolar archaeol (diether) lipids, or a monolayer if made exclusively from bipolar caldarchaeol (tetraether) lipids, or a combination of monolayers and bilayers if made from caldarchaeol lipids in addition to archaeol lipids or standard bilayer-forming phospholipids [13, 17, 23]. Variations to these standard lipid cores include archaeols with increased number of C5 units or caldarchaeols with cyclopentane rings within the C40-phytanil chains. Recently, a large number of new core lipid structures have been elucidated (e.g., isocaldarchaeol, H-shaped aldarchaeol, macrocyclic archaeol with one or two cyclopentane rings in the byphytanediyl chain and alkyl ether type core lipids) [20].

Tetraether-type lipids mainly consist of macrocycles composed of two membrane-spanning alkyl chains that link the two glycerol backbones. These atypical bipolar lipids play a key role to the adaptation of Archaea to extreme habitats by optimizing membrane organization and properties in direct response to the growth conditions of the organisms. The maintenance of membrane fluidity, transport functions, intracellular solute concentrations, chemiosmotic gradients and membrane protein stability are a few examples of the functions of these lipids [19].

The polar head groups of archaeal lipids are phospholipids or glycosides that are connected to one of the core lipids. The polar moieties are similar to those (phospho, glyco, polyol, amino, hydroxyl groups) encountered in ester lipids, but phosphatidylcholine is rarely present in archaeal lipids [2].

The most common phospholipids are phosphoserin, phosphoinositol, phosphoglycerol, phosphoetanolamin, and many phosphoglyicolipids; among them, the most common carbohydrates found among archaeal lipids are glucose, gulose, mannose, galactose, inositol, and N-acetylglucosamine, which can form mono-, di-, or oligosaccharides on one or both sides of caldarchaeol. Phosphoglycolipids with two polar head groups on both sides of the caldarchaeol may have glycerophosphate as the phosphoester moiety on one side and glucose alone or glucose and mannose, which form mono-, di-, or oligosaccharides as the sugar moiety on the other side. Replacement of one glycerol moiety of the core lipid backbone by a nonitol has also been observed [20].

The primary and fundamental physiological function of polar lipids is to form a cell membrane, which encloses a cell and provides a permeability barrier for various essential solutes for life. One of the essential features required for lipid membrane to fulfill biological functions is that they are in the liquid crystalline phase. Archaeal lipid membranes have, in contrast, much lower phase transition temperature than fatty acyl ester lipids. While membranes made of fatty acyl ester lipids are in the gel phase or in the liquid crystalline phase depending mostly on their fatty acid composition, archaeol- and caldarchaeol-based polar lipid membranes of archaea are assumed to be in the liquid crystalline phase at a wide temperature range of 0–100°C [20].

In view to optimize the performance of archaeosomes, specific archaeal lipid membrane properties have to be considered: (i) the ether linkages are more stable than esters over a wide range of pH, high temperature and the branching methyl groups help to reduce crystallization (membrane lipids in the liquid crystalline state at ambient temperature) and membrane permeability (steric hindrance of the methyl side groups); (ii) the saturated alkyl chains would impart stability towards oxidative degradation; (iii) the unusual stereochemistry of the glycerol backbone (opposite to mesophilic organisms) would ensure resistance to attack by phospholipases released by other organisms; (iv) the bipolar lipids span the membranes and enhance their stability properties; and (v) the addition of cyclic structures (in particular five-membered rings) in the transmembrane portion of the lipids appears to be a thermoadaptive response, resulting in enhanced membrane packing and reduced membrane fluidity [24].

More specific functions of tetraether lipids have also to be considered: (1) the tetraether structure is thought to span the membrane from the inner to the outer side to form a monolayer membrane organization instead of the standard bilayer model. Thus, a covalent linkage operates in the middle of the lipid layer: Each bipolar tetraether molecule is completely stretched and spans its entire thickness. This monolayer organization would rigidify the membrane especially at the high growth temperatures of thermophilic, methanogens and of thermoacidophiles; (2) the presence of a covalent bond between the alkyl chains in H-shaped tetraethers might protect the cells against membrane lysis at high temperature by reinforcing the strength of the monolayer membrane; (3) high proportions of glycosyl polar head groups both in methanogenic and thermoacidophilic lipids may further stabilize the membrane structure by interglycosyl headgroup hydrogen bonding. The presence of large sugar heads towards the convex surface of the membrane may assume an asymmetric orientation, making therefore the monolayer organization easier. Owing to such originalities, natural and synthetic archaeal lipids are promising tools for medical, biological, and biotechnological applications. Interest in these areas is stimulated primarily by the tetraether lipid self-assembling properties leading to supramolecular structures such as lipid films or liposomes [19].

These unique lipid structures in general conferred archaeosomes higher stability and less permeable than conventional liposomes that frequently need up to 33% of cholesterol to improve their stability. Thanks to archaeal lipid properties, archaeosomes can be formed at any temperature in the physiological range or lower, thus making it possible to encapsulate thermally labile compounds. Moreover, they can be prepared and stored in the presence of air/oxygen without any degradation [10, 12], they do not fuse or aggregate during storage at 4°C over a period of 4 months; therefore, they are a valuable resource in the development of novel biotechnological processes [20].

9.2.1 Biotechnological Applications of Archaeosomes

Although archaeosomes are a recent technology, they have already proven to be a safe delivery system for bioactive compounds. Their pH and thermal stability, tissue distribution profiles, as well as the adjuvant activity indicate that archaeosome formulations may offer a superior alternative for several biotechnological applications, including delivery systems for drugs, genes, or cancer imaging agents. As an ideal vector they should be highly efficient in delivering the drug in a target-specific manner (various molecules can be incorporated into archaeosomes or associated with the vesicles to target specific tissues), stable *in vitro* as well as *in vivo*, non-toxic and non-immunogenic. In this sense, archaeosomes delivery systems answer most of these conditions [12, 20].

Their composition and preparation can be chosen to achieve desired features such as selection of lipid, charge, size, size distribution, entrapment, and location of antigens or adjuvants. Depending on the chemical properties, water-soluble antigens (proteins, peptides, nucleic acids, carbohydrates, haptens) are entrapped within the aqueous inner space of liposomes, whereas lipophilic compounds (lipopeptides, antigens, adjuvants, linker molecules) are intercalated into the lipid bilayer and antigens or adjuvants can be attached to the liposome surface by adsorption or stable chemical linking. These carriers provide adjuvant activity by enhancing antigen delivery or by activating innate immune responses [25].

First, since liposomes and archaeosomes naturally target cells of the mononuclear phagocytic system, they would be ideally suited for the delivery of antigens, as carrier systems and/or directly as adjuvants that stimulate the immune system [7]. This was supported by two important observations, namely (i) as archaeosomes are phagocytosed effectively, they may efficiently target antigen to antigen presenting cells (APC), facilitating a strong humoral and cell-mediated (Th1/Th2) immune reactions to soluble entrapped proteins and consequent long-term memory [26]; and (ii) more importantly, archaeosomes induce CD8+ cytotoxic T lymphocyte responses (critical for protection against cancer and intracellular infections) to entrapped proteins [25].

The immunity stimulating effects of the archaeosomes should be highly useful by oral administration in order to mobilize the mucosal immune system for both systemic and mucosal immune responses. One indicator of effective cell mediated immune response is the responding ability against antigen *in vitro* of lymphocyte populations from immunized mice. In a study, the potency of archaeosomes facilitating proliferative responses of CD8+ T cells via oral immunity route may be explained by distinctive features of the tetraether lipid structures. It seems possible that rather stable archaeosomes with high tetraether lipid content may facilitate MHC class I cross presentation resulting in the activation of potent specific cell mediated immunity including CD8+ T cell responses [12].

It is also showed that archaeosomes promote the recruitment and activation of professional APC *in vivo* and deliver the antigen to both MHC class I and II pathways for antigen presentation, without eliciting overt inflammatory responses. Furthermore, depending on the archaeal lipid composition, activation signals to dendritic cells may be provided by archaeosomes [27].

Krishnan *et al.* studied the adjuvant activity of archaeosome that was prepared from lipid extracted from *Methanobrevibacter smithii.* They reported that archaeosomes could show adjuvant activity and induced T helper and cytotoxic T lymphocyte responses to entrapped antigen. Additional work in this way is suggested to trace the archaeosome in cancer cells for evaluating the potential application in delivery of molecules [28]. Instead, Réthoré *et al.* studied the potential application of archaeosomes as gene delivery system for plasmid DNA expressing the luciferase reporter gene, showing that the developed archaeobacterial-like lipids, usable as cationic lipids or co-lipids for *in vitro* gene transfection, represent a new approach for modulating the lipidic membrane fluidity of the complexes they form with plasmid DNA [29].

Li *et al.* prepared archaeosomes, made of the polar lipid fraction E (PLFE) extracted from *Sulfolobus acidocaldarius,* entrapping insulin as a model peptide and they explored their potential application as a vehicle for oral delivery of the peptide drug. Their study showed that the archaeosomes exhibit significant high stability in simulated gastric and intestinal fluids. As a carrier of oral insulin, archaeosomes are superior in reducing the blood glucose levels in diabetic mice, compared with conventional liposomes [13]. Carrer *et al.* produced ultradeformable archaeosomes (UDA), a variety of UDL containing total polar archaeolipids (TPA), extracted from the extreme halophile archaea *Halorubrum tebenquichense*, soybean phosphatidylcholine and sodium cholate (3:3:1 w:w), for topical delivery of ovalbumin [14].

Lastly, Archaeal lipids have also been proposed as monomers for bioelectronics for several reasons. Their bipolar tetraether structure offers novel opportunities for protein-lipid interactions, which are of interest for the assembly of electronic devices based on redox proteins or enzymes. Some other applications include use of archaeal lipids as taxonomic markers and as novel lubricants for engines, since they are chemically stable at high temperatures, lubricate, either with or without a carrier lubricant, and provide a friction coefficient of less than 0.1 [20]. In Table 9.1, some examples of compounds already loaded in archaeosome formulations are listed.

Table 9.1 Examples of compounds loaded in archaeosomes

Archaea	Major phospholipids	Compound	Ref.
Sulfolobus acidocaldarius	Phosphatidylglycerol, diphosphatidylglycerol, phosphatidylethanolamine lipids containing glycerol ethers	Methylene blue	[11]
Halorubrum tebenquichense	Total polar lipids	Ovalbumin	[14]
Sulfolobus acidocaldarius	Polar lipid fraction E (a mixture of tetraether lipids)	Ovalbumin	[12]
Sulfolobus acidocaldarius	Polar lipid fraction E (a mixture of tetraether lipids)	Insulin	[13]
Aeropyrum pernix	Polar-lipid methanol fraction	Fluorescence probes	[21]
Halorubrum tebenquichense	Total polar lipids	Ovalbumin	[30]
—	PEGylated tetraether lipid	Carboxyfluorescein	[24]
Aeropyrum pernix	—	Calcein, listeriolysin, keratin 14, plasmid DNA	[18]

9.2.2 Preparation and Physicochemical Characterization of Archaeosomes

The first step, before archaeosome preparation, consists in either the lipid extraction from selected archaea species or the synthesis of archaeal-like lipids. Concerning natural archaeal lipids, the total lipid extract, composed of TPL and neutral lipids, is usually obtained by chloroform/methanol/water extraction from frozen thawed biomass coming from the selected archaea species. TPL and neutral lipids are then separated by precipitation of TPL in acetone. These TPL made from isoprenoid ether lipids of opposite sn-2,3 stereochemistry can be stored as chloroform or chloroform/methanol (2/1) solutions without special conditions. Pure archaeal lipids can be further obtained by chromatography, either column chromatography or preparative thin-layer chromatography [7].

In this regard, in the procedure applied by Benvegnu *et al.*, archaea were grown under their optimal living environmental conditions before undergoing extensive organic solvent extraction (chloroform/methanol/water mixtures). TPL were finally obtained by precipitation with acetone from a chloroform/methanol (2:1: v/v) solution. The resulting extracts were analyzed by thin- layer chromatography and mass spectrometry and were used as such or were further purified by preparative thin-layer chromatography to isolate pure polar lipid components [7].

The characterizations of the fractionated lipids were conducted in another work on thin-layer chromatography, mass spectral, infrared spectroscopy, and nuclear magnetic resonance by Li *et al.* [12]. Barbeau *et al.* used, instead, high-performance thin-layer chromatography, an advanced form of thin-layer chromatography that can analyze mixtures by separating the compounds and determining the number of components in a mixture, to assess the lipid extracted [24].

The resulting pure archaeal lipids can be chemically modified in order to introduce specific head groups. Starting from natural, chemically modified, or synthetic archaeal lipids, it was possible to prepare archaeosome formulations and encapsulate/associate hydrophilic or hydrophobic compounds using standard methods developed for the preparation of conventional liposomes, e.g., hydrated film submitted to sonication, extrusion or detergent/

dialysis, at any temperature in the physiological range or lower. For instance, archaeosomes can be prepared by the detergent/dialysis method. However, the presence of a detergent could be a major drawback for pharmaceutical applications of the corresponding archaeosomes. In addition, the detergent/dialysis method can result in poor entrapment due to the leakage of loaded molecules during the dialysis process. A well-developed technique for preparing archaeosomes consists of the hydration of archaeal lipid film followed by size reduction using sonication or extrusion through polycarbonate membranes. Archaeosomes can be also prepared by the reverse phase evaporation method combined or not with bath sonication. Other methods, such as pressure extrusion, freeze, and thaw [12, 13], as well as dehydration-hydration, are applicable for archaeosome preparation. The selected biologically active molecules can be encapsulated or associated with the prepared archaeosomes either during their formulation, the molecule to be encapsulated contained in the hydration media, or after their formulation where the molecule (especially DNA) to be encapsulated is added in the archaeosome aqueous dispersion [7].

Regarding ultradeformable archaeosomes, the work described by Carrer *et al.* reports the production of these carriers by the thin-film hydratation method. Briefly, appropriate amounts of soya phosphatidylcholine in chloroform and TPA and NaChol in chloroform:methanol (1:1, v:v) were mixed in round bottom flasks. Solvents were rotary evaporated at 40°C until elimination and the lipid films obtained were flushed with N_2. Lipid films were hydrated by the addition of an aqueous phase (10 mM Tris–HCl buffer plus 0.9% w/v NaCl, pH 7.5 (Tris–HCl buffer)) up to a final concentration of 43 mg of phospholipids/mL. The resultant suspensions were sonicated (45 min with a bath type sonicator 80 W, 40 kHz) and extruded 15 times through three stacked 0.2, 0.1 and 0.1 μm pore size polycarbonate filters using a 100 mL Thermobarrel extruder. The drug was dissolved in the aqueous phase used for the hydration of the thin film at 1.15 mg/mL. After extrusion, they were submitted to five freeze–thaw cycles between –70 and 40°C. Finally, vesicles were purified on Sephadex G-75 using the minicolumn centrifugation technique [14].

From the characterization point of view, empty and/or drug-loaded archaeosomes can be characterized by methods used for conventional liposomes (Table 9.1).

Dynamic light scattering (DLS) allows the measurement of the average diameter and size distribution of the corresponding vesicles, phase analysis light scattering may be applied for zeta potential evaluation [14]; transmission electron microscopy (TEM) [12], cryo-TEM and TEM after freeze-fracturing of the preparation, contrast and fluorescence microscopy, give access to the shape, diameter and structure of the corresponding nanocarriers.

Atomic Force Microscope (AFM) allows to obtain some images concerning liposomal shape for morphology evaluation [13]. Archaeosomes may be studied also by using electron paramagnetic resonance and fluorescence emission spectrometry [21].

Differential scanning calorimetry (DSC) is used to investigate changes in their thermal behavior and is a great tool which can be used to investigate the interaction between liposomes and drug molecules. It can be realized that whether the drug is loaded in the carrier [24].

Phase contrast light microscopy (PCLM) and scanning electron microscopy (SEM) allow the observation of structures of archaeosomes and aggregates obtained by mixing archaeosomes with multivalent cations or organic cationic compounds, named archaeal lipid mucosal vaccine adjuvant and delivery (AMVAD) systems as described by Patel *et al.* [31]. Using the previously cited techniques it has been ascertained that archaeosomes had a small, individual and spherical structure whereas their AMVAD presented a large spherical structure aggregated into clumps. Patel *et al.* confirmed these observations by fluorescence microscopy carried out with 5(6)-carboxyfluorescein (CF) encapsulated archaeosomes and AMVAD [31].

Other authors used phase contrast microscopy and/or fluorescence microscopy with CF-encapsulated formulations to characterize the structure of their archaeosomes [32].

Besides archaeosome size, size distribution, shape and structure characterizations, studies to evaluate vesicle stabilities under various conditions have to be performed. Choquet *et al.* tested the stability of archaeosomes prepared from TPL of various archaea under different conditions (presence of phospholipases, temperature, pH) by fluorescence measurements (CF release), by radioactivity leakage ([14C]-sucrose leakage) and by vesicle size measurements. The results demonstrated that archaeobacterial ether lipids formed archaeosomes able to resist enzymatic attack. Studies on CF release as a function of temperature and on [14C]-sucrose leakage as a function of pH suggested that: (1) the presence of tetraether lipids is a stabilizing factor of vesicles at high temperature; (2) the ether bonds are partly responsible for the enhanced stability of archaeosomes at high temperature and extreme pH. Moreover, oxidation degradation is limited because of high stability of fully saturated phytanyl chains [33]. Similar studies were also carried out by Benvegnu *et al.* with vesicles based on synthetic archaeal lipid analogues. They showed that increasing amounts of synthetic archaeal lipids in archaeosome formulations exhibited a better stability in the presence of surfactants reproducing the behavior of bile salts, calf serum mimicking the blood medium and in acid pH conditions as found in the stomach [7].

Electrospray ionization mass spectrometry (ESI-MS) technique is an extremely soft method for molecular ionization in the gaseous phase, which induces the formation of molecular ions (protonated, deprotonated species, and adducts). In the field of lipidomics, ESI-MS has proven useful in lipids identification and quantification. Elaborated studies based on this technique has been applied to quantify different types of non-covalent associations such as electrostatic interactions (e.g., salt bridges), dipolar interactions (e.g., hydrogen bonds), and van der Waals interactions (e.g., hydrophobic interactions), between protein–protein, protein–small molecule, and protein–DNA complexes [14].

In addition, two important technical parameters that should be taken into account are (i) measurement of deformability (D), and the (ii) determination of encapsulation efficiency (EE).

The deformability (D) value of liposome-derivatives may be calculated according to the following equation:

$$D = J \times \left(\frac{rv}{rp}\right) \times 2,$$

where J is the rate of penetration through a permeability barrier, rv is the post extrusion liposomal radii and rp is the extrusion membrane pore radii. This Van der Bergh assay is a simple method used to identify highly deformable nanoliposomes. In there the nanoliposomal flux under a low extrusion pressure across membranes of pore radii half of the liposomal radii (usually 100 nm mean diameter) and the mean size post-extrusion are measured. As a result, the relative deformability—which is inversely proportional to the absolute elastomechanic parameter Young's modulus—is determined. For example, in Carrer *et al.* study in order to measure *J*, nanoliposomes were extruded through two stacked 50 nm *rp* membranes at 0.8 MPa using a Thermobarrel extruder. Extruded volume was collected every minute along 15 min, phospholipid was quantified in each fraction and *J* was calculated as the area under the curve of the plot of phospholipid recovered as a function of time. The average post extrusion liposomal radii (*rv*) was measured by DLS [14].

The encapsulation efficiency (*EE*) may be assessed by UV spectrophotometer, indirectly by an ultracentrifugation method. To quantify the drug content, the supernatant is withdrawn and subsequently analyzed. The encapsulated amount of drug is calculated by subtracting the free amount of drug from the total amount in the dispersion, applying the following equation:

$$EE = \left(\frac{W_a - W_s}{W_a}\right) \times 100,$$

where W_a is the weight of drug added in the formulation and W_s is the weight of drug analyzed in the supernatant (after separation of lipid and aqueous phases by ultracentrifugation) [34].

Additional studies that should be carried out in order to complete the characterization and the evaluation of archaeosomes as drug delivery systems include the pharmacokinetics and biodistribution assays, which can be carried out via (i) *in vitro* drug release studies, usually performed by using dialysis membrane method and a specially designed Franz diffusion cell; (ii) *in vitro*/*in vivo* skin permeation; (iii) drug release kinetics. The release mechanism is evaluated using different kinetic models including zero-order, first-order, Higuchi, and Korsmeyer–Peppas (also called power law) (the latter is dependent on the fraction of drug released at time, rate constant, and release exponent) [11], in order to better characterize the drug release behavior:

$$\frac{\%R_t}{\%R_\infty} = 1 - e^{-K_1 \times t} \text{ (first order)}$$

$$\frac{\%R_t}{\%R_\infty} = KH \times t^{1/2} \text{(Higuchi model)}$$

$$\frac{\%R_t}{\%R_\infty} = K \times t^n \text{ (Korsmeyer-Peppas model)},$$

where $\%R_t$ is the percentage drug released at time t, $\%R_\infty$ is the total percentage drug released, $\%R_t/\%R_\infty$ is the fraction of drug released at time t, K is the release rate constant, n is the diffusion release exponent that could be used to characterize the different release mechanism $n < 0.5$ (Fickian diffusion), $0.5 < n < 1.0$ (anomalous transport), and >1.0 (case II transport; i.e., zero-order release) [35].

Permeation studies are performed by using vertical amber glass Franz-type diffusion cells. The excised skin is mounted between donor and receptor compartment of Franz diffusion cell avoiding bubbles formation, with SC side facing the donor compartment and the dermis facing the receptor medium receiver compartment, filled with ethanol/transcutol P1 (50:50; v/v) allowing sink conditions to be maintained. Samples are applied to the donor compartment. Temperature is kept at 32°C to mimic *in vivo* conditions by a circulating-water jacket and stirred with a magnetic rotor at a speed of 700 rpm. Aliquots of some hundred mL are withdrawn at regular time intervals and replaced with the same volume of doubled distilled water after each collection with great care to avoid trapping air beneath the dermis. The samples are until analyzed by HPLC [34].

For skin permeation parameters, the cumulative amount of drug permeated (Q_t) through excised human skin is plotted as function of time and determined based on the following equation:

$$Q_t = \frac{V_r \times C_t + \sum_{i=0}^{t=1} V_S \times C_i}{S},$$

where C_t is the drug concentration of the receiver solution at each sampling time, C_i the drug concentration of the i sample, and V_r and V_s the volumes of the receiver solution and the sample, respectively, S represents the skin surface area. Drug fluxes (J, m g/cm^2/h) through the skin, from the assayed formulations is calculated from the slope of linear portion of the cumulative amounts permeated through the human skin per unit surface area versus time plot by regression [34].

At the end of the permeation study, the amount of drug remaining on the skin is quantified in order to know if there are differences between formulations assayed. Skin mounted on the diffusion cell is carefully taken off, cleaned with gauze soaked in a 0.05% solution of dodecyl sulfate and washed in distilled water. The permeation area of the skin is cut and weighed. Drug contended in the skin is extracted with ethanol/transcutol P1 (90:10) during 20 min under cold sonication in an ultrasound bath. The resulting solution is centrifuged and measured by HPLC, yielding the amount of drug retained in the skin (expressed as microgram per gram square centimeter) [34].

Another important study is the histological evaluation. For instance, in Clares' work the other part of the skin samples coming from permeation studies is harvested and fixed in 10% neutral-buffered formalin and then dehydrated by ethanol solution in different concentrations, then embedded in paraffin wax. Formalin-fixed, paraffin-embedded samples are cut in 5 mm-thick sections on a microtome with a disposable blade and conventionally stained with hematoxylin-eosin. The samples are then examined by light microscope and compared with the control sample [34].

Homogenization is an important factor in the archaeosomes preparation, and their size control. It is believed that homogenization probably prevents the aggregation and fusion of particles. Reduction in the particle size is important parameter for improving the performance of poorly soluble drugs. Also, the results of a study demonstrated that homogenization time can significantly affect the particle size of vesicles. It has been accepted that decreasing the particle size of vesicles may cause an increase in the penetration of encapsulated drugs into the deeper skin layers [11]. As is the case of liposomes, it is possible to incorporate ligands such as polymers to archaeosomes. In Fact, incorporation of PEG and coenzyme Q10 into archaeosomes improve the tissue distribution profiles of intravenously administered vesicles [2, 22]. It is suggested that cholesterol may increase the stability and modify the fluidity of archaeosomes prepared from lipid extracted by archaea. Cholesterol has been employed as helper lipid in the liposome formulation to improve further stability and rigidity and decrease leakage of the encapsulated drugs. Also, it has been widely used to modify drug release, improve physical stability, and prolong circulation half-life of liposomal drug *in vivo*. Cholesterol interacts with fatty acids of liposomes via hydrogen binding, enhancing the cohesiveness and mechanical strength of the membrane [24].

9.3 Other Ultradeformable (Elastic) Liposomes

The current challenge of drug delivery is liberation of drug agents at the right time in a safe and reproducible manner to a specific target site. A number of novel drug delivery systems have emerged encompassing various routes of administration, to achieve controlled and targeted drug delivery. The success of liposomes as drug carriers has been reflected in a number of liposome-based formulations, which are commercially available or are currently undergoing clinical trials. Also, novel lipid carrier-mediated vesicular systems are originated [36].

The vesicular systems are highly ordered assemblies of one or several concentric lipid bilayers, when certain amphiphilic building blocks are confronted with water. The term "synthetic bilayers" alludes to the non-biological origin of such vesiculogenes. In recent years, vesicles have become the vehicle of choice in drug delivery, playing a role in modeling biological membranes, in transport and targeting of active agents. They reduce the cost of therapy improved bioavailability of medication, especially in case of poorly soluble drugs. They delay drug elimination or rapidly metabolized drugs and function sustained release systems. This system solves the problems of drug insolubility, instability, and rapid degradation [22].

Various types of vesicular systems, over liposomes and archaeosomes, have been developed such as ethosomes, transfersomes, among others which are fully described in Chapters 8 and 10.

9.4 Applications for Skin Injuries

Topical delivery can be defined as the application of a drug containing formulation to the skin to directly treat cutaneous disorders like acne or the cutaneous manifestations of a general disease like psoriasis with the intent of containing the pharmacological or other effects of the drug to the surface of the skin or within the skin. A successful topical dermatological formulation can be considered to be one that satisfies the target product profile and is: (i) physically and chemically stable having adequate shelf-life; (ii) releases drug from the formulation and delivers it into the skin as required for the target indication; (iii) is cosmetically elegant and acceptable to patients; (iv) contains only excipients that are necessary, FDA-approved or acceptable from a regulatory perspective, and acceptable for the disease state; (v) is easy to apply and compatible with the desired packaging, and (vi) can be manufactured with a process that is scalable to commercial levels [5].

Skin acts as a key target as well as a principle barrier for topical/transdermal bioactive delivery. This drug administration avoids the hepatic first pass effect, provides continuous drug delivery, decreases side effects, and improves patient compliance. A major obstacle in TT delivery is low percutaneous penetration. Several approaches have been used to weaken skin barrier and improve topical/transdermal delivery [2].

One of the many physiological functions of the skin is protection, but being a vital organ, the skin must be nourished as the other organs of the body. Therefore, percutaneous absorption of a drug without allowing other environmental agents in and without damaging the skin is a challenge [34].

Briefly, the transdermal delivery system offers many advantages over conventional injection and oral methods, namely the following: (i) It is convenient method and requires only once weekly application. Such a simple dosing regimen can aid in patient adherence to drug therapy; (ii) transdermal drug delivery can be used as an alternative route of administration to accommodate patients who cannot tolerate oral dosage forms; (iii) it is of great advantage in patients who are nauseated or unconscious; (iv) drugs that cause gastrointestinal upset can be good candidates for transdermal delivery because this method avoids direct effects on the stomach and intestine; (v) drugs that are degraded by the enzymes and acids in the gastrointestinal system may also be good targets; (vi) first pass metabolism, an additional limitation to oral drug delivery, can be avoided with transdermal administration; (vii) drugs that require relatively consistent plasma levels are very good candidates for transdermal drug delivery [37].

However, probable disadvantages are: (i) possibility of local irritation at the site of application; (ii) erythema, itching, and local edema can be caused by the drug, the adhesive, or other excipients in the patch formulation; (iii) may cause allergic reactions; (iv) a molecular weight less than 500 Da is essential; (v) sufficient aqueous and lipid solubility, a log P (octanol/water) between 1 and 3 is required for permeate to transverse SC and underlying aqueous layers [37].

The pH of dermal formulations is an important factor to avoid skin irritation or make the skin susceptible to bacterial infection. Natural acidity of the skin ranges from 4 to 6, depending on the skin area and the age of the individual, due to the buffer system in the skin that is able to absorb small quantities of acid or alkali material applied to reduce irritation [34].

With transdermal delivery systems, the active pharmaceutical ingredient makes contact with the target site before entering the systemic circulation. Systemic side effects can be greatly reduced, e.g., in circumventing hepatic first-pass metabolism and major fluctuations of plasma levels caused by oral administration. The biggest challenge for the topical/transdermal delivery systems is to ensure the effective penetration of drugs, for example, across the scar skin and play the anti-scar role. Besides that, problems of short-term effect and requirement of frequent administration existed widely in the current topical treatment of hypertrophic scar and keloid. For adult tissues, healing of deep wounds often leads to excessive scarring of hypertrophic scars and keloids. With frequent application, many adverse effects occurred such as atrophic changes, steroid atrophy, telangiectasia, striae, purpura, stellate pseudoscars, ulceration, easy bruising, infections, masked microbial infections (*tinea incognito*), and aggravation of cutaneous candidiasis, herpes or demodex. Thus, sustained drug release efficacy and the permeation enhancement are fairly important in topical drug delivery system. With the sustained release and enhanced permeation could reduce the incidence of adverse effects caused by drugs. Recently, several topical delivery systems such as liposomes and other vesicular systems have been reported for the advantages of biocompatibility, protecting drugs from degradation by external conditions, sustained drug release efficacy and the permeation enhancement for the treatment of hypertrophic scars and keloids [4].

It is well known that the SC, the outer layer of epidermis, represents the main physical barrier of the skin, so that for a substance permeating across the skin, diffusion through the SC is the rate limiting step [38]. In fact, only lipophilic drugs with molecular weight < 500 Da can pass through it [5].

In spite of this, novel drug delivery technology, such as some of those previously reported, and penetration enhancers may help to obviate some of these objections.

Although topical treatment of cutaneous diseases represents an important challenge, a way to overcome this problem is encapsulation of drugs in liposomal vesicles, such as miconazole nitrate, an antifungal agent. According to the results, liposomal formulation of the drug may increase percutaneous permeability if compared to conventional formulations [39]. Topical application of liposome vesicles has many advantages over the conventional dosage forms. It has been previously shown that the liposomal gel of lidocaine HCL may perform therapeutically better effects than the conventional formulation. It has been suggested that the formulated liposomes may be applied onto the skin as gel [40]. Using different enhancers, enhancement permeability of gentamicin sulfate through shed snake skin and liposomal membranes has been studied. The results indicated that direct effect of surfactants on shed snake skin and liposomal membranes is responsible for their enhancing effects [41]. Fluconazole is a synthetic antifungal agent that is used in the treatment of esophageal, vulvovaginal disorder and is effective against dermatophytoses. Fluconazole has adverse effects such as bloating, nausea, abdominal problems and vomiting. Mitkari *et al.* have formulated and evaluated a topical liposomal gel of fluconazole. Their investigations showed that the entrapment efficiency was 57.78–66.64% and liposomal gel was increased the skin permeation and deposition of fluconazole compared to control. Therefore, in this way the risk of gastrointestinal side effect may decrease [42].

The physicochemical properties and the nature of drug and absorption enhancers are very important factors in enhancing skin permeation of the drugs. For example, methylene blue is a polar compound and, as mentioned before, SC represents the main restriction barrier for transdermal delivery of polar compounds. It is believed that follicular shunt route is responsible for the permeation of polar molecules and drugs. Nevertheless, it is accepted that these routes comprise a fractional area for permeation of approximately 0.1% of total permeation. Consequently, penetration enhancement techniques have been concentrated on increasing transport across the SC rather than via the appendages. According to some findings, archaeosome can be employed as penetration enhancer for polar compounds and drugs. Many studies indicated that archaeal lipids are good sources for preparation of liposomes due to their remarkable thermostability and safety [11]. Delivery of various materials through the skin is highly important in different areas particularly in cosmetics and skin care. For transdermal delivery of bioactive agents using carrier systems, the bioactive compounds must be associated with specifically designed vehicles, in the form of highly deformable particles, and applied on the skin non-occlusively. The inventors claim that transferosomes are ultradeformable and squeeze through pores less than one-tenth of their diameter. Therefore, 200–300 nm-sized transfersomes are claimed to penetrate intact skin. Their penetration works best under *in vivo* conditions and requires a hydration gradient from the skin surface towards the viable tissues. It has been suggested that transfersomes can respond to external stresses by rapid shape transformations requiring low energy. This high deformability allows them to deliver drugs across barriers, including skin [43].

Another peculiarity of the skin is that to be rich in potent APC, that are not readily accessible to parenteral vaccination excepting by intradermal route, which is difficult to practice. In this sense, the topical route is an attractive alternative enabling a much closer access to these skin APC. Topical vaccination takes the advantages offered by the avoidance of injectables, such as the increased patient compliance, the reduced potential re-infection by contaminated material and waste of disposable material, as well as the need for specialized trained personal, sterilized material and maintenance of cold chain [30].

Particulate and especially nanoparticulate materials, when phagocytosed by immature APC, induce stronger antigen-specific systemic humoral, cellular and memory responses as well as mucosal immunity, than soluble antigen. However, the SC is the main impairment for topical delivery of nanoparticulate material to skin APC, and the previous disruption of a substantial SC area by mechanical, chemical or physical means, such as hydration, tape stripping, electroporation or abrasion is required. Besides, upon disruption keratinocytes induce pro-inflammatory cytokines that recall an adaptive immune response by stimulating the maturation and migration of Langerhans cells. Topical liposomes-based derivatives have rendered successful pre-clinical results. Some of these approaches are already in advanced clinical studies. Higa *et al.* prepared UDA that are vesicles made of soybean phosphatidylcholine, sodium cholate and total polar lipids from *Halorubrum tebenquichense* (3:1:3 wt/wt). While UDL, made of soybean phosphatidylcholine and sodium cholate at 6:1 wt/wt, and UDA were neither captured nor caused cytotoxicity on keratinocytes, UDA was avidly captured by macrophages, being their viability reduced 60% down to 25%. Instead, UDL were poorly captured and caused no toxicity. Both *in vitro* and *in vivo*, archaeosomes are more avidly internalized by macrophages and APC than liposomes, showing to be suitable as topical adjuvants [30].

Similar results are reported by Carrer *et al.* confirming UDA as promising topical adjuvants whose deformability is an essential property for intact skin penetration up to the viable epidermis/dermis [14].

Liposomes loaded with drugs have been investigated as the carriers for topical drug delivery to the scar sites for treatment because of their efficiency in facilitating skin penetration, releasing drugs slowly, and keeping the moisture of scar skin, etc. Being similar to biological membranes, liposomes can navigate water soluble and lipophilic substances in different phases or domains. Liposomal preparations are reported to reduce the skin roughness because of their interaction with the corneocytes and intercellular lipids resulting in skin softening and smoothening. However, liposomes are metastable systems and their pharmaceutical use is often limited by instability. Instability can be due to leakage of the vesicles, change in vesicle size due to aggregation or fusion, as well as ester hydrolysis and formation of oxidation products [4].

Vogt *et al.* reported that polyvinyl pyrrolidone-iodine (PPI) was carried by a new liposome hydrogel formulation (Betasom hydrogel) and was applied to the patients with meshed skin grafts after burns [44]. Clinical assessment indicated that better antiseptic condition and wound healing quality has been received with the Betasom hydrogel treatment. Compared to wounds treated with a conventional antiseptic chlorhexidine-gauze, PPI liposomal can provide higher moisture to the wound surface, release at a low rate, and target the substance exactly by interaction with the cells surface [44]. Yang *et al.* reported that the topical treatment of liposome-encapsulated hydroxycamptothecin could significantly reduce the epidural scar, compared with that of saline control group [45].

Umalkar *et al.* formulated and evaluated ultradeformable vesicles gel of diflorasone diacetate, a corticosteroid, for topical application, prepared by using soya phosphatidylcholine and four different surfactants, i.e., Span 80, Tween 80, Sodium cholate, Span 60 by the lipid film hydration method, showing effectiveness in the treatment of psoriasis [46].

Clares *et al.* developed liposomes for the delivery of retinyl palmitate oil, the most stable form of vitamin A, widely known for playing an important role in cellular differentiation and carcinogenesis prevention, and largely employed in anti-aging formulations. It is also thought to induce thickening of the epidermis and to be effective for the treatment of skin diseases. MLV were obtained following the thin-film hydration method at a molar ratio of drug to phospholipids (1:1) [34].

In recent years, niosomes, ethosomes and highly flexible liposomes (also called transfersomes) have been studied to take the place of conventional liposomes. Niosomes offer higher chemical stability, lower costs, and great availability of surfactant classes when compared to conventional liposomes. The ethosomal system is composed of ethanol, phospholipid and water. The effect of ethanol on stratum corneum lipids and on vesicle fluidity as well as a dynamic interaction between ethosomes and the stratum corneum may contribute to the enhancement of skin permeation. Transfersomes that follow the transepidermal water activity gradient in the skin can enhance the transdermal bioavailability of drugs [4].

They can incorporate molecules within the hydrophilic core, within the lipophilic or hydrophilic areas of the bilayer walls, or attached to the surface. By adding surfactants as edge activators or ethanol (ethosomes) to the formulation highly deformable or ultraflexible liposomes are received. In both ways the membrane fluidity is increased, which allows the carrier to squeeze through pores that are smaller than the nominal carrier diameter when under pressure. Flexible liposomes are pulled across the SC via the inter-corneocyte lipid channels by the trans-epithelial hydration gradient. This mechanism is supposed to depend on non-occlusive conditions. During the passage the lipid bilayers of the vesicles remain intact so that the cargo is protected from potential degradation or undesired distribution. Also this mechanism brings them rapidly towards the deeper layers of the SC, close to the epidermis, so that they possibly get accessible to LCs sampling the environment. In addition, when using ethosomes, ethanol may work as a fluidizing agent on the intercellular lipid bilayers of the SC, which may further enhance their transport across the SC. In contrast, when conventional liposomes are applied to the skin they spread on the skin surface, dehydrate and coalesce without entering the lipid bilayers. Flexible liposomes have been used to improve skin penetration of a wide range of active ingredients and have demonstrated their superiority compared with conventional liposomes [47].

Skin ageing is associated with physiological processes and is inevitable; however, there is an exogenous ageing which is caused by extrinsic harmful environment factors and can be prevented. One of the most harmful extrinsic factors for skin is UV radiation, either A or B radiation. Exposition to these radiations leads the formation of reactive oxygen species (ROS), which are involved in phototoxicity reactions and inflammation, aggravating previous skin diseases and being involved in the development of malignant tumors. Among the approaches used to protect skin from these degenerative effects, use of antioxidant has been adopted as an important strategy, being incorporated in cosmetic and pharmaceutical formulations to scavenge free radicals in skin. However, delivery of these antioxidants using the conventional dosage forms is a challenge due to various reasons like poor solubility and permeability, instability and extensive first pass metabolism before reaching systemic circulation. To overcome these disadvantages of antioxidant orally administration, topically administration could be a more effective alternative because of the accumulation of the antioxidant compounds just in the area which needs to be protected. However, by applying an antioxidant directly on the skin surface, it will find a number of difficulties to penetrate through the stratum corneum, and therefore the use of a delivery system, such as liposomes, could have a doubly beneficial effect. On the one hand, it will protect the antioxidant compound of phenomena that can alter its chemical structure or biological activity, and on the other, it will facilitate the passage through the skin and the formation of a deposit in deeper layers of the stratum corneum. Despite the many advantages of liposomes, their instability constitutes one of the main drawbacks associated with their therapeutic application. In this sense, archaeal membrane lipids are an excellent raw material to form liposomes [35].

In this regard, González-Paredes *et al.* carried out a study to evaluate the incorporation of antioxidant phenolic compounds, recovered from olive mill waste, in archaeosomes formulation, in a proper excipient, Carbopol® and Pluronic® gels, in order to obtain a dosage form for topical application, namely, archaeosomes hydrogels [35]. The result showed that archaeosomes were to be more versatile than conventional liposomes for incorporation into gels.

9.5 Conclusions

Recent advances in pharmaceutical research have focused on novel delivery systems utilizing new devices to achieve modification of delivery time, targeting, as well as improve the *in vivo* solubility and hence bioavailability of poorly soluble drugs. Generally, most lipid drug delivery systems used as drug carriers have high stability, high carrier capacity, feasibility of incorporating both hydrophilic and hydrophobic substances and feasibility of variable routes of administration, including oral, topical, parenteral routes.

The enormous versatility of liposomes, archaeosomes and other vesicular systems, make them valuable carrier systems for bioactive compounds. It is predictable that these delivery systems will be increasingly applied in the near future with success, leading to major improvements in the therapeutic field.

Acknowledgments

The authors wish to acknowledge the financial support from the Portuguese Science and Technology Foundation, Ministry of Science and Education (FCT/MEC) through national funds, and co-financed by FEDER, under the Partnership Agreement PT2020, project references M-ERA-NET/0004/2015-PAIRED and UID/QUI/50006/2013.

References

1. Souto EB, Severino P, Basso R, Santana MH. Encapsulation of antioxidants in gastrointestinal-resistant nanoparticulate carriers. *Methods in Molecular Biology*. 2013; 1028: 37–46.
2. Khosravi-Darani K, Mozafari M. Nanoliposome potentials in nanotherapy: A concise overview. *International Journal of Nanoscience and Nanotechnology*. 2010; 6: 3–13.
3. Moghimipour E, Handali S. Liposomes as drug delivery systems: Properties and applications. *Research Journal of Pharmaceutical, Biological and Chemical Sciences*. 2013; 4: 169–185.
4. Chen X, Lihua P, Jianqing P. Novel topical drug delivery systems and their potential use in scars treatment. *Asian Journal of Pharmaceutical Sciences*. 2012; 7: 155–167.
5. Tadwee IK, Gore, S, Giradkar, P. Advances in topical drug delivery system: A review. *International Journal of Pharmaceutical Research and Allied Sciences*. 2012; 1: 14–23.
6. Puri D, Bhandari A, Sharma P, Choudhary D. Lipid nanoparticles (SLN, NLC): A novel approach for cosmetic and dermal pharmaceutical. *Journal of Global Pharma Technology*. 2010; 2: 1–15.
7. Benvegnu T, Lemiegre L, Cammas-Marion S. New generation of liposomes called archaeosomes based on natural or synthetic archaeal lipids as innovative formulations for drug delivery. *Recent Patents in Drug Delivery and Formulation*. 2009; 3: 206–220.
8. Sharma V, Mishra D, Sharma A, Srivastava B. Liposomes: Present Prospective and Future Challenges. *International Journal of Current Pharmaceutical Research*. 2010; 1: 7–16.

9. Patel C, Satyanand Tyagi P, Patel Tushar P, Patel Priyanka H. A Novel Pharmaceutical Microscopic Vesicle: Liposome. *International Journal of Current Trends in Pharmaceutical Research*. 2014; 2: 555–561.
10. Çaðdaþ M, Sezer, DA, and Bucak, S. Liposomes as potential drug carrier systems for drug delivery. In (Sezer AD, ed), *Application of Nanotechnology in Drug Delivery*: InTech; 2014. pp. 1–50.
11. Moghimipour E, Kargar M, Ramezani Z, Handali S. The potent *in vitro* skin permeation of archaeosome made from lipids extracted of Sulfolobus acidocaldarius. *Archaea*. 2013; 2013: 782012.
12. Li Z, Zhang L, Sun W, Ding Q, Hou Y, Xu Y. Archaeosomes with encapsulated antigens for oral vaccine delivery. *Vaccine*. 2011; 29: 5260–5266.
13. Li Z, Chen J, Sun W, Xu Y. Investigation of archaeosomes as carriers for oral delivery of peptides. *Biochemical and Biophysical Research Communications*. 2010; 394: 412–417.
14. Carrer DC, Higa LH, Tesoriero MV, Morilla MJ, Roncaglia DI, Romero EL. Structural features of ultradeformable archaeosomes for topical delivery of ovalbumin. *Colloids and Surfaces B Biointerfaces*. 2014; 121: 281–289.
15. Benvegnu T, Rethore G, Brard M, Richter W, Plusquellec D. Archaeosomes based on novel synthetic tetraether-type lipids for the development of oral delivery systems. *Chemical Communications (Camb)*. 2005; 44: 5536–5538.
16. Chime AS, and Onyishi IV. Lipid-based drug delivery systems (LDDS): Recent advances and applications of lipids in drug delivery. *African Journal of Pharmacy and Pharmacology*, 2013; 7: 3034–3059.
17. Patel B, Chen W. Archaeosomes as drug and vaccine nanodelivery systems. In (Mozafari MR, ed), *Nanocarrier Technologies: Frontiers of Nanotherapy*, Springer; 2006. pp. 1–16.
18. Zavec AB, Ota, A, Zupancic, T, Komel, R, Ulrih, NP, Liovic, M. Archaeosomes can efficiently deliver different types of cargo into epithelial cells grown *in vitro*. *Journal of Biotechnology*. 2014; 192(Pt A): 130–135.
19. Jacquemet A, Barbeau J, Lemiegre L, Benvegnu T. Archaeal tetraether bipolar lipids: Structures, functions and applications. *Biochimie*. 2009; 91: 711–717.
20. Ulrih NP, Gmajner D, Raspor P. Structural and physicochemical properties of polar lipids from thermophilic archaea. *Applied Microbiology and Biotechnology*. 2009; 84: 249–260.

21. Ota A, Gmajner D, Sentjurc M, Ulrih N. Effect of growth medium pH of Aeropyrum pernix on structural properties and fluidity of archaeosomes. *Archaea*. 2012; 2012: 285152.

22. Gangwar M, Singh R, Goel R, Nath G. Recent advances in various emerging vescicular systems: An overview. *Asian Pacific Journal of Tropical Biomedicine*. 2012; 2: S1176–S1188.

23. Sustar V, Zelko J, Lopalco P, Lobasso S, Ota A, Poklar Ulrih N, *et al.* Morphology, biophysical properties and protein-mediated fusion of archaeosomes. *PLoS One*. 2012; 7: e39401.

24. Barbeau J, Cammas-Marion S, Auvray P, Benvegnu T. Preparation and characterization of stealth archaeosomes based on a synthetic PEGylated archaeal tetraether lipid. *Journal of Drug Delivery*, 2011; 2011: 396068.

25. Schwendener RA. Liposomes as vaccine delivery systems: A review of the recent advances. *Therapeutic Advances in Vaccines*. 2014; 2: 159–182.

26. Krishnan L, Sad S, Patel GB, Sprott GD. Archaeosomes induce long-term CD8+ cytotoxic T cell response to entrapped soluble protein by the exogenous cytosolic pathway, in the absence of CD4+ T cell help. *Journal of Immunology*. 2000; 165: 5177–5185.

27. Sprott GD, Sad S, Fleming LP, Dicaire CJ, Patel GB, Krishnan L. Archaeosomes varying in lipid composition differ in receptor-mediated endocytosis and differentially adjuvant immune responses to entrapped antigen. *Archaea*. 2003; 1: 151–164.

28. Krishnan L, Sad S, Patel GB, Sprott GD. The potent adjuvant activity of archaeosomes correlates to the recruitment and activation of macrophages and dendritic cells *in vivo*. *Journal of Immunology*. 2001; 166: 1885–1893.

29. Rethore G, Montier T, Le Gall T, Delepine P, Cammas-Marion S, Lemiegre L, *et al.* Archaeosomes based on synthetic tetraether-like lipids as novel versatile gene delivery systems. *Chemical Communications (Camb)*. 2007: 2054–2056. DOI: 10.1039/b618568a.

30. Higa LH, Schilrreff P, Perez AP, Iriarte MA, Roncaglia DI, Morilla MJ, *et al.* Ultradeformable archaeosomes as new topical adjuvants. *Nanomedicine*. 2012; 8: 1319–1328.

31. Patel G, Ponce A, Zhou H, Chen W. Structural characterization of archaeal lipid mucosal vaccine adjuvant and delivery (AMVAD) formulations prepared by different protocols and their efficacy upon intranasal immunization of mice. *Journal of Liposome Research*. 2008; 18: 127–143.

32. Sprott G, Tolson D, Patel G. Archaeosomes as novel antigen delivery systems. *FEMS Microbiology Letters*. 1997; 154: 17–22.

33. Choquet C, Patel G, Sprott G, Beveridge T. Stability of pressure-extruded liposomes made from archaeobacterial ether lipids. *Applied Microbiology and Biotechnology*. 1994; 42: 375–384.

34. Clares B, Calpena AC, Parra A, Abrego G, Alvarado H, Fangueiro JF, *et al.* Nanoemulsions (NEs), liposomes (LPs) and solid lipid nanoparticles (SLNs) for retinyl palmitate: Effect on skin permeation. *International Journal of Pharmaceutics*. 2014; 473: 591–598.

35. Gonzalez-Paredes A, Clares-Naveros B, Ruiz-Martinez MA, Durban-Fornieles JJ, Ramos-Cormenzana A, Monteoliva-Sanchez M. Delivery systems for natural antioxidant compounds: Archaeosomes and archaeosomal hydrogels characterization and release study. *International Journal of Pharmaceutics*. 2011; 421: 321–331.

36. Kumar D, Sharma D, Singh G, Singh M, Rathore MS. Lipoidal soft hybrid biocarriers of supramolecular construction for drug delivery. *ISRN Pharmaceutics*. 2012; 2012: 474830.

37. Dhiman S, Singh, GT, and Rehni, KA. Transdermal patches: A recent approch to new drug delivery system. *International Journal of Pharmacy and Pharmaceutical Sciences*. 2011; 3: 26–34.

38. Prow TW, Grice JE, Lin LL, Faye R, Butler M, Becker W, *et al.* Nanoparticles and microparticles for skin drug delivery. *Advanced Drug Delivery Reviews*. 2011; 63: 470–491.

39. Agarwal R, Katare O. Miconazole nitrate–loaded topical liposomes. *Pharmaceutical Technology*. 2002; 26: 48–60.

40. Glavaš-Dodov M, Simonoska M, Goračinova K. Formulation and characterization of topical liposome gel bearing lidocaine HCl. *Macedonian Journal of Chemistry and Chemical Engineering*. 2005; 24: 59–65.

41. Moghimipour E, Sajadi Tabassi S, Ramezani Mohammad L. Enhanced permeability of gentamicin sulfate through shed snake-skin and liposomal membranes by different enhancers. *Iranian Journal of Basic Medical Sciences*. 2003; 6: 9–19.

42. Mitkari B, Korde S, Mahadik K, Kokare C. Formulation and evaluation of topical liposomal gel for fluconazole. *Indian Journal of Pharmaceutical Education & Research*. 2010; 44: 324–333.

43. Mozafari MR, and Khosravi-Darani, K. An overiew of liposome-derived nanocarrier technologies. In (Mozafari MR, ed), *Nanomaterials and Nanosystems for Biomedical Applications,* Springer; 2007. pp. 113–124.

44. Vogt PM, Hauser J, Rossbach O, Bosse B, Fleischer W, Steinau HU, *et al.* Polyvinyl pyrrolidone-iodine liposome hydrogel improves epithelialization by combining moisture and antisepis. A new concept in wound therapy. *Wound Repair Regen.* 2001; 9: 116–122.

45. Yang J, Ni B, Liu J, Zhu L, Zhou W. Application of liposome-encapsulated hydroxycamptothecin in the prevention of epidural scar formation in New Zealand white rabbits. *The Spine Journal.* 2011; 11: 218–223.

46. Umalkar D, Rajesh K. Preparation and characterization of ultra deformable vesicles gel for treatment of psoriasis. *Indo American Journal of Pharmaceutical Research.* 2013; 3: 6291–6298.

47. Hansen S, Lehr C. Nanoparticles for transcutaneous vaccination. *Microbial Biotechnology.* 2012; 5: 156–167.

Chapter 10

Ethosomes: A Novel Carrier for Dermal or Transdermal Drug Delivery

Dinesh Kumar Mishra,[a] Neelam Balekar,[a] Vinod Dhote,[b] and Pradyumna Kumar Mishra[c]

[a]*IPS Academy, College of Pharmacy, Indore, Madhya Pradesh, India*
[b]*Truba Institute of Pharmacy, Bhopal, Madhya Pradesh, India*
[c]*Department of Molecular Biology,*
National Institute for Research in Environmental Health,
Indian Council of Medical Research, Bhopal, Madhya Pradesh, India

dineshdops@gmail.com

10.1 Introduction

Transdermal delivery of drugs and vaccines is an effective alternative to oral and parenteral routes of administration. Through transdermal drug delivery, it is feasible to evade "first-pass" inactivation by the liver, diminish the likelihood of gastrointestinal irritation, offer steady absorption of medication over long periods of time, reduce the frequency of dosing, which in turn improves adherence, and lower the frequency of adverse effects through the avoidance of high serum drug peaks [1]. Following critical examination of numerous transdermal formulations that are currently available in the market and from emerging evidence of several ongoing clinical trials for debilitating human ailments,

Carrier-Mediated Dermal Delivery: Applications in the Prevention and Treatment of Skin Disorders
Edited by Andreia Ascenso, Sandra Simões, and Helena Ribeiro

ISBN 978-981-4745-58-1 (Hardcover), 978-1-315-36447-6 (eBook)
www.panstanford.com

it is amply clear this route of delivery is highly cost-effective. Besides, safety and effectiveness, increased acceptability of transdermal products in patients have led to an unprecedented interest in novel drug delivery technologies to ensure maximum drugs are amenable to be administered by this route [2, 3]. To design a better transdermal delivery system, it is extremely important to understand the structure of skin. The human skin offers protective coverage for the internal structure and organs. It covers a total surface area of approximately 1.8 m^2 at an average thickness of 0.00394 in (0.1 mm), accounting for between 15–18% of the total body weight. The skin has three layers—the epidermis, dermis, and fat layer (also called the subcutaneous layer). It protects the organism against environmental factors and regulates heat and water loss from the body (Fig. 10.1) [4].

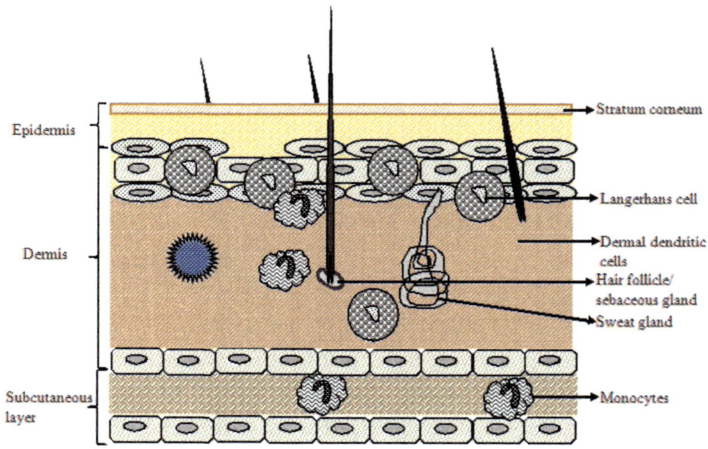

Figure 10.1 Structure of skin.

Epidermis is the outermost layer of skin and composed of four strata: the stratum germinativum (or basal layer), the stratum granulosum (the malpighian layer), the stratum lucidum (the granular layer), and stratum corneum (the horny layer). The stratum corneum (SC) is the heterogeneous outermost layer of the epidermis and is approximately 10–20 μm thick. Within the epidermis there are distinct types of cells: keratinocytes, melanocytes, and dendritic cells (DCs). DCs are also present in dermis layer participating in the recognition of invading pathogens. The surface lipid film and the SC form the interface with the environment providing the site for the topical application of medications [5, 6].

Dermis is 10–40-fold thicker than the epidermis, depending on the area of the body. It is a matrix of loose connective tissue composed of polysaccharides and protein (collagen and elastin) and metabolically less active than the epidermis. This matrix contains nerves, blood vessels, hair follicles, sebaceous and sweat glands [7, 8]. Mast cells, macrophages, leukocytes, and endothelial cells of the blood vessels are also located in the dermis. The function of the dermis is to nourish the epidermis and anchor it to the subcutaneous tissue. Subcutaneous tissue serves as a receptacle for the formation and storage of fat. It acts as both heat regulator and shock absorber [2].

Transdermal route has received wide attention for many decades owing to its huge surface area and properties that promote drug delivery [3, 6, 9]. However, skin tends to be the strongest barrier for the entry of drug entities, and hence it is essential to design the drug delivery system in the most appropriate manner which includes the selection of a vehicle to deliver the medicament into the skin layers (cutaneous delivery), or through the skin and into the systemic circulation (percutaneous absorption) [10]. Different types of drug delivery systems have been designed for application on the skin. In dermatopharmacotherapy, the primary purpose is to apply drugs to the skin for inducing local effects at the site of application [11].

To achieve this target in dermatopharmacotherapy, selective delivery systems were developed to enhance penetration of active ingredients at the site of action. However, the poor skin penetration by most of drugs is one of the major problems in dermatopharmacotherapy, since only a small portion of dose finally reaches the sites of action producing limited local activity [6, 12]. Moreover, few drugs which can easily penetrate the skin are quickly removed by blood circulation, producing systemic effect rather than local effects. In order to deliver drugs through the skin, most compounds necessitate various degrees of permeation enhancement [13].

Classic enhancement methods focus primarily on chemical enhancement or modulation of interactions between the drug and the vehicle [14]. Recent literature indicates the use of innovative vesicular carriers, electrically assisted delivery, various micro invasive methods, and some incorporating technologies from other fields [15–17]. The best avenue to improve drug penetration and/or localization is obviously by manipulating the vehicle or using a drug carrier concept. Dermatological and cosmetic preparations frequently contain active pharmaceutical ingredients, which can only act when they penetrate at least the outermost layer of the skin [18]. On the other hand, the efficacy of these actives is suboptimal since the transport into the skin is leisurely owing the resistance offered by outermost layer of the skin. To conquer these restrictions, various approaches are undertaken to increase permeation (Fig. 10.2) [19].

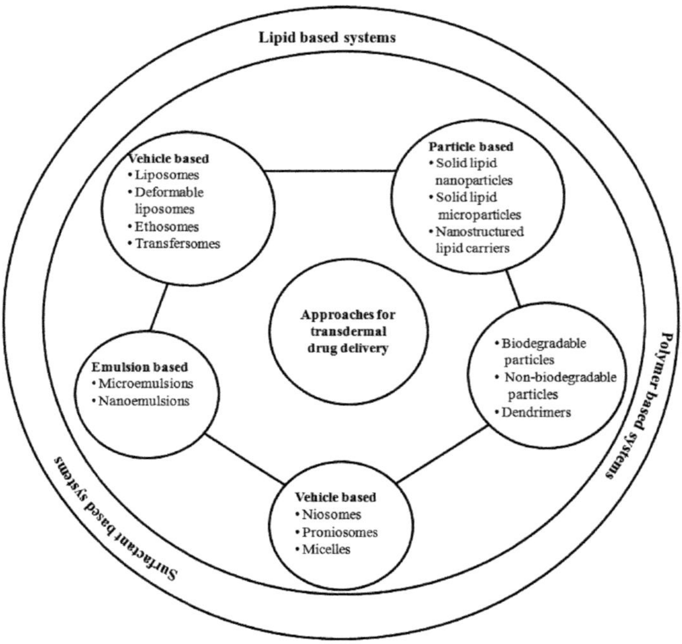

Figure 10.2 Various approaches for permeation enhancement through dermal and transdermal route.

10.2 Novel Carriers as Tools for Modulation of Skin Permeability

Skin is considered as a remarkably efficient barrier designed to keep our insides in and the outside out. The modulation of this barrier property, including permeability to drugs, chemicals, and bioactive agents is the prime requirement as well as target for transdermal delivery. Therefore, several approaches have been attempted to enhance the permeation rate of various agents. One of the approaches is the application of novel carriers that are capable of providing controlled release, delivery of drug at predetermined rate and targeted delivery. This may lead to improved efficacy, safety, and better patient compliance. Novel delivery systems or carriers for effective dermal and transdermal delivery are microparticles, nanoparticles, liposomes, elastic liposomes, niosomes, ethosomes, etc. The ideal delivery systems should be safe, non-toxic as well as non-immunogenic and also economical. Moreover, such systems should be pharmaceutically acceptable, stable, biocompatible, and amenable to the patient [20–23]. The main delivery systems are discussed in the following paragraphs.

10.2.1 Microparticles/Nanoparticles

Micro/nanoparticles are suitable vehicles for delivery of the drug, proteins, and peptides. These carriers offer numerous advantages like their potential ability for controlled release of encapsulated agents and long lasting responses, efficient phagocytosis due to their particulate nature, and also capacity to induce cellular immune responses. Various nanoparticles constituted by gelatin, PLGA, chitosan, and lipid have been successfully used for enhanced transdermal delivery [24, 25].

10.2.2 Liposomes

Liposomes are colloidal particles containing concentric bimolecular layers and possess ability to encapsulate both polar and non-polar drugs. These lipid vesicles are usually made up of phospholipids and cholesterol. They can be prepared with various sizes, number of lamellae, structure, and different payloads. They are a versatile tool for delivery of large array of bioactives ranging from drugs to high molecular weight proteins and peptides. The amphipathic nature of liposomes may allow them to be widely used as a non-invasive delivery agent [21, 26].

It is generally reported that topical application of liposomes tends to accumulate in the SC, upper skin layers and in the appendages, with minimum or no penetration into deeper skin or systemic circulation. However, the results obtained so far by various researchers about the efficacy of liposomal system for facilitating transdermal delivery are quite conflicting. In fact, the composition of the vesicles influences their physico-chemical characteristics such as size, charge, thermodynamic status, lamellarity, and bilayer elasticity [27, 28]. Thus by modifying the structure and composition of liposomes we can have more versatile, safe and efficient tool for transdermal delivery.

10.2.3 Elastic Liposomes

A novel class of modified liposomes, containing an optimum amount of edge activator with highly elastic nature, has been developed: Transfersome®. These carriers were firstly described by Cevc and Blume (1992), and subsequently they have been the subject of numerous patents and literature reports [29]. Elastic liposomes (Transfersome®) are specially optimized since ultradeformable lipid vesicles have been claimed to penetrate and permeate the skin layers as intact vesicles to reach the systemic circulation. The elasticity possessed by these vesicles is the consequence of an edge activator (often a single-chain surfactant that enhances the deformability via lipid bilayer destabilization) incorporated within the phospholipid-based system. In most cases, the phospholipid and edge activator contents have been optimized to attain the desired deformable nature of the elastic liposomes, to increase elasticity and penetrability. These novel carriers are applied under non-occlusive conditions in the form of semi-diluted suspension to offer an efficient dermal and transcutaneous drug delivery of high and low-molecular-weight substances [23, 30, 31].

10.2.4 Niosomes

Niosomes are non-ionic surfactant based vesicles similar to liposomes formed by hydration of synthetic non-ionic surfactants and cholesterol. These vesicles are similar to liposomes in terms of physical properties and prepared in the same way. Niosomes alleviate the demerits associated with classic liposomes such as chemical instability, variable purity of phospholipids and high cost. In addition, they have more penetrating capability than the classic liposomes [32, 33].

10.2.5 Ethosomes

Ethosomes are soft, malleable ethanolic phospholipid vesicles tailored for enhanced delivery of active agents. The high concentration of ethanol causes disturbance of skin lipid bilayer organization; hence, when incorporated into a vesicular membrane, it improves the vesicle's capability to go through the SC [34]. Although the lipid membrane is less tightly packed than conventional vesicles, it has equivalent stability and improves drug distribution ability in SC lipids [28]. This extended non-invasive delivery of drug molecules of variant size is useful for delivery of cultured cells and microorganisms as well. Enhanced delivery of these bioactive molecules through the skin and cellular membranes by means of an ethosomal carrier offers numerous challenges and prospects for the research and future development of novel improved therapies [35–37].

Ethanolic liposomes, or ethosomes, are novel lipid based and non-invasive delivery carriers that enable biologically active agents to reach deeper skin layers and/or systemic circulation. These systems are mainly composed of phospholipids, a relatively high concentration of ethanol (20–50%) and water (Fig. 10.3) [38]. Previously, it was generally thought that a high alcohol concentration lead to the destruction of lipid vesicular structure, owing to the interdigitating effect of alcohol on lipids. Afterward Touitou *et al.* (1997) demonstrated the coexistence of phospholipid vesicles with a high concentration of ethanol, leading to the formation of soft, malleable, highly fluid vesicles [39]. The size of ethosomes can be modulated from tens of nanometers to microns [40, 41].

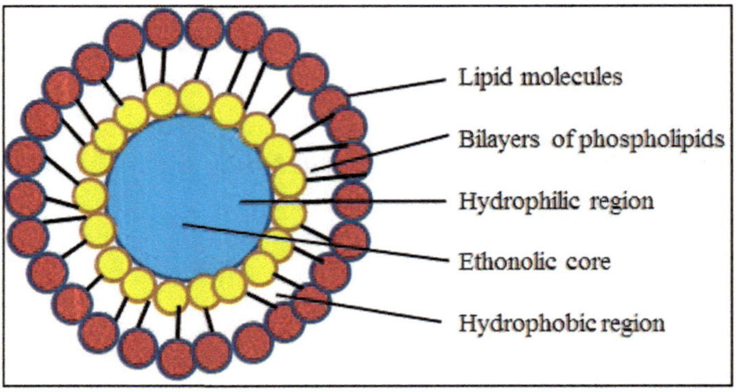

Figure 10.3 Structural feature of ethosomes.

10.2.5.1 Ethosomes composition

Ethosomes are vesicular carriers that comprise hydroalcoholic or hydro/alcoholic/glycolic phospholipids in which the concentration of alcohols or their combination is relatively high [35]. The various type of additives used in the ethosomes preparations are represented in Table 10.1 [42–48].

Table 10.1 Different additives employed in formulation of ethosomes

Additives	Uses	Examples
Phospholipid	Vesicles forming component	Soya phosphatidyl choline, egg phosphatidyl choline, dipalmityl phosphatidyl choline, distearyl phosphatidyl choline
Polyglycol	Skin penetration enhancer	Propylene glycol, transcutol
Cholesterol	Stabilizer	Cholesterol
Alcohol	For providing the softness for vesicle membrane and as a penetration enhancer	Ethanol, isopropyl alcohol
Vehicle	As a gel former	Carbopol 934
Dye	For characterization study	6-Carboxy fluorescence, rhodamine-123, rhodamine red, fluorescene isothiocynate

Ethosomes mainly consists of phospholipids with various chemical structures like phosphatidylcholine (PC), hydrogenated PC, phosphatidylethanolamine (PE), and others alcohol (ethanol or isopropyl alcohol), water and propylene glycol (or other glycols). Ethanol binds to the lipid polar heads and increases the fluidity of the liquid crystalline state. Cholesterol at concentrations ranging between 0.1–1% can be also added to the preparation. By employing ^{31}p NMR, the polar head group motions of PC were found to be restricted and anisotropic, and the existence of bilayers at 20–45% ethanol range was also confirmed [49, 50]. The permeability of phospholipid dispersions at 30–45% ethanol was revealed by paramagnetic-ion NMR technique, demonstrating phospholipid bilayers in the form of closed vesicles. The ethosomes composed of PC: ethanol (2:30%) are claimed to be multilamellar vesicles, with evenly spaced lamellas extending to the core of the vesicles [51]. As the ethanol content approaches to 45%, a small soluble fraction of phospholipids is mixed with the closed vesicles [52].

A unique property of ethosomes lies on the possibility of controlling their size as a function of the ethanol content. For example, the size of ~200 nm diameter vesicles composed of 2% PC can be reduced by half as the ethanol concentration is increased from 20% to 45%. On the contrary, the ethosomes size has a limited dependence on the PC concentration. An eight-fold increase in PC concentration from 0.5% to 4% results in a two-fold increase in ethosomes size (120 to 250 nm) [53–55]. Moreover, the high ethanol content is also responsible for the negative zeta potential. The high negative zeta potential confers higher colloidal stability than its liposomal counterparts. As already mentioned, the phospholipids in ethosomes are less tightly packed compared to conventional liposomal bilayers, and the membrane presents higher permeability for hydrophilic/ionic solutes. Hence, ethosomes could be not suited enough to entrap certain hydrophilic solutes [50, 56]. In spite of the high ethanol concentration, the average size and size distribution of ethosomes usually remain constant for at least two years at room temperature. A key difference with transfersomes is that ethosomes can be successfully applied under occlusive conditions employing patches [57, 58].

10.2.5.2 Mechanism of skin penetration

In terms of the enhanced potential for transdermal delivery of biologically active agents—loaded ethosomal carriers, the exact mechanism of skin permeability modulation remains somehow speculative. According to Touitou *et al.* (2000) a synergistic mechanism between ethanol, vesicles and skin lipids exists, leading to an improved permeation profile. The proposed mechanism of ethosomal skin modulation lies in the interaction of ethanol with lipid molecules in the polar head region, resulting in a reduction of the transition temperature (T_m) of SC lipids, thus enhancing their fluidity and leading to a disordered form. This transition provides a potential site for soft, malleable ethosomes to penetrate more easily within the skin layers (Fig. 10.4) [1, 17].

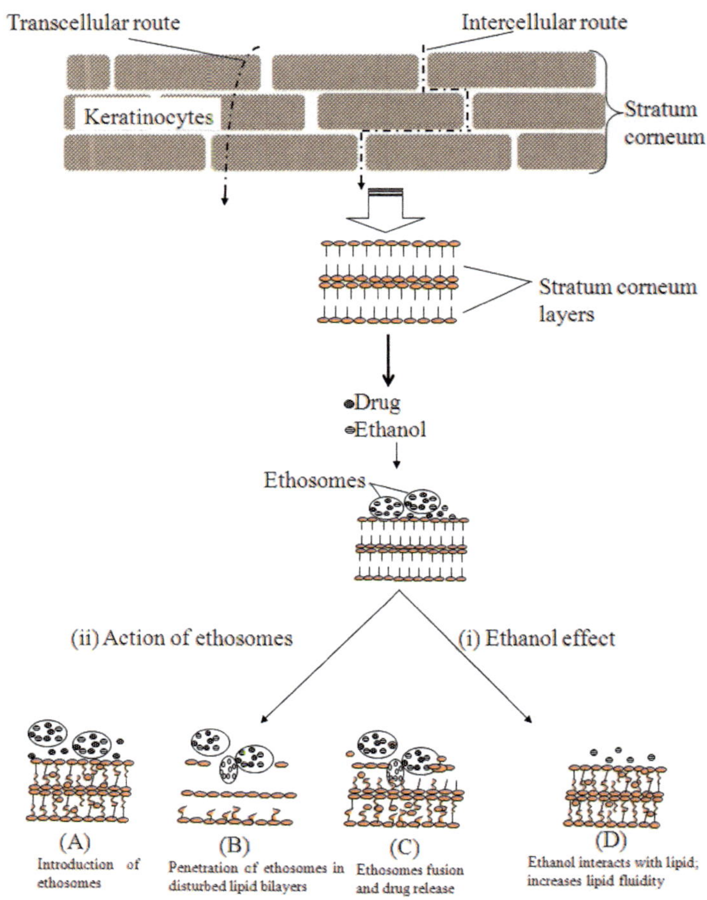

Figure 10.4 Proposed mechanisms for skin delivery via ethosomes.

After topical application, the permeation enhancement from ethosomes is much greater than would be expected from just pure ethanol, suggesting a synergic mechanism between ethanol, vesicles and skin lipids. Ethosomes are generally more effective permeation enhancers than ethanol, aqueous ethanol, or ethanolic phospholipid solutions. It is hypothesized that ethosomes might act as enhancers of drug permeation and as drug carriers through the SC [50]. Ethanol may increase the solubility of the drug in the vehicle, disturb the organization of the SC lipid bilayer and enhance its lipid fluidity. The subsequent mixing of phospholipids with SC lipids of the intercellular layers was observed to enhance the skin permeability [51–53].

10.2.5.3 Advantages and limitations

Current available drug delivery systems lack in achieving all the lofty goals, but sincere attempts have been made to achieve them through novel approaches in drug delivery context. Ethosomes have emerged encompassing various routes of administration, to achieve controlled and targeted drug delivery through the mechanism mentioned above [28]. Encapsulation of the drug in its vesicular structures can be predicted to prolong the existence of the drug in systemic circulation, and reduce the toxicity if a selective uptake can be achieved. It offers numerous advantages over conventional drug delivery systems, as depicted in Fig. 10.5 [17, 54].

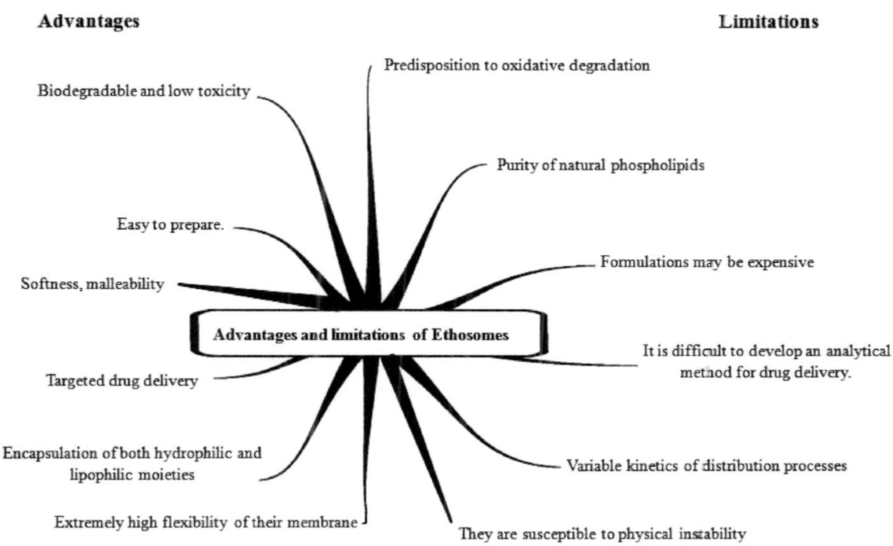

Figure 10.5 Advantages and limitations associated with ethosomes.

10.2.5.4 Methods of preparation

Ethosomes preparation grounds on simple and easy scale up techniques without entailment of any sophisticated instruments at both pilot and industrial levels. Basic methods used for preparation of these vesicular carriers are as follows:

10.2.5.4.1 *Cold method*

This is one of the most extensively used methods for preparation of ethosomes consisting of two basic and simple steps. In the first step, phospholipid or other lipid material is dissolved in ethanol by vigorous stirring in a covered vessel and with continuous addition of polyols as propylene glycol in a water bath at 30°C. Water is heated up to 30°C in a separate vessel and slowly added in a fine stream to the above mixture [59]. The drug can be dissolved in water or ethanol depending on its hydrophilic/hydrophobic properties. Stirring is continued for another 5 min and the resultant vesicle suspension will be cooled at room temperature. The vesicle size of ethosomal formulation can be reduced to desired one using sonication or extrusion method (Fig. 10.6). Finally, the formulation should be stored under refrigerated conditions [55].

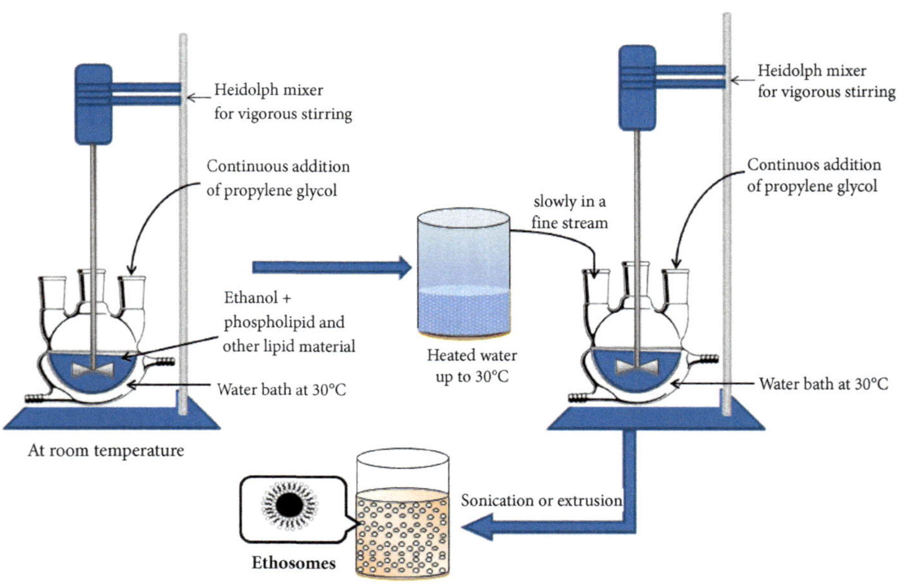

Figure 10.6 Cold method.

10.2.5.4.2 Hot method

In this method, the phospholipid is dispersed in water in a water bath at 40°C until a colloidal solution is obtained. In a separate vessel, ethanol and glycols are mixed and heated up to 40°C. As temperature of both mixtures reaches 40°C, the organic phase is added to the aqueous phase. The final procedure is quite similar to cold method (Fig. 10.7) [51, 59].

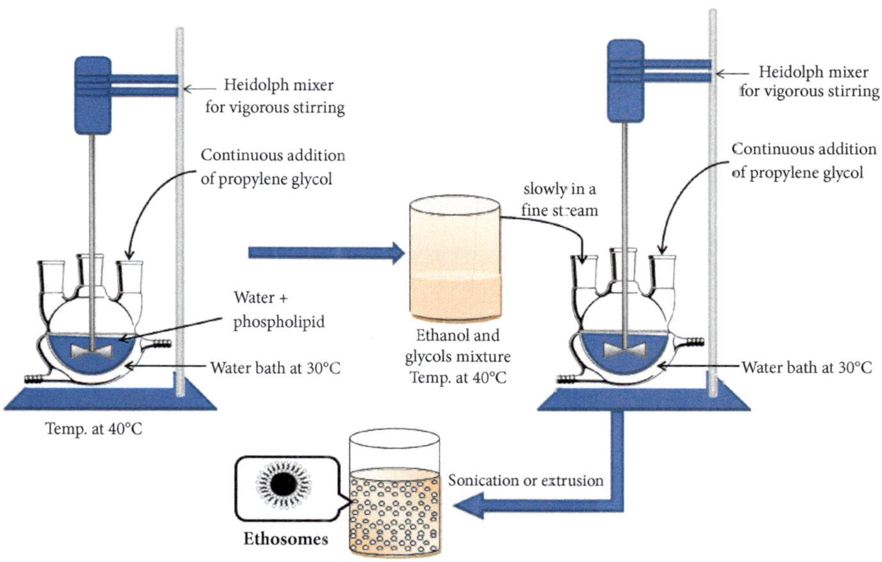

Figure 10.7 Hot method.

10.2.5.4.3 Classic mechanical dispersion method

In this method, phospholipid is dissolved in an organic solvent or mixture of organic solvents in a round-bottom flask. Organic solvent is further removed using a rotary vacuum evaporator above lipid transition temperature to form a thin lipid film on the wall of the flask [45]. Traces of the solvent should be removed from the deposited lipid film by leaving the contents under vacuum overnight. The lipid film is then hydrated with drug hydroethanolic solution by rotating the flask at suitable temperature with or without intermittent sonication. Finally, the resultant ethosomal suspension is cooled at room temperature (Fig. 10.8) [24, 34].

These methods have been extensively used by researchers to investigate efficacious dermal/transdermal delivery of different actives.

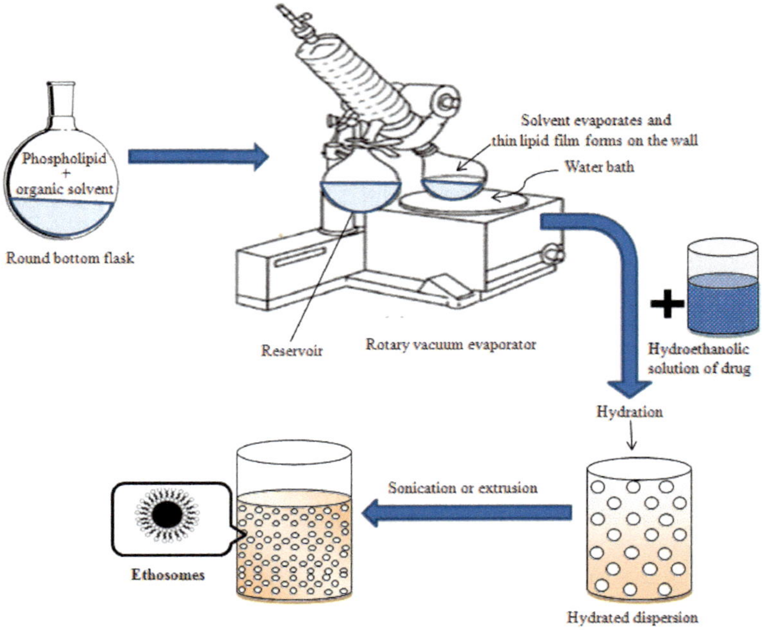

Figure 10.8 Classic mechanical dispersion method.

10.2.5.5 Characterization of ethosomes

Characterization of any dosage form or delivery system is very important from the manufacturing as well as therapeutic efficacy point of view, being also needed to get a reproducible product. The characterization parameters of ethosomes are quite similar to other vesicular carrier systems (Fig. 10.9). Vesicle shape and surface morphology can be studied by scanning electron microscopy (SEM) and transmission electron microscopy (TEM) in which spherical or nearly spherical shaped structure and unilamellar or multilamellar vesicles can be observed [36]. Vesicles size and size distribution can be measured by dynamic light scattering (DLS) technique. Zeta potential measured by laser Doppler anemometry (LDA) is the electric potential of the vesicles including its ionic atmosphere, which affects vesicular properties such as stability and skin-vesicle interaction [46]. High zeta potential whether positive or negative improves the vesicles physical stability by preventing their aggregation owing to electrostatic repulsion and increasing the inter bilayer distance. The entrapment efficiency of drug-loaded ethosomes can be measured by different methods after the purification step of ultracentrifugation, size exclusion chromatography or dialysis. Entrapment efficiency gives an idea about the amount of drug in possible three areas of the vesicles: the quantity adsorbed in the vesicular membrane, the quantity incorporated into the membrane bilayer and the quantity included in the internal core phase [48–53]. The entrapment efficiency is affected by both lamellarity of vesicles and drug solubility in the ethosomal system.

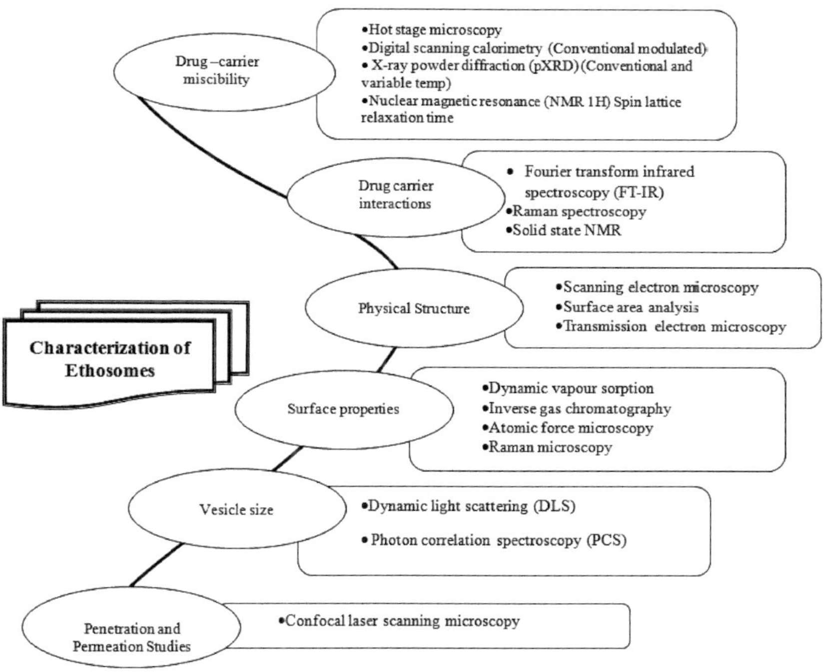

Figure 10.9 Characterization of ethosomal formulations.

Vesicles softness can be indicated by measurement of transition temperature of the lipids present in the vesicular system [49]. The extent or depth of skin penetration can be evaluated by confocal laser scanning microscopy (CLSM) with application of various dyes-loaded vesicles [24, 34, 35].

Finally, the ethanol content is a significant parameter of ethosomes and can be estimated using enzymatic diagnostic kit based upon oxidation of alcohol to aldehyde. The phospholipid content can be quantified by Barlett assay by colorimetric estimation at 830 nm [26].

10.2.5.6 Stability of ethosomes

Technically stability is defined as the capacity of formulation in a specific system to remain within its physical, chemical, microbiological, therapeutic, and toxicological limits. Stability is a major concern for the development of marketed preparations. *In vitro* stability is related with the issues prior to formulation administration while *in vivo* stability is evaluated after administration of formulation via different routes to the biological fluids. Ethosomes offer enhanced stability in contrast to other vesicular carriers such as liposomes, vesosomes, etc. [60]. In case of stored liposomes, they are liable to fuse and grow into larger vesicles which promote the breakage of liposomal vesicles. In fact, drug leakage from the vesicles represents a significant problem during storage. The lack of electrostatic repulsion is expected to account for the propensity of neutral liposomes to amassed, whereas in ethosomes, ethanol causes a variation of the net charge of the system (e.g., impart negative charge) and confers some degree of steric stabilization leading to increased stability of vesicles against agglomeration and drug leakage from vesicles. Increasing the concentration of ethanol from 20% to 45% improves the entrapment efficiency owing to an increase in the fluidity of the membranes [61–63]. However, a further increase in the ethanol concentration (>45%) destabilizes the vesicles membrane leading to a decrease in entrapment efficiency. Another difficulty remains on the purity of phospholipids and phospholipids containing unsaturated fatty acids that endure oxidation whose reaction products can cause permeability changes in the ethosomal bilayers [64]. Oxidative degradation can be diminished by certain procedures, such as protecting the lipid preparation from light; using phospholipids which contain saturated fatty acids, and adding antioxidants as α-tocopherol. Additionally, hydrolysis of lipids leads to the formation of lyso-lecithin. The presence of lyso-lecithin in lipid bilayers greatly enhances the permeability of ethosomes; hence, it causes leakage of drug from ethosomal vesicles. Thus, it is important to start with phospholipids which are free of lyso-lecithin and any phospholipases.

10.3 Applications

Ethosomal systems are a modified version of well-established and popular dosage form i.e. liposomes. They are considered as non-invasive type of carriers that deliver drugs to deep skin layers and/or systemic circulation. Ethosomes are sophisticated, safe, and effective systems and easily prepared. Once any moiety is entrapped within the system, it is not exposed to metabolic degradation and dilution. These types of systems have a prolonged drug release improving the therapeutic benefit. Some of important applications of ethosomes are briefly discussed below and summarized in Fig. 10.10.

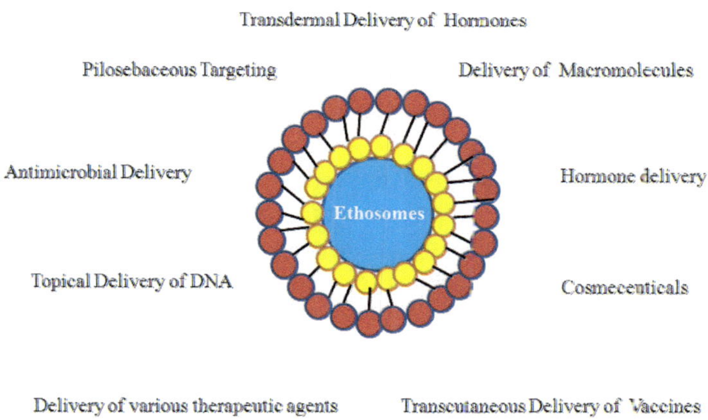

Figure 10.10 Applications of ethosomes in drug delivery.

10.3.1 Pilosebaceous Targeting

Hair follicles and sebaceous glands are recognized as potential components in the percutaneous drug delivery. Minoxidil-loaded ethosomal formulation was investigated for the treatment of baldness. It was observed that conventional topical formulation had very poor skin permeation as well as poor skin retention properties compared to ethosomal formulation. This study showed the utility of ethosomal carriers for pilosebaceous targeting of minoxidil to achieve better clinical efficacy [17, 65, 66]. In addition, pilosebaceous targeting using ethosomes represented a better option for several dermatological disorders like eosinophilic pustular folliculitis (EPF) and Kaposi sarcoma [67].

10.3.2 Hormones Delivery

Hormones delivery by means of oral delivery is associated with low bioavailability, hepatic first pass metabolism, dose related side effects and low patient compliance. Researchers enumerated different approaches for the transdermal delivery of hormones by ethosomes. Touitou *et al.* investigated the efficacy of ethosomes for transdermal delivery of testosterone over marketed testosterone transdermal patch. When both formulations were compared, a greater skin permeation of testosterone-loaded ethosomes was reported. Both *in vitro* as well as *in vivo* studies demonstrated increased skin permeation and bioavailability of testosterone from ethosomal formulation [68, 69].

10.3.3 Antimicrobial Delivery

In case of microbial infection, the microbes tend to be localized intracellularly hindering the treatment with antimicrobials. Ethosomes could be a better option for antimicrobial delivery as it was realized by many researchers. Touitou *et al.* demonstrated better intracellular delivery of bacitracin-loaded ethosomes with application of CLSM and fluorescence activated cell sorting (FACS) techniques in different cell lines [10, 43]. Greater cellular uptake of anti-HIV drugs also suggested ethosomes as an attractive alternative for anti-HIV therapy [34].

10.3.4 DNA Delivery

Skin is now considered as an excellent protective barrier against environmental pathogens, immunologically active and able to express certain genes. Accordingly, ethosomes were also used for topical delivery of DNA molecules to express genes in skin cells. Touitou *et al.* incorporated the green fluorescent protein (GFP)-CMV-driven transfecting construct into ethosomal formulation which was applied on the dorsal skin. CLSM revealed effective delivery of GFP via ethosomes and expression of genes in skin cells [70, 71].

10.3.5 Macromolecules Delivery

Macromolecules such as proteins or peptides still face some challenges for oral delivery due to degradation in gastrointestinal tract. Non-invasive delivery could be a better option for overcoming the problems associated with oral delivery. Effective insulin delivery using ethosomes proved its utility for macromolecules delivery [66].

10.3.6 Vaccines Delivery

Transcutaneous immunization offers a new method for the delivery of vaccines that relies on the application of antigen with adjuvant onto the outer layer of the skin, and subsequent delivery to underlying Langerhans cells (LCs) that serve as antigen-presenting cells (APCs). Non-invasive mode of vaccine delivery could decrease the incidence of needle-borne diseases, reduce the complications associated to physical skin penetration, and improve access to vaccination by eliminating the need of trained personnel and sterile equipment. Ethosomes are interesting and innovative carriers that present an ample opportunity to transport active molecules more efficaciously through the SC into deeper skin layers than conventional formulations. In another study, HBs Ag-loaded ethosomes were observed to cause a protective immune response due to their ability to traverse and target the immunological milieu of the skin. This suggested the potential of ethosomal carrier system in development and delivery of vaccines [24].

10.3.7 Cosmeceuticals

The concept of better skin distribution ability of nanocarrier ethosomes was explored to serve useful cosmeceutical purpose. Esposito *et al.* formulated an ethosomal gel of an anti-keratinizing agent (azelaic acid) for the treatment of acne and compared *in vitro* release with conventional liposomes. The release rate was found to be better in case of ethosomal systems. Apart from dermatological skin treatment, ethosomes can be also effectively used to deliver anti-aging agents [65, 72–75]. Considering that antioxidants are usually not stable and can be degraded under light exposure, Koli *et al.* developed antioxidant-loaded ethosomes for topical delivery including the synergistic mixture of vitamin A palmitate, Vitamin E, and Vitamin C. The results suggested that synergistic interaction of Vitamin C in the aqueous core and vitamin A and E in the lipid bilayer provided complete protection against oxidation in ethosomal system [76].

10.4 Marketed Products Based on Ethosomes

Following the research and technology advancement, some of the products based on ethosomal system are already available in the market. Table 10.2 shows some of these marketed products [46, 55, 58, 77].

Table 10.2 Marketed products based on ethosomal drug delivery system

Product	Uses	Manufacturer
Cellutight EF®	Topical cellulite cream that contains a powerful combination of ingredients to increase metabolism and break down fat.	Hampden Health, USA
Decorin cream®	Anti-aging cream for treating, repairing, and delaying the visible aging signs of the skin including wrinkle lines, sagging, age spots, loss of elasticity and hyperpigmentation.	Genome Cosmetics, Pennsylvania, US
Nanominox®	First minoxidil-loaded ethosomes product. Contains 4% minoxidil, a well-known hair growth promoter that must be metabolized by sulfation	Sinere, Germany
Noicellex®	Topical anti-cellulite cream	Novel Therapeutic Technologies, Israel
Skin genuity®	Powerful cellulite buster that reduces orange peel.	Physonics, Nottingham, UK
Supravir cream®	Formulation of acyclovir for the treatment of herpes virus. It has a long shelf life with no stability problems for at least three years at 25°C. Skin permeation experiments showed that this cream retained its initial penetration enhancing properties even after three years.	Trima, Israel

10.5 Translational Perspective

Delivery of injectable or oral drugs by adopting a transdermal approach has a significant contribution to clinical practice, but it has yet to fully achieve its potential as an effective alternative to oral route and hypodermic injections. Whereas absence of first-pass metabolism and less frequent dosing regimens offer a distinct advantage compared with the oral route; non-invasive administration and minimal risk of disease transmission are better prospects than injectable route. Easy access to a relatively large surface area for better absorption without adding any additional costs makes this approach highly patient compliant. Despite these advantages, hydrophilic drugs are less amenable to be administrated by this route. Nevertheless, delivery of vaccines through this route seems highly promising as they could generate heightened immune responses by targeting epidermal LC and dermal DCs residing in the skin at much lower doses than deeper injection. In recent years through optimizing preparation, characterization, safety, and efficacy criteria, several novel vesicular carrier systems have demonstrated great potential in proof-of-concept studies. Ethosomes consisting of ethanol, propylene glycol, phospholipids and water offer a superior prospect as a therapeutic molecular vector as compared to a spectrum of other nanocarriers due to its higher permeating ability through skin. Although the precise mechanism of drug delivery by ethosomes remains less defined, a synergistic mechanism between ethanol, vesicles and skin lipids allows them to easily penetrate into deeper skin layers. The permeation enhancing ability of ethosomes has been largely attributed to the complex formation between phospholipid vesicles and ethanol which rapidly penetrates into the skin and influences the bilayer structure of the SC. The presence of ethanol at high concentration makes the lipid membrane less rigid which in turn helps the ethosomes to squeeze through small spaces such as the openings created by disturbing the SC lipid. The introduction of ethosomes marks a new view in vesicular research for topical drug delivery. The successful commercialization of ethosomal products has affirmed their position in market. While majority of the launched products are cosmeceuticals, exploration of this system for pharmaceuticals holds immense promise. As ethosomes can be tailored to improve the delivery of a number of molecules to the cellular membranes, it offers a good opportunity for non-invasive delivery of small-, medium-, and large-sized drug molecules. The ethosomal technology can be broadened to expose the agents to cultured cells and microorganisms for developing bio-engineering approaches. The delivery of hydrophilic drugs and quantum dots through human skin via ethosomes holds good promise. Through a series of investigative studies, ethosomes have proved to be an effective molecular vector offering superior safety, efficacy, long-term stability, simplified industrial manufacture as well as better patient compliance as compared to conventional transdermal permeation enhancement techniques. However, the clinical evaluation of intracellular targeting via ethosomes is still in its "infancy." Future refinements pertaining to issues in composition, preparatory methods, molecular mechanisms of skin permeation, characterization, and stability at product development stage will open new avenues for ethosomes as a potential molecular vector for clinical translation.

References

1. Touitou E, Godin B, Weiss C. Enhanced delivery of drugs into and across the skin by ethosomal carriers. *Drug Dev Res*, 2000; 50: 406–415.
2. Verma P, Pathak K. Nanosized ethanolic vesicles loaded with Econazole nitrate for the treatment of deep skin infections through topical gel formulation. *Nanomed Nanotechnol Biol Med*, 2012; 8: 489–496.
3. Zhang Z, Wo Y, Zhang Y, Wang D, He R, Chen H *et al. In vitro* study of ethosome penetration in human skin and hypertrophic scar tissue. *Nanomed Nanotechnol Biol Med*, 2012; 8: 1026–1033.
4. Cevc G. Lipid vesicles and other colloids as drug carriers on the skin. *Adv Drug Del Rev*, 2004; 56: 675–711.
5. Dayan N, Touitou E. Carriers for skin delivery of trihexyphenidyl HCl: Ethosomes vs. liposomes. *Biomaterials*, 2000; 21: 1879–85.
6. Elsayed MM, Abdallah OY, Naggar VF, Khalafallah NM. Deformable liposomes and ethosomes: Mechanism of enhanced skin delivery. *Int J Pharm*, 2006; 322: 60–66.
7. Elsayed MM, Abdallah OY, Naggar VF, Khalafallah NM. Liposomes: Novel lipid vesicles for skin delivery of drugs. *J Pharm Pharmacol*, 2007; 59: 1447–1450.

8. Fang YP, Huang YB, Wu PC, Tsai YH. Topical delivery of 5-aminolevulinic acid-encapsulated ethosomes in a hyperproliferative skin animal model using the CLSM technique to evaluate the penetration behaviour. *Eur J Pharm Biopharm*, 2009; 73: 391–398.

9. Galic B. Boroxine composition for removal of skin changes. 2012; US 8278289.

10. Godin B, Touitou E. Mechanism of bacitracin permeation through the skin and cellular membranes from an ethosomal carrier. *J Control Release*, 2004; 94: 365–379.

11. Godin B, Touitou E, Rubinstein E, Athamma A, Athamma M. A new approach for treatment of deep skin infections by an ethosomal antibiotic preparation: An in-vivo study. *J Antimicrobial Chemother*, 2005; 55: 989–994.

12. Mitragotri S. Modeling skin permeability to hydrophilic and hydrophobic solutes based on four permeation pathways. *J Control Release*, 2003; 86: 69–92.

13. Rattanpak T, Young K, Rades T, Hook S. Comparative study of liposomes, transfersomes, ethosomes and cubosomes for transcutaneous immunisation: Characterisation and *in vitro* skin penetration. *J Pharm Pharmacol*, 2012; 64: 1560–1569.

14. Rao Y, Zheng F, Zhang X, Gao J, Liang W. In-vitro percutaneous permeation and skin accumulation of finasteride using vesicular ethosomal carriers. *AAPS Pharm Sci Tech*, 2008; 9: 860–865.

15. Singh A, Rathore P, Shukla M, Nayak S. Comparative studies on skin permeation of miconazole using different novel carriers. *Int J Pharm Sci Res*, 2010; 1: 61–66.

16. Touitou E. Bergelson L. Products for preventing penetration into the skin. 2010; US 7736630.

17. Touitou E, Dayan N, Bergelson L, Godin B, Eliaz M. Ethosomes-novel vesicular carriers for enhanced delivery: Characterization and skin penetration properties. *J Control Release*, 2000; 65: 403–418.

18. Touitou E. Compositions for applying active substances to or through the skin. 1998; US005716638.

19. Touitou E. Compositions for applying active substances to or through the skin. 1996; US005540934A.

20. Xu DH, Zhang Q, Feng X, Xu X, Liang WQ. Synergistic effects of ethosomes and chemical enhancers on enhancement of naloxone permeation through human skin. *Pharmazie*, 2007; 62: 316–318.

21. Elsayed MM, Abdallah OY, Naggar VF, Khalafallah NM. Lipid vesicles for skin delivery of drugs: Reviewing three decades of research. *Int J Pharm*, 2007; 332: 1–16.

22. Dubey Vaibhav, Mishra D, Nahar M, Jain NK. Vesicles as tools for modulation of skin permeability. *Expert Opin Drug Deliv*, 2007; 4(6): 579–593.

23. Benson HAE. Transfersomes for transdermal drug delivery. *Expert Opin Drug Deliv*, 2006; 3(6): 727–737.

24. Mishra D, Mishra PK, Dubey V, Nahar M, Dabadghao S, Jain NK. Systemic and mucosal immune response induced by transcutaneous immunization using Hepatitis B surface antigen loaded modified liposomes. *Eur J Pharm Sci*, 2008; 33: 424–433.

25. Mishra D, Dhote V, Mishra PK. Transdermal immunization: Biological framework and translational perspectives. *Exp Opin Drug Deliv*, 2013; 10(2): 183–200.

26. Cortesi R, Romagnoli R, Drechsler M, Menegatti E, Zaid AN, Ravani L, *et al.* Liposomes and ethosomes associated distamycins: A comparative study. *J Liposome Res*, 2010; 20: 277–285.

27. El Maghraby GM, Williams AC, Barry BW. Can drug-bearing liposomes can penetrate intact skin. *J Pharm Pharmacol*, 2006; 58: 415–429.

28. Honeywell-Nguyen PL, Bouwstra JA. Vesicles as a tool for transdermal and dermal delivery. *Drug Discovery Today*, 2005; 2(1): 67–74.

29. Cevc G, Blume G. Lipid vesicles penetrate into intact skin owing to the transdermal osmotic gradients and hydration force. *Biochim Biophys Acta*, 1992; 1104: 226–232.

30. Cevc G. Transfersomes, liposomes and other lipid suspension on the skin: Permeation enhancement, vesicle penetration, and transdermal drug delivery. *Crit Rev Ther Drug Carrier Syst*, 1996; 13(3–4): 257–388.

31. Honeywell-Nguyen PL, Bouwstra JA. The *in vitro* transport of pergolide from surfactant-based elastic vesicles through human skin: A suggested mechanism of action. *J Control Release*, 2003; 86: 145–156.

32. Uchegbu IF, Vyas SP. Non-ionic surfactant based vesicles (niosomes) in drug delivery. *Int J Pharm*, 1998; 172: 33–70.

33. Fang JY, Hong CT, Chiu WT, Wang YY. Effect of liposomes and niosomes on skin permeation of enoxacin. *Int J Pharm*, 2001; 219: 61–72.

34. Dubey V, Mishra D, Nahar M, Jain V, Jain NK. Enhanced transdermal delivery of an anti-HIV agent via ethanolic liposomes. *Nanomed Nanotech Biol Med*, 2010; 6: 590–596.

35. Dubey V, Mishra D, Dutta T, Nahar M, Saraf DK, Jain NK. Dermal and transdermal delivery of an anti-psoriatic agent via ethanolic liposomes. *J Control Release*, 2007; 123: 148–154.

36. Dubey V, Mishra D, Jain NK. Melatonin loaded ethanolic liposomes: Physicochemical characterization and enhanced transdermal delivery. *Eur J Pharm Biopharm*, 2007; 67: 398–405.

37. Jain S, Umamaheshwari RB. Bhadra D, Jain NK. Ethosomes: A novel vesicular carrier for enhanced transdermal delivery of an Anti-HIV agent. *Ind J Pharm Sci*, 2004; 66: 72–81.

38. Maurya SD, Prajapati SK, Gupta AK, Saxena GK, Dhakar RC. Formulation development and evaluation of ethosome of stavudine. *Ind J Pharm Educ Res*, 2010; 44: 102–108.

39. Touitou E, Alkabes M, Dayan N. Ethosomes: Novel lipid vesicular system for enhanced delivery. *Pharm Res*, 1997; S14: 305–306.

40. Maurya SD. Enhanced transdermal permeation of Indinavir sulphate through stratum corneum via. Novel permeation enhancers: Ethosomes. *Der Pharm Lettre*, 2010; 2: 208–220.

41. Zhou Y, Wei Y, Liu H, Zhang G, Wu X. Preparation and *in vitro* evaluation of ethosomal total alkaloids of Sophora alopecuroides loaded by a transmembrane pH-gradient method. *AAPS Pharm Sci Tech*, 2010; 11: 1350–1358.

42. Prasanthi D, Lakshmi PK. Development of ethosomes with taguchi robust design-based studies for transdermal delivery of alfuzosin hydrochloride. *Int Curr Pharm J*, 2012; 1(11): 370–375.

43. Touitou E, Godin B, Dayan N, Weiss C, Piliponsky A, Levi-Schaffer F. Intracellular delivery mediated by an ethosomal carrier. *Biomaterials*, 2001; 22: 3053–3059.

44. Zhang JP, Wei YH, Zhou Y, Li YQ, Wu XA. Ethosomes, binary ethosomes and transfersomes of terbinafine hydrochloride: A comparative study. *Arch Pharm Res*, 2012; 35:109–117.

45. Zhaowu Z, Xiaoli W, Yangde Z. Preparation of matrine ethosomes, its percutaneous permeation *in vitro* and anti-inflammatory activity *in vivo* in rats. *J Liposomes Res*, 2009; 19: 155–162.

46. Zhou Y, Wei YH, Zhang GQ, Wu XA. Synergistic penetration of ethosomes and lipophilic prodrug on the transdermal delivery of acyclovir. *Arch Pharm Res*, 2010; 33: 567–574.

47. Akhtar N, Pathak K. Cavamax W-7 composite ethosomal gel of clotrimazole for improved topical delivery: Development and comparison with ethosomal gel. *AAPS Pharm Sci Tech*, 2012; 13: 344–345.

48. Bendas ER, Tadros MI. Enhanced transdermal delivery of salbutamol sulphate via ethosomes. *AAPS Pharm Sci Tech*, 2007; 8: 1–15.
49. Bhalaria MK, Naik S, Mishra AN. Ethosomes: A novel delivery system for antifungal drugs in the treatment of topical fungal disease. *Ind J Exp Biol*, 2009; 47: 368–437.
50. Charyulu RN, Harish NM, Sudhakar CK, Udupa G. Formulation and evaluation of clotrimazole ethosomes for topical delivery. *AAPS Annual Meeting Expo*, 2009; 7–12.
51. Cortesi R, Ravani L, Zaid AN, Menegatti E, Romagnoli R, Drechsler M, Esposito E. Ethosomes for the delivery of anti-HSV-1 molecules: Preparation, characterization and in-vitro activity. *Pharmazie*, 2010; 65: 743–751.
52. Chen JG, Yu F, Gai TW. Preparation and anti-inflammatory activity of triptolide ethosomes in an erythema model. *J Liposome Res*, 2010; 20, 297–303.
53. Dave V, Kumar D, Lewis S, Paliwal S. Ethosomes for enhanced transdermal drug delivery of aceclofenac. *Int J Drug Deliv*, 2010; 2: 81–92.
54. Dubey V, Mishra D, Jain NK, Nahar M. Miconazole loaded ethosomes for effective management of topical fungal infection. *AAPS Annual Meeting & Exposition* 2007; 74: 9–15.
55. Girhepunje K, Pal R, Gevariya H, Behera A. Ethosomes: A novel vesicular carrier for enhanced dermal delivery of Ciclopiroxolamine. *Der Pharmacia Lettre*, 2010; 2: 360–367.
56. Got A, Patel V. Ethosomally entrapped clotrimazole: A view to improve therapeutic response of antifungal drug. *AAPS Nat Bio Con*, 2007; 82: 23–28.
57. Sheer A, Chauhan M. Ethosomes as vesicular carrier for enhanced transdermal delivery of Ketoconazole: Formulation and evaluation. *IJPI's J Pharma Cosmetol*, 2011; 1: 1–14.
58. Bahia AP, Azevedo EG, Ferreira LAM, Frézard F. New insights into the mode of action of ultradeformable vesicles using calcein as hydrophilic fluorescent marker. *Eur J Pharm Sci*, 2010; 39: 90–96.
59. Pandey V, Golhani D, Shukla R. Ethosomes: Versatile vesicular carriers for efficient transdermal delivery of therapeutic agents. *Drug Deliv*, 2014 Posted online on March 3, 2014 (doi:10.3109/10717544.2014.889777).
60. Touitou E. Compositions and methods for intracellular delivery. 1995; EP 0804160.
61. Touitou E. Compositions and methods for intracellular delivery. 2010; US20100298420 A1.

62. Vander MD, Riviere JE. Comparative studies on the effects of water, ethanol and water ethanol mixtures on chemical partitioning into porcine stratum corneum and silastic membrane. *Toxicol in vitro*, 2005; 19: 69–77.

63. Wilson NAL. Transdermal administration of phycotoxins. 2013; US008377951B2.

64. Zadini F, Zadini G. Dissolution of arterial plaque. 2012; US 8304383.

65. Esposito E, Menegatti E, Cortesi R. Ethosomes and liposomes as topical vehicles for azelaic acid: A preformulation study. *J Cosmet Sci*, 2004; 55: 253–264.

66. Horwitz E, Pisanty S, Czerninski R, Helser M, Eliav E, Touitou E. A clinical evaluation of a novel liposomal carrier for acyclovir in the topical treatment of recurrent herpes labialis. *Oral Surg Oral Med Oral Pathol Oral Radiol Endod*, 1999; 87: 700–705.

67. Hsu PJ, Huang CJ, Wu MT. Pathergy in a typical eosinophilic pustular folliculitis. *Int J Dermatol*, 2005; 44: 203–205.

68. Dkeidek, I., Touitou, E. Transdermal absorption of polypeptides. *AAPS Pharm Sci*, 1999; 1: S202.

69. Ainbinder D, Touitou E. Testosterone ethosomes for enhanced transdermal delivery. *Drug Deliv*, 2005; 12(5): 297–303.

70. Kumar KP, Radhika PR, Sivakumar T. Ethosomes-a priority in transdermal drug delivery. *Int J Adv Pharm Sci*, 2010; 1: 111–121.

71. Godin B, Touitou E. Ethosomes: New prospects in transdermal delivery. *Crit Rev Ther Drug Carrier Syst*, 2003; 20(1): 63–102.

72. Harris RA, Burnett R, McQuiilkin S, McCloud A, Simon FR. Effect of ethanol on membrane order: Fluorescence studies. *Ann NY Acad Sci*, 1987; 492: 125–133.

73. Lodzki M, Godin B, Rakou L, Mechoulam R, Gallily R, Touitou E. Cannabidiol-transdermal delivery and anti-inflammatory effect in a murine model. *J Control Release*, 2003; 93: 377–87.

74. Modi P. Stabilized compositions for topical administration and methods of making same. 2010; US 7727537.

75. Scimeca JV, Zimmerman AC, Mettler MF, Kudo A, Kawasaki Y. Cosmetic treatment system and methods. 2010; US 7758878.

76. Koli JR, Lin S. Develpement of antioxidant ethosomes for topical delivery utilizing the synergistic properties of Vit A palmitate, Vit E and Vit C. *AAPS Pharm Sci Tech*, 2009; 11: 1–8.

77. Touitou E. Terbinafine compositions for onchomycosis treatment; 2010: WPO Patent app WO/ 2010/086723.

Chapter 11

Lipid-Based Nanocarriers for the Treatment of Infected Skin Lesions

Sandra Simões, Manuela Carvalheiro, and Maria Manuela Gaspar

Research Institute for Medicines (iMed.ULisboa),
Faculty of Pharmacy, University of Lisbon,
Avenida Professor Gama Pinto, 1649-003 LISBOA, Portugal
ssimoes@ff.ulisboa.pt

11.1 Skin Infections

The skin is the first line of defense between the body and the environment, protecting the body from toxic chemicals, ultraviolet radiation, microorganisms, and mechanical trauma [1, 2]. However, if skin barrier is destroyed as a result of any accident such as burns, wounds, bites, ulceration, or laceration, infections may occur. Skin infections constitute a vast array of conditions that range in severity from benign to life threatening. Bacterial, mycobacterial, parasitic, fungal, and viral skin infections represent a primary category of dermatologic illnesses contributing annually for substantial health and economic burden. These pathologies are a common cause of hospital admissions leading to significant morbidity. The clinical presentation of infectious skin diseases varies with the type of pathogen involved, the skin layers and structures damaged and the underlying medical conditions of patients [3]. Numerous antimicrobial drugs have been considered

Carrier-Mediated Dermal Delivery: Applications in the Prevention and Treatment of Skin Disorders
Edited by Andreia Ascenso, Sandra Simões, and Helena Ribeiro

ISBN 978-981-4745-58-1 (Hardcover), 978-1-315-36447-6 (eBook)
www.panstanford.com

for the treatment of bacteria and parasites induced diseases and disorders. However, inefficient delivery of these drugs might result in inadequate therapeutic effect and in severe side effects.

Currently, progress and challenges in constructing nanocarrier platforms for delivering various antimicrobial drugs for infected skin lesions have been attempted. In the present review problems related to mycobacterial and parasitic skin infections will be particularly focused as well as new alternative strategies for treatment of these pathologies.

11.1.1 Mycobacterial Skin Infections

Mycobacterial diseases may be divided in three categories: tuberculosis, leprosy, and non-tuberculous mycobacteria (NTM) [4]. Mycobacteria responsible for tuberculosis and leprosy are sensitive to several existing antimycobacterial drugs. However, NTM has become more important in developed countries primarily due to the emergence of AIDS. NTM are also called atypical mycobacterial infections being capable of affecting lungs, lymphatic system, bone tissues, or skin. In particular, skin NTM depends on host immunity as well as environmental and living conditions.

The mycobacteria species that may cause localized infections of the skin and subcutaneous tissue are *M. chelonae*, *M. fortuitum*, *M. abscessus*, *M. marinum*, and *M. ulcerans*. Rapid growth of *M. chelonae*, *M. fortuitum*, or *M. abscessus* may occur due to localized drainage or abscess formation at the site of puncture wounds, open traumatic injuries or fractures, surgical wound infections, post-injection abscesses [5]. Occasionally, these infections may also spread to lymphatics [6]. *M. marinum* represents one of the most frequent cutaneous atypical mycobacterial infections. The usual source is water from lakes, swimming pools, or aquariums [6]. Buruli ulcer (BU), caused by *M. ulcerans*, is the third most widespread mycobacterial infection in the world after tuberculosis and leprae being the least understood. It has been recently recognized by the World Health Organization (WHO) as an emerging disease [7, 8]. The causative organism is from the family of bacteria which causes tuberculosis and leprae; however, BU has received less attention than these diseases. BU has been reported in over 30 countries with tropical and subtropical climates affecting mainly poor rural communities in Africa,

North America (Mexico), South America, Southeast Asia, and Oceania. Currently, Bellarine Peninsula in Victoria is known to be the highest endemic area in Australia [9]. In light of this emerging disease able to cause great human suffering, this review will focus on current therapies used, drawbacks associated, and new alternative lipid nanocarriers for fighting this mycobacterial skin infection.

11.1.1.1 Buruli ulcer

BU is an emerging infectious disease of the skin caused by *M. ulcerans*. The natural reservoir of *M. ulcerans* still continues to be an open question after decades of investigation. Nevertheless, early studies in Africa demonstrated that the infection occurred near stagnant water or areas surrounded by water [10]. Usually humans get infected by traumatic introduction of *M. ulcerans* into the skin. Contamination of damaged skin may result from direct exposition to slow-moving water, aerosols arising from ponds and swamp surfaces. Lesions may also be developed at the site of antecedent traumas namely land mine wound, snakebite, human or insect bite [11, 12]. However, the role of aquatic insect bites for transmission of BU is still under investigation [13].

In Fig. 11.1 is shown a scheme of possible progression of BU infection after exposure to *M. ulcerans* in the environment. In some conditions, people exposed to *M. ulcerans* do not develop BU disease or even lesions may heal spontaneously depending on the host immunity. In case of progression, the infection is primarily related with two properties of *M. ulcerans*: the low optimal temperature of growth, favoring the development of lesions in cooler tissues particularly the skin and subcutaneous tissue, and production of a macrolide toxin, the mycolactone, being responsible for the toxic effects observed in BU lesions [14]. Mycolactones are secreted and diffuse into infected tissues. BU presents different clinical forms that can be dramatic if untreated. The first clinical presentation is non-ulcerative characterized by papules (skin lesions with a diameter inferior to 1 cm), nodules (extension into subcutaneous tissue, diameter inferior to 2 cm), plaques (lesions with diameter superior to 2 cm) and edema forms distributed mainly on limbs [15]. Figure 11.2 shows a schematic representation of BU pathogenesis progression from pre-ulcerative to ulcerative laceration stage.

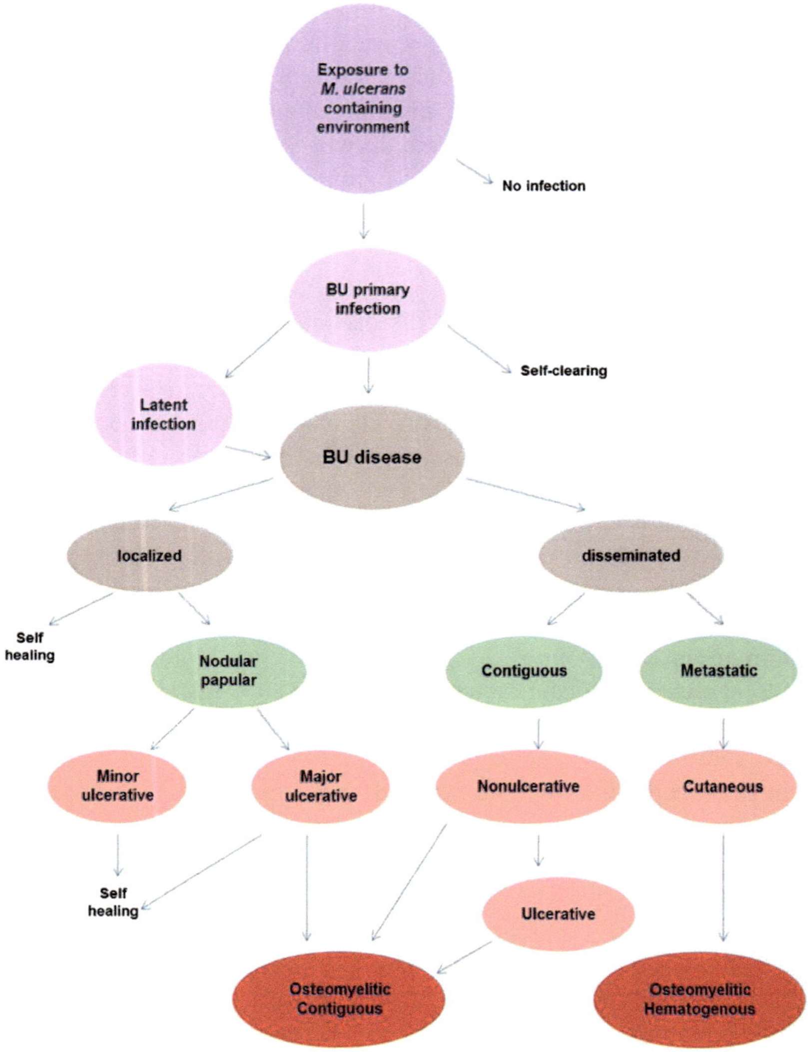

Figure 11.1 Possible progression of untreated BU infection. Adapted with permission from Portaels *et al.* [18].

The disease progresses without pain or fever, which may partially explain why infected persons often do not seek prompt treatment. In more severe clinical forms of BU, lesions lead to larger ulcerations with undetermined edges; they may destroy nerves, appendages, and blood vessels and eventually invade bone causing painful osteomyelitis or reactive osteitis [16]. Infected people not early treated often suffer long-term functional disability, such as restriction of joint movements as well as the obvious cosmetic problems. Metastatic lesions may occur in skin, soft tissue, or bone via spread through the vasculature or lymphatics. In regions where BU is endemic, patients with severe osteomyelitis may undergo amputation and other crippling disabilities [17].

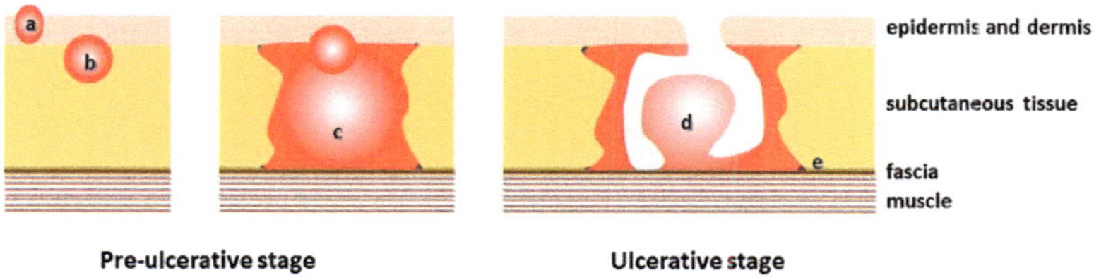

Figure 11.2 Pathogenesis of BU: progression from pre-ulcerative stages to ulcerative laceration. Adapted with permission from van der Werf *et al.* [19].

11.1.2 Parasitic Skin Infections

A parasite is an organism that grows, feeds, and is sheltered from or at the expense of its host. Some parasites do not affect the host while others make the host sick resulting in a parasitic infection [20].

A wide range of parasites infect humans through the skin causing an infection that may be confined to the skin or may have skin involvement as part of its pathology [21]. The symptoms of parasitic infections vary widely, depending on the type of organism and the human health condition. Skin lesions, allergic rashes, and papules are the most common superficial signs but subcutaneous lesions, nodules or cysts may also occur [20].

Arthropods, helminthes, and protozoans are the major groups of organisms responsible for skin infections in humans [22].

Arthopods comprise insects, mites, ticks, and spiders that live and feed of human skin or blood. Their bite or sting results in a skin lesion either due to tissue damage or to an immune reaction to the injected saliva or venom [23]. Pathogens responsible for infectious diseases may also be transmitted by the byte of various arthropods due to their role as intermediate hosts in the life cycle of parasites. Vector-borne infectious diseases with skin manifestations transmitted by arthropod include Rocky Mountain spotted fever, Lyme disease, and leishmaniasis [22].

Helminths or parasitic worms are the most common infectious agents of humans in developing countries causing a wide variety of diseases. Important examples include schistosomiasis, cysticercosis, lymphatic filariasis, onchocerciasis, and loiasis. Infections may occur through the skin by contact with contaminate water or after ingestion of the parasite eggs in contaminated soils or foods [22].

Protozoans are microscopic unicellular eukaryotic organisms that can multiply inside mammals causing the development of serious infections. Examples of protozoan parasites that cause severe skin infections include *Trypanosoma cruzi* and *T. brucei* responsible for trypanosomiasis, known as sleeping sickness and Chagas disease, respectively. Toxoplasmosis, another important skin infection is caused by the protozoan specie *Toxoplasma gondii*. Common symptoms of such infections include extreme itching and appearance of nodules that get ulcerated within few months [24]. The protozoan parasites of the genus *Leishmania* are the causative agents of a group of diseases called Leishmaniasis. These parasitic infections are transmitted to mammals via the bite of female phlebotomine sandflies. In the mammalian host, these obligate intracellular parasites proliferate within several cells of the phagocytic mononuclear system, including macrophages [25]. Leishmaniasis has been classified in three different clinical forms according to parasite tropism: cutaneous (CL), mucocutaneous (ML) and visceral (VL). These different forms can cause a wide spectrum of disorders, ranging from self-healing skin ulcers, to severe, life-threatening manifestations depending on the virulence of the infecting species and the immune response of the host [26–28]. It is endemic in more than 80 low-income countries throughout the world and represents a serious public health problem in tropical and sub-tropical regions.

11.1.2.1 Cutaneous leishmaniasis

CL is considered the most serious protozoan skin infection in many developing countries with an annual incidence estimated between 0.7 and 1.2 million cases. Ninety percent of CL cases occur in Middle East and South American countries [29]. As shown in Fig. 11.3, the clinical CL manifestations may range from single or disseminated, nodular or ulcerative skin lesions to the destruction of mucosal tissue [27, 30]. In most cases CL is limited to a single or few lesions that are localized at the site of parasite inoculation within dermal macrophages on exposed parts of the body [31]. The infection begins as an erythema that develops into a nodule and then enlarges and matures into an ulcerated lesion over 1–3 months (Fig. 11.3) [27, 32]. These lesions may heal spontaneously after months or even years. Less common CL presentations include disseminated and diffuse (very rare) forms that origin multiple skin lesions over large body areas [31, 32]. ML, which usually occurs after an initial cutaneous infection, involves the mucous membranes of the upper respiratory tract and oral cavity (nose, mouth, and throat) [33]. Post kala-azar dermal leishmaniasis is yet another form of CL that may arise after visceral infection caused by *L. donovani* [34]. In the more severe forms of CL, the lesions do not self-cure triggering significant disfigurement and social stigmatization [31].

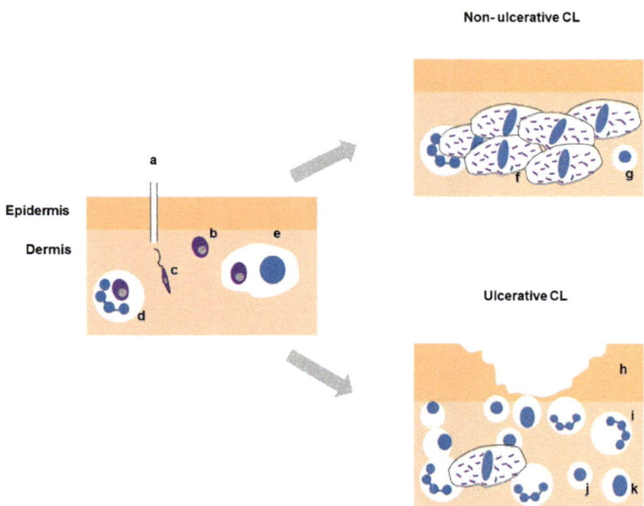

Figure 11.3 Cutaneous Leishmaniasis (CL) pathogenesis: progression to non-ulcerative or ulcerative forms: (a) sand fly mouth; (b) amastigote; (c) promastigote; (d) infected neutrophil; (e) macrophage; (f) infected macrophage; (g) T cell; (h) epidermal hyperplasia; (i) neutrophil; (j) B cell; (k) T cell. Adapted with permission from Rethi and Eidsmo [35].

Several *Leishmania* species give rise to the cutaneous disease. According to the geographic location, CL can be divided into Old World CL (Asia, the Middle East, Africa or Europe) and New World CL (Mexico, Central and South America). The majority of Old World CL is caused by one of two species of parasites; *L. major* (in desert areas) and *L. tropica* (in urban areas) and presents essentially localized wet or dry lesions, respectively, that tend to resolve within 2 to 15 months [34, 36]. The New World CL is characterized by the presence of ulcerative skin lesions and the potential of spreading to lymph nodes and other cutaneous sites. The main causative agents are *L. mexicana*, *L. amazonensis*, and *L. chagasi* and also species of the subgenera *L. viannia* (*L. braziliensis*, *L. panamensis* and *L. guyanensis*), responsible for the development of ML due to systemic dissemination of parasites to nasopharyngeal mucosa [25, 34].

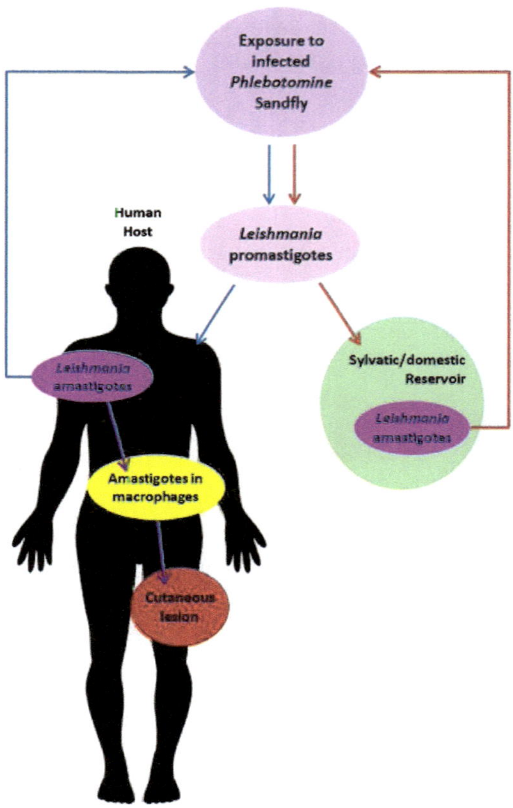

Figure 11.4 Epidemiology and pathogenesis of CL. *Leishmania* promastigotes are injected into the skin of the host and differentiate into amastigotes within macrophages. In humans, the infection can result in a cutaneous lesion. The red arrows represent the zoonotic cycle and the blue arrows the anthroponotic cycle. Adapted with permission from Goto and Lindoso [37].

The epidemiology of CL has changed in the last years with new foci emerging in non-endemic areas and with the concomitant increase in the number of reported cases. Many factors contribute to this expansion; some rely on changes in the environment or vector habitats, while others are related with human activity such as population migration trends due to urbanization or political conflicts or the increasing presence of international travelers and military personnel in endemic areas [38, 39].

11.1.3 Current Therapies

Despite some research efforts, to date, there are no effective vaccines against either BU or CL. These infectious diseases are recognized by the WHO as emerging health problems primarily due to their frequent disabling and stigmatizing complications. While medically diverse, these two neglected diseases share features that allow them to persist in conditions of poverty, often in remote rural areas, urban slums or in conflict zones (Table 11.1). Most of currently existing therapies have been introduced for other pathologies. Efforts regarding the development of alternative therapies have been initiated within last decade.

The following topics will describe the current therapies for these neglected diseases, problems associated and new strategies for improving treatments.

Table 11.1 Main features of BU and CL

Skin disease	Risk factors	Environmental stages	Transmission	Dependence on macrophages	Keys for pathogenesis
Buruli ulcer[1]	Geographical	Soil, water, aquatic plants	Indirect or insect vector	No (early stages)	Low Temperature growth (30–32°C); mycolactone production; extracellular/ intracellular
Cutaneous leishmaniasis[2]	Geographical	n.a.	Insect vector	Yes	Species dependent; obligate intracellular

(1) mycobacteria; (2) protozoa; n.a.: not applicable.

11.1.3.1 Buruli ulcer

Historically, the method of choice for the management of BU consists on removal of all infected tissue followed by skin grafting or amputation. However, this treatment is very drastic, involving long hospitalization. In addition, recurrence rates ranging from 16% to 47% have been reported when extended excision in healthy appearing surrounding tissues was not accomplished [19, 40, 41]. Different medical approaches have been explored: chemical topical treatments such as nitrites, phenytoin powders, or clay; physical topical treatments through application of heat to accelerate healing after surgery, or systemic treatments using hyperbaric oxygen, heparin, or antibiotics. Chemical tropical treatments evidenced some clinical improvements but only in young people presenting ulcers of less than 30 cm [42, 43].

Local heat has been applied to patients with ulcers of less than 10 cm leading to complete healing without surgery [44]. However, this treatment is not a very good option particularly in remote areas due to the need of confinement in bed for at least two weeks and absence of infrastructures such as electricity. Systemic treatments with antimycobacterial molecules constitute another alternative. Several antibiotics have demonstrated high *in vitro* antimycobacterial activity against *M. ulcerans*. However, the good *in vitro* results were not translated into therapeutic benefits when administered in humans. This might be attributed to irreversible tissue damage and necrosis or even due to poor irrigation in necrotizing tissues obstructing the penetration of drugs into tissues, where *M. ulcerans* proliferates [43]. Rifamycins (rifampicin (RFP) and rifapentine), aminoglycosides (streptomycin (SM), clarythromicin, and amikacin), fluoroquinolones (sparfloxacin, moxifloxacin, and aminofloxacin) are examples of tested antibiotics either in animal models or in clinical trials alone or in combination [43]. Since 2004 the WHO recommended daily subcutaneous administration of SM and oral intake of RFP for two months as first-line treatment regimen for BU [45]. This antibiotic combination presents several drawbacks, namely the following: It is not effective for extensive lesions, being surgery the only option; it has long treatment duration, associated with adverse side effects; and involves subcutaneous painful administration of SM, requiring health skilled personnel. Overall, these facts lead to poor patient compliance. Moreover, SM toxicity is cumulative and thus special attention must be taken into account for patients previously treated with aminoglycoside [45]. Recently, Almeida and co-workers have tested the antimycobacterial activity of another rifamycin, the rifapentine, with longer circulation half-life in comparison with RFP [46]. These *in vivo* studies in a murine model of *M. ulcerans* demonstrated that rifapentine was faster on reducing the bacterial loads. However, this faster bactericidal activity was not correlated into better prevention of relapses [46].

Altogether, no antibiotic combination has proven to be effective in the management of BU up to now. There is no vaccine currently effective against BU and so other strategies to fight this skin mycobacterial infection are needed.

The phage therapy represents one alternative therapeutic strategy as it involves the targeted application of bacteriophages. Bacteriophages are viruses that attack bacteria being able to replicate within them causing disruption of bacterial cells [47]. As their lytic action is highly specific, this approach can be applied for treatment of bacterial infections that do not respond to available antibiotics. Nevertheless, for the successful application of phages, the knowledge of their host range is a prerequisite. Particularly, a set of virulent phages for *M. ulcerans* has been reported [48]. Among this set of phages, the therapeutic effect of the mycobacteriophage D29 was evaluated in a *M. ulcerans* murine footpad model. This phage was subcutaneously administered in the infected footpad and a single injection was able to decrease the bacterial load in footpad in comparison with infected and untreated mice [49].

Topical application of phages in the treatment of ulcerative lesions represents an alternative strategy to the subcutaneous administration. This approach has been already used in the treatment of multi-drug-resistant *Staphylococcus aureus* infections using a biodegradable formulation able to promote a sustained release [50]. Considering that *M. ulcerans* is an extracellular mycobacteria, topical application of lytic phages allows immediate access to skin infection.

The same approach can be applied for treatment of BU either by incorporating already approved antibiotics such as RFP or SM in adequate lipid carriers or even new molecules obtained from chemical synthesis or extracted from natural products [11].

Taking into account the secretion by *M. ulcerans* of mycolactone the design of inhibitors to this toxin may constitute an alternative strategy for developing new therapies [51].

In any case, the early diagnosis and beginning of treatment are vital in preventing the spread of infection. Drug topical application has for the moment limited value in BU management but it should be considered as an alternative unexplored strategy particularly in the early stages of the disease.

11.1.3.2 Cutaneous leishmaniasis

Therapeutic goals vary with the clinical form of leishmaniasis. In CL, a non-fatal condition, treatment aims to prevent mucosal invasion, to accelerate the healing of skin lesions and to avoid disfiguring scars [27, 52].

The search for a vaccine has been the subject of extensive efforts, taking into account the knowledge that recovery from natural or deliberate infection CL results in the development of long-term immunity to reinfection.

Regrettably, these efforts have not yet been completely successful and vaccination is still under experimental and clinical trials [53, 54]. In the absence of a vaccine, chemotherapy remains the most efficient way to treat CL. However, the therapeutic arsenal is very limited and there is no single optimal treatment for all forms of this disease. The parenteral administration of pentavalent antimonials such as sodium stibogluconate (Pentostam®) and meglumine antimoniate (Glucantime®) have been the first-line treatment for over 50 years. For example, WHO recommends the administration of 20 mg/kg per day for 20 days of antimonials for New World CL [55–57]. Despite their widespread usage, these drugs have proved inconsistency in their effectiveness across different *Leishmania* species [58]. In addition, antimonial treatments are accompanied by painful administration, need of long course treatments and severe adverse reactions, such as cardiac and renal toxicity [59–61].

Second-line treatments include amphotericin B deoxycholate (Fungizone®) and pentamidine, efficient drugs also limited by severe side effects and the need for parenteral administration. Originally developed as a systemic antifungal, amphotericin B (AmB) is also an efficient antileishmanial drug targeting sterols within the Leishmania membrane inhibiting the binding of promastigotes to the macrophage membrane. It also causes the formation of pores in the parasite membrane leading to cell lysis [62]. Despite its high efficiency, this drug requires prolonged hospitalization and close monitoring due to severe side effects, in particular significant nephrotoxicity [34, 63]. Currently, AmBisome® an amphotericin B liposomal formulation (L-AmB) that minimizes these side effects without affecting the drug activity is the preferred treatment for VL. However, the number of reports on its use for CL is still limited. In a small study by Solomon *et al.*, 3 mg/kg/day of L-AmB given for five consecutive days and a sixth dose administered on day 10 caused complete clinical cure in patients infected with *L. braziliensis* [64]. Large controlled studies are needed to assess the therapeutic efficacy of L-AmB in the treatment of CL. In addition, its use remains very limited in low-income regions due to very high cost.

Pentamidine, an aromatic diamidine, has been used as a second-line treatment for VL and CL for over 40 years mainly in cases of antimonials intolerance or resistance [59, 65]. Its clinical use has been accompanied by frequent adverse effects with moderate morbidity including an unusually high rate of hyperglycemia [57, 65]. Nevertheless, when the causative agent is *L. guyanensis*, specie particularly resistant to antimonials, pentamidine isethionate (Pentam 300) becomes the first-line treatment with the advantage of presenting a short time course [34, 36].

Other alternative systemic agents include azoles such as ketoconazole, fluconazole, or itraconazole. These antifungal drugs block ergosterol biosynthesis which in turn affects the permeability of the parasite membrane [66]. They are well-tolerated drugs but have been reported as having variable cure rates against different *Leishmania* species [67]. Most of these compounds showed better results when applied in oral or topical therapy [61, 68].

Miltefosine (hexadecylphosphocholine), a structural analogue of alkyllysophospholipids, originally developed as an anticancer drug, is the first efficient oral drug for the treatment of leishmaniasis, essentially used in VL treatment [61, 69]. More recently, it has been reported in the treatment of several New World CL species with variable efficacies [66, 70]. In Colombia, where *L. panamensis* is the more frequent specie, the oral administration of miltefosine, at a dose of 2.5 mg/kg/day for 28 days, resulted in over 91% cure rate [25]. Most studies were conducted in Colombia, where the drug is registered for CL treatment. However, in studies conducted in areas with a predominance of *L. braziliensis* and *L. mexicana,* cure rates were lower ranging from 35% to 70% [70, 71]. Miltefosine is well tolerated and has a half-life of about 8 days after oral administration. Nevertheless, a general concern about severe side effects, including teratogenicity, and the easy appearance of resistant mutants exists [66].

Topical or local treatment of CL represents an exciting alternative for overcoming problems associated with systemic administration of antileishmanial drugs. Advantages include lower drug toxicity, decreased adverse side effects, reduced treatment cost, and ease of administration [72]. This approach is particularly advantageous for simple localized forms of CL or when disease dissemination is not expected to occur [31]. Local treatment modalities include physical methods (cryotherapy, localized controlled heat and carbon dioxide laser therapy), topical paromomycin (PRM), intralesional antimonial injections, and ethanolic AmB solutions [31]. The local intralesional injection of antimonials compounds is still the method of choice for many patients. WHO recommends, as standard treatment of CL infections caused by *L. major*, *L. tropica*, *L. braziliensis*, or *L. panamensis,* an antimonial injection under the edges of the lesions until the surface has blanched [36, 73].

PRM is an aminoglycoside that inhibits the mitochondrial activity of *Leishmania* parasites. It was registered in 2006 in India as an effective, well-tolerated, and affordable treatment for VL at a dose of 11 mg/kg/day for 21 days [26, 56, 74]. In the treatment of CL, it has been used clinically with irregular success since 1985 [73]. Two topical PRM formulations are commercially available to treat CL; both have demonstrated good efficacy against *L. major* infections. One contains 15% PRM and 12% methylbenzethonium chloride and the other combines a 15% PRM ointment in 10% urea [75]. The main shortcoming of the first is the occurrence of skin irritation due to the methylbenzethonium chloride [76]. The second shows no toxicity problems but presents variable efficacies with cure rates similar to placebo as reported in clinical trials performed in Tunisia and Iran [72]. Overall results so far show that topical PRM treatment may be a useful alternative mainly for Old World CL. There is, however, some concern in using this approach to treat New World CL due to the potential risk of mucosal dissemination [65]. Nevertheless, in regions where the risk is very low, topical PRM proved to be effective in the treatment of infections caused by *L. braziliensis* [32]. WR279396, a topical formulation of PRM (15%) and gentamicin (0.5%) in a hydrophilic base, was developed to address the need of an effective and safe treatment of New World CL. This expanded-spectrum aminoglycoside ointment was found to be effective in treating *L. major* infections presenting also a satisfactory safety profile [77]. In a small clinical trial in patients with *L. panamensis* infection, WR279396 was well tolerated and more effective than PRM alone [78].

Combination therapy is a new strategy in the treatment of CL that may contribute to reduce the total drug dose administered, to shorten the treatment duration; to limit the emergence of drug resistance; and to enhance cure rates. Currently clinical trials are testing different combination therapy regimens involving the most promising drugs (PRM, miltefosine, antimonials, and L-AmB) in the treatment of CL [73, 79].

11.2 Advanced Drug Delivery Systems in Topical Therapy

Topical therapy combines the advantage of limiting the systemic drug adverse effects [80] as long as the drug is exposed only to affected skin, and permitting self-administration, attaining good patient compliance [81]. In the case of skin diseases, the interest of treating directly the skin via topical application is even increased [82]. Skin conditions play an important role in the outcome of topical treatment of skin infections. Besides the structure of the biological membrane, in order to achieve a formulation suitable for topical delivery, drug log *P* and molecular mass are important factors to be considered. Poor solubility and poor permeation are challenging aspects for formulators. New drug delivery systems have been extensively explored to enhance skin drug bioavailability and have also gained interest as adjuvants for vaccines [53, 83, 84]. In the case of skin infections, a great number of studies are related with fungal infections [85]. For other skin infections, namely those with ulcerative forms, the investigation is not so abundant, despite the promising role of advanced carriers.

The great interest received by drug carriers in the case of neglected diseases treatment is based on the possibility of a sustained release diminishing toxicity and dosage numbers [86]. Efficient but toxic drugs may be reformulated in nanocarriers and long treatment periods could be shortened avoiding therapy abandonment.

11.2.1 The Skin Barrier

The skin is a multilayer functional unit capable of performing multiple and fundamental roles. It exhibits a complex structure and a great capacity of regeneration. The skin acts as a barrier, avoiding the penetration of foreign bodies and simultaneously maintaining the body's ingredients, including water, inside.

The skin structure varies somewhat from site to site. The architecture of the skin exhibits two main layers: the epidermis (external) and the dermis (Fig. 11.5) [87]. A third layer, the subcutaneous tissue, the deepest, is essentially composed of fatty material and differs in thickness and composition between body sites, individuals and species.

The epidermis is non-vascular and contains no nerve endings. It is crossed by cutaneous annexes (hair follicles, sebaceous and sweat glands) and is thinner (0.04–0.1 mm) than the dermis (Fig. 11.5). The dermis contains an extensive collagen and elastin matrix, produced by fibroblasts, but few cells.

The epidermis is composed by various layers that represents different stages of cell differentiation, and consequently can be classified in *strata*: from the internal to external epidermal surface the *stratum basale*, the *stratum spinosum*, the *stratum granulosum*, the *stratum lucidum*, and the *stratum corneum* (SC). Keratinocytes are 95% of total cells in the epidermis, the others being Langerhans cells and melanocytes. The viable epidermis has *ca* 60–75% water content in contrast with the 10–15% water of the SC. The dermis contains an extensive collagen and elastin matrix, produced by fibroblasts, but few cells. This skin layer also shelters blood capillaries, nerve endings and the roots of cutaneous appendages, which covers 0.1% of skin surface.

The skin permeability barrier to the percutaneous absorption of compounds as well as to water loss is a consequence of the anatomical and biochemical organization of the SC. This outermost region of the epidermis consists of inert corneocytes embedded into lamellae of lipids organized in a very special lipid matrix. These lamellae are stacked in multiple layers between corneocytes (name for the SC keratinized cells) [88, 89].

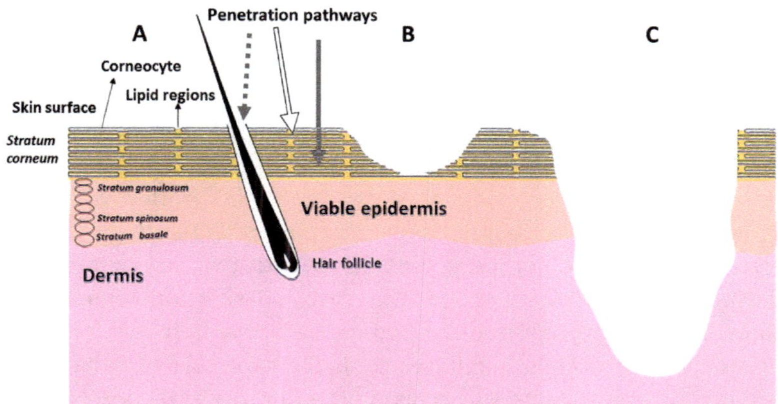

Figure 11.5 *Stratum corneum* (SC) model scheme ("brick-and-mortar") with the lipid intercellular matrix (yellow) and the possible penetration pathways for intact skin (A) (intercellular: white arrow; transcellular: gray arrow; annexial: dashed arrow). In the basal layer of the epidermis cells proliferate, differentiate and migrate in the surface direction. Viable cells are transformed into dead keratin filled cells (corneocytes). These are surrounded by a cell envelope. Substances permeate mainly along the tortuous pathway in the intercellular lamellar regions which are oriented parallel to the corneocyte surface. In skin infections the SC may not exist (B) as a consequence of the disease or the lesion can affect epidermis and dermis (C).

The SC has an exceptional lipid composition with long chain ceramides, free fatty acids, and cholesterol as the main lipid classes. Due to its composition, the lipid phase behavior is different from that of other biological membranes [90, 91]. The SC was described as having a "brick-and-mortar" structure (Fig. 11.5A) but this is a too simplified picture. Hydrophilic pathways in the skin are located between cell envelopes and intercellular lipids [88]. The mean thickness of SC is 15–25 μm; however, its composition and architecture assure a tightly packed structure. Some studies have demonstrated that by sequential removal of SC cell layers, a decrease on skin electric resistivity is observed [92].

The knowledge of skin composition is not only important for understanding some diseases where SC barrier is compromised, but also for designing skin permeation enhancers. It is important to realize that SC is not an inert wall resulting from epidermal differentiation. Rather than this, SC is a complex, heterogeneous barrier with persistent biochemical activity interacting with the underlying tissues in terms of injury responses and pathophysiology [93].

The relative importance of the potential routes of entry from the skin surface into the sub-epidermal tissue has been vastly investigated for percutaneous penetration prediction and formulation development. In the case of skin infections, the skin structure is often disturbed or destroyed. In such cases, the dependence of hair follicles with their associated sebaceous glands or sweat ducts to act as shunts, or the transport across the continuous SC between these appendages might become irrelevant.

Intercellular transport is energetically favorable, passive, and based on diffusion, occurring within the matrix between the corneocytes. The diffusion based molecular transfer through the skin is controlled by partition and diffusion. Many transport enhancement strategies are focused on the modification of these parameters to improve drug delivery through the skin. Paradoxically, Fick's laws of diffusion can describe diffusion of most studied low MW and hydrophobic drugs through the complex and heterogeneous barrier of SC.

11.2.2 Overcoming the Skin Barrier

Topical application using conventional or more complex drug delivery systems has been approved and became widely accepted for local therapy. The benefits of topical application are definitely recognized promoting the local deposition of drugs in the site of action. The outermost layer of the skin, the SC, prevents, however, skin permeation of most drugs at clinically useful rates. In order to facilitate the delivery of such entities, several strategies were developed. These include the use of chemicals, supramolecular structures, or physical techniques to enhance molecular diffusion through the SC.

11.2.2.1 Strategies for intact skin

Drug delivery across intact skin permits noninvasive transport or at least minimal invasive application, contributing for injection pain reduction and consequently patient compliance. Moreover, it improves drug pharmacokinetics and targeted drug delivery [94]. For these reasons, the development of advanced delivery systems has been a challenging area of research involving both optimization of drug permeation and deposition, as well as physical and chemical stability of the drugs and their formulation.

The anatomy and biochemistry of intact skin determine the barrier function of this organ. In recent years, researchers have studied a variety of methodologies for improving the permeation of poorly absorbable drugs through the skin with the final aim to develop effective dermal and transdermal delivery systems. Essentially, there are four types of transport enhancement: the chemical approach, the carrier-mediated transport, the physically facilitated transport, and the SC/epidermis by pass (Table 11.2).

Dermal drug delivery may be either passively or actively facilitated. Depending on the external energy source used to facilitate skin permeation, different techniques have been developed (Table 11.2). Passive methods are technologically easy to produce in a large-scale approach, and economically more advantageous. This aspect is an important issue to consider when thinking in developing countries markets. These methods are based on chemical enhancement and thus on permeability improvement or in carrier-mediated transport. In this latter case, concentration gradient driven transport or penetration of supramolecular structures (deformable liposomes) may be involved [94]. Examples of chemical enhancers are fatty acids and fatty esters, solvents, and surfactants [95]. Chemical enhancers became very useful as they can be employed in dermal regional delivery or incorporated in transdermal delivery systems. The chemical enhancer mechanism of action includes drug modification by increasing drug solubilization and drug partitioning into the skin and/or skin modification by fluidizing the crystalline structure of the SC, or dissolving skin lipids

[96]. Combinations of chemical enhancers have resulted in skin delivery increment [96].

Table 11.2 Strategies to overcome skin barrier and deliver drugs into superficial and deep skin layers.

Strategy	Method
Chemical enhancement	Chemical enhancers Prodrugs Peptides Hydration Supersaturation
Carrier-mediated	Microemulsions Liposomes Transfersomes Ethosomes Lipid nanoparticles Cyclodextrins
Physically assisted	Iontophoresis Low-frequency ultrasound Electroporation
SC/epidermis bypass	Delivery through hair follicles Tape-stripping Thermal or radio frequency wave-mediated ablation Microneedles Liquid-jet injections Powder injection

11.2.2.2 Permeation in infected skin

Targeted delivery to the skin depends on skin physiology and on formulation (Fig. 11.6). In the case of skin infections, the clinical evolution plays an important role on skin targeting [97]. It is generally accepted that percutaneous absorption is increased when SC is defective or when SC is disrupted by physical or chemical means [2]. Unfortunately, there is lack of comprehensive studies on the role of diseased skin on drug absorption.

Drug formulations can be applied either to thickened lesions, to open lesions or even to ulcers. In each case the integrity of the skin barrier differs. Therefore, formulations for topical application in skin lesions should be able to be applied into intact and damaged skin situations. In the case of intact barrier, formulations should promote deep skin drug delivery, as long as infective agents are located in the dermis, as observed in many infectious diseases.

In the case of skin infections, topical delivery of antimicrobials may have different targets as the drug needs to be delivered in the SC, in the epidermis or in the dermis. In parallel, the vehicle should guarantee enough drug stability and promote drug retention within the skin. For example, within the infected host, mycobacteria are capable of existing both intracellularly and extracellularly. For the treatment of mycobacterial infections, drugs must be able to penetrate host macrophages, in addition to the cell envelope of the pathogenic organism [98].

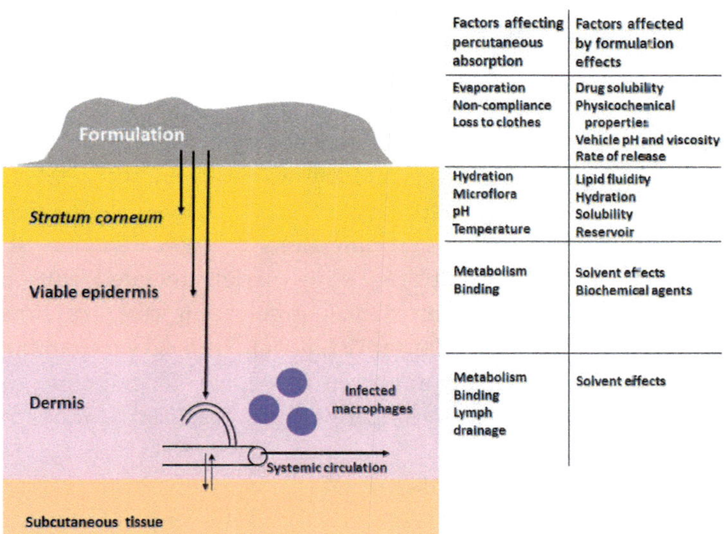

Figure 11.6 Percutaneous absorption from a topical formulation on infected skin.

In infectious diseases like CL, the permeability of the skin is highly variable in comparison to the normal skin. Characteristic lesion of CL appears with the loss of epidermis and part of the dermis, thus the absorption of drugs, especially hydrophilic molecules, used in topical treatment can be enhanced [99]. CL lesions develop differently according to the *Leishmania* species.

In some clinical conditions spontaneously healing of lesion may occur, as above described. Drug skin penetration may become hindered in healed lesions and if keratotic nodules and scar tissues are developed.

11.2.3 Lipid-Based Nanocarriers Applied in the Treatment of Buruli Ulcer and Cutaneous Leishmaniasis

Conventional formulations cannot provide sufficient drug targeting and sustained delivery. Advanced nanostructured carriers, especially lipid-based systems have proven efficient drug targeting [100], sustained drug release [82] and protection from degradation [101]. These carriers are versatile structures able to incorporate both hydrophilic and hydrophobic molecules. However, more important is the fact that most microorganisms are intracellularly located. Targeted drugs must penetrate the phagocytic cells and thus the use of carriers may facilitate drug accumulation and antimicrobial cell penetration [102].

Prabhu and co-workers recently reviewed the role of nanocarriers in topical delivery of antiinfectives, highlighting the challenges in treating skin infections by topical route [103]. However, most of the studies regarding carrier-mediated transport of antimicrobial agents for topical application refer to fungal infections therapies [104]. Pevaryl® Lipogel, a topical gel containing ecanazole was the first approved liposome formulation that proved the feasibility of liposomal systems prepared at industrial scale [105].

BU is, as previously described, a cutaneous and soft skin disease belonging to the category of difficult-to-treat infections [106]. Regarding its treatment, little investigation on drug delivery systems has been done. The management of the infection, as above mentioned, has been essentially surgical or by antibiotics association, not necessarily successful [107]. Alternative therapeutic regimens (rifampicin-quinolones and rifampin-clarithromycin) have been reported [108]. However, the use of carrier-mediated transport of already known molecules or of new drug candidates for treating *M. ulcerans* infection remains an opportunity in drug delivery. To the best of our knowledge, no reports were found in the treatment of BU using a carrier-mediated approach.

In the case of CL, specific therapy investigation is marginal as most strategies used are based on the current therapies of VL and for that liposomal formulations are already available. In particular, the successful use of intravenous AmBisome® and a topical liposomal formulation of PRM (WR279, 963), have been reported in the treatment of patients with localized CL [109, 110]. However, the evaluation of different treatment protocols, new drugs and new formulations for CL is mandatory.

The following sections present the most relevant lipid-based carriers investigated for topical delivery especially regarding antimycobacterial and antiparasitic drugs. One must have in mind that the knowledge acquired on carrier-mediated application in other skin diseases should be considered when developing formulations for BU and CL treatments.

11.2.3.1 Liposomes

Much has been written about the ability of liposomes and other vesicular colloidal carriers to penetrate the SC [91, 111]. Liposomes are vesicles constituted by one or more lipid bilayers separated by aqueous compartments. The hydrophilic part of lipids, most often phospholipids, is oriented to the inner and to the outer bilayer surface. The hydrophobic acyl chains are aligned inward avoiding water [100, 112, 113]. Hydrophilic substances are encapsulated in aqueous space while hydrophobic compounds are integrated into the bilayer.

Liposomes are characterized in terms of size, lamellarity, lipid composition, charge, surface morphology, and their properties may be modulated by the preparation method. Almost 50 years of research produced a mature technological platform to develop different classes of liposomes according to the therapeutic objectives, resulting in a real improvement of drugs pharmaceutical properties. These facts attracted the pharmaceutical industry that has developed methods for industrial scale production and sterilization of liposomal products [114, 115].

Mezei and Gulasekharam were the first to report that liposomes loaded with triamcinolone acetonide facilitate accumulation within the epidermis and dermis with low systemic levels [116].

The years following 1989 were particularly prolific in liposome research for topical application. Some of the most relevant studies include phospholipid interactions with the skin, lipid vesicle transport through SC and dermatologically pertinent studies with liposomes [117]. Many of these studies revealed interactions between lipid vesicles and the skin, e.g., via high surface absorption, lipid incorporation or deposition between the intercellular lipid lamella (between corneocytes but not so evident between keratinocytes). Other possibilities include deposition between the poorly organized skin lipid regions, or even more likely, fusion with intercellular lipids [118]. In fact, liposomes may affect SC permeability. The basic mechanisms by which this occurs are still uncertain. Two possibilities are proposed: (1) Molecules penetrate into the skin associated with intact liposomes; (2) molecules cross the skin, as a result of their interaction with intact liposomes or their constituents after disruption. In any case, liposomal constituents are incorporated into SC lipid bilayers and thus facilitate drug transport by changing the lipid layers properties [119, 120]. When or if this happens, liposomal constituents act as chemical enhancers either via hydration increase, after fusion with the skin, or through the action of enhancer-like components (degraded), such as hydrolyzed fatty acids.

Weiner and collaborators were also the first to report the use of liposomes incorporating interferon to treat skin infections [121]. They found that liposomes were not able to enhance permeation although they provided the needed physicochemical environment for interferon to be transferred into the skin.

In the 1980s the work of Setterstrom and co-workers demonstrated that the microencapsulation of antibiotics once applied topically to open contaminated wounds released drug at a controlled rate over an extended period [122]. After these studies, there was a large gap in this area of research; only after 2000 the search for targeted topical delivery of antibiotics reappeared. To the best of our knowledge, there are no reports on the use of drug-loaded liposomes to treat BU skin lesions. However, liposomal formulations of several molecules were developed and successfully used to treat other mycobacterioses [123, 124] and VL [125–127] in appropriated murine models.

In these studies formulations were administered by parenteral route. The exploitation of these liposomal formulations for topical application appears as a good strategy to be followed in either BU or CL treatments.

This approach was already accomplished against methicillin-resistant *Staphylococcus aureus* (MRSA), where cationic epigallocatechin gallate (EGCG) was incorporated in positive, negative and neutral surface charged liposomes [128]. These systems were tested *in vitro* for determination of minimum inhibitory concentration and killing rates, compared with free form. Additionally, EGCG formulations were topically applied *in vivo* in burned mouse skin infected by MRSA. The results have shown the effectiveness of cationic EGCG liposomes in the studied model [128].

An interesting recent study of local delivery of antibiotics concerned the use of mupirucin liposomes where drug release was affected by the vesicle size [129]. Microbiological evaluation confirmed its antimicrobial potential against *Staphylococcus aureus* and *Bacillus subtilis*. The system exhibited superior bioadhesiveness and sustained mupirocin release profiles than marketed products.

Also very recently, Gao *et al.* reported the use of small pH-responsive gold nanoparticle adsorbed at liposome surface as a strategy to stabilize liposomes against fusion yet allowing them to fuse preferentially with the bacteria upon arrival at the infection sites [130]. This system constructed for the incorporation of antimicrobial agents, represents a promising platform for topical application in various infected skin lesions.

Comparing the two selected skin infections discussed in this review, both in the frame of neglected diseases, CL has gained much more interest and research investment worldwide than BU [25]. Actually, the use of liposomes to incorporate antileishmanial drugs has been largely explored and different routes of administration have been tested. Either in *in vitro* studies [131] or in different animal models of CL [132, 133] or in humans patients [134], liposomes as drug delivery system is a strategy based on a robust scientific support.

Specific CL treatment should be decided upon identification of the *Leishmania* species involved. The route of administration may vary according to the available treatment in endemic regions, the number and dissemination of cutaneous lesions, and the physicochemical properties of the therapeutic agent. In parallel to topical application, intralesional injections have been tested despite the pain associated [36, 135].

Liposomal systems containing pentavalent antimonial drugs to treat CL were also tried, but the routes of administration were intravenous or subcutaneous [136]. The therapeutic effect of these liposomal formulations was considered minimal, causing a suppression of lesion growth rather than a cure. Topical formulations containing antimonials did not show also satisfactory activity against CL, in part due to the high water solubility of the drug and its low permeability coefficient across the skin. The use of appropriate carriers may be an opportunity to deliver these drugs.

AmB was evaluated *in vitro* for its stability and transdermal absorption following incorporation in positively and negatively charged liposomes. L-AmB was found to be more stable than the free drug. Moreover, L-AmB stability and transdermal absorption were lipid composition dependent [137].

In humans, successful treatment of CL with AmBisome® has been reported [134, 138, 139]. However, CL *caused by L. infantum* was unresponsive to intravenous AmBisome®, a failure not common for this *Leishmania* specie. Subsequently, the patient responded to topical imiquimod and had no relapse [140].

Yardley and Croft reported the influence of the route of administration of L-AmB in the treatment of CL in a BALB/c *L. major* model [141]. When L-AmB was intravenously administered, once a day on six alternate days, a dose-response effect was achieved. Conversely, subcutaneous administration, close to a lesion, had no significant activity, as well as the free drug at non-toxic doses. These authors referred AmBisome® as a therapeutic alternative against CL despite the higher doses needed in comparison to those used for VL.

A new system comprising Pan-DR-binding epitope (PADRE)-derivatized-dendrimer (PDD) complexed with L-AmB was tested in a *L. major* mouse model and the therapeutic efficacy of low-dose PDD/L-AmB versus full PDD/L-AmB was evaluated. The *in vivo* studies showed a reduction of the effective dose and toxicity of L-AmB and elicitation of strong parasite specific T-cell responses [142].

PRM a highly hydrophilic aminoglycoside with relative lipid insolubility has been used in the treatment of CL. As previously referred, PRM has been formulated in several conventional vehicles [143, 144]. The more representative example is a cream formulation (WR279,396) containing two aminoglycosides (PRM and gentamicin). This product was found to be safe and effective in treating Old World CL infections. Moreover, authors discussed the advantages of simple, easily applicable, and inexpensive treatments for this neglected disease [110].

Advanced lipid-based systems have been explored for PRM topical delivery. This drug was encapsulated in unilamellar liposomes resulting in improved skin permeation and retention across the intact skin [99]. In addition, controlled topical delivery across stripped skin was compared with intact skin. As PRM is an hydrophilic drug, its transport through intact skin is usually a problem due to its inability to penetrate into SC. Lipid drug carriers have demonstrated the ability to increase drug penetration into and across skin. As a result, when applied to the stripped skin, liposomes demonstrated targeted and sustained topical delivery. Since in infected skin the barrier is often compromised, this finding could be of extreme importance and liposomes could play an important role in sustained release.

Topical liposomal PRM sulfate was applied in *L. major* infected BALB/c mice [145]. Authors observed that eight weeks after the beginning of the treatment, every mouse treated with PRM liposomes was completely cured irrespective of the dose tested. In addition, no spleen parasite burden was observed in mice treated with liposomal PRM.

PRM liposomes were confirmed to be a promising alternative for the topical treatment of CL by Carneiro and co-workers [146]. In fact, PRM incorporation in soybean phosphatidylcholine (SPC) liposomes improved therapeutic efficacy in *L. Major* experimentally infected mice, in comparison with the free drug. They have also studied the influence of the lipid composition on skin PRM permeation and penetration. PRM penetration from fluid liposomes (SPC liposomes) was significantly higher than the one observed from rigid vesicles: 4 and 2 times higher, respectively, in comparison with PRM solution. Drug permeation across stripped skin was also improved for the liposomal PRM, compared to the free drug and more evident for fluid than for rigid liposomes [25].

11.2.3.2 Transfersomes

To improve the transdermal drug delivery, specialized lipid-based nanocarriers have been developed. Some of these designed carriers have claimed the ability to cross the skin intact and deliver the loaded drugs into the systemic circulation, being at the same time responsible for the percutaneous absorption of the drug within the skin. To differentiate them from conventional liposomes they were named as deformable vesicles. Deformability character enables the carriers to penetrate the narrow gaps between the cells of the SC while assuring the delivery of loaded or associated material.

The first deformable vesicles were called Transfersomes (a trademark of IDEA, AG) being composed of highly flexible membranes as a result of combining into a single structure phospholipids and an edge-active component to increase the bilayer flexibility [147] and consequently the vesicle deformability necessary for through-the-skin passagework [148]. Transfersomes are much more flexible and deformable than liposomes being able to move spontaneously against water concentration gradient in the skin. The vesicles have typically 150 ± 50 nm. Additional requirement for successful skin penetration is non-occlusive application which maintains the driving force across the skin. The natural hydration then forces carriers through weak junctions between lipid structures in intercellular space in SC.

Despite the successful applications resulting from incorporating in Transfersomes innumerous drugs, and the specificity of these carriers for topical application, only sparse investigation was made with these vesicles for the treatment of skin infections [149].

Li *et al.* recently published the preparation of Transfersomes with the name of flexible nanoliposomes containing daptomycin. *In vitro* and *in vivo* studies demonstrated that daptomycin was able to permeate the skin efficiently exerting powerful antibacterial action and activity against biofilms [150].

Recent published data showed that PRM Transfersomes made of SPC and sodium cholate with or without ethanol might be useful candidates for the topical treatment of CL, as observed *in vitro* and in a *L. major* murine model [151].

Photodynamic therapy (PDT) is a promising strategy used in neoplastic skin diseases where a photosensitizer in the presence of molecular oxygen is excited with visible light inducing the formation of reactive oxygen species, highly toxic to targeted tissues. Since 2007 PDT has been evaluated against CL using Transfersomes as carriers for different photosensitizers constituting an interesting strategy for treating CL [152]. However, the need of special medical equipment, namely lamps, may hinder the use of PDT to treat CL especially in developing countries. Montanari *et al.* determined the *in vitro* leishmanicidal activity both in the darkness and upon sunlight irradiation of Transfersomes incorporating a photosensitizer [153]. In addition, these authors have also evaluated the penetration profile through intact skin of the developed system.

Transfersomes represent, thus, a great potential to be explored in skin infectious diseases as long as they can enhance the delivery of drugs to deep skin regions where pathogens are located.

11.2.3.3 Ethosomes

Ethosomes are a special kind of deformable vesicles in which ethanol, present in the formulation, turns lipid bilayers very fluid and soft thus improving the delivery of various molecules into deep skin layers. The results that have been published indicate a preferential incorporation of hydrophobic and low MW drugs. Ethosomes share with Transfersomes many characteristics: unilateral lamellarity, 150 nm mean size, composed essentially of SPC, and instead of detergent molecules, ethosomal suspensions may contain up to 45% ethanol without vesicle disruption [154]. The main difference is that upon topical application, under non-occlusive conditions, the agent responsible for vesicle's deformability, the ethanol, evaporates from skin surface.

In what concerns infectious diseases, a large number of studies with ethosomes as drug carriers for topical application is reported in literature. However, their main application has been described for the delivery of molecules with antifungal activity [155].

Godin and Touitou incorporated erythromycin in ethosomes aiming to facilitate the transport of antibiotics through the SC and through bacterial membrane [156]. Authors found that after topical application on infected mice, erythromycin ethosomal formulations efficiently delivered the antibiotic in the deep skin *strata* where bacteria proliferate. In another study, the same authors investigated the dermal and intracellular delivery of bacitracin incorporated in ethosomes and concluded that they could be a highly beneficial approach for the treatment of a number of skin infections, requiring intracellular delivery of antibiotics [157]. Unexpectedly, no studies with topical application of ethosomes for treatment of *M. ulcerans* skin lesions are reported. Nevertheless, preliminary results have been already published regarding *in vitro* studies of a rifamycin, incorporated in ethosomes [158]. Penetration and release studies through both synthetic and biological membranes revealed that ethosomal formulations of this rifamycin present an adequate profile to be used in a topical application and thus a new alternative for BU cutaneous lesions.

The use of permeation enhancers like ethanol associated to marketed AmB formulations, used for intravenous administration, has been tested for CL. When topically administered the presence of ethanol in these formulations was able to improve skin-carrier interactions and killing localized parasites, using very low total drug concentrations. However, no transdermal delivery was achieved [159].

The use of a colloidal dispersion of AmB (Amphocil®) dispersed in 5% ethanol for topical application was reported to be used successfully in a pediatric case not responding to repeated courses with PRM ointment [160].

As previously concluded for Transfersomes, ethosomes are another excellent alternative for drug delivery by topical route. Unexpectedly, few studies have been published regarding topical application of antimicrobials associated to ethosomes for treatment of BU or CL.

11.2.3.4 Other

Archaeosomes are vesicles enclosed by one or more bilayers prepared with Total Polar Lipids extracted from microorganisms that belong to the domain *Archaea*. Ultradeformable archaeosomes are vesicles made of SPC, sodium cholate and polar lipids from *Halorubrum tebenquichense* being reported to be avidly captured by macrophages [161]. These systems were proposed as vehicles for transdermal delivery with improved skin penetration.

Nanostructured lipid carrier is a second generation of solid lipid nanoparticles, a drug delivery system with a solid matrix at room temperature [162]. These systems received great interest as colloidal lipid nanocarriers for topical application due to their ability to protect chemically labile ingredients, to improve skin hydration and to modulate drug release [163]. An important aspect is the fact that they could be used on damaged or inflamed skin as they are based on non-irritant and non-toxic lipids being easily sterilized without loss of their physicochemical properties [126, 164].

11.3 Conclusions

Infectious skin diseases represent a major group of pathologies that contribute annually for significant health economic expenses. In the present review special attention was paid on two neglected tropical skin infections, the BU and the CL, and on the important role of topical application of antiinfectives. The delivery of drugs through the skin provides a convenient route of administration as can be self-administered resulting in a better patient compliance. However, due to their physicochemical properties, some drugs present low penetration into the skin. The advances in nanomedicine, in particular by the use of nanostructured lipid carriers, have improved the dermal and transdermal delivery of drugs. In the present review, a huge number of *in vitro* and *in vivo* studies have demonstrated the usefulness of lipid-based carriers for the treatment of skin infections. Despite these encouraging results, few lipid nanocarriers have reached clinical trials. Regarding CL therapy some promising *in vivo* studies revealed the importance of lipid-based nanocarriers for improving the topical application of antileishmanial drugs. On the other hand, for BU the development of topical formulations still remains a great challenge for scientists.

Therefore we are in the beginning of the development of a stimulating nanotechnological field and the clinical possibilities arising from the use of lipid nanocarriers for topical skin infection treatments are very attractive. To accomplish this line of research the interest from governments and industries should also be promoted.

References

1. B. G. Amsden and M. F. A. Goosen, Transdermal delivery of peptide and protein drugs—an overview. *Aiche Journal*, 41, 1972–1997 (1995).
2. C. Hwa, E. A. Bauer, and D. E. Cohen, Skin biology. *Dermatologic Therapy*, 24, 464–470 (2011).
3. A. L. Dawson, R. P. Dellavalle, and D. M. Elston, Infectious skin diseases: A review and needs assessment. *Dermatologic Clinics*, 30, 141–151 (2012).
4. G. Hautmann, A. Katsambas, and T. Lotti, Non-tuberculous mycobacterial skin infections. *Journal of the European Academy of Dermatology and Venereology*, 9, 1–35 (1997).
5. American Thoracic Society, Diagnosis and treatment of disease caused by nontuberculous mycobacteria. This official statement of the American Thoracic Society was approved by the Board of Directors, March 1997. Medical Section of the American Lung Association. *American Journal of Respiratory and Critical Care Medicine*, 156, S1–S25 (1997).
6. J. M. Jackson, Emerging infections in dermatology. *Seminars in Cutaneous Medicine and Surgery*, 25, 201–206 (2006).
7. V. Prendki, P. Germaud, P. Bemer, A. Masseau, and M. Hamidou, Non tuberculous mycobacterial infections. *Revue de Medecine Interne*, 29, 370–379 (2008).
8. K. Asiedu and S. Etuaful, Socioeconomic implications of Buruli ulcer in Ghana: A three-year review. *American Journal of Tropical Medicine and Hygiene*, 59, 1015–1022 (1998).
9. S. C. Boyd, E. Athan, N. D. Friedman, A. Hughes, A. Walton, P. Callan, A. McDonald, and D. P. O'Brien, Epidemiology, clinical features and diagnosis of Mycobacterium ulcerans in an Australian population. *Medical Journal of Australia*, 196, 341–344 (2012).
10. F. Portaels, Epidemiology of mycobacterial diseases. *Clinics in Dermatology*, 13, 207–222 (1995).
11. A. Yemoa, J. Gbenou, D. Affolabi, M. Moudachirou, A. Bigot, S. Anagonou, F. Portaels, J. Quetin-Leclercq, and A. Martin, Buruli ulcer: A review of *in vitro* tests to screen natural products for activity against mycobacterium ulcerans. *Planta Medica*, 77, 641–646 (2011).

12. D. S. Walsh, F. Portaels, and W. M. Meyers, Buruli Ulcer: Advances in understanding Mycobacterium ulcerans infection. *Dermatologic Clinics*, 29, 1–8 (2011).

13. E. Marion, S. Eyangoh, E. Yeramian, J. Doannio, J. Landier, J. Aubry, A. Fontanet, C. Rogier, V. Cassisa, J. Cottin, A. Marot, M. Eveillard, Y. Kamdem, P. Legras, C. Deshayes, J. P. Saint-Andre, and L. Marsollier, Seasonal and regional dynamics of M. ulcerans transmission in environmental context: Deciphering the role of water bugs as hosts and vectors. *Plos Neglected Tropical Diseases*, 4, e731 (2010).

14. K. M. George, D. Chatterjee, G. Gunawardana, D. Welty, J. Hayman, R. Lee, and P. L. C. Small, Mycolactone: A polyketide toxin from Mycobacterium ulcerans required for virulence. *Science*, 283, 854–857 (1999).

15. A. C. Chany, C. Tresse, V. Casarotto, and N. Blanchard, History, biology and chemistry of Mycobacterium ulcerans infections (Buruli ulcer disease). *Natural Product Reports*, 30, 1527–1567 (2013).

16. D. S. Walsh, F. Eyase, D. Onyango, A. Odindo, W. Otieno, J. N. Waitumbi, W. D. Bulimo, D. C. Schnabel, W. M. Meyers, and F. Portaels, Short report: Clinical and molecular evidence for a case of buruli ulcer (mycobacterium ulcerans infection) in kenya. *American Journal of Tropical Medicine and Hygiene*, 81, 1110–1113 (2009).

17. F. Portaels, K. Chemlal, P. Elsen, P. D. R. Johnson, J. A. Hayman, J. Hibble, R. Kirkwood, and W. M. Meyers, Mycobacterium ulcerans in wild animals. *Revue Scientifique et Technique de l Office International des Epizooties*, 20, 252–264 (2001).

18. F. Portaels, M. T. Silva, and W. M. Meyers, Buruli ulcer. *Clinics in Dermatology*, 27, 291–305 (2009).

19. T. S. van der Werf, W. T. A. van der Graaf, J. W. Tappero, and K. Asiedu, Mycobacterium ulcerans infection. *Lancet*, 354, 1013–1018 (1999).

20. B. Alberts, A. Johnson, J. Lewis, M. Raff, K. Roberts, and P. Walter, Pathogens, infection and innate immunity, in *Molecular Biology of the Cell*, Garland Science, Taylor & Francis Group, New York, pp. 1485–1538 (2005).

21. D. Modi, Parasites and the skin. *C M E*, 254–260 (2009).

22. J. M. Goldsmid and W. Melrose, Parasitic infections of the skin, in *Primer of Tropical Medicine*, Australian College of Tropical Medicine, pp. 1–18 (2005).

23. J. Goddard, Medical conditions caused by arthropod stings or bites, in *Infectious Diseases and Arthropods* (J. Goddard, ed.), Humana Press, pp. 201–208 (2008).

24. O. Lupi, B. L. Bartlett, R. N. Haugen, L. C. Dy, A. Sethi, S. N. Klaus, J. M. Pinto, F. Bravo, and S. K. Trying, Tropical dermatology: Tropical diseases caused by protozoa. *Journal of the American Academy of Dermatology*, 60, 897–925 (2009).

25. G. Carneiro, M. G. Aguiar, A. P. Fernandes, and L. A. M. Ferreira, Drug delivery systems for the topical treatment of cutaneous leishmaniasis. *Expert Opinion on Drug Delivery*, 9, 1083–1097 (2012).

26. S. Antinori, L. Schifanella, and M. Corbellino, Leishmaniasis: New insights from an old and neglected disease. *European Journal of Clinical Microbiology & Infectious Diseases*, 31, 109–118 (2012).

27. R. Reithinger, J. C. Dujardin, H. Louzir, C. Pirmez, B. Alexander, and S. Brooker, Cutaneous leishmaniasis. *Lancet Infectious Diseases*, 7, 581–596 (2007).

28. M. Gramiccia and L. Gradoni, The current status of zoonotic leishmaniases and approaches to disease control. *International Journal for Parasitology*, 35, 1169–1180 (2005).

29. J. Alvar, I. D. Velez, C. Bern, M. Herrero, P. Desjeux, J. Cano, J. Jannin, and M. den Boer, Leishmaniasis worldwide and global estimates of its incidence. *Plos One*, 7, e35671 (2012).

30. A. E. P. Mondolfi, G. B. Duffey, L. E. Horton, M. Tirado, O. R. Jaimes, A. Perez-Alvarez, and O. Zerpa, Intermediate/borderline disseminated cutaneous leishmaniasis. *International Journal of Dermatology*, 52, 446–455 (2013).

31. T. Garnier, A. Mantyla, T. Jarvinen, M. J. Lawrence, M. B. Brown, and S. L. Croft, Topical buparvaquone formulations for the treatment of cutaneous leishmaniasis. *Journal of Pharmacy and Pharmacology*, 59, 41–49 (2007).

32. H. W. Murray, J. D. Berman, C. R. Davies, and N. G. Saravia, Advances in leishmaniasis. *Lancet*, 366, 1561–1577 (2005).

33. A. Strazzulla, S. Cocuzza, M. R. Pinzone, M. C. Postorino, S. Cosentino, A. Serra, B. Cacopardo, and G. Nunnari, Mucosal leishmaniasis: An underestimated presentation of a neglected disease. *Biomed Research International*, 2013, http://dx.doi.org/10.1155/2013/805108 (2013).

34. M. Ameen, Cutaneous leishmaniasis: Therapeutic strategies and future directions. *Expert Opinion on Pharmacotherapy*, 8, 2689–2699 (2007).

35. B. Rethi and L. Eidsmo, FasL and TRAIL signaling in the skin during cutaneous leishmaniasis–implications for tissue immunopathology and infectious control. *Frontiers in Immunology*, 3(163), 1–8 (2012).

35. L. Eidsmo, C. Fluur, B. Rethi, S. E. Ygberg, N. Ruffin, A. De Milito, H. Akuffo, and F. Chiodi, FasL and TRAIL induce epidermal apoptosis and skin ulceration upon exposure to Leishmania major. *American Journal of Pathology*, 170, 227–239 (2007).

36. P. Minodier and P. Parola, Cutaneous leishmaniasis treatment. *Travel Medicine and Infectious Disease*, 5, 150–158 (2007).

37. H. Goto and J. A. L. Lindoso, Cutaneous and mucocutaneous leishmaniasis. *Infectious Disease Clinics of North America*, 26, 293–307 (2012).

38. M. Ameen, Cutaneous leishmaniasis: Advances in disease pathogenesis, diagnostics and therapeutics. *Clinical and Experimental Dermatology*, 35, 699–705 (2010).

39. R. D. Pearson and A. D. Sousa, Clinical spectrum of Leishmaniasis. *Clinical Infectious Diseases*, 22, 1–11 (1996).

40. M. Debacker, J. Aguiar, C. Steunou, C. Zinsou, W. M. Meyers, and F. Portaels, Buruli ulcer recurrence, Benin. *Emerging Infectious Diseases*, 11, 584–589 (2005).

41. J. M. Kanga, Epidemiology of Buruli ulcer in Cote d'Ivoire: Results of a national survey. *Bulletin de la Societe de Pathologie Exotique*, 94, 46–51 (2001).

42. E. Y. Klutse, O. Adjei, E. Ampadu, and L. Arthur, Management of Buruli ulcer cases with topical application of phenytoin powder. Report of the 6th WHO Advisory Group Meeting on Buruli ulcer, 103–112 (2003).

43. V. Sizaire, F. Nackers, E. Comte, and F. Portaels, Mycobacterium ulcerans infection: Control, diagnosis, and treatment. *Lancet Infectious Diseases*, 6, 288–296 (2006).

44. W. M. Meyers, W. M. Shelly, and D. H. Connor, Heat-treatment of mycobacterium-ulcerans infections without surgical excision. *American Journal of Tropical Medicine and Hygiene*, 23, 924–929 (1974).

45. World Health Organization, Provisional guidance on the role of specific antibiotics in the management of Mycobacterium ulcerans disease (Buruli ulcer) (2004).

46. D. V. Almeida, P. J. Converse, S. Y. Li, S. Tyagi, E. L. Nuermberger, and J. H. Grosset, Bactericidal activity does not predict sterilizing activity: The case of rifapentine in the murine model of mycobacterium ulcerans disease. *Plos Neglected Tropical Diseases*, 7, e2085 (2013).

47. B. Weber-Dabrowska, M. Mulczyk, and A. Gorski, Bacteriophage therapy of bacterial infections: An update of our institute's

experience. *Archivum Immunologiae et Therapia Experimentalis*, 48, 547–551 (2000).

48. J. Rybniker, S. Kramme, and P. L. Small, Host range of 14 mycobacteriophages in Mycobacterium ulcerans and seven other mycobacteria including Mycobacterium tuberculosis-application for identification and susceptibility testing. *Journal of Medical Microbiology*, 55, 37–42 (2006).
49. G. Trigo, T. G. Martins, A. G. Fraga, A. Longatto, A. G. Castro, J. Azeredo, and J. Pedrosa, Phage therapy is effective against infection by Mycobacterium ulcerans in a murine footpad model. *Plos Neglected Tropical Diseases*, 7, 2183 (2013).
50. D. Jikia, N. Chkhaidze, E. Imedashvili, I. Mgaloblishvili, G. Tsitlanadze, R. Katsarava, G. J. Morris, and A. Sulakvelidze, The use of a novel biodegradable preparation capable of the sustained release of bacteriophages and ciprofloxacin, in the complex treatment of multidrug-resistant Staphylococcus aureus-infected local radiation injuries caused by exposure to Sr90. *Clinical and Experimental Dermatology*, 30, 23–26 (2005).
51. C. Deshayes, S. K. Angala, E. Marion, I. Brandli, J. Babonneau, L. Preisser, S. Eyangoh, Y. Delneste, P. Legras, C. De Chastellier, T. P. Stinear, M. Jackson, and L. Marsollier, Regulation of Mycolactone, the Mycobacterium ulcerans Toxin, Depends on Nutrient Source. *Plos Neglected Tropical Diseases*, 7, e2502 (2013).
52. M. Balasegaram, K. Ritmeijer, M. A. Lima, S. Burza, G. O. Genovese, B. Milani, S. Gaspani, J. Potet, and F. Chappuis, Liposomal amphotericin B as a treatment for human leishmaniasis. *Expert Opinion on Emerging Drugs*, 17, 493–510 (2012).
53. A. Badiee, V. H. Shargh, A. Khamesipour, and M. R. Jaafari, Micro/nanoparticle adjuvants for antileishmanial vaccines: Present and future trends. *Vaccine*, 31, 735–749 (2013).
54. I. Okwor and J. Uzonna, Vaccines and vaccination strategies against human cutaneous leishmaniasis. *Human Vaccines*, 5, 291–301 (2009).
55. N. Beheshti, S. Soflaei, M. Shakibaie, M. H. Yazdi, F. Ghaffarifar, A. Dalimi, and A. R. Shahverdi, Efficacy of biogenic selenium nanoparticles against Leishmania major: *In vitro* and *in vivo* studies. *Journal of Trace Elements in Medicine and Biology*, 27, 203–207 (2013).
56. T. S. Tiuman, A. O. Santos, T. Ueda-Nakamura, B. P. Dias, and C. V. Nakamura, Recent advances in leishmaniasis treatment. *International Journal of Infectious Diseases*, 15, E525–E532 (2011).

57. L. F. Oliveira, A. O. Schubach, M. M. Martins, S. L. Passos, R. V. Oliveira, M. C. Marzochi, and C. A. Andrade, Systematic review of the adverse effects of cutaneous leishmaniasis treatment in the New World. *Acta Tropica*, 118, 87–96 (2011).

58. S. L. Croft and P. Olliaro, Leishmaniasis chemotherapy-challenges and opportunities. *Clinical Microbiology and Infection*, 17, 1478–1483 (2011).

59. S. L. Croft, K. Seifert, and V. Yardley, Current scenario of drug development for leishmaniasis. *Indian Journal of Medical Research*, 123, 399–410 (2006).

60. C. R. Davies, P. Kaye, S. L. Croft, and S. Sundar, Leishmaniasis: New approaches to disease control. *British Medical Journal*, 326, 377–382 (2003).

61. M. Ouellette, J. Drummelsmith, and B. Papadopoulou, Leishmaniasis: Drugs in the clinic, resistance and new developments. *Drug Resistance Updates*, 7, 257–266 (2004).

62. N. Singh, M. Kumar, and R. K. Singh, Leishmaniasis: Current status of available drugs and new potential drug targets. *Asian Pacific Journal of Tropical Medicine*, 5, 485–497 (2012).

63. J. A. P. C. Tavares, A. Ouaissi, and A. Cordeiro-da-Silva, Therapy and further development of anti-leishmanial drugs. *Current Drug Therapy*, 3, 204–208 (2008).

64. M. Solomon, S. Baum, A. Barzilai, A. Scope, H. Trau, and E. Schwartz, Liposomal amphotericin B in comparison to sodium stibogluconate for cutaneous infection due to Leishmania braziliensis. *Journal of the American Academy of Dermatology*, 56, 612–616 (2007).

65. P. Mitropoulos, P. Konidas, and M. Durkin-Konidas, New World cutaneous leishmaniasis: Updated review of current and future diagnosis and treatment. *Journal of the American Academy of Dermatology*, 63, 309–322 (2010).

66. V. K. Mahajan and N. L. Sharma, Therapeutic options for cutaneous leishmaniasis. *Journal of Dermatological Treatment*, 18, 97–104 (2007).

67. P. Le Pape, Development of new antileishmanial drugs–current knowledge and future prospects. *Journal of Enzyme Inhibition and Medicinal Chemistry*, 23, 708–718 (2008).

68. U. Gonzalez, M. Pinart, L. Reveiz, and J. Alvar, Interventions for Old World cutaneous leishmaniasis. *Cochrane Database of Systematic Reviews*, 8, CD005067. doi: 10.1002/14651858.CD005067.pub3(2008).

69. F. J. Perez-Victoria, M. P. Sanchez-Canete, K. Seifert, S. L. Croft, S. Sundar, S. Castanys, and F. Gamarro, Mechanisms of experimental resistance of Leishmania to miltefosine: Implications for clinical use. *Drug Resistance Updates*, 9, 26–39 (2006).

70. B. Monge-Maillo and R. Lopez-Velez, Therapeutic options for old world cutaneous leishmaniasis and new world cutaneous and mucocutaneous leishmaniasis. *Drugs*, 73, 1889–1920 (2013).

71. O. L. S. Almeida and J. B. Santos, Advances in the treatment of cutaneous leishmaniasis in the new world in the last ten years: A systematic literature review. *Anais Brasileiros de Dermatologia*, 86, 497–506 (2011).

72. T. Garnier and S. Croft, Topical treatment for cutaneous leishmaniasis. *Current Opinion in Investigational Drugs*, 3, 538–544 (2002).

73. A. Khatami, A. Firooz, F. Gorouhi, and Y. Dowlati, Treatment of acute Old World cutaneous leishmaniasis: A systematic review of the randomized controlled trials. *Journal of the American Academy of Dermatology*, 57, 335–346 (2007).

74. J. V. Richard and K. A. Werbovetz, New antileishmanial candidates and lead compounds. *Current Opinion in Chemical Biology*, 14, 447–455 (2010).

75. S. A. Ben, H. Zakraoui, A. Zaatour, A. Ftaiti, B. Zaafouri, A. Garraoui, P. L. Olliaro, K. Dellagi, and I. R. Ben, A randomized, placebo-controlled trial in Tunisia treating cutaneous leishmaniasis with paromomycin ointment. *The American Journal of Tropical Medicine and Hygiene*, 53, 162–166 (1995).

76. J. Elon, S. Halevy, M. H. Grunwald, and L. Weinrauch, Topical treatment of old-world cutaneous leishmaniasis caused by leishmania-major—a double-blind control study. *Journal of the American Academy of Dermatology*, 27, 227–231 (1992).

77. H. Lecoeur, P. A. Buffet, G. Milon, and T. Lang, Early curative applications of the aminoglycoside WR279396 on an experimental leishmania major-loaded cutaneous site do not impair the acquisition of immunity. *Antimicrobial Agents and Chemotherapy*, 54, 984–990 (2010).

78. N. Sosa, Z. Capitan, J. Nieto, M. Nieto, J. Calzada, H. Paz, C. Spadafora, M. Kreishman-Deitrick, K. Kopydlowski, D. Ullman, W. F. McCarthy, J. Ransom, J. Berman, C. Scott, and M. GROGL, Randomized, Double-blinded, phase 2 trial of WR 279,396 (paromomycin and gentamicin) for cutaneous leishmaniasis in panama. *American Journal of Tropical Medicine and Hygiene*, 89, 557–563 (2013).

79. T. P. C. Dorlo, M. Balasegaram, J. H. Beijnen, and P. J. de Vries, Miltefosine: A review of its pharmacology and therapeutic efficacy in the treatment of leishmaniasis. *J Antimicrobial Chemother*, 67, 2576–2597 (2012).

80. M. P. Cramer and S. R. Saks, Translating safety, efficacy and compliance into economic value for controlled-release dosage forms. *Pharmacoeconomics*, 5, 482–504 (1994).

81. D. F. Archer, V. Cullins, G. W. Creasy, and A. C. Fisher, The impact of improved compliance with a weekly contraceptive transdermal system (Ortho Evra (R)) on contraceptive efficacy. *Contraception*, 69, 189–195 (2004).

82. S. C. Raposo, S. D. Simoes, A. J. Almeida, and H. M. Ribeiro, Advanced systems for glucocorticoids' dermal delivery. *Expert Opinion on Drug Delivery*, 10, 857–877 (2013).

83. K. S. Paudel, M. Milewski, C. L. Swadley, N. K. Brogden, P. Ghosh, and A. L. Stinchcomb, Challenges and opportunities in dermal/transdermal delivery. *Therapeutic Delivery*, 1, 109–131 (2010).

84. R. H. Guy, Transdermal drug delivery. *Handbook of Experimental Pharmacology*, 197, 399–410 (2010).

85. I. P. Kaur and S. Kakkar, Topical delivery of antifungal agents. *Expert Opinion on Drug Delivery*, 7, 1303–1327 (2010).

86. N. Duran, P. D. Marcato, Z. Teixeira, M. Duran, F. T. M. Costa, and M. Brocchi, State of the art of nanobiotechnology applications in neglected diseases. *Current Nanoscience*, 5, 396–408 (2009).

87. E. J. Wood and I. R. Harris, Reconstructed human skin: transplant, graft or biological dressing. *Essays in Biochemistry*, 29, 65–85 (1995).

88. A. Schatzlein and G. Cevc, Non-uniform cellular packing of the stratum corneum and permeability barrier function of intact skin: A high-resolution confocal laser scanning microscopy study using highly deformable vesicles (Transfersomes). *British Journal of Dermatology*, 138, 583–592 (1998).

89. P. M. Elias, Lipids and the epidermal permeability barrier. *Archives of Dermatological Research*, 270, 95–117 (1981).

90. P. W. Wertz, D. C. Swartzendruber, W. Abraham, K. C. Madison, and D. T. Downing, Essential fatty-acids and epidermal integrity. *Archives of Dermatology*, 123, 1381–1384 (1987).

91. J. A. Bouwstra, P. L. Honeywell-Nguyen, G. S. Gooris, and M. Ponec, Structure of the skin barrier and its modulation by vesicular formulations. *Progress in Lipid Research*, 42, 1–36 (2003).

92. C. Pailler-Mattei, C. Guerret-Piecourt, H. Zahouani, and S. Nicoli, Interpretation of the human skin biotribological behaviour after tape stripping. *Journal of the Royal Society Interface*, 8, 934–941 (2011).

93. P. W. Wertz and D. T. Downing, Stratum Corneum: Biological and biochemical considerations, in *Transdermal Drug Delivery: Developmental Issues and Research Initiatives* (J. Hadgraft and R. H. Guy, eds.), Marcel Dekker, New York, pp. 1–22 (1989).

94. S. I. Simões, T. C. Delgado, R. M. Lopes, S. Jesus, A. A. Ferreira, J. A. Morais, M. E. M. Cruz, M. L. Corvo, and M. B. F. Martins, Developments in the rat adjuvant arthritis model and its use in therapeutic evaluation of novel non-invasive treatment by SOD in Transfersomes. *Journal of Controlled Release*, 103, 419–434 (2005).

95. J. Hadgraft and M. E. Lane, Skin: The ultimate interface. *Physical Chemistry Chemical Physics*, 13, 5215–5222 (2011).

96. J. T. Huzil, S. Sivaloganathan, M. Kohandel, and M. Foldvari, Drug delivery through the skin: Molecular simulations of barrier lipids to design more effective noninvasive dermal and transdermal delivery systems for small molecules, biologics, and cosmetics. *Wiley Interdisciplinary Reviews-Nanomedicine and Nanobiotechnology*, 3, 449–462 (2011).

97. M. S. Roberts, Targeted drug delivery to the skin and deeper tissues: Role of physiology, solute structure and disease. *Clinical and Experimental Pharmacology and Physiology*, 24, 874–879 (1997).

98. W. W. Barrow, Treatment of mycobacterial infections. *Revue Scientifique et Technique de l Office International des Epizooties*, 20, 55–70 (2001).

99. L. S. Ferreira, G. A. Ramaldes, E. A. Nunan, and L. A. M. Ferreira, *In vitro* skin permeation and retention of paromomycin from liposomes for topical treatment of the cutaneous leishmaniasis. *Drug Development and Industrial Pharmacy*, 30, 289–296 (2004).

100. M. E. M. Cruz, S. I. Simões, M. L. Corvo, M. B. F. Martins, and M. M. Gaspar, Formulation of NPDDS for macromolecules, in *Drug Delivery Nanoparticles: Formulation and Characterization* (Y. Pathak and D. Thassu, eds.), Informa Healthcare, New York, pp. 35–49 (2009).

101. A. Ascenso, M. Cruz, C. Euleterio, F. A. Carvalho, N. C. Santos, H. C. Marques, and S. Simoes, Novel tretinoin formulations: A drug-in-cyclodextrin-in-liposome approach. *Journal of Liposome Research*, 23, 211–219 (2013).

102. E. Briones, C. I. Colino, and J. M. Lanao, Delivery systems to increase the selectivity of antibiotics in phagocytic cells. *Journal of Controlled Release*, 125, 210–227 (2008).

103. P. Prabhu, V. Patravale, and M. Joshi, Nanocarriers for effective topical delivery of anti-infectives. *Current Nanoscience*, 8, 491–503 (2012).

104. M. Y. Ning, Y. Z. Guo, H. Z. Pan, X. L. Chen, and Z. W. Gu, Preparation, *in vitro* and *in vivo* evaluation of liposomal/niosomal gel delivery systems for clotrimazole. *Drug Development and Industrial Pharmacy*, 31, 375–383 (2005).

105. R. Naeff, Feasibility of topical liposome drugs produced on an industrial scale. *Advanced Drug Delivery Reviews*, 18, 343–347 (1996).

106. F. Alcaide and J. Esteban, Cutaneous and soft skin infections due to non-tuberculous mycobacteria. *Enfermedades Infecciosas y Microbiología Clínica*, 28 Suppl 1, 46–50 (2010).

107. M. T. Ruf, A. Chauty, A. Adeye, M. F. Ardant, H. Koussemou, R. C. Johnson, and G. Pluschke, Secondary Buruli ulcer skin lesions emerging several months after completion of chemotherapy: Paradoxical reaction or evidence for immune protection? *Plos Neglected Tropical Diseases*, 5, e1252 (2011).

108. P. D. R. Johnson, J. A. Hayman, T. Y. Quek, A. M. Janet, G. A. Jenkin, J. A. Buntine, E. Athan, M. Birrell, J. Graham, and C. J. Lavender, Consensus recommendations for the diagnosis, treatment and control of Mycobacterium ulcerans infection (Bairnsdale or Buruli ulcer) in Victoria, Australia. *Medical Journal of Australia*, 186, 64–68 (2007).

109. G. Wortmann, M. Zapor, R. Ressner, S. Fraser, J. Hartzell, J. Pierson, A. Weintrob, and A. Magill, Lipsosomal amphotericin b for treatment of cutaneous leishmaniasis. *American Journal of Tropical Medicine and Hygiene*, 83, 1028–1033 (2010).

110. A. Ben Salah, P. A. Buffet, G. Morizot, N. Ben Massoud, A. Zaatour, N. Ben Alaya, N. B. H. Hamida, Z. El Ahmadi, M. T. Downs, P. L. Smith, K. Dellagi, and M. Grogl, WR279,396, a third generation aminoglycoside ointment for the treatment of Leishmania major cutaneous leishmaniasis: A phase 2, randomized, double blind, placebo controlled study. *Plos Neglected Tropical Diseases*, 3, e432 (2009).

111. G. Cevc, A. G. Schatzlein, H. Richardsen, and U. Vierl, Overcoming semipermeable barriers, such as the skin, with ultradeformable mixed lipid vesicles, transfersomes, liposomes, or mixed lipid micelles. *Langmuir*, 19, 10753–10763 (2003).

112. G. Gregoriadis ed., *Liposome Technology: Liposome Preparation and Related Techniques*, Informa Healthcare USA Inc., New York (2007).

113. M. Eugenia, M. Cruz, M. M. Gaspar, M. B. F. Martins, and M. L. Corvo, Liposomal superoxide dismutases and their use in the treatment of experimental arthritis. *Liposomes, Pt e*, 391, 395–413 (2005).

114. O. R. Justo and A. M. Moraes, Economical feasibility evaluation of an ethanol injection liposome production plant. *Chemical Engineering & Technology*, 33, 15–20 (2010).

115. A. N. Lukyanov and V. P. Torchilin, Autoclaving of liposomes. *Journal of Microencapsulation*, 11, 669–672 (1994).

116. M. Mezei and V. Gulasekharam, Liposomes—A selective drug delivery system for the topical route of administration. 1. Lotion dosage form. *Life Sciences*, 26, 1473–1477 (1980).

117. W. Wohlrab and J. Lasch, The effect of liposomal incorporation of topically applied hydrocortisone on its serum concentration and urinary excretion. *Dermatol Monatsschr*, 175, 348–352 (1989).

118. N. Weiner, L. Lieb, S. Niemiec, C. Ramachandran, Z. Hu, and K. Egbaria, Liposomes—A novel topical delivery system for pharmaceutical and cosmetic applications. *Journal of Drug Targeting*, 2, 405–410 (1994).

119. J. Duplessis, N. Weiner, and D. G. Muller, The influence of in-vivo treatment of skin with liposomes on the topical absorption of a hydrophilic and a hydrophobic drug in-vitro. *International Journal of Pharmaceutics*, 103, R1–R5 (1994).

120. M. Schaller and H. C. Korting, Interaction of liposomes with human skin: The role of the stratum corneum. *Advanced Drug Delivery Reviews*, 18, 303–309 (1996).

121. N. Weiner, N. Williams, G. Birch, C. Ramachandran, C. Shipman, and G. Flynn, Topical delivery of liposomally encapsulated interferon evaluated in a cutaneous herpes guinea-pig model. *Antimicrobial Agents and Chemotherapy*, 33, 1217–1221 (1989).

122. J. A. Setterstrom, T. R. Ticet, and W. E. Myers, Development of encapsulated antibiotics for topical administration to wounds. *Recent Advances in Drug Delivery Systems*, 185–198 (1984).

123. M. M. Gaspar, S. Neves, F. Portaels, J. Pedrosa, M. T. Silva, and M. E. M. Cruz, Therapeutic efficacy of liposomal rifabutin in a Mycobacterium avium model of infection. *Antimicrobial Agents and Chemotherapy*, 44, 2424–2430 (2000).

124. M. M. Gaspar, A. Cruz, A. E. Penha, J. Reymao, A. C. Sousa, C. V. Eleuterio, S. A. Domingues, A. G. Fraga, A. L. Filho, M. E. M. Cruz, and J. Pedrosa, Rifabutin encapsulated in liposomes exhibits increased therapeutic activity in a model of disseminated tuberculosis. *International Journal of Antimicrobial Agents*, 31, 37–45 (2008).

125. M. Carvalheiro, M. A. Esteves, D. Santos-Mateus, R. M. Lopes, M. A. Rodrigues, C. V. Eleutério, E. Scoulica, G. Santos-Gomes, M. E. M. Cruz, Hemisynthetic trifluralin analogues incorporated in liposomes for the treatemnt of leishmanial infections. *European Journal of Pharmaceutics and Biopharmaceutics*, 93, 346–352 (2015).

126. M. M. Gaspar, S. Calado, J. Pereira, H. Ferronha, I. Correia, H. Castro, A. M. Tomás, M. E. M. Cruz, Targeted delivery of paromomycin in murine infectious diseases through association to nano lipid systems. *Nanomedicine: NBM*, 11, 1851–1860 (2015).

127. R. M. Lopes, M. M. Gaspar, J. Pereira, C. V. Eleuterio, M. Carvalheiro, A. J. Almeida, and M. E. M. Cruz, Liposomes versus lipid nanoparticles: comparative study of lipid-based systems as oryzalin carriers for treatment of leishmaniasis. *Journal of Biomedical Nanotechnology*, 10, 1–11 (2014).

128. A. Gharib, Z. Faezizadeh, and M. Godarzee, Therapeutic efficacy of epigallocatechin gallate-loaded nanoliposomes against burn wound infection by methicillin-resistant staphylococcus aureus. *Skin Pharmacology and Physiology*, 26, 68–75 (2013).

129. J. Hurler, O. A. Berg, M. Skar, A. H. Conradi, P J. Johnsen, and N. Skalko-Basnet, Improved burns therapy: Liposomes-in-hydrogel delivery system for mupirocin. *Journal of Pharmaceutical Sciences*, 101, 3906–3915 (2012).

130. W. Gao, D. Vecchio, J. Li, J. Zhu, Q. Zhang, V. Fu, J. Li, S. Thamphiwatana, D. Lu, and L. Zhang, Hydrogel containing nanoparticle-stabilized liposomes for topical antimicrobial delivery. *ACS Nano*, 8, 2900–2907 (2014).

131. S. E. T. Borborema, R. A. Schwendener, J. A. Osso, H. F. de Andrade, and N. do Nascimento, Uptake and antileishmanial activity of meglumine antimoniate-containing liposomes in Leishmania (Leishmania) major-infected macrophages. *International Journal of Antimicrobial Agents*, 38, 341–347 (2011).

132. N. B. Barros, V. Migliaccio, V. A. Facundo, P. Ciancaglini, R. G. Stabeli, R. Nicolete, and I. Silva-Jardim, Liposomal-lupane system as alternative chemotherapy against cutaneous leishmaniasis: Macrophage as target cell. *Experimental Parasitology*, 135, 337–343 (2013).

133. C. B. Panosian, M. Barza, F. Szoka, and D. J. Wyler, Treatment of experimental cutaneous leishmaniasis with liposome-intercalated amphotericin-b. *Antimicrobial Agents and Chemotherapy*, 25, 655–656 (1984).

134. F. Butsch, M. Faulde, A. Debus, C. Bogdan, and E. von Stebut, Two cases of successful treatment of multilesional cutaneous leishmaniasis

with liposomal amphotericin B. *Journal der Deutschen Dermatologischen Gesellschaft*, 11, 83–85 (2013).

135. H. Neuber, Leishmaniasis. *Journal der Deutschen Dermatologischen Gesellschaft*, 6, 754–764 (2008).

136. F. Frezard and C. Demicheli, New delivery strategies for the old pentavalent antimonial drugs. *Expert Opinion on Drug Delivery*, 7, 1343–1358 (2010).

137. A. Manosroi, L. Kongkaneramit, and J. Manosroi, Stability and transdermal absorption of topical amphotericin B liposome formulations. *International Journal of Pharmaceutics*, 270, 279–286 (2004).

138. M. Solomon, F. Pavlotzky, A. Barzilai, and E. Schwartz, Liposomal amphotericin B in comparison to sodium stibogluconate for Leishmania braziliensis cutaneous leishmaniasis in travelers. *Journal of the American Academy of Dermatology*, 68, 284–289 (2013).

139. M. Ono, K. Takahashi, K. Taira, H. Uezato, S. Takamura, and S. Izaki, Cutaneous leishmaniasis in a Japanese returnee from West Africa successfully treated with liposomal amphotericin B. *Journal of Dermatology*, 38, 1062–1065 (2011).

140. J. A. Hervas, A. Martin-Santiago, D. Hervas, E. Rojo, A. Mena, V. Rocamora, and J. Duenas, Old world leishmania infantum cutaneous leishmaniasis unresponsive to liposomal amphotericin b treated with topical imiquimod. *Pediatric Infectious Disease Journal*, 31, 97–100 (2012).

141. V. Yardley and S. L. Croft, Activity of liposomal amphotericin B against experimental cutaneous leishmaniasis. *Antimicrobial Agents and Chemotherapy*, 41, 752–756 (1997).

142. P. M. Daftarian, G. W. Stone, L. Kovalski, M. Kumar, A. Vosoughi, M. Urbieta, P. Blackwelder, E. Dikici, P. Serafini, S. Duffort, R. Boodoo, A. Rodriguez-Cortes, V. Lemmon, S. Deo, J. Alberola, V. L. Perez, S. Daunert, and A. L. Ager, A Targeted and adjuvanted nanocarrier lowers the effective dose of liposomal amphotericin B and enhances adaptive immunity in murine cutaneous leishmaniasis. *Journal of Infectious Diseases*, 208, 1914–1922 (2013).

143. J. Soto, P. Fuya, R. Herrera, and J. Berman, Topical paromomycin methylbenzethonium chloride plus parenteral meglumine antimonate as treatment for American cutaneous leishmaniasis: Controlled study. *Clinical Infectious Diseases*, 26, 56–58 (1998).

144. J. Elon, R. Livshin, Z. Evenpaz, D. Hamburger, and L. Weinrauch, Topical treatment of cutaneous leishmaniasis. *Journal of Investigative Dermatology*, 87, 284–288 (1986).

145. M. R. Jaafari, N. Bavarsad, B. S. F. Bazzaz, A. Samiei, D. Soroush, S. Ghorbani, M. M. L. Heravi, and A. Khamesipour, Effect of topical liposomes containing paromomycin sulfate in the course of leishmania major infection in susceptible BALB/c mice. Antimicrobial Agents and Chemotherapy, 53, 2259–2265 (2009).

146. G. Carneiro, D. C. M. Santos, M. C. Oliveira, A. P. Fernandes, L. S. Ferreira, G. A. Ramaldes, E. A. Nunan, and L. A. M. Ferreira, Topical delivery and *in vivo* antileishmanial activity of paromomycin-loaded liposomes for treatment of cutaneous leishmaniasis. *Journal of Liposome Research*, 20, 16–23 (2010).

147. G. Cevc and G. Blume, Lipid vesicles penetrate into intact skin owing to the transdermal osmotic gradients and hydration force. *Biochimica et Biophysica Acta*, 1104, 226–232 (1992).

148. G. Cevc, A. Schatzlein, and H. Richardsen, Ultradeformable lipid vesicles can penetrate the skin and other semi-permeable barriers unfragmented. Evidence from double label CLSM experiments and direct size measurements. *Biochimica et Biophysica Acta: Biomembranes*, 1564, 21–30 (2002).

149. G. Cevc, Transfersomes, liposomes and other lipid suspensions on the skin: Permeation enhancement, vesicle penetration, and transdermal drug delivery. *Critical Reviews in Therapeutic Drug Carrier Systems*, 13, 257–388 (1996).

150. C. Li, X. L. Zhang, X. L. Huang, X. Y. Wang, G. J. Liao, and Z. B. Chen, Preparation and characterization of flexible nanoliposomes loaded with daptomycin, a novel antibiotic, for topical skin therapy. *International Journal of Nanomedicine*, 8, 1285–1292 (2013).

151. N. Bavarsad, B. S. F. Bazzaz, A. Khamesipour, and M. R. Jaafari, Colloidal, *in vitro* and *in vivo* anti-leishmanial properties of transfersomes containing paromomycin sulfate in susceptible BALB/c mice. *Acta Tropica*, 124, 33–41 (2012).

152. I. P. Hernandez, J. Montanari, W. Valdivieso, M. J. Morilla, E. L. Romero, and P. Escobar, *In vitro* phototoxicity of ultradeformable liposomes containing chloroaluminum phthalocyanine against New World Leishmania species. *Journal of Photochemistry and Photobiology B-Biology*, 117, 157–163 (2012).

153. J. Montanari, C. Maidana, M. I. Esteva, C. Salomon, M. J. Morilla, and E. L. Romero, Sunlight triggered photodynamic ultradeformable liposomes against Leishmania braziliensis are also leishmanicidal in the dark. *Journal of Controlled Release*, 147, 368–376 (2010).

154. E. Touitou, N. Dayan, L. Bergelson, B. Godin, and M. Eliaz, Ethosomes—novel vesicular carriers for enhanced delivery: Characterization

and skin penetration properties. *Journal of Controlled Release*, 65, 403–418 (2000).

155. M. K. Bhalaria, S. Naik, and A. N. Misra, Ethosomes: A novel delivery system for antifungal drugs in the treatment of topical fungal diseases. *Indian Journal of Experimental Biology*, 47, 368–375 (2009).

156. B. Godin and E. Touitou, Erythromycin ethosomal systems: Physicochemical characterization and enhanced antibacterial activity. *Current Drug Delivery*, 2, 269–275 (2005).

157. B. Godin and E. Touitou, Mechanism of bacitracin permeation enhancement through the skin and cellular membranes from an ethosomal carrier. *Journal of Controlled Release*, 94, 365–379 (2004).

158. B Gaspar, A Khan, M. M. Gaspar, and S. I. Simões, Ethosomes as carriers for rifabutin topical application. *Biomedical and Biopharmaceutical Research*, 10, 115–125 (2013).

159. S. Frankenburg, D. Glick, S. Klaus, and Y. Barenholz, Efficacious topical treatment for murine cutaneous leishmaniasis with ethanolic formulations of amphotericin B. *Antimicrobial Agents and Chemotherapy*, 42, 3092–3096 (1998).

160. A. Zvulunov, E. Cagnano, S. Frankenburg, Y. Barenholz, and D. Vardy, Topical treatment of persistent cutaneous leishmaniasis with ethlanolic lipid amphotericin B. *Pediatric Infectious Disease Journal*, 22, 567–569 (2003).

161. L. H. Higa, P. Schilrreff, A. P. Perez, M. A. Iriarte, D. I. Roncaglia, M. J. Morilla, and E. L. Romero, Ultradeformable archaeosomes as new topical adjuvants. *Nanomedicine-Nanotechnology Biology and Medicine*, 8, 1319–1328 (2012).

162. S. Doktorovova and E. B. Souto, Nanostructured lipid carrier-based hydrogel formulations for drug delivery: A comprehensive review. *Expert Opinion on Drug Delivery*, 6, 165–176 (2009).

163. M. A. Iqbal, S. Md, J. K. Sahni, S. Baboota, S. Dang, and J. Ali, Nanostructured lipid carriers system: Recent advances in drug delivery. *Journal of Drug Targeting*, 20, 813–830 (2012).

164. S. A. Wissing, O. Kayser, and R. H. Muller, Solid lipid nanoparticles for parenteral drug delivery. *Advanced Drug Delivery Reviews*, 56, 1257–1272 (2004).

Chapter 12

Liquid Crystals as Drug Delivery Systems for Skin Applications

Marilisa Guimarães Lara

Faculdade de Ciências Farmacêuticas de Ribeirão Preto, Universidade de São Paulo, Ribeirão Preto, SP, Brazil

mlara@fcfrp.usp.br

12.1 Liquid Crystals

Liquid crystals have been widely used as drug delivery systems for several drugs and several routes of administration, including skin applications. These systems are formed by amphiphilic molecules in the presence of water and show properties of both liquid and solid states. Specifically, they can flow like a liquid and simultaneously exhibit a long-ranged orientational order of molecules. Liquid crystals form when amphiphilic molecules are mixed with a solvent (usually water) and form aggregates via a self-assembly process. Their internal structures consist of nanostructured hydrophilic and hydrophobic domains separated by a lipid bilayer, providing a matrix of complex geometry [1–4].

Liquid crystals can be classified in lyotropic or thermotropic liquid crystalline phases when the formation of the systems

Carrier-Mediated Dermal Delivery: Applications in the Prevention and Treatment of Skin Disorders
Edited by Andreia Ascenso, Sandra Simões, and Helena Ribeiro

ISBN 978-981-4745-58-1 (Hardcover), 978-1-315-36447-6 (eBook)
www.panstanford.com

is induced by the addition of a solvent or by temperature, respectively. For drug delivery purposes, the most used systems are the lyotropic liquid crystalline phases, which are formed by polar lipids in contact with water [2, 4, 5].

Lyotropic liquid crystals are formed when amphiphilic molecules arrange themselves into different ordered arrays in the presence of water and form several phases with distinct structures and geometries, which are called mesophases. The main mesophases studied in the development of drug delivery systems are lamellar (L), cubic (Q), and hexagonal phases (H). Due to their highly ordered structures, which can modulate drug release, these systems have been extensively explored as matrices for drug delivery [1, 2, 4, 6–9]. The presence of hydrophilic and hydrophobic domains provides an ideal vehicle for drugs with different solubilities, sizes, and polarities [7, 8]. Therefore, these systems can enclose hydrophilic, lipophilic, and amphiphilic drugs and their complex structures can provide a diffusion pathway for controlled release of drugs [2, 9–12]. Liquid crystals are also thermodynamically stable and possess similar structures than physiological lipid membranes. This similarity may account for interactions with membranes at the site of application. In addition, liquid crystalline systems can also incorporate and protect some drugs (peptides, proteins, and nucleic acids) from chemical and physical degradation, which is desirable for macromolecule drug delivery [2, 7]. Liquid crystalline systems can also improve the solubility of drugs in the formulation. The hydrophobic domains of the liquid crystalline structure provide a medium for enhancing the solubility of poorly soluble drugs and hence alter the thermodynamic activity of drug in the formulation, which may affect drug release and skin permeation [9, 12–15].

Liquid crystalline systems are promising vehicles for skin drug delivery because their structures are similar to the structural organization of the lipid components of the stratum corneum, which favors interactions of these systems with skin and improves drug permeation [2, 9, 10].

The therapeutic effect of percutaneous preparations depends not only on the action of the drug itself but also on other factors related to the structure of the vehicle [16]. Formulation of a vehicle can determine the kinetic of release and/or percutaneous absorption and, therefore, influence biological response of a drug applied on skin. The choice of appropriate vehicles is a strategy to improve skin delivery of drugs [17]. Drug delivery systems can be designed to control drug release and consequently influence the amount of drug available to permeate the skin and the therapeutic effect of this drug. The vehicle can also interact with skin and alter its barrier function, which in turn affects the amount of drug permeated. Interaction of formulations with skin surface plays an important role in determining percutaneous absorption [17]. In this way, the drug delivery systems can significantly affect the therapeutic efficacy of percutaneous preparations, either by controlling drug release or by modifying skin barrier properties. Liquid crystals can modulate drug release and interact with skin to improve drug permeation. Thus, they constitute a very interesting class of vehicles for skin drug delivery.

12.2 Liquid Crystal-Forming Lipids

Lyotropic liquid crystalline systems can be formed by several amphiphilic lipids, such as monoglycerides, phospholipids, and glycolipids, which can spontaneously self-assemble in water to form liquid crystals. Some examples of materials commonly used to form liquid crystals are glyceryl monooleate (GMO), glyceryl monolinoleate (GML), glyceryl monooleyil ether (GME), phytantriol (PT), oleyl glycerate (OG), phytanyl glycerate (PG), monolinolein, monoelaidin, phosphatidylethanolamine, oleoylethanolamide, phospholipids, vitamin E TPGS, PEGylated phospholipids, alkyl glycerates and glycolipids. GMO, OG, and PT are the most widely studied lipids that form liquid crystals for drug delivery purposes [2, 7, 8, 12, 15, 18, 19].

GMO is a polar unsaturated monoglyceride containing a *cis* double bond at C9 and an acyl chain attached to a glycerol backbone by an ester bond. It is a nontoxic, biodegradable, and biocompatible material that is classified as GRAS (generally recognized as safe). GMO can form a lamellar phase at low water contents and a cubic phase with an excess of water. It is also considered a skin permeation enhancer because it can temporarily and reversibly disrupt stratum corneum lipids, promote ceramide extraction and enhance the lipid fluidity of this skin layer. Therefore, liquid crystalline systems formed by GMO are a very interesting vehicle for skin delivery due to their permeation enhancer properties and the ability to control drug delivery [2, 4, 5, 20, 21].

OG presents a similar structure to that of GMO, but it can form a reverse hexagonal phase in an excess of water at physiological conditions, whereas GMO forms a cubic phase. The structural difference between the cubic and reverse hexagonal phase results in distinct release rates for the same drug when released from systems based on OG and GMO [9].

The phase behavior of PT is similar to that of GMO, mainly with respect to the formation of a bicontinuous cubic phase in an excess of water at room temperature and a reverse hexagonal phase at higher temperatures. Therefore, it is also widely used in drug delivery systems. However, the structure of PT does not include an ester functional group, which confers an improved stability to PT compared with GMO, as the ester bonds within the GMO headgroup are susceptible to enzymatic breakdown (esterase-catalyzed hydrolysis) to form oleic acid and glycerol. PT can also improve the moisture retention of skin and consequently improve drug permeation [2, 8, 9, 22].

12.3 Liquid Crystalline Macroscopic Forms

Liquid crystals have been used as drug delivery systems in several forms, such as bulk liquid crystalline gels, liquid crystalline precursors and dispersed systems (cubosomes or hexosomes). Bulk liquid crystalline gels are easily prepared by simply mixing the lipid and aqueous phase, which may contain additives [2, 9, 10]. Drugs can be dissolved in the lipid or aqueous phases depending on their solubilities.

Cubic phase gels have been commonly used as drug delivery systems for several routes of administration, but the stiffness and viscous nature of the bulk phase of these gels usually hinders their spreading during application. Therefore, some drug delivery systems are formulated as cubic phase precursors, which are less viscous but can transform into a cubic phase at the site of application, mainly by water absorption (*in situ* gelification upon contact with water from body fluids). This strategy is often employed in the development of drug delivery systems that contact body fluids, like oral, subcutaneous, and mucosal delivery systems [9, 10, 15].

Dispersed systems can also be prepared when the bulk liquid crystalline structure is dispersed in an excess of water and forms nanoparticles. In this way, cubic phase gels can be fragmented and, with the addition of stabilizers, form stable colloidal dispersions, which are termed cubosomes. In the same way, hexagonal phases can also be dispersed and form structures named hexosomes [2, 10]. These nanostructured particles consist of a unique drug delivery system with interesting properties, such as an increased surface area generated by their inherent nanostructure. An important feature of these particles is that they retain the inner structures of liquid crystalline systems, which confers sustained release properties [9, 12, 21–23].

Cubosomes and hexosomes can be prepared via several fragmentation methods of liquid crystalline systems. Some methods involve high-energy input to fragment the systems. Specifically, a mixture of the structure-forming lipid and stabilizers is hydrated to self-assemble in a viscous bulk phase. The bulk is then dispersed upon the input of high-level energy (high-pressure homogenization, ultrasonication) to form cubosomes or hexosomes [2, 9]. Other methods of preparation include the reconstitution of dispersions from dried lipid/stabilizer films and precipitation upon diluting lipids in the presence of solutions containing an aqueous phase or upon dialyzing a mixed micellar solution to form nanostructured systems [9]. Cubosomes or hexosomes can be formed by the controlled addition of aqueous medium, which rapidly reduces the lipid solubility and results in particle formation [2]. Most systems require a dispersing agent or stabilizers (bile salts, amphiphilic proteins or block copolymers) because these nanostructured particles tend to aggregate [9]. The properties of these nanoparticles (such as their size, structure and stability) can be affected by their internal composition, dispersion polymer concentration and processing conditions [11].

Liquid crystals can also be formed in emulsions formulations. Some surfactants used to stabilize emulsions may form liquid crystalline structures in the continuous aqueous phase when used in excess, extending as a three-dimensional network. Lamellar liquid crystalline phases also can be formed surrounding the oil droplets of an emulsion formulation, maintaining an ordered structure and a low interfacial tension. Emulsions containing liquid crystals are suggested to have advantages, such as improved stability (the network formed can slow the movement of emulsion droplets and decrease the coalescence of the dispersed phase), prolonged skin hydration, sustained drug release and improved skin penetration [14, 16].

12.4 Liquid Crystalline Mesophases

The use of surfactants able to form liquid crystalline systems is explored in the development of topical drug delivery systems [24, 25]. Liquid crystalline mesophases are formed when amphiphilic molecules spontaneously form thermodynamically stable, self-assembled structures in the presence of water, and they exhibit several arrangements, such as lamellar, cubic, hexagonal phases. Figure 12.1 presents several arrangements of crystal, liquid crystal, and fluid phases formed by GMO and water systems [1, 6, 12, 19, 26].

The lamellar phase (L) consists of planar lipid bilayers stacked in a one-dimensional lattice separated by layers of water. The polar head groups of the lipid molecules are facing each other and come in direct contact with the thin lamellae, whereas the hydrophobic tails are oriented away from the water molecules in tightly packed bilayers. This system can flow because the parallel layers can slide over each other during shear. The texture is streaky or mosaic-like and can be easily identified by polarizing light microscopy due to a characteristic anisotropic pattern [3, 4, 9].

Cubic phase structures consist of two continuous but nonintersecting water channels separated by a lipid bilayer. This bilayer is curved and extends in three dimensions. This mesophase exhibits the most complex spatial organization of the liquid crystalline phases, which allows it to be used as a slow release matrix for several drugs. Cubic phases present different lattice arrangements, which can be classified into three types: double-diamond cubic phase (Pn3m), body-centered cubic phase (Im3m), and gyroid cubic phase (Ia3d). These arrangements can be identified by X-ray crystallographic studies. Cubic phases are thermodynamically stable, isotropic by polarizing light microscopy and consist of a transparent, very viscous, and rigid gel [1, 2, 4, 7, 9, 10].

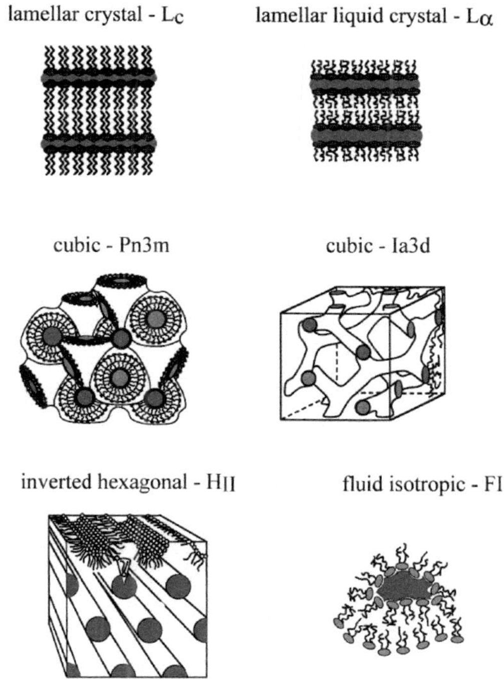

Figure 12.1 Schematic representation of phases formed by GMO systems. Reproduced with permission from Qiu, Caffrey, 2000 [26].

The hexagonal phases present micellar columnar structures. A normal hexagonal phase (HI) consists of long cylindrical micelles separated by water, whereas an inverse hexagonal phase (HII) consists of long cylindrical cores of water arranged on a hexagonal lattice. The inverse hexagonal phase consist of aggregates arranged as cylinders in a continuous matrix consisting of long rod-like micelles lying parallel to each other in a hexagonal array, and the long-range order is two-dimensional. These systems are anisotropic under polarizing light microscopy [1, 3, 6, 9].

Another mesophase that can be formed by amphiphilic molecules is the sponge phase (L3), which is obtained in the presence of solvents. This mesophase is obtained and characterized in liquid crystalline systems formed by GMO when a solvent is added to a GMO/water cubic phase in sufficient amounts for the solvent molecules to slightly penetrate the lipid bilayer, making it more flexible and flat. The solvents used to prepare sponge phases can be propylene glycol (PG), dimethyl sulfoxide (DMSO), ethanol, *N*-methyl alpha pyrrolidone (NMP) and polyethylene glycol (PEG). The internal structure of this phase is one congruent lipid bilayer that extends in three dimensions and is surrounded on both sides by a mixture of water and solvent. This mesophase is described as a molten cubic phase because it shares the properties of a bicontinuous cubic phase but does not exhibit a long-arranged order [1, 27, 28].

12.5 Identification of Mesophases

Liquid crystalline phases can be identified by several techniques, such as polarizing light microscopy, differential scanning calorimetry (DSC), small-angle X-ray scattering (SAXs), low-frequency dielectric spectroscopy, nuclear magnetic resonance (NRM), Raman scattering spectroscopy, cryo-transmission electron microscopy (Cryo-TEM), or cryo-field emission scanning electron microscopy (Cryo-FESEM) [4, 9].

The simplest technique to identify liquid crystals is polarizing light microscopy, which identifies the mesophases based on the optical birefringence of the systems. Because of the anisoptropic molecular arrangement of the lamellar and hexagonal phases, they can be easily identified by specific patterns under polarizing light microscopy. The lamellar phase is identified by the presence of oily streaks and Maltese crosses, whereas the hexagonal phase exhibits a fan-shaped or cone-shaped mosaic. Typical photomicrographs of lamellar and hexagonal phases are shown in Fig. 12.2. Nevertheless, the cubic phase is isotropic under polarizing light microscopy, i.e., its molecular arrangement is nonbirefringent, showing a dark field under a polarizing light microscope. Polarizing light microscopy is a simple and useful method to identify liquid crystalline systems, but it only indicates the phase of the liquid crystal. Consequently, other methods are necessary to confirm the crystallographic structure of the systems [4, 9].

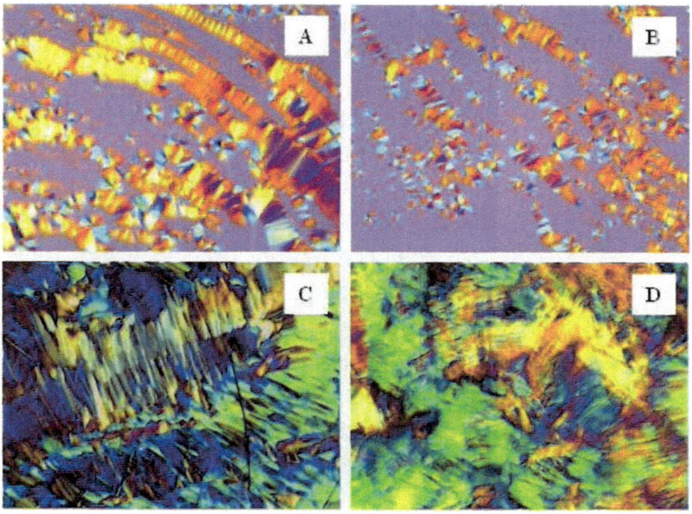

Figure 12.2 Representative photomicrographs of liquid crystalline systems of GMO and water (A and B lamellar phase) and GMO, water and oleic acid (C and D hexagonal phase).

12.6 Factors That Affect the Formation of Liquid Crystals

The formation of liquid crystalline phases by a polar lipid and water is affected by several factors, such as the molecular structure of the lipid molecule, water content of the system, temperature, pH, pressure, and addition of a third substance (drugs or additives) [2, 15, 21].

Several phase diagrams describe the mesophases formed under certain conditions of temperature and water contents, mainly for GMO/water systems, which are the most studied liquid crystalline systems proposed for drug delivery. Briefly, the phase behavior of GMO can be described as micellar (reverse micelles L2) at low water contents. As the water content increases, a phase transition to the lamellar phase is observed. As the hydrocarbon chain disorder increases due to either heating or increases in the water content, there is a transition from the lamellar to a cubic phase, which is stable in an excess of water. GMO is known as a lipid that exhibits a wide cubic phase region in the phase diagram, including gyroid and diamond cubic phase structures. Therefore, this lipid is a very interesting material for drug delivery purposes. A reverse hexagonal phase (HII) is found at higher temperatures (higher than 80°C). Therefore, the water content and temperature are determining factors in the formation of liquid crystalline mesophases [4–7, 26, 29].

To form a liquid crystal, amphiphilic molecules arrange themselves to minimize the free energy of the system when in contact with water, exposing the hydrophilic regions to the aqueous environment and packing the hydrophobic domains to minimize interfacial forces [3]. A lamellar phase is usually obtained at low water contents and consists of planar bilayers separated by layers of water [4, 9]. These bilayers present no curvature (zero curvature), although nonlamellar phases are characterized by a nonzero curvature of the lipid bilayer. A lipid bilayer tends to curl to relieve stress within the layers, and this tendency depends on the composition and interactions between the lipids and water [6, 21]. The presence of unsaturations in the molecule is also reported to favor the formation of nonlamellar phases by increasing the tendency of a bilayer to curve.

The molecular geometry of the forming lipid plays an important role in the determination of phase behavior [2, 10]. The effect of molecular shape on interfacial curvature has been described in terms of molecular packing using the concept of the crucial packing parameter (CPP), which is defined as CPP = $v/a_0 l$, where v is the hydrophobic chain volume, a_0 is the cross-sectional area of the surfactant headgroup and l is the hydrophobic chain length. The CPP is useful to predict the phases that are preferentially formed by a given lipid in a liquid crystalline system. When CPP = 1, a lamellar phase is favored, whereas oil-in-water structures, such as normal micelles and normal cubic and hexagonal phases (HI), are formed when CPP is smaller than 1. When CPP is higher than 1, water in oil structures are formed, such as reverse cubic phase, reverse hexagonal phases (HII), and reversed micelles [10, 12, 29].

Temperature also influences the formation of mesophases because the increase in temperature increases chain disorder, which leads to a phase transition. Temperature and pressure, as well as the salt concentration and pH value, were reported to induce phase transitions between cubic and hexagonal phases.

Phase transitions from a lamellar to a cubic phase were observed at low concentrations of Ca^{2+} [2], whereas the formation of a hexagonal phase was favored at low pH values [30].

The presence of compounds other than the lipid and water in a liquid crystalline system can influence the packing parameter of the lipid and consequently the formation of the liquid crystalline phases [2, 31]. These compounds may be drugs added to the systems to be delivered or additives added with a particular aim, such as permeation enhancers, viscosity modifiers, bioadhesive polymers, and others. Thus, the presence of drugs or additives can influence the structure of the mesophases and induce phase transitions, which can also affect the drug release profiles. Phase transitions in GMO liquid crystalline systems have been observed with several drugs, such as lidocaine, chlorpheniramine maleate, diltiazem, propranolol HCl, and vitamin K [2, 7, 32, 33, 38]. The polarity and molecular structure of the additive determine its location inside the liquid crystalline structure (location at the polar region or the nonpolar domain of the lipid bilayer), which may influence phase behavior. Therefore, the additive affects the interfacial curvature of the systems in different manners, which can lead to different phase transitions. Generally, the addition of hydrophilic drugs favors a lamellar phase, whereas lipophilic additives induce phase transitions from cubic to hexagonal phases [7, 29, 34]. Hydrophobic additives dissolve in the lipophilic domain and increase the hydrocarbon chain volume, which increases the CPP and consequently favors the formation of nonlamellar phases [12, 34]. Hydrophilic additives decrease the availability of water to the mesophase and favors lamellar phases [34]. For example, the addition of oleic acid to cubic phase systems induced a transition to a hexagonal phase. Propylene glycol and polyethylene glycol favor the formation of lamellar and cubic phases and, at certain concentrations, may form a sponge phase [20, 27–29, 31, 35].

Additives can also change the transition temperatures of liquid crystalline systems, and this approach can be used to manipulate the phase structure exhibited by systems and improve their properties as drug delivery systems. The addition of hydrophobic molecules, such as oleic acid, oils, or co-surfactants, to liquid crystalline systems promotes phase transitions to obtain hexagonal phases, which could be obtained only at high temperatures. This approach allows the use of hexagonal phases as drug delivery systems by overcoming the hindering high temperatures typically required to obtain this mesophase [3]. Hexagonal phases are interesting for skin drug delivery because they are less viscous than cubic phase and can be easily applied on skin. The addition of vitamin E acetate to phytantriol bulk phases alters the transition temperatures of the systems and the phase transitions [22]. The addition of amphiphilic additives can alter the transitions temperatures between a highly diffusive (cubic phase) and a less diffusive mesophase (reverse hexagonal, lamellar), which presents different release properties. A temperature-induced transition from cubic to hexagonal can be obtained and therefore, drug release can be modulated at physiological temperatures. Switching the applied temperature can induce phase transitions and consequently control the release rate of drugs [3].

Additives also can be used to alter the viscosity of liquid crystalline systems. An example of this approach is the use of solvents to reduce the viscosity of cubic phase systems. Specifically, cubic phases produce very rigid, stiff gels, which can complicate their practical applicability and limit their use as drug delivery systems. The addition of solvents, such as ethanol, PG, or PEG 400, influences the phase diagrams and reduces the viscosity of the systems, forming soft gels whose flow properties make them easier to spread. This behavior is desirable for some routes of administration. Considering topical application of these systems, the resultant soft gels retain a liquid crystalline structure and can be easily spread on the skin. For other routes of administration, such as injectable or mucosal drug delivery, this strategy can be used to develop liquid crystalline precursors that can undergo a rapid transition to the highly viscous cubic phase at the site of application via water absorption [7, 15, 36–38].

12.7 Liquid Crystals as Drug Release Systems

Liquid crystalline systems can be used as carriers for several drugs and can provide a slow drug release matrix for different routes of administration [2, 9, 10]. The presence of hydrophilic and hydrophobic domains in the liquid crystalline structure allows the incorporation of hydrophilic, lipophilic, and amphiphilic drugs. Hydrophilic drugs will be located in the aqueous domain, whereas lipophilic drugs will be loaded in the lipid bilayer, and amphiphilic drugs will be accommodated close to the interface (Fig. 12.3) [2, 9, 10].

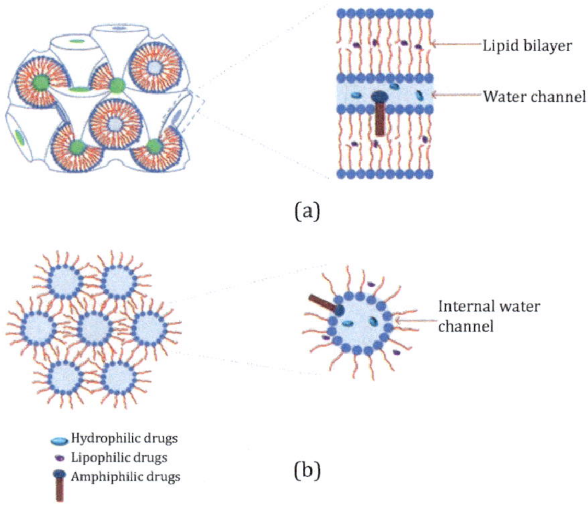

Figure 12.3 Structures of (a) reversed bicontinuous cubic and (b) hexagonal phases and possible localization of drugs. For simplicity, only parcial lattice is represented. Reproduced with permission from Chen *et al.*, 2014 [2].

The release rate is controlled by several factors related to the physicochemical properties of drug and the release properties of the system. The proposed mechanism for drug release is the diffusion of drug in the aqueous channels of the liquid crystalline structure, which are considered the principal route for a drug to be released. The diffusion of a drug in the liquid crystalline system can be hindered by the physical restriction of the drug motion due to the complex internal structure of the systems and/or interactions with the systems, resulting in a slow release [7, 13, 15].

Many factors influence drug release, such as the type of liquid crystalline phase, initial water content, swelling capacity, interactions between drugs and lipid bilayers, drug solubility, diffusion coefficient, drug load, temperature, pH, and ionic strength of the dissolution media [2, 15, 39]. Drug release depends on the physicochemical properties of the drug, which will define the location of the drug inside the system. Because drugs are released by diffusion in the aqueous channels of the cubic phase, the drug will be readily released if it resides in these channels. However, if a drug is very lipophilic and located in the lipid bilayer, partitioning into the aqueous channels may become the rate-limiting step. Therefore, different drugs may be released from liquid crystalline systems in different manners, depending on their chemical nature, dimensions, solubilities, and location of the drug inside the systems [7, 37].

The release of drugs from liquid crystalline systems also significantly depends on their mesophase structure [3]. Therefore, all above-discussed factors that affect phase transitions will influence drug release. The release of both hydrophilic and hydrophobic drugs use to be greater from the cubic phase than the hexagonal phase due to differences in the aqueous domains of theses mesophases. In cubic phases, the aqueous channels are likely open to the external media, whereas the aqueous compartments are closed extended micellar columnar structures in the hexagonal phase. Therefore, the aqueous domain has easier access to the external solution in a cubic phase, which favors drug release [8, 15].

The release profiles of sodium diclofenac and celecoxib from the cubic and lamellar phases of GMO/ethanol/water systems have been studied. Specifically, the cubic phase released larger amounts of drug than the lamellar phase, indicating that sodium diclofenac release depended on the mesophase structure. However, celecoxib release was not affected by the mesophase structure because both the cubic and lamellar phases presented similar drug release profiles. These differences can be attributed to the diffusion pathways of drugs: Whereas sodium diclofenac likely diffuses through the aqueous channels, celecoxib seems to diffuse through the lipid continuous region [37].

However, factors other than the mesophase of the system may account for drug release. Drug release from liquid crystalline systems may also be influenced by the material that comprises the system. Systems prepared with different lipids but presenting the same mesophase can provide different release profiles for the same drug. This difference can be attributed to differences in the arrangement of lipids, which may vary in terms of curvature, the lattice parameter of liquid crystalline structures and possible hydrophobic interactions between drug molecules and the lipid bilayer. These factors can affect drug diffusion through the systems and consequently influence drug release [13, 15]. Interactions between drug molecules and lipid bilayer via hydrogen bonding or ion-pairing have been observed for several drugs and are related to an incomplete release of some drugs from liquid crystalline systems. These interactions involve specific interactions between drug molecules and lipid components that may account for the incorporation of the drug into the lipid domain of the system and influence drug release [4, 15, 33].

Importantly, mesophases presented by liquid crystalline systems are related to the water content of this system. As mentioned above, lamellar phases are favored at low water contents. In the presence of excess of water, liquid crystalline system matrices absorb water, swell, and go on a phase transition to a cubic phase. When swelling is significant, it can also influence drug release because the drug will be released by simultaneous processes of matrix swelling (water migration into the systems and increase in the diffusional path of the drug) and drug diffusion through the matrix [36, 39].

The influence of the initial water content of liquid crystalline systems on their release properties is controversial. Some investigations showed that the initial water content directly correlates with the rate of drug release due the increased degree of hydration of the matrices. This phenomenon was observed for the release of several drugs (salicylic acid [39], the peptidic drug (DADLE) [40], celecoxib [37]). In matrices exhibiting a higher initial water content, the hydrophilic channels available for the release of drugs are expanded, which accelerates diffusion and release compared with less swollen matrices [8]. Additives can also hydrate the lipid headgroups and swells the water channels [41]. The addition of hydration-enhancing surfactants enhances the hydrophilic domain, which can control the dimensions of water channels and is correlated with the drug diffusion rates [3]. However, some investigations found that the initial water content did not affect the release rates due to the rapid formation of a cubic phase [8, 42]. In the presence of excess of water, liquid crystalline matrices rapidly absorb water to form a cubic phase. Because this phase transition is very rapid, the incorporated drug will be released primarily from the cubic phase formed after water absorption, resulting in similar release profiles independent of the initial water content [7, 8]. Given these different results, the influence of the initial water content on drug release likely depends on the drug properties, their diffusion pathway, and the presence of significant drug partition in the lipid bilayer [37, 39]. Drug load can also affect drug release from liquid crystalline systems, but this influence also depends on the drug properties. Specifically, drug release may or may not depend on the drug load for various studied drugs [39, 43, 44].

The pH value of the surrounding media can also influence drug release because it influences drug solubility and ionization. Changes in the pH values can alter drug solubility in the dissolution media, which may affect the drug release process. The pH value also affects drug ionization and consequently will account for the degree of possible ionic interactions between charged drugs and the lipid bilayer, which can lead to the binding of the drug to the mesophases, incorporation of drug into the lipid bilayer, and formation of less soluble ion-pairs. Consequently, these interactions may influence drug release [4, 30, 33, 44].

12.8 Liquid Crystals and Skin Permeation

Liquid crystals should be proposed as vehicles for topical/transdermal drug delivery for several reasons. As mentioned before, these systems present several properties that allow them to be used as drug delivery systems, such as thermodynamic stability, lipid and aqueous domains (which can dissolve drugs with different polarities), and the ability to improve the solubility of poorly water-soluble drugs. Their complex structure can also provide controlled drug release, and the aforementioned factors that affect drug release can also influence skin permeation by altering the amount of drug available to permeate skin. In addition, their particular structure favors interaction with skin lipids, which can improve drug permeation [2, 9, 10].

The stratum corneum is considered to be the rate-limiting barrier in transdermal drug delivery. Lipids form lamellar structures in the intercellular spaces of the stratum corneum, which contributes to the barrier properties of the skin layer [17, 45]. Liquid crystals present a structural organization very similar to that found in the lipid components of the stratum corneum, which would provide the high affinity of these systems for skin. The interaction of the liquid crystalline phases with the intercellular lipid packing of the stratum corneum results in a disruption of the lamellar structure of the lipid bilayer in this skin layer. Of particular interest for this route of administration are cubic and hexagonal phases, which are the most studied mesophases for skin drug delivery. Cubic and hexagonal phases can reportedly penetrate the stratum corneum due to their structural similarity with skin lipids. Therefore, these systems are interesting vehicles for topical/transdermal drug delivery. This disruption leads to more fluid and permeable lipid packing, which facilitates drug permeation [10, 20]. Liquid crystalline systems promote a particular distribution of drugs when applied on skin. This distribution differs from those obtained with common vehicles used for topical drug delivery. The distribution of a fluorescent model drug in human skin was studied in four different formulations (water, ointment, and cubic phases formed by either GMO or PT), and the fluorescence was measured by two-photon microscopy. This fluorescence correlated with the local concentration of drug, which facilitates the evaluation of drug penetration and distribution inside the skin. The distribution pattern when using cubic phases differed from those of common ointment and water. The intercellular pathway for drug penetration seems to be predominant when using water vehicle and ointment, whereas an intercluster pathway dominated for cubic phases. The fluorescence from cubic phase systems mainly localized to micro-fissures caused by the microscopic clustering of keratinocytes in the skin. These findings indicate that cubic phases were able to penetrate into micro-fissures. This route contributes to drug delivery and clearly shows that the penetration pathway of the drug into skin differs between cubic phases and water or ointments vehicles. Cubic phase systems also yield threadlike patterns of fluorescence in inner skin layers, whereas a more homogeneous pattern was found for water vehicles. The threadlike pattern observed for cubic phase systems indicates the diffusion of the formulation into the lipid matrix between cell layers, which suggests interactions between the lipid bilayers of the liquid crystals and cellular lipid matrix [46].

Another interesting feature of liquid crystalline systems regarding skin drug delivery is that some commonly used lipid components, such as GMO and PT, are recognized as permeation enhancers. Thus, the lipid that comprises the lipid domain of the liquid crystalline system can improve drug permeation when in contact with skin by increasing stratum corneum fluidity and extracting lipids. In addition to structural similarity with skin lipids, the presence of these lipids with permeation enhancer properties result in synergistic effects that improve the permeation of skin by the drug [9, 10, 20, 32]. Liquid crystalline systems also allow the incorporation of others permeation enhancers, such as oleic acid or isopropyl myristate, which can be dissolved in the lipid domain of the liquid crystal structure. Oleic acid has been used as permeation enhancer in liquid crystalline systems for skin delivery of cyclosporine-A [20, 31]. It is also possible to incorporate other substances with the aim of improvement of drug permeation. Liquid crystalline drug delivery systems containing cell-penetrating peptides, such as penetratin, TAT, and RALA peptide, have been studied to enhance the penetration of nonsteroidal anti-inflammatory drugs (sodium diclofenac and celecoxib). Cell-penetrating peptides are a kind of peptides with functions of penetration into living cells that can be used to enhance delivery through skin in the development of drug delivery systems. They can overcome the stratum corneum barrier and improve drug permeation [41, 47, 48].

Liquid crystalline systems can be applied to the skin in several forms, such as bulk cubic or hexagonal phases, nanodipersions or gel formulations that contain liquid crystalline nanodispersions. Cubosomes and hexosomes have been proposed for topical and/or transdermal drug delivery due to the similarity of their inner structure and the epithelium cells as well as their good permeability. They present some advantages over bulk phase gels, such as high fluidity and a large superficial area of the dispersions in contact with skin, which might improve the interaction with skin. For example, nanodispersions of a hexagonal phase increased the delivery of vitamin K to skin compared with a hexagonal bulk gel [9, 10, 12, 31, 32].

The type of mesophase presented by the systems can also influence drug permeation. Different liquid crystalline structures present differences in the internal structure, viscosity, and drug release properties, which may account for differences in the skin permeation profiles. Therefore, developing drug delivery systems with appropriate mesophases that maximize drug diffusion and/or skin permeation is important to optimize skin drug delivery [20, 48].

The topical delivery of cyclosporine A was evaluated by applying cubic and hexagonal phases made of GMO or GMO/oleic acid, respectively. Both systems increased the skin penetration of cyclosporine A *in vitro* and *in vivo*. The cubic phase increased the drug penetration in the stratum corneum and (epidermis + dermis) but did not exhibit percutaneous delivery, whereas the reverse hexagonal phase increased drug penetration in (epidermis + dermis) and percutaneous delivery of the drug. Therefore, the hexagonal phase appeared to increase drug penetration in deeper skin layers and through skin. These differences must be due to the internal structure of the systems (cubic and hexagonal phase), that provide different drug release profiles. However, differences in the components of the systems (the presence of oleic acid in hexagonal phase) should also affect the permeation profiles [20].

The cubic and lamellar phases of GMO/ethanol and water mixtures containing a cell-penetrating peptide (TAT) as a permeation enhancer were evaluated. TAT-enhanced drug permeation from the cubic phases to a remarkable extent, although the effect on drug permeation from the lamellar phase was very small. The structures of cubic and lamellar phases can likely control the diffusion of the studied drugs. In addition, there was a binding of TAT and GMO in the lamellar phase, and the diffusion of the peptide was decreased within the lamellar system, but not in the cubic phase system. Different levels of drug and peptide released from cubic and lamellar phases lead to different permeation profiles [48].

The influence of the mesophase presented by the systems on skin drug delivery was also reported for the transdermal delivery of paclitaxel, which was studied using BRIJ-based liquid crystalline phases presenting hexagonal and lamellar mesophases. The hexagonal phase showed a lower release and cutaneous delivery of paclitaxel than the lamellar phase, which was attributed to the higher viscosity of the hexagonal phase, which hinders drug diffusion through the system [24].

Different mesophases clearly present different drug release properties. Therefore, the drug release will determine the amount of drug available to permeate skin and consequently, skin permeation. Thus, differences between drug permeation from different mesophases may be related to the drug release properties of each mesophase that yields a particular drug release profile. However, drug release is not the sole factor determining drug permeation, as skin interactions can also be the rate-limiting step for skin permeation. In this case, different mesophases may yield different drug release profiles and even present similar skin permeation rates. Lamellar and cubic phases presented similar permeation rates for sodium diclofenac, although cubic phase showed a higher amount of drug released. The explanation for this behavior is that, in this case, skin interactions have more influence in skin permeation than drug release [48].

The structure of drug molecules can also influence their release/permeation profiles from liquid crystalline systems. The *in vitro* skin permeation of diclofenac salts from liquid crystalline systems was studied, and different salt derivatives of diclofenac exhibited distinct permeation profiles. The different physicochemical properties of each salt may influence their permeability through skin due to an influence at the ability to diffuse through the system and possible chemical interactions with the lipids of the liquid crystalline structure, which influence drug release [13].

Additives can also be used to modify the properties of liquid crystalline systems and improve drug permeation. Additives can modify the mesophase presented by the systems, which, as described before, may affect drug release and skin permeation [20, 31]. The effect of additives on drug permeation may also be related to increases in drug release. The addition of co-solvents or surfactants may increase drug release due to an increase in drug solubility in the media [40]. The effect of different additives on the skin permeation of finasteride from a liquid crystalline system was studied and showed different influences depending on the molecule added to the system. Additives that affect the interfacial curvature of the systems (propylene glycol, glycerol, or polyethylene glycol) increased drug release by allowing water pores to swell. This effect resulted in increased drug permeation. The addition of surfactants can increase drug solubility and change the thermodynamic activity of drugs, and, thus, increase drug permeation [14]. Surfactants also improve drug permeation by accelerating drug release via the effect of the surfactants on the partitioning into the interfacial domain of the liquid crystals. In addition, higher concentrations of surfactants may increase the fluidity of the membrane and eventually increase drug permeation. In contrast, addition of oleic acid showed to reduce drug permeation, likely because this hydrophobic molecule increased the hydrophobic volume of the lipid domain and promoted a transition from cubic to hexagonal phase. Despite being a permeation enhancer, oleic acid decreased drug release and, due to its hydrophobicity, was retained in the lipid domain of the nanoparticles and not released to directly act on skin [49, 50]. The effect of additives on drug permeation may also be related to a direct effect on the skin. Specifically, additives may promote the disruption of the skin barrier and consequently improve drug permeation. The addition of monoglycerides with acyl chains of different lengths to lamellar phase systems affected the delivery of paclitaxel to skin and an inverse relationship between paclitaxel penetration and the acyl chain length of the monoglyceride was demonstrated. The effect of the monoglycerides was not related to an increased drug release but to a more pronounced disruption of the skin barrier [25].

Liquid crystalline systems may be modified with a defined aim as a strategy for topical delivery. Liquid crystalline nanoparticles whose surfaces have been modified by the binding of mucoadhesive polymer (chitosan) are proposed to enhance the topical delivery of drugs. Due to its positive charge, chitosan nanoparticles may act as a driving force for skin permeation of drugs entrapped [51].

Several studies have shown that liquid crystalline systems can improve the skin permeation of drugs. Table 12.1 presents several reports of the use of liquid crystalline systems for skin drug delivery. These studies describe the development of liquid crystalline systems in several forms (bulk liquid crystalline phases or nanodispersions) and the characterization of these systems by different techniques, such as polarizing light microscopy, small-angle X-ray scattering, Fourier transform infrared spectroscopy, cryogenic-transmission electron microscopy, rheological studies, *in vitro* drug release studies, *in vitro* drug permeation studies using Franz diffusion cells, evaluation of skin irritation, and *in vivo* drug activity. Most of these studies show that liquid crystalline systems can control drug diffusion, sustain drug release, and increase the drug permeability coefficient. Therefore, these systems improve the topical/transdermal delivery of several drugs. The mechanism of the enhanced drug permeation is attributed to possible interactions of the liquid crystalline systems with stratum corneum lipids and the disruption of its organized structure [10, 14, 25, 52, 53].

Table 12.1 Studies using liquid crystalline systems as drug delivery systems for skin applications

Type of system	Composition	Drug	Ref.
Bulk lamellar phase	Vitamin E TPGS/isopropyl myristate/propylene glycol	quercetin	[18]
Bulk lamellar phase	BRIJ/monocaprylin/ monomyristolein	paclitaxel	[25]
Bulk cubic phase	GMO/propylene glycol	capsaicin	[36]
Bulk cubic phase	GMO/water	paeonol	[52]
Bulk cubic phase	GMO/water	δ-aminolevulinic acid (ALA)	[54]
Bulk cubic phase	GMO or PT/water	δ-aminolevulinic acid (ALA)	[55]
Bulk hexagonal phase	GMO/tricaprylin/water + RALA	Sodium diclofenac	[41]
Bulk hexagonal phase	GMO/tricaprylin/water + penetratin	Sodium diclofenac	[47]
Bulk lamellar and cubic phases	GMO/ethanol/water + TAT	Sodium diclofenac/ celecoxib	[48]
Bulk hexagonal and cubic phases	GMO/water/oleic acid	cyclosporine A	[20]
Bulk lamellar and hexagonal phases	BRIJ/medium chain mono-diglycerides	paclitaxel	[24]
Bulk lamellar and cubic phases	GMO/ethanol/water	diclofenac salts	[13]
Bulk lamellar and cubic phases	GMO/water	celecoxib	[56]

Bulk sponge phase + iontophoresis	GMO/propylene glycol	δ-aminolevulinic acid (ALA)and methyl-esters	[28]
Cubosomes	GMO/Poloxamer 407	indomethacin	[53]
Cubosomes	GMO/Poloxamer 407	α-lipoic acid	[57]
Liquid crystalline nanoparticles	GMO/Poloxamer 407/surfactants	finasteride	[50]
Liquid crystalline nanoparticles	GMO/Poloxamer 407/additives	finasteride	[49]
Liquid crystalline nanodispersions (hexagonal phase)	GMO/oleic acid/ Poloxamer 407/ cationic polymers	siRNA	[58]
Liquid crystalline nanoparticles	GME/hydrogenated lecithin/1,3 butylene glycol/Poloxamer 407	flurbiprofen	[59]
Liquid crystalline nanoparticles	GMO/Poloxamer 407/ chitosan	finasteride/ dutasteride	[51]
Liquid crystalline nanodispersion (hexagonal phase)	GMO/oleic acid/ Poloxamer 407	cyclosporine A	[31]
Liquid crystalline nanoparticles and bulk hexagonal phase	GMO/water and GMO/water/ Poloxamer 407	vitamin K	[32]
Emulsion with liquid crystals	Oil/water/surfactants	itraconazole	[16]
Emulsion with liquid crystals	Oil/water/surfactants	hydroquinone/ salicylic acid	[14]
Bulk thermoresponsive gel	Cholesteryl cetyl carbonate/lauryl alcohol	indomethacin	[60]
Liquid crystalline pharmacogel (lamellar phase)	Synthetic prodrug	propranolol hydrochloride	[61]

The studies presented in Table 12.1 suggest that liquid crystalline systems are considered potential vehicles in topical drug delivery and cosmetic formulations [62].

Lamellar liquid crystals or liquid crystalline nanodispersions have been investigated for use in cosmetics and skin care products due to their similarity in structure to the intercellular lipids of the stratum corneum, which exhibit a lamellar arrangement. The structural similarity with lipids confers a high affinity to skin and enables the system to act as a specific modulator of water-retention in the stratum corneum [45, 62]. Liquid crystalline systems (bulk gels or nanodispersions) showed better skin care characteristics with regard to permeability, skin hydration, and skin occlusion as well as clinical benefits, such as a reduction in facial lines and improvements in skin color and texture without irritation or other adverse effects [45, 57, 62].

Skin delivery of drugs is proposed to optimize drug administration and reduce side effects associated with oral administration. Liquid crystalline systems can be considered a promising approach to deliver drugs to skin as they can sustain drug release and improve drug penetration in the skin layers [13, 56].

Several studies use liquid crystalline systems to deliver finasteride for the treatment of androgenic alopecia as a alternative strategy to avoid drug-related side effects [49–51]. In the same way, these systems are proposed for the delivery of anti-inflammatory drug for the treatment of severe pain and others inflammatory conditions. The chronic use of nonsteroidal anti-inflammatory drugs (NSAIDs) might be accompanied with severe side effects that limit their use by long-term oral administration. Nevertheless, other routes of administration such as skin delivery have been explored. Liquid crystalline systems could be a promising approach to delivering NSAIDs to the skin as they can sustain drug release, improve drug penetration into the skin layers, and minimize side effects [41, 47, 48, 56]. Another interesting feature of liquid crystalline systems is the potential for topical delivery of prodrugs and photosensitizers used in photodynamic therapy (PDT) with promising results. There are some difficulties related to the administration of these drugs by conventional systems such as chemical instability of drugs and limited transdermal and/or tumor penetration. Liquid crystalline systems are a strategy to overcome these difficulties and optimize PDT [28, 54, 55]. Liquid crystalline nanodispersions modified with cationic polymers have been proposed as carriers for the topical delivery of siRNA, with promising results. The topical delivery of siRNAs can modulate local gene expression in a variety of cutaneous disorders while avoiding systemic side effects. Nevertheless, the skin barrier is an obstacle to topical nucleic acid delivery and is a particular hindrance to the topical delivery of siRNAs. The aforementioned systems present interesting properties to overcome these obstacles, such as the presence of lipid-forming liquid crystals with permeation enhancer properties (such as GMO) that facilitate siRNA penetration into the skin and the ability to protect siRNAs from physical and enzymatic degradation. These systems enable the dermal delivery of siRNA and can form a deposit on the skin surface, resulting in prolonged release [58, 63, 64].

Novel liquid crystalline systems, such stimuli-responsive systems that undergo phase transitions upon different stimuli, have been explored. These transitions can modulate drug release in optimized conditions for a given route of administration. Several strategies are proposed to manipulate drug transport by different stimuli, such as pH-responsive systems, the use of ionic interactions to the control entrapment and release of drugs, the modulation of hydrophobic interactions between drug and lipid bilayers, the modulation of the mesophase water channel sizes, thermo-responsive systems, light-responsive nanosctructured mesophases, or the modulation of drug transport with externally applied magnetic fields [3]. These different attempts offer novel drug delivery systems for different routes of administration, and some of them can be used to improve systems designed for topical delivery.

References

1. Koynova R., Tenchov B. Recent patents on nonlamellar liquid crystalline lipid phases in drug delivery. *Recent Pat. Drug Deliv. Formul.*, 7 (2013), pp. 165–173.
2. Chen Y., Ma P., Gui S. Cubic and hexagonal liquid crystals as drug delivery systems. *BioMed Res. International.*, 2014 (2014), p. 815981.
3. Zabara A., Mezzenga R. Controlling molecular transport and sustained drug release in lipid-based liquid crystalline mesophases. *J. Control. Release,* 188 (2014), pp. 31–43.

4. Milak S., Zimmer A. Glycerol monooleate liquid crystalline phases used in drug delivery. *Int. J. Pharm.*, 478 (2015), pp. 569–587.
5. Clogston J., Rathman J., Tomasko D., Walker H., Caffrey M. Phase behavior of a monoacylglycerol: (Myverol 18-99 K)/water system. *Chem. Phys. Lipids.*, 107 (2000), pp. 191–220.
6. Tate M. W., Eikenberry E. F., Turner D. C., Shyamsunder E., Gruner S. M. Nonbilayer phases of membrane lipids. *Chem. Phys. Lipids,* 57 (1991), pp. 147–164.
7. Shah J. C., Sadhale Y., Chilukuri D. M. Cubic phase gels as drug delivery systems. *Adv. Drug Deliv. Rev.*, 47 (2001), pp. 229–250.
8. Rizwan S. B., Hanley T., Boyd B. J., Rades T., Hook S. Liquid crystalline systems of phytantriol and glyceryl monooleate containing a hydrophilic protein: Characterisation, swelling and release kinetics. *J. Pharm. Sci.*, 98 (2009), pp. 4191–4204.
9. Pan X., Han K., Peng X., Yang Z., Qin L., Zhu C., *et al.* Nanostructured cubosomes as advanced drug delivery system. *Curr. Pharm. Des.*, 19 (2013), pp. 6290–6297.
10. Guo C., Wang J., Cao F., Lee R. J., Zhai G. Lyotropic liquid crystal systems in drug delivery. *Drug Discov. Today,* 15 (2010), pp. 1032–1040.
11. Amar-Yuli I., Wachtel E., Shoshan E. B., Danino D., Aserin A., Garti N. Hexosome and hexagonal phases mediated by hydration and polymeric stabilizer. *Langmuir,* 23 (2007), pp. 3637–3645.
12. Hirlekar R., Jain S., Patel M., Garse H., Kadam V. Hexosomes: A novel drug delivery system. *Curr. Drug Deliv.*, 7 (2010), pp. 28–35.
13. Yariv D., Efrat R., Libster D., Aserin A., Garti N. *In vitro* permeation of diclofenac salts from lyotropic liquid crystalline systems. *Colloids Surf. B: Biointerfaces,* 78 (2010), pp. 185–192.
14. Otto A., Wiechers J. W., Kelly C. L., Dederen J. C., Hadgraft J., du Plessis J. Effect of emulsifiers and their liquid crystalline structures in emulsions on dermal and transdermal delivery of hydroquinone, salicylic acid and octadecenedioic acid. *Skin Pharmacol. Physiol.,* 23 (2010), pp. 273–282.
15. Boyd B. J., Whittaker D. V., Khoo S.-M., Davey G. Lyotropic liquid crystalline phases formed from glycerate surfactants as sustained release drug delivery systems. *Int. J. Pharm.*, 309 (2006), pp. 218–226.
16. Nesseem D. I. Formulation and evaluation of itraconazole via liquid crystal for topical delivery system. *J. Pharm. Biomed. Anal.*, 26 (2001), pp. 387–399.

17. Shaefer H., Redelmeier T. E. Factors effecting percutaneous absorption. In: *Skin Barrier. Principles of Percutaneous Absorption* (Shaefer H., Redelmeier T. E., eds.), Karger AG (1996), pp. 153–212.

18. Vicentini F. T. M. C., Casagrande R., Verri Jr W. A., Georgetti S. R., Bentley M. V. L. B., Fonseca M. J. V. Quercetin in lyotropic liquid crystalline formulations: Physical, chemical and functional stability. *AAPS PharmSciTech*, 9 (2008), pp. 591–596.

19. Phan S., Fong W. K., Kirby N., Hanley T., Boyd B. J. Evaluating the link between self-assembled mesophase structure and drug release. *Int. J. Pharm.*, 421 (2011), pp. 176–182.

20. Lopes L. B., Lopes J. L. C., Oliveira D. C. R., Thomazini J. A., Garcia M. T. J., Fantini M. C. A., *et al.* Liquid crystalline phases of monoolein and water for topical delivery of cyclosporin A: Characterization and study of *in vitro* and *in vivo* delivery. *Eur. J. Pharm. Biopharm.*, 63 (2006), pp. 46–55.

21. Fong C., Le T., Drummond C. J. Lyotropic liquid crystal engineering-ordered nanostructured small molecule amphiphile self-assembly materials by design. *Chem. Soc. Rev.*, 41 (2012), pp. 1297–1322.

22. Dong Y. D., Larson I., Hanley T., Boyd B. J. Bulk and dispersed aqueous phase behavior of phytantriol: Effect of vitamin E acetate and F127 polymer on liquid crystal nanostructure. *Langmuir*, 22 (2006), pp. 9512–9518.

23. Boyd B. J., Whittaker D. V., Khoo S.-M., Davey G. Hexosomes formed from glycerate surfactants—Formulation as a colloidal carrier for irinotecan. *Int. J. Pharm.*, 318 (2006), pp. 154–162.

24. Hosmer J. M., Shin S. H., Nornoo A., Zheng H., Lopes L. B. Influence of internal structure and composition of liquid crystalline phases on topical delivery of paclitaxel. *J. Pharm. Sci.*, 100 (2011), pp. 1444–1455.

25. Hosmer J. M., Steiner A. A., Lopes L. B. Lamellar liquid crystalline phases for cutaneous delivery of paclitaxel: Impact of the monoglyceride. *Pharm. Res.*, 30 (2013), pp. 694–706.

26. Qiu H., Caffrey M. The phase diagram of the monoolein/water system: Metastability and equilibrium aspects. *Biomaterials*, 21 (2000), pp. 223–234.

27. Alfons K., Engstrom S. Drug compatibility with the sponge phases formed in monoolein, water, and propylene glycol or poly(ethylene glycol). *J. Pharm. Sci.*, 87 (1998), pp. 1527–1530.

28. Merclin N., Bender J., Sparr E., Guy R. H., Ehrsson H., Engström S. Transdermal delivery from a lipid sponge phase–iontophoretic and passive transport *in vitro* of 5-aminolevulinic acid and its methyl ester. *J. Control. Release,* 100 (2004), pp. 191–198.

29. Caboi F., Amico G. S., Pitzalis P., Monduzzi M., Nylander T., Larsson K. Addition of hydrophilic and lipophilic compounds of biological relevance to the monoolein/water system. I. Phase behavior. *Chem. Phys. Lipids,* 109 (2001), pp. 47–62.

30. Sallam A. S., Khalil E., Ibrahim H., Freij I. Formulations of an oral dosage form utilizing the properties of cubic liquid crystalline phases of glyceryl monooleate. *Eur. J. Pharm. Biopharm.,* 53 (2002), pp. 343–352.

31. Lopes L. B., Ferreira D. A., de Paula D., Garcia M. T. J., Thomazini J. A., Fantini M. C. A., Bentley M. V. L. B. Reverse hexagonal phase nanodispersion of monoolein and oleic acid for topical delivery of peptides: *In vitro* and *in vivo* skin penetration of cyclosporine A. *Pharm. Res.,* 23 (2006), pp. 1332–1342.

32. Lopes L. B., Speretta F. F., Bentley M. V. L. B. Enhancement of skin penetration of vitamin K using monoolein-based liquid crystalline systems. *Eur. J. Pharm. Sci.,* 32 (2007), pp. 209–215.

33. Chang C. M., Bodmeier R. Binding of drugs to monoglyceride-based drug delivery systems. *Int. J. Pharm.,* 147 (1997), pp. 135–142.

34. Shah M. H., Paradkar A. Effect of HLB of additives on the properties and drug release from the glyceryl monooleate matrices. *Eur. J. Pharm. Biopharm.,* 67 (2007), pp. 166–174.

35. Bender J., Jarvoll P., Nydén M., Engström S. Structure and dynamics of a sponge phase in the methyl delta-aminolevulinate/monoolein/water/propylene glycol system. *J. Colloid Interface Sci.,* 317 (2008), pp. 577–584.

36. Peng X., Wen X., Pan X., Wang R., Chen B., Wu C. Design and *in vitro* evaluation of capsaicin transdermal controlled release cubic phase gels. *AAPS PharmSciTech,* 11 (2010), pp. 1405–1410.

37. Cohen-Avrahami M., Shames A. I., Ottaviani M. F., Aserin A., Garti N. On the correlation between the structure of lyotropic carriers and the delivery profiles of two common NSAIDs. *Colloids Surf. B Biointerfaces,* 122 (2014), pp. 231–240.

38. Chang C.-M., Bodmeier R. Low viscosity monoglyceride-based drug delivery systems transforming into a highly viscous cubic phase. *Int. J. Pharm.,* 173 (1998), pp. 51–60.

39. Lara M. G., Bentley M. V. L. B., Collett J. H *In vitro* drug release mechanism and drug loading studies of cubic phase gels. *Int. J. Pharm.*, 293 (2005), pp. 241–250.

40. Lee J., Kellaway I. W. Combined effect of oleic acid and polyethylene glycol 200 on buccal permeation of [D-Ala2, D-Leu5]enkephalin from a cubic phase of glyceryl monooleate. *Int. J. Pharm.*, 204 (2000), pp. 137–144.

41. Cohen-Avrahami M., Aserin A., Garti N. HII mesophase and peptide cell-penetrating enhancers for improved transdermal delivery of sodium diclofenac. *Colloids Surf. B Biointerfaces*, 77 (2010), pp. 131–138.

42. Burrows R., Collett J. H., Attwood D. The release of drugs from monoglyceride-water liquid crystalline phases. *Int. J. Pharm.*, 111 (1994), pp. 283–293.

43. Geraghty P. B., Attwood D., Collett J. H., Dandiker Y. The *in vitro* release of some antimuscarinic drugs from monoolein/water lyotropic liquid crystalline gels. *Pharm. Res.*, 13 (1996), pp. 1265–1271.

44. Chang C.-M., Bodmeier R. Effect of dissolution media and additives on the drug release from cubic phase delivery systems. *J. Control. Release*, 46 (1997), pp. 215–222.

45. Iwai H., Fukasawa J., Suzuki T. A liquid crystal application in skin care cosmetics. *Int. J. Cosmet. Sci.*, 20 (1998), pp. 87–102.

46. Bender J., Simonsson C., Smedh M., Engström S., Ericson M. B. Lipid cubic phases in topical drug delivery: Visualization of skin distribution using two-photon microscopy. *J. Control. Release*, 129 (2008), pp. 163–169.

47. Cohen-Avrahami M., Libster D., Aserin A., Garti N. Penetratin-induced transdermal delivery from HII mesophases of sodium diclofenac. *J. Control. Release*, 159 (2012), pp. 419–428.

48. Cohen-Avrahami M., Shames A. I., Ottaviani M. F., Aserin A., Garti N. HIV-TAT enhances the transdermal delivery of NSAID drugs from liquid crystalline mesophases. *J. Phys. Chem. B*, 118 (2014), pp. 6277–6287.

49. Madheswaran T., Baskaran R., Thapa R. K., Rhyu J. Y., Choi H. Y., Kim J. O., *et al.* Design and *in vitro* evaluation of finasteride-loaded liquid crystalline nanoparticles for topical delivery. *AAPS Pharm Sci Tech*, 14 (2013), pp. 45–52.

50. Madheswaran T., Baskaran R., Yong C. S., Yoo B. K. Enhanced topical delivery of finasteride using glyceryl monooleate-based liquid crystalline nanoparticles stabilized by cremophor surfactants. *AAPS PharmSciTech*, 15 (2014), pp. 44–51.

51. Madheswaran T., Baskaran R., Sundaramoorthy P., Yoo B. K. Enhanced skin permeation of 5α-reductase inhibitors entrapped into surface-modified liquid crystalline nanoparticles. *Arch. Pharm. Res.*, 38 (2015), pp. 534–542.

52. Luo M., Shen Q., Chen J. Transdermal delivery of paeonol using cubic gel and microemulsion gel. *Int. J. Nanomed.*, 6 (2011), pp. 1603–1610.

53. Esposito E., Cortesi R., Drechsler M., Paccamiccio L., Mariani P., Contado C., *et al.* Cubosome dispersions as delivery systems for percutaneous administration of indomethacin. *Pharm. Res.*, 22 (2005), pp. 2163–2173.

54. Turchiello R. F., Vena F. C. B., Maillard P., Souza C. S., Bentley M. V. B. L., Tedesco A. C. Cubic phase gel as a drug delivery system for topical application of 5-ALA, its ester derivatives and m-THPC in photodynamic therapy (PDT). *J. Photochem. Photobiol. B Biol.*, 70 (2003), pp. 1–6.

55. Bender J., Ericson M. B., Merclin N., Iani V., Rosén A., Engström S., *et al.* Lipid cubic phases for improved topical drug delivery in photodynamic therapy. *J. Control. Release*, 106 (2005), pp. 350–360.

56. Estracanholii E. A., Praça F. S., Cintra A. B., Pierre M. B., Lara M. G. Liquid crystalline systems for transdermal delivery of celecoxib: *In vitro* drug release and skin permeation studies. *AAPS Pharm Sci Tech*, 15 (2014), pp. 1468–1475.

57. Sherif S., Bendas E. R., Badawy S. The clinical efficacy of cosmeceutical application of liquid crystalline nanostructured dispersions of alpha lipoic acid as anti-wrinkle. *Eur. J. Pharm. Biopharm.*, 86 (2014), pp. 251–259.

58. Vicentini F. T. M. D. C., Depieri L. V., Polizello A. C. M., Del Ciampo J. O., Spadaro A. C. C., Fantini M. C. A., *et al.* Liquid crystalline phase nanodispersions enable skin delivery of siRNA. *Eur. J. Pharm. Biopharm.*, 83 (2013), pp. 16–24.

59. Uchino T., Murata A., Miyazaki Y., Oka T., Kagawa Y. Glyceryl monooleyl ether-based liquid crystalline nanoparticles as a transdermal delivery system of flurbiprofen: Characterization and *in vitro* transport. *Chem. Pharm. Bull.*, 63 (2015), pp. 334–340.

60. Aeinleng N., Songkro S., Noipha K., Srichana T. Physicochemical performances of indomethacin in cholesteryl cetyl carbonate liquid crystal as a transdermal dosage. *AAPS PharmSciTech*, 13 (2012), pp. 513–521.

61. Namdeo A., Jain N. K. Liquid crystalline pharmacogel based enhanced transdermal delivery of propranolol hydrochloride. *J. Control. Release*, 82 (2002), pp. 223–236.

62. Yamada K., Yamashita J., Todo H., Miyamoto K., Hashimoto S., Tokudome Y., *et al.* Preparation and evaluation of liquid-crystal formulations with skin-permeation-enhancing abilities for entrapped drugs. *J. Oleo. Sci.,* 60 (2011), pp. 31–40.

63. Zhen G., Hinton T. M., Muir B. W., Shi S., Tizard M., McLean K. M., Hartley P. G. Glycerol monooleate-based nanocarriers for siRNA delivery *in vitro. Mol. Pharm.,* 9 (2012), pp. 2450–2457.

64. Vicentini F. T., Borgheti-Cardoso L. N., Depieri L. V., de Macedo Mano D., Abelha T. F., Petrilli R., Bentley M. V. Delivery systems and local administration routes for therapeutic siRNA. *Pharm. Res.,* 30 (2013), pp. 915–913.

Chapter 13

Cyclodextrins and Skin Disorders: Therapeutic and Cosmetic Applications

Oluwatomide Adeoye,[a,b] **Ana Figueiredo,**[a] **and Helena Cabral Marques**[a]

[a]*Research Institute for Medicines (iMed.ULisboa), Faculdade de Farmácia da Universidade de Lisboa, Avenida Prof. Gama Pinto, 1649-003 Lisboa, Portugal*
[b]*Faculty of Pharmacy, Obafemi Awolowo University, Ile-Ife, Nigeria*

hcmarques@ff.ulisboa.pt

13.1 Introduction

Skin disorders currently affect millions of people worldwide with physical and often traumatic psychological implications for quality of life [1, 2]. The primitive topical application of mud, urine, oils, plant extracts and animal products gives evidence to the knowledge of ancient societies on the need to maintain a healthy skin and treat skin disorders [3, 4]. This knowledge has evolved with the convergence of clinical and scientific research and development to produce an overlap between traditional pharmaceutical approaches and cosmetic principles [3]. Many polymers with mainstream pharmaceutical and dermatological applications are being used in cosmetic science. One of such

Carrier-Mediated Dermal Delivery: Applications in the Prevention and Treatment of Skin Disorders
Edited by Andreia Ascenso, Sandra Simões, and Helena Ribeiro

ISBN 978-981-4745-58-1 (Hardcover), 978-1-315-36447-6 (eBook)
www.panstanford.com

polymers is a group of cyclic oligosaccharides called cyclodextrins (CyDs). They are made up of six to eight dextrose units linked by glycosidic bonds. Today, many dermal products formulated with CyD are available for the prevention and treatment of skin disorders. Classifying these products as either cosmetic or therapeutic is often problematic, and complicated by the aetiology and description of many skin disorders. However, we know that many skin disorders will benefit from the use of cosmetics products and do not necessarily require dermatological interventions with dermal drugs. Many skin diseases can also be prevented by the use of cosmetic products [5]. The controversial definition of cosmeceuticals as a broad range of product between cosmetics and drugs has not resolved this classification problem, since many pharmaceutical and clinical scientists still consider dermal products as either cosmetic or pharmaceutical [3, 6, 7].

Considering this overlap, and within the context of this book, this chapter aims to provide an overview of CyDs in the prevention and treatment of skin disorders from both the cosmetic and therapeutic perspective. It will describe CyDs and their application as drug carriers for dermal therapeutics and also review the bases for its numerous cosmetic uses in the prevention of skin disorders and/or maintenance of a healthy skin.

13.2 Cyclodextrins: Historical Background and Description

Cyclodextrins were first isolated and reported by the French scientist Villiers in 1891, as the product of bacteria (*Bacillus amylobacter*)-induced degradation of starch. Villiers named this product "cellulosine" because of its similarity to cellulose, and described its non-reducing and acid hydrolysis resistant properties [8]. During the first decade of the 20th century, Schardinger initiated the study of CyDs chemistry (Fig. 13.1). He described them as cyclic oligosaccharides and identified both alpha- (α-CyD) and beta-cyclodextrin (β-CyD) [9–11]. By the 1950s, Freudenberg and co-workers had chemically identified CyD molecules as repeating units of glucopyranose linked by

α-(1,4) glycosidic bonds in a ring formation. They also isolated gamma-cyclodextrin (γ-CyD) and successfully synthesized the first pure CyD [11–17].

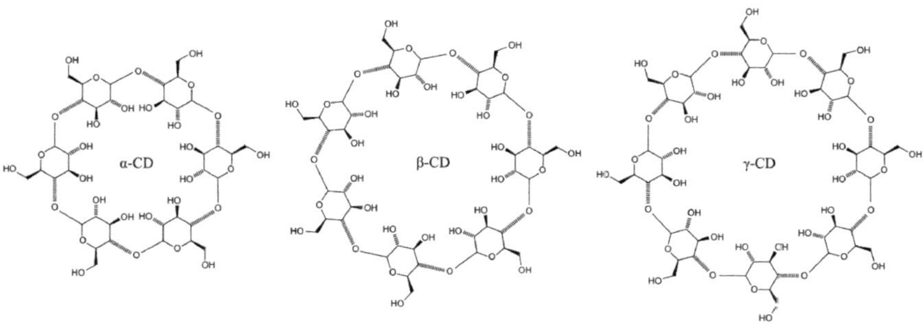

Figure 13.1 Schematic representation of the chemical structure of α-cyD, β- CyD and γ–CyD. Adapted from [29].

The biosynthesis of CyD is facilitated by an enzymatic product of bacteria known as cyclodextrin glucosyltransferase enzyme (CGtase). The CGtase enzyme can be produced from different types of microorganisms such as strains of *Bacillus* sp. (*B. amylobacter, B. macerans, B. circulans, B. subtilis, B. megaterium,* alkalophilic *Bacillus* sp.), *Klebsiella* (*K. pneumoniae, K. oxytoca*), *Thermoanaerobacter* sp., *Clostridium* sp. [16]. During the biosynthesis of CyD, CGtase induces the intramolecular transglycosylation reaction to degrade the amylose helix fraction of starch, and facilitate the subsequent cyclization of the product to form toroidal or doughnut shaped molecules (Fig. 13.2). This toroidal shape is due to the stabilization of the C2–C3 hydroxyl groups by hydrogen bonds, and its rigid non-rotating structure. The most important characteristic of the CyD toroids is the hydrophilic exterior surfaces and hydrophobic interior cavities, which confers the ability to clathrate various molecules, thus forming inclusion complexes. This toroidal characteristic is due to the orientation of the hydroxyl groups of the glucose residues towards the external surfaces of the toroid with the primary (C-6) and secondary (C-2 and C-3) hydroxyl groups at the narrow and wider edges, respectively. On the other hand, the interior cavity is lined by skeletal carbons and ethereal oxygens, which give it a lipophilic character [15, 18–21]. Many types of starch have been used as substrates for CyDs biosynthesis. However, potato starch is still the most commonly used due to high yield [22].

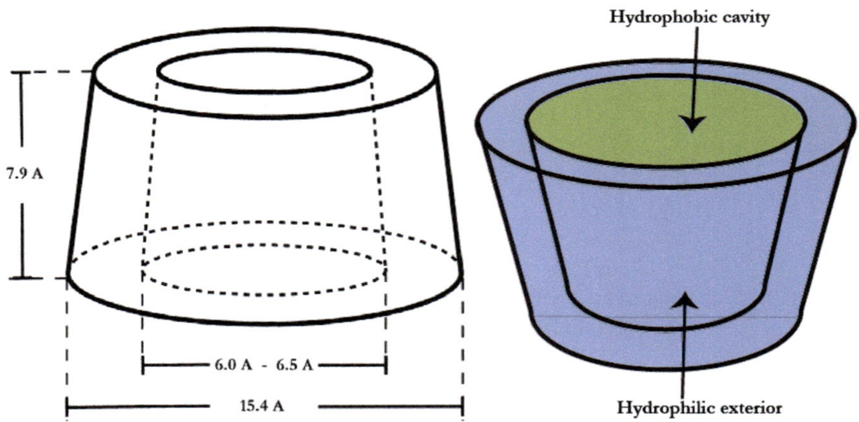

Figure 13.2 Toroidal structure of β-cyclodextrin.

The natural CyDs, i.e. α-CyD, β-CyD and γ-CyD, are commonly referred to as the first-generation CyD and are formed by six, seven, and eight glucopyranose units, respectively [11]. While CyD with fewer glucose units cannot be formed due to steric hindrances, homologs with nine or more glucose units have been identified. French and co-workers [23] identified delta (δ-CyD) and epsilon (ε-CyD). Thoma and Stewart [24] identified zeta (ζ-CyD) and eta (η-CyD). Unlike the natural CyDs, these larger CyDs have small biosynthetic yields, they are difficult to purify and also have poor molecular inclusion properties [16, 25, 26]. Natural CyDs are crystalline, homogeneous, and non-hygroscopic substances. Their melting point is not clearly defined as they start to decompose at about 270°C and produce a sharp endothermic endpoint (differential scanning calorimetry) indicating CyD decomposition at 300°C [16]. Some other important physicochemical properties [27, 28] of natural CyD are presented in Table 13.1.

Table 13.1 Physicochemical properties of natural cyclodextrins

Properties	α	β	γ
No. of glucose units	6	7	8
Molecular weight	972	1135	1297
solubility in water (g/100 ml), room T	14.5	1.85	23.2
$[R]_D$ 25°C	150.0 ± 0.5	162.5 ± 0.5	177.4 ± 0.5
Inner cavity diameter (Å)	4.7–5.3	6.0–6.5	7.5–8.3
Height of Torus (Å)	7.9 ± 0.5	7.9 ± 0.5	7.9 ± 0.5
Diameter of outer periphery (Å)	14.6 ± 0.4	15.4 ± 0.4	17.5 ± 0.4
Approximate cavity volume ($Å^3$)	174	262	427
Approximate cavity volume in 1 mol CyD (ml)	104	157	256
Approximate cavity volume in 1 g CyD (ml)	0.1	0.14	0.2
Crystalline forms (from water)	Hexagonal plates	Monoclinic parallelograms	Quadratic prisms
Crystal water content (% wt)	10.2	13.2–14.5	8.13–17.7
Water molecules in cavity	6	11	17
Solubility in water (room temp., g/100 ml)	14.5	1.85	23.2
Diffusion constant at 40°C	3.443	3.224	3
Hydrolysis by *A. oryzae* α-amylose	Negligible	Slow	Rapid
pK (by potentiometry) at 25°C	12.332	12.202	12.081
Surface tension (MN/m)	71	71	71
Partial molar volumes in solution (ml/mol)	611.4	703.8	801.2
Adiabatic compressibility in aqueous solutions ml (mol^{-1} bar^{-1}) × 10^4	7.2	0.4	–5

Source: Adapted from [30].

The developments of biotechnological and industrial chemistry processes of the 20th century led to the synthetic or semi-synthetic modification of many natural molecules. These modification processes were carried out to enhance desired physicochemical properties while reducing the unwanted ones. These technologies also led to the synthesis of many CyD derivatives by molecular substitution (amination, esterifications, etherifications) of their primary and secondary hydroxyl groups. Depending on the nature and degree of substitution, the hydrophobic molecular cavity volume, affinity for guest molecules, solubility, and stability of natural CyD were modified [31–34].

For instance, while natural CyD are hydrophilic, their aqueous solubility is limited. This aqueous solubility which is dependent on hydrogen bonding between hydroxyl groups is dramatically increased by substitution of the hydroxyl groups. Even substitution by hydrophobic moieties such as methoxy functions can increase aqueous solubility [35, 36]. The use of hydropropyl derivative of β-CyD (HP-β-CyD) while increasing aqueous solubility also reduces the drug binding ability of the CyD cavity. However, the drug binding ability of these parent CyD (β-CyD) can be enhanced by partial methylation of the hydroxyl groups 2-, 3- and 6-positions at the risk of increasing its toxicity [37]. Another group of derivatives are the amphiphilic CyDs, formed by primary and/or secondary side chain substitution with long alkyl and fluoroalkyl groups. They are important in dermal drug delivery due to their ability to form monolayers, micelles and bilayer vesicles [38]. While more than a hundred derivatives of CyDs have been synthesized, toxicological and safety concerns have limited the approval and application of many as drug carriers for therapeutics [39]. Some common CyD derivatives are listed in Table 13.2.

Table 13.2 Physicochemical properties of natural cyclodextrins

Cyclodextrin derivatives		Molecular weight	Solubility in water (mg/ml at 25°C)
Alpha-cyclodextrin	α-CyD	972	145
Beta-cyclodextrin	β-CyD	1135	18.5
2-Hydroxylpropyl-beta-cyclodextrin	HP-β-CyD	1400	>600

Cyclodextrin derivatives		Molecular weight	Solubility in water (mg/ml at 25°C)
Dimethyl-beta-cyclodextrin	DM-β-CyD	1331	—
Randomly methylated-beta-cyclodextrin	RM-β-CyD	1312	>500
Sulphobutylether-beta-cyclodextrins	SBE-β-CyD	2163	>500
Gamma-cyclodextrin	γ-CyD	1297	232
2-Hydroxylpropyl-gamma-cyclodextrin	HP-γ-CyD	1576	>500

Source: Adapted from [11].

13.3 Cyclodextrin-Guest Molecule Complexes

CyDs are versatile drug carriers due to their ability to form inclusion complexes with guest hydrophobic molecules in an aqueous medium. They are also capable of forming non-inclusion complexes and complex aggregates [28, 30, 40]. Thus, they are able to modify the physicochemical and pharmacokinetic properties of drug molecules for efficient drug delivery. These properties such as drug solubility, bioavailability, membrane permeability, enzymatic and chemical stability, photostability, physical state, taste and odour, skin irritability and intrinsic volatility have all been modified by complexation with CyD [19]. Drug-CyD inclusion complexes are formed by non-covalent bonds or simply by the clathration of the guest molecule into the toroidal hydrophobic cavity of CyD (Figs. 13.3 and 13.4). During inclusion complex formation, whole drug molecule or some lipophilic moiety of a compound with poor aqueous solubility replaces enthalpy-rich water molecules thus reducing the energy of the system. The equilibrium between the guest molecule and free CyD facilitates inclusion complex formation. The efficiency of this inclusion process depends on the polarity, steric dimensions, molecule size, electronic effects, temperature, pH and method of inclusion. Other factors like the hydrogen bonding, van der Waals, surface tension, molar ratio (1:1, 1:2, 2:1) have also been reported to play some role in the formation of inclusion complexes. It is important to note that not all drug molecules form stable inclusion complexes. Molecules with high aqueous solubility tend to interact with the hydrophilic surfaces to form association compounds or non-inclusion complexes [20, 21].

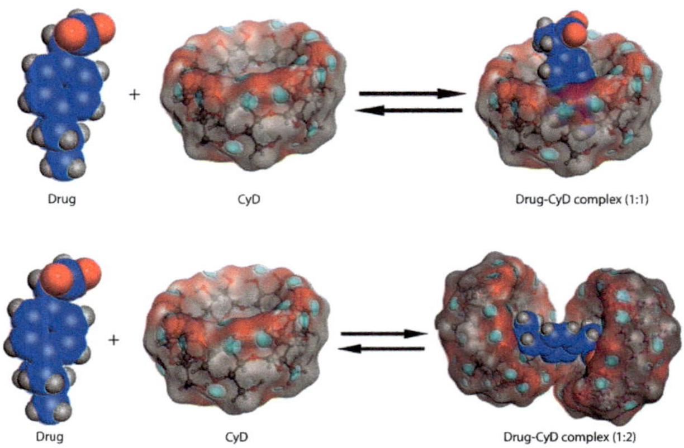

Figure 13.3 Formation of drug–CyD stoichiometric complexes.

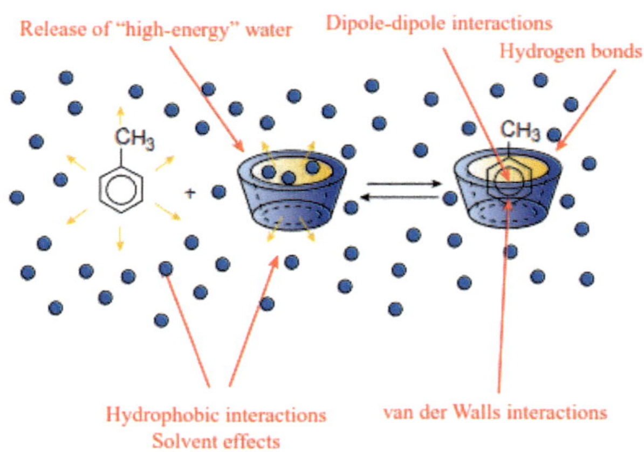

Figure 13.4 Driving force for complex formation. Adapted from [16].

Since their discovery by Villiers, CyDs have attracted enormous research attention and scientific exploration [17]. It has evolved into an important material for the pharmaceutical, chemistry, cosmetics, agriculture, environmental, textile industries, etc. While the first patent on CyD was issued to Freudenberg and co-workers in 1953 [11], industrial application was limited due to poor production methods. The biotechnological development of the early 1980s enabled the genetic engineering of different

CGTase enzymes with specific affinity for α-, β- or γ-CyD; and the subsequent large-scale production of low cost CyDs for industrial application [41]. Many review articles and books are available with enormous details on CyD chemistry, technology and application [11]. This chapter will discuss the application of CyD as drug carriers for the prevention and treatment of skin diseases.

13.4 Skin Safety and Toxicity Considerations of Cyclodextrins

CyDs' safety like all other excipients remains a major concern when seeking approval for pharmaceutical formulations. The biological fate and toxicological profile of the CyDs and all other excipients are required to be fully defined. This is particularly important as contradictory results of CyDs' skin safety have been reported in literature [11]. Piel *et al.* [42] demonstrated the skin compatibility of CyDs using corneoxenometry techniques on human *stratum corneum*. Generally, most natural CyDs and their derivatives are considered non-irritant and safe for dermal drug delivery because they do not induce the disruption of *stratum corneum* layer. Lipophilic CyDs such as the methylated derivatives are perhaps the only group that alters the barrier properties of the skin through component extraction and fluidization [43, 44]. This ability is, however, dependent on CyD concentration in the formulation [40]. A study performed by Bentley and co-workers [45], using a hairless mouse model, suggests that HP-β-CyD can increase the permeability of the *stratum corneum* possibly as a result of extraction of lipids. However, it is important to note that these studies were conducted with empty CyD. The encapsulation of drug molecules in CyDs' cavities is likely to reduce its ability for skin component extraction.

13.5 Cyclodextrins and Dermal Drug Delivery

In the treatment of any skin disorder, the efficacy of dermal therapy is dependent on the efficient delivery of the drug to the target sites of drug action. Typically, topical medications are for local prophylactic or therapeutic effect on the skin surface; cutaneous penetration to treat disorders within the skin; and percutaneous absorption for contact with non-target sites in systemic circulation [46]. The dose of the drug delivered to these target and/or non-target sites determines the desired pharmacologic and adverse effects. In dermal drug therapeutics, the goal is to situate the drug within the epidermis or dermis and prevent systemic absorption [47]. This is difficult to achieve in a controlled and reproducible way because the skin provides a formidable barrier and determines the efficacy of topically applied medicines. The primary function of the skin is to protect the body from environmental toxic agents and excessive water loss. The *stratum corneum* serves this protective function being the rate-limiting barrier for percutaneous drug transport [4]. Generally, a molecule that will penetrate the skin is expected to navigate the physical, chemical/biochemical (lipids, acids, hydrolytic enzymes, anti-microbial peptides and macrophages) and the adaptive immunological barriers of the epidermis in sufficient quantity [48]. Such a molecule must be sufficiently hydrophilic and lipophilic for adequate availability at the site of absorption and *stratum corneum* penetration, respectively [49]. Substances that are very hydrophobic cannot pass easily through the *stratum corneum* and viable epidermis because they are not available in sufficient concentration at the site of absorption to facilitate passive diffusion. It is also expected to have small molecular weight (<500 Da) [49–51]. While (per)cutaneous delivery of lipophilic drugs is majorly facilitated by dissolution into intercellular lipids in the *stratum corneum*, hydrophilic drugs can be absorbed through the pores of the hair follicles and sebaceous glands [52]. How the degree of interaction between these properties affect percutaneous transport is often difficult to define. Essentially, carrier-mediated dermal drug delivery is used to manipulate the drug and/or the target site in order to optimize drug formulation effectiveness, dose availability (bioavailability) and therapeutic efficiency.

Thus, CyDs' ability to modify the physicochemical property of drug molecules without changing their intrinsic abilities has found application in the design and optimization of dermal formulations and cosmetic products for the prevention and treatment of skin disorders. There are also many reports of

CyDs' ability to modify the anatomical structure of the skin. Some of the capabilities of CyDs as drug carriers for dermal delivery are discussed below.

13.5.1 Enhancement of Dermal Drug Absorption and/or Penetration

Skin penetration enhancers are a group of agents used in dermal formulations to facilitate the disruption of physiological barrier system of the skin for enhanced drug penetration. They decrease the ability of the barrier system by distorting the intracellular lipid configurations, changing solvent nature, increasing hydration and enhancing the drug partition into membranes. Thus, lipophilic and hydrophilic drug molecules that will ordinarily be prevented from (per)cutaneous penetration are allowed to permeate the membranes irrespective of the delivery system. CyDs have been used as penetration enhancers in dermal formulation. However, because of their large molecular weight (> 1000 Da), they do not penetrate the skin to disrupt the barrier system. They act by affecting different aspects/levels of dermal flux (according to Fick's law of diffusion [42] governs permeation across the *stratum corneum* in dermal drug absorption [41, 53].

$$J = \Delta D \frac{KD}{h}, \tag{13.1}$$

where J is the drug flux across stratum corneum, K the drug partition coefficient between the skin and the formulation, D the drug diffusion coefficient or diffusivity of drug molecules; ΔD the concentration difference between donor phase and the receptor side of the membrane, and h the thickness of the stratum corneum.

Generally, drug dissociation from the inclusion complex by dilution in aqueous fluids is impossible with dermal formulations. This makes tissue uptake the likely mechanism of absorption. CyDs become effective penetration enhancers by increasing the concentration gradient (ΔD) to achieve the required sink condition of tissue uptake. By increasing the aqueous solubility of the dermal formulation, CyD increases drug availability on the skin surface for higher tissue uptake and absorption [49]. Using hairless mouse skin as a model, Lopez *et al.* [54] reported an enhanced dexamethasone permeation from β-CyD and HP-β-CyD inclusion complexes due to increased drug concentration on the skin surfaces. Ventura *et al.* [55] also described the enhanced celecoxib drug flux across the *stratum corneum* due to an increase in the dissolution rate. Both inclusion complex with DM-β-CyD and non-inclusion complex with HP-β-CyD positively influenced the aqueous solubility of celecoxib. Ascenso *et al.* [56] incorporated a complex of tretinoin and DM-β-CyD in a hydrogel and reported an enhanced skin retention of the drug compared to the free drug, due to an increase in the availability of the drug on the membrane surface.

Furthermore, the incorporation of dermal drugs into topical formulation can be increased without compromising the physicochemical stability by utilizing inclusion complexes of drug with enhanced solubility properties. This increases the concentration of the drug delivered to the skin surface for aqueous solubilization and tissue uptake. Higher doses of tenoxicam gels have been formulated with Me-β-CyD as complexes. The increase in the concentration of the drug in vehicle and partitioning behaviour positively influenced the observed percutaneous penetration of tenoxicam [57].

Contrary to this generally accepted mechanism, some studies have suggested the paracellular route of drug absorption for CyD complexes [55, 58]. Other studies have also described the CyD facilitated penetration enhancement by *stratum corneum* disruption. This is reported to occur by the extraction of some lipophilic constituents of the skin into the CyD hydrophobic cavity and the subsequent diffusion of the drug molecules through the disrupted barrier (reduction of h from Eq. 13.1) [59]. This ability is often dependent on the derivative of natural CyD used. Methylated CyDs have been shown to have higher ability to induce membrane damage by the extraction of cholesterol [60, 61]. While this ability has been used in some cosmetic formulations (empty CyDs), lipophilic extraction is unlikely to occur if the hydrophobic cavities are already occupied by hydrophobic drug molecules [53, 55].

The partition coefficient (K in Eq. 13.1) of the drug is also an important parameter for optimal dermal absorption. Drug solubility and thermodynamic activity have been reported to affect the partition coefficient. Thus, CyDs' ability to increase drug solubility positively affects the partition coefficient for enhanced drug absorption [62]. The prevention of skin metabolism or bio-conversion of some drugs have also been described as the possible way by which CyDs increase drug availability and subsequent drug penetration of the *stratum corneum*. For instance, a 35% reduction in dexamethasone degradation was observed when β-CyD and HP-β-CyD complexes were used instead of the free drug [54].

13.5.2 Enhancement of Drug Tolerability

In the treatment of many skin disorders, adherence to therapy and subsequent therapeutic efficacy are often dependent on the tolerability of the drug formulation and the dosage form. Many drugs elicit topical adverse reactions when in direct contact with the skin or when the formulation is inadequate to prevent such reactions. Adverse topical reactions like erythema, burning sensations, desquamation, xerosis and local irritation have all been reported due to the corrosive nature of many drugs. These can be reduced or eliminated by incorporating such drugs into CyDs by inclusion complexes formation which reduces the drug molecules contact with the skin. The skin irritability of drugs like retinoic acid, prochlorperazine and celecoxib has been successfully reduced by complexation with CyDs [55, 63, 64]. Also because drug solubility is increased by CyD complexation, lower quantity of free drug is often required in such formulations to achieve defined therapeutic effects. This reduction of drug dose can significantly reduce skin exposure and susceptibility to adverse drug reactions.

13.5.3 Enhancement of Drug Stability in Dermal Formulations

Physicochemical and enzymatic stability of drug molecules in formulation is required for the therapeutic effectiveness of such formulation. Due to poor solubility, some drug molecules are liable to precipitation from aqueous formulation during manufacture, storage or use, while others may undergo various forms of chemical instability such as hydrolysis, oxidation and photodecomposition. Enzymatic degradation of drug molecules due to the presence of esterases, dehydrogenases, etc., on the skin can lead to bio-conversion and inactivation of such drugs. Many studies have demonstrated the enhancement of physicochemical and enzymatic stability of many dermal drug molecules when formulated as inclusion complexes of CyDs. For instance, the crystallization of sericoside from topical formulations was eliminated by using HP-γ-CyD and γ-CyD complexes of this drug [65].

Most dermal formulations are dispersed systems with two immiscible phases stabilized by a surface acting agent. The stabilization of these systems with particles such as CyDs (Pickering emulsions) instead of emulsifiers has also been shown to enhance the tolerability and effectiveness of formulations [66]. CyDs form low soluble complexes at oil/water interface thus decreasing interfacial tension and modifying the interfacial rheology to stabilize the system. Natural CyDs-stabilized emulsions have been reported to have long-term stability in various storage conditions [66–68]. Linoleic acid, a substance that plays a vital role in the maintenance of skin barrier functions, has been applied in skin care and skin restorative cosmetic/dermatological products. However, it is susceptible to oxidation and rancidification which limits its use in such products. Reversible complexation with CyDs as described by Regiert [69] has ensured its stability to oxygen, light, and temperature. Many other studies have demonstrated the enhancement of drug stability with CyDs [70, 71].

13.5.3.1 Encapsulation and controlled delivery of volatile compounds

Many active pharmaceutical ingredients of interest in the treatment of skin disorders exist as essential oils and volatile liquids. The encapsulation of the agents can reduce their volatility, protect them from environmental degradation and prolong skin contact [16, 19, 20, 72]. Arana-Sanchez *et al.* [73] microencapsulated oregano oil (*Lippia graveolens*) by spray drying with β-CyD to preserve the anti-microbial activity and improve the anti-oxidant activity of the active compound. Daletos *et al.* [74] observed a reduction in the odour and tastes of the essential oils: thymol and carvacrol by the forming inclusion complexes with β-CyD.

13.6 Application of Cyclodextrin in Dermatologic Products

Skin disorders encompass a large number of skin diseases ranging from the acute and benign to chronic and life threatening disorders. Their aetiology is linked to numerous infectious pathogens and inflammatory conditions. Therapeutic approaches to resolving these diseases are often based on the pathogenic origin of the disease and the skin structure that is affected [75, 76]. Gupta *et al.* [4] provide a pictorial chart (Fig. 13.5) of the basic classification of skin disorders.

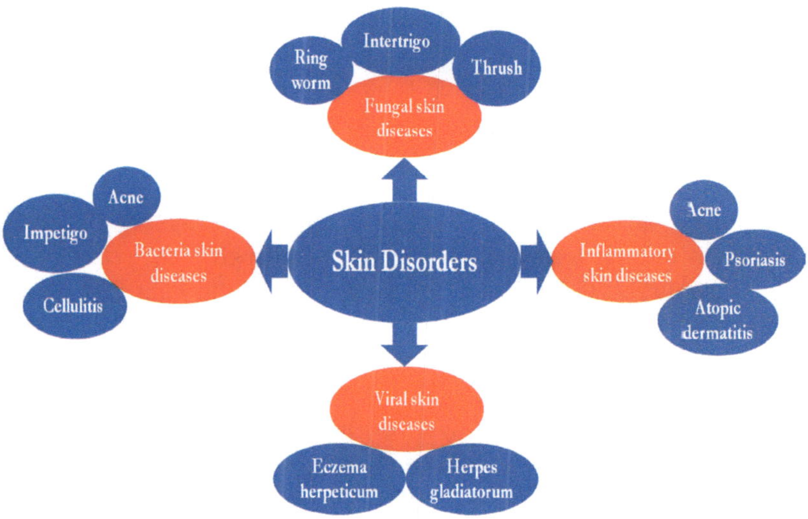

Figure 13.5 Schematic representation of various skin disorders.

Dermatologic drugs are only effective when they are able to reach these target sites (epidermis, keratinous tissues and hair follicles; dermis and subcutaneous tissues, etc.) within the skin in adequate concentrations. As stated earlier, effective delivery of these drugs within these skin structures is limited by the *stratum corneum* [77] and CyDs are suitable drug carriers for enhancing the therapeutic efficacy of these drugs. Despite the fact that very few CyDs containing dermal pharmaceutical products are currently being marketed [78], a review of literature shows numerous promising possibilities for CyDs enhanced therapeutic efficacy of many dermatologic drugs. These possibilities/applications in the treatment of some skin disorders are discussed below.

13.6.1 Anti-Acne

Acne vulgaris is a chronic inflammatory condition of the pilosebaceous unit characterized by facial eruption of comedones which are sebum and keratin containing plugs lodged in the follicular ducts. The aetiology of acne includes increased sebum production and hyper-cornification of the pilosebaceous ducts (follicular keratinization); and colonization by the bacterial *Propionibacterium acnes.* Due to the multidimensional nature of acne's aetiology, therapeutic interventions involve several factors implicated in its pathogenesis such as androgenic hormonal stimulation responsible for sebum hyper-secretion, faulty occlusion of the follicular orifice, *P. acnes* colonization and immunological inflammatory response [79–81].

Keratolytic agents such as topical vitamin A acid (tretinoin, all-trans retinoic acid) and its synthetic analogues (retinoids) increase the turnover of follicular epithelial cells, thus normalizing keratinization, preventing follicular occlusion and inhibiting comedone formation [80–82]. However, the therapeutic use of tretinoin dermal formulations is limited by high skin irritability, low aqueous solubility, and high instability to air, light and heat [83]. Numerous studies have reported the use of CyD derivatives as drug carriers to ameliorate these topical tretinoin drawbacks. Anadolu *et al.* [64] observed that the treatment efficacy of retinoic acid (tretinoin)-β-CyD formulated as topical products was higher than those of commercial products containing twice the amount of retinoic acid. This is probably due to the enhanced water solubility and stability of the CyD complex of the drug. They also reported a reduction in the side effects of the drugs. This reduction in skin irritability and side effects has also been reported by Amdidouche *et al.* [84]. In a series of studies, Ascenso and co-workers [56, 85, 86] demonstrated the ability of DM-β-CyD to enhance the physicochemical and dermatological properties of tretinoin as mentioned before.

A US patent by At and Mallard [87] described the use of CyDs in the preparation of a stable topical adapalene for the treatment of acne and other dermatological diseases related to the keratinization disorders (comedones, papulopustular, etc.). Beyond the randomly methylated β-CyD (RAMEB) facilitated increase in drug solubility, its stability in the formulation prior to dermal application was achieved by preparing complexes of adapalene in an excess CyD complex solution. This prevented formulation-induced de-complexation of the adapalene complex in dermal products and made the complex amenable to formulation in various topical galenic forms such as aqueous, aqueous-alcoholic or oily dispersions, dispersions of lotion type, aqueous, anhydrous or lipophilic gels, etc. De-complexation only occurred upon application to the skin with enhanced skin penetration of the drug and beneficial effect on comedolytic activity.

13.6.2 Psoriasis

Psoriasis is a chronic, genetically determined immune-mediated inflammatory disease characterized by thick, red, scaly plaque on the skin, scalp and joints [88]. Topical therapy, which is the first line treatment for psoriasis [89] can prevent severe psoriasis and its negative impact on the quality of life [90]. Vitamin D analogues like calcipotriol can inhibit the proliferation and differentiation of keratinocytes thereby increasing the amount of anti-inflammatory cytokines while decreasing the amount of pro-inflammatory cytokines. Poor aqueous solubility and skin irritability have limited their topical administration in the management of psoriasis. Emulgels containing calcipotriol complex with the anionically charged CyD (Captisol®) have been shown to increase drug release compared to those containing pure drug [88]. Singh *et al.* [91] studied the skin permeation and deposition of colchicine, a drug used in the treatment of psoriasis due to its immunosuppressive and anti-inflammatory activity. A 3.4-fold increase in skin permeation was observed with colchicine-β-CyD complex compared to colchicine drug solution. When formulated as elastic liposomes, colchicine-β-CyD complex facilitated a 12.4-fold increase in skin deposition compared the drug solution, while the formulation of colchicine elastic liposomes was only able to achieve 8.4-fold increase. Because skin accumulation is required for effective topical therapy with colchicine, an observed reduction in rate of skin permeation and systemic absorption in formulations containing β-CyD is indicative of CyD's ability to mediate controlled topical delivery of colchicine in the topical treatment of mild psoriasis.

13.6.3 Dermatitis

CyDs mediate skin delivery and have been used to optimize the efficacy of numerous anti-inflammatory drugs applied topically to the skin for various forms of dermatitis. Some of these drugs are non-steroidal anti-inflammatory drugs (NSAIDs) whose anti-skin cancer activities have also been demonstrated in cell culture and animal studies [92]. Skin accumulation of these drugs from topical therapy can cause localized cyclooxygenase inhibition and subsequent anti-inflammatory response to radiation-induced skin damage [93]. Thus, the objective of CyD carrier-mediated delivery is to optimize skin accumulation and prevent system absorption from the transdermal route. Topical formulations of ibuprofen at the maximum amount soluble in 10% HP-β-CyD resulted in a higher degree of skin accumulation and photo-protection, with huge potentials in the management of skin cancers [93]. Also, liquid crystalline gels containing lornoxicam-β-CyD showed a higher reduction in the degree of pain and inflammation *ex vivo* due to higher skin accumulation and avoidance of systemic absorption [94]. In another study, the coating of topical flurbiprofen PLGA and PLGA-PEG nanospheres with HP-β-CyD resulted in higher *in vivo* topical anti-inflammatory efficiency despite similar *ex vivo* skin accumulation data with uncoated nanospheres [95]. It is possible that CyDs can modulate enhanced anti-inflammatory activity by other means apart from skin accumulation.

In atopic dermatitis (eczema), immunologically mediated leukocyte infiltration into the skin significantly contributes to the pathogenesis of the disease. Often characterized by itching, swelling, redness, blisters, crusting, ulceration, pain, scaling, cracking, etc., therapeutic agents like corticosteroids have been administered both topically and systemically. CyD complexes of these drugs have been found to enhance/mediate skin delivery and topical efficacy [54, 96, 97]. Drug-induced contact dermatitis has also been alleviated by complexation of such drugs with CyDs [63, 98]. Xanthohumol, a prenylated chalcone found in hops (*Humulus lupulus*, Cannabaceae), has been reported to reduce contact dermatitis [99]. It has also been the subject of a recent CyD-related patent [100] in which the CyD-mediated increase in aqueous solubility was important for the treatment of skin disorders such as contact, atopic, and seborrhoeic dermatitis, eczema and rosacea, acne, etc.

13.6.4 Microbial Skin Diseases

The skin is a natural host for many microorganisms such as bacteria, fungi and virus. Many factors such as age, hormonal status, local humidity, amount of sebum and sweat, and anatomical location affect the type and density of these organisms [101]. These organisms are often commensal or symbiotic but can become pathogenic if they are able to circumvent normal host defence mechanisms. For instance, a cluster of 100,000 fungi will only have a few pathogens capable of causing dermatomycoses [102]. While these organisms in their non-pathogenic state are generally tolerated on the skin surface, they are able to induce inflammation when present in the dermis [103]. Recurrent bacterial infections have also been known to complicate the prognosis of diseases like atopic dermatitis and psoriasis [104].

While the treatment protocol for microbial infections of the skin is often by oral therapy, topical application of antibiotic and antifungal agents are more effective as they are able to localize the drug at the superficial and cutaneous sites of disease, while also reducing systemic side effects. CyDs have been used to overcome numerous drawbacks on the topical use of these agents. Tenjarla and co-workers [105] prepared various complexes of miconazole with natural CyDs and their derivatives; and evaluated the ability of these CyDs to enhance the aqueous solubility and topical bioavailability. Skin permeation studies revealed a 2.6-fold increase in skin retention of miconazole due to higher solubility (54.6-fold) and/or the effect of α-CyD on the skin. Other drugs like topical doxepin have also been used for the treatment of pruritus and pain symptoms associated with herpes zoster-induced post-herpetic neuralgia and atopic dermatitis. Its localized analgesic activity demonstrated in the transcutaneous delivery of doxepin-HP-β-CyD was ascribed to HP-β-CyD-induced sustained drug release, prolonged analgesic activity, and a reduction in the systemic absorption/side effects [106].

The high risk of nephrotoxicity from the parenteral administration of amphotericin B, a poorly soluble, broad antifungal and leishmanicidal drug with low incidence of resistance; limits its use in the treatment of cutaneous infections. The topical formulation of this drug will thus prevent systemic toxicity and localize the drug at the site of required action. Ruiz *et al.* [107] prepared creams and gels containing amphotericin B-γ-CyD inclusion complexes and evaluated their topical antifungal and anti-leishmanial efficacy. The amphotericin B-γ-CyD gel formulation exhibited a greater antifungal activity against all the strains tested *in vitro* compared to other formulations and clotrimazole creams. They hypothesized that the higher activity obtained with the amphotericin B-γ-CyD formulation was due to the greater drug solubility and physicochemical stability. The amphotericin B-γ-CyD gel was also effective in the topical treatment of cutaneous leishmaniosis as it was able to penetrate the epidermis and kill leishmania parasites localized in the dermis.

13.6.5 Wound Healing (Pressure Ulcers)

Pressure ulcers, which are localized lesions of the skin or underlying tissues, arise from reduced blood flow to the skin and subsequent skin death especially in hospitalized patients. This can have high pathological implications for underlying tissues and the body. It is characterized by skin blisters and open sores. Insulin has been known to promote wound healing due to its ability to facilitate re-epithelialization by epidermal differentiation and the formation of junctions between the epidermis and the dermis [108]. However, the aggregation of the hydrophobic portions of proteins such as insulin can lead to reduced activity, thermal denaturing and degradation. The incorporation of these hydrophobic portions into CyD cavities can have positive effects on the thermal stability of insulin even under high temperature conditions [109, 110]. The CyDs' ability to increase the stability of disulphide bonds, which are important to the conformation of insulin, as observed in HP-β-CyD has also been proposed as a mechanism of preventing insulin degradation [109]. Valentini and co-workers [111] formulated insulin-HP-β-CyD complexes in water-soluble Carbopol®-based gels and evaluated its clinical use in healing pressure ulcers. They observed that the CyD-containing gels promoted better healing of the ulcers (reduction of the pressure ulcer size, absence of necrotic tissues, and revitalization of the tissues). They also observed the absence of systemic hypoglycaemia in the patients, which was indicative of the localized insulin effect.

13.7 Application of Cyclodextrins in Cosmetics

Cosmetic products play important roles in the prevention and treatment of skin disorders. For instance, the impairment of *stratum corneum* barrier functions observed in cutaneous disorders such as psoriasis, ichthyosis, atopic skin and contact dermatitis has been linked to skin dryness [112]. Environmental chemicals, low humidity, low temperature, normal aging process and psychological stress all modulate skin dryness and barrier homeostasis causing high susceptibility to irritancy and secondary bacterial infections [113]. By inducing superficial and deep layer changes, cosmetic products like moisturizing creams have been used to facilitate *stratum corneum* hydration in order to prevent skin dryness and also as adjuncts in the treatment of some skin disorders [112, 114, 115]. The use of sunscreens is also important in the prevention of ultraviolet-radiation-induced skin damage and cancers [116].

The past three decades have witnessed the publication of over 40,000 articles in scientific journals with specific interest on the application of CyDs to the field of cosmetics. These applications are based on the scientific knowledge of CyDs' ability to modulate the physicochemical properties of these products, especially the aqueous solubility of lipophilic water-insoluble active compounds [117, 118]. Various cosmetic formulations containing CyDs have been designed, developed and approved for use. Marques [16] lists the basic functions of CyD in cosmetics as follows:

(i) prevention of the destruction of certain flavours, colours or vitamins associated with certain ingredients by processing or on storage
(ii) increasing water solubility of lipophilic materials
(iii) conversion of liquid or oily materials to powder form
(iv) increasing the physical and chemical stability of guest molecules by protecting them against decomposition, oxidation, hydrolysis and loss by evaporation

(v) reducing or preventing skin irritation
(vi) providing controlled release of active ingredients
(vii) prevention of interactions between various formulation ingredients
(viii) increasing or decreasing the absorption of various compounds into skin

Some of these functions in specific dermal products are discussed below.

13.7.1 Deodorants and Formulations for Odour Control

The offensive nature of body odour has been recognized for centuries [119]. Bacterial microflora present on the skin is implicated in the biogenesis of body odour. They facilitate the metabolism of non-odorous precursors in apocrine and sebaceous secretions to produce odorous volatile fatty acids, steroids, and sulphur compounds [120–123]. Several methods including increased personal hygiene, application of fragrances and antimicrobial agents, etc., have all been used to reduce or eradicate these odours with varying degrees of success [119]. Empty CyDs are able to extract and form inclusion complexes with gas and other volatile molecules generated by the microbial degradation of sweat [118]. These inclusion complexes immobilize the odorous molecules and odour precursors, thereby decreasing their vapour pressure below the olfactory perception threshold [119]. Many antiperspirants and deodorants target these microbial agents on the skin surface for their activity. CyDs molecules like β-CyD, HP-β-CyD, etc., have been used for this purpose [123]. Natural and chemically modified CyDs are also able to clathrate and deodorize unsaturated aldehydes responsible for aging odour in the middle age and elderly population [122]. The patent literature is replete with examples of CyDs used in the preparation of cosmetic deodorants [124–131].

CyDs have also been used to deodorize many cosmetic products containing active substances with unpleasant base odours. When CyD complexes of such substances are used, it is possible to reduce or completely eliminate such odour. For example, gluthathione an inhibitor of melanin pigment formation is an active ingredient of many de-pigmentation agents, skin-improving and/or whitening cosmetic products. However, its offensive odour often limits its cosmetic acceptability and use. The formation of a gluthathione-CyD complex eliminates the unpleasant odour and enhances its acceptability [132]. When dihydroxyacetone-CyD complexes are used in the manufacture of the tanning agent Ultrasun©, the unpleasant odour produced by the reaction of the free drug with the skin is eliminated [133].

Phytocide, an essential oil extracted from *Thujopsis dolabrata* timber, has been formulated in aqueous solutions and CyD as a functional sea water for bathing application with the ability to suppress body odour, improve health and prevent other skin diseases [134].

13.7.2 Fragrances

Fragrances are employed in cosmetic formulations to mask unpleasant base odour of a cosmetic product and/or provide a pleasant odour and product identity to meet consumer requirements. Perfuming processes with fragrances are, however, very difficult to achieve since they exist as poorly soluble or insoluble compounds in liquid state. The use of surfactants in their solubilization has resulted in other drawbacks such as skin irritation, sensitization to light, cloudiness and turbidity of erstwhile transparent products, and loss of fragrance constituents on storage due to high volatility and poor stability [15]. Long-lasting fragrances have been produced by forming inclusion complexes with CyDs. Numanoglu *et al.* [135] described the conversion of linalool and benzyl acetate fragrances to powders and the increased stability (decreased volatility) and aqueous solubility through the formation of inclusion complexes with β-CyD and 2-HP-β-CyD. The formation of cosmetic gels with these inclusion complexes ensured the control release of the fragrance over longer period to time.

13.7.3 Sunscreens

The increase in ultraviolet radiation to the earth due to the progressive stratospheric ozone layer damage, and the deleterious effects on the skin has been a major public health concern. Ultraviolet radiation induces damage of the skin's immunological system, skin cancers, erythema, cutaneous photoaging, and other skin diseases [136, 137]. This has led to the increased use of cosmetic sunscreens. The active components of these sunscreens are organic chemicals capable of absorbing UV radiations and attenuating their effects on the skin. They are often referred to as UV filters and are formulated as cosmetic lotions, creams and sprays. Light-induced decomposition and generation of skin toxic degradation products is a major concern in the formulation of these agents [138, 139]. One of the formulation approaches to ensure the photo-stability of these UV filters is the formation of CyD inclusion complexes sunscreens. The HP-β-CyD inclusion complexes with phenylbenzimidazole sulphonic acid, butyl-methoxydibenzoylmethane and 2-ethylhexyl-p-dimethylaminobenzoate showed enhanced photo-stability [98, 140, 141]. As observed in β-CyD and *trans*-2-ethylhexyl-*p*-methoxycinnamate inclusion complexes, CyD can also enhance the chemical stability of sunscreen agents by limiting adverse interactions with other formulation ingredients [142]. By controlling or reducing membrane diffusion of the active agents, CyD inclusion complexes can reduce or eliminate skin penetration, interaction and irritability of UV filters like oxybenzone [143]. Scalia and co-workers [98, 140, 141] also reported a reduction in their skin irritation potential of UV filters when formulated as inclusion complexes. Sulphobutylether-β-CyD (SBE-β-CyD) was more effective than HP-β-CD in reducing the epidermal penetration and potential irritation of butyl-methoxydibenzoylmethane and oxybenzone [144–146]. Also, CyD facilitated increase in the concentration of the UV filters at the *stratum corneum* by the formation of drug reservoirs resulting into enhanced photoprotective effects [137, 144].

13.7.4 Skin Cleansers and Scrubs

The ability of empty CyDs to clathrate molecules such as fatty acids and sebum has led to their application as skin cleaners in some acne cosmetic formulations [53]. Cosmetic products containing β-CyD and Me-β-CyD have been described in patent literature by Saeki and Morifuji [147] and Matsuda and Ito [148], respectively. Also, due to their mild scrubbing ability, water-insoluble CyD polymer beads consisting of un-complexed epichlorohydrin treated β-CyD has been employed in the formulation of facial cleansers [149]. The low solubility of salicylic acid limits its use as a skin cleanser despite its antibacterial and anti-keratolytic properties. When inclusion complexes of salicylic acid and HP-β-CyD are prepared, its aqueous solubility increases leading to the enhancement of its disinfectant, bacteriostatic, and keratolytic properties [150].

13.8 Conclusions and Prospects for the Future

While CyDs have been available for over a century, their application in the pharmaceutical and cosmetic industry is just about four decades old. Currently, there are many CyD-containing cosmetic and cosmeceutical products approved for use. However, there seems to be a gap between research data and patents on CyD-mediated dermal delivery of drugs and the translation of such information into pharmaceutical products. This may be due to the increased focus of the industry on the formulation of dermal drugs as cosmeceuticals. The design, development and marketing strategies of cosmeceuticals ensure access to the less rigorous approval procedures compared to pharmaceutical products. Also, because the claim of drug efficacy is not expressly required, a larger client base can be easily targeted with these products [151]. With numerous calls for a stricter regulation of the cosmeceuticals and the classification of about forty percent of all marketed pharmaceutical active ingredients as practically insoluble, the potential of CyD-mediated delivery in the treatment of dermal diseases is huge [151, 152]. Also, recent synergistic approaches of combining CyD's drug delivery ability with other drug delivery technologies such as nanotechnology, for enhanced therapeutic efficacy of dermal drugs is expected to expand the scientific use of CyD in the treatment of dermal disorders [4, 153, 154].

References

1. Evers A, Lu Y, Duller P, Van Der Valk P, Kraaimaat F, Van De Kerkhof P. Common burden of chronic skin diseases? Contributors to psychological distress in adults with psoriasis and atopic dermatitis. *British Journal of Dermatology*. 2005; 152(6): 1275–1281.

2. Bickers DR, Lim HW, Margolis D, Weinstock MA, Goodman C, Faulkner E, *et al.* The burden of skin diseases: 2004: A joint project of the American Academy of Dermatology Association and the Society for Investigative Dermatology. *Journal of the American Academy of Dermatology*. 2006; 55(3): 490–500.
3. Roberts M, Walters K. Skin structure, pharmaceuticals, cosmetics, and the efficacy of topically applied agents. In: (Walters K, Roberts M, eds.), *Dermatologic, Cosmeceutic, and Cosmetic Development*. New York: Informa Healthcare; 2008. pp. 1–10.
4. Gupta M, Agrawal U, Vyas SP. Nanocarrier-based topical drug delivery for the treatment of skin diseases. *Expert Opinion on Drug Delivery*. 2012; 9(7): 783–804.
5. Wallach D. The field of cosmetic dermatology: The need for a patient-centred approach. *Journal of Cosmetic Dermatology*. 2002; 1(3): 137–141.
6. Kilgman A. Cosmeceuticals: A broad-spectrum category between cosmetics and drugs. In: (Elsner P, Maibach H, eds), *Cosmeceuticals and Active Cosmetics Drug versus Cosmetics*. 2nd ed: Taylor & Francis; 2005. pp. 1–9.
7. Ramos-e-Silva M, Celem LR, Ramos-e-Silva S, Fucci-da-Costa AP. Anti-aging cosmetics: Facts and controversies. *Clinics in Dermatology*. 2013; 31(6): 750–758.
8. Villiers A. Sur la transformation de la fécule en dextrine par le ferment butyrique. *Comptes Rendus de l'Académie des Sciences*. 1891; 112: 435–438.
9. Schardinger F. Acetongärung. *Wiener klinische Wochenschrift*. 1904; 17: 207–209.
10. Schardinger F. Bildung kristallisierter polysaccharide (dextrine) aus stärkekleister durch microben. *Zentralbl Bakteriol Parasitenk Abt II*. 1911; 29: 188–197.
11. Loftsson T, Duchêne D. Cyclodextrins and their pharmaceutical applications. *International Journal of Pharmaceutics*. 2007; 329(1–2): 1–11.
12. Freudenberg K, Meyer-Delius M. Über die Schardinger-Dextrine aus Stärke. *Berichte der deutschen chemischen Gesellschaft (A and B Series)*. 1938; 71(8): 1596–1600.
13. Freudenberg K, Plankenhorn E, Knauber H. Schardinger dextrins-derived from starch. *Chemistry & Industry*. 1947(48): 731–735.
14. Cramer F. *Einschlussverbindungen*: Springer; 1954.

15. Marques HC. Structure and properties of cyclodextrins. Inclusion complex formation. *Revista portuguesa de farmácia*. 1994; 44(2): 77–84.

16. Marques HMC. A review on cyclodextrin encapsulation of essential oils and volatiles. *Flavour and Fragrance Journal*. 2010; 25(5): 313–326.

17. André Sá Couto PS, Marques HC. Cyclodextrins. In: (Ramawat KG, Mérillon J-M, ed.), *Polysaccharides: Bioactivity and Biotechnology*, Cham: Springer International Publishing; 2015. pp. 247–288.

18. Davis ME, Brewster ME. Cyclodextrin-based pharmaceutics: Past, present and future. *Nature Reviews Drug Discovery*. 2004; 3(12): 1023–1035.

19. Valle EMM. Cyclodextrins and their uses: A review. *Process Biochemistry*. 2004; 39: 1033–1046.

20. Cal K, Centkowska K. Use of cyclodextrins in topical formulations: Practical aspects. *European Journal of Pharmaceutics and Biopharmaceutics*. 2008; 68(3): 467–478.

21. Duchêne D. Cyclodextrins and their inclusion complexes. In: (Bilensoy E, ed), *Cyclodextrins in Pharmaceutics, Cosmetics, and Biomedicine: Current and Future Industrial Applications*. New Jersey; 2011. pp. 1–18.

22. Szerman N, Schroh I, Rossi AL, Rosso AM, Krymkiewicz N, Ferrarotti SA. Cyclodextrin production by cyclodextrin glycosyltransferase from Bacillus circulans DF 9R. *Bioresource Technology*. 2007; 98(15): 2886–2891.

23. French D. The schardinger dextrins. *Advances in Carbohydrate Chemistry*. 1957; 12: 189–260.

24. Thoma J, Stewart L. Cycloamyloses. *Starch: Chemistry and technology*. 1965; 1: 209–249.

25. Ueda H. Large ring cyclodextrins-Recent progress. *FABAD Journal of Pharmaceutical Sciences*. 2004; 29: 27–38.

26. Taira H, Nagase H, Endo T, Ueda H. Isolation, Purification and characterization of large-ring cyclodextrins (CD36~~CD39). *Journal of Inclusion Phenomena and Macrocyclic Chemistry*. 2006; 56(1–2): 23–28.

27. Dodziuk H. *Molecules with Holes—Cyclodextrins*. 2006. 1–30 pp.

28. Kurkov SV, Loftsson T. Cyclodextrins. *International Journal of Pharmaceutics*. 2013; 453(1): 167–180.

29. Skowron S. Cyclodextrin.svg 2006 [cited 2015 November, 6]. Available from: https://en.wikipedia.org/wiki/Cyclodextrin#.

30. Salústio PJ, Pontes P, Conduto C, Sanches I, Carvalho C, Arrais J, *et al.* Advanced technologies for oral controlled release: Cyclodextrins for oral controlled release. *AAPS PharmSciTech.* 2011; 12(4): 1276–1292.

31. Szejtli J. The properties and potential uses of cyclodextrin derivatives. *Journal of Inclusion Phenomena and Molecular Recognition in Chemistry.* 1992; 14(1): 25–36.

32. Szente L, Szejtli J. Highly soluble cyclodextrin derivatives: Chemistry, properties, and trends in development. *Advanced Drug Delivery Reviews.* 1999; 36(1): 17–28.

33. Mura P, Furlanetto S, Cirri M, Maestrelli F, Corti G, Pinzauti S. Interaction of naproxen with ionic cyclodextrins in aqueous solution and in the solid state. *Journal of Pharmaceutical and Biomedical Analysis.* 2005; 37(5): 987–994.

34. Buchanan CM, Buchanan NL, Edgar KJ, Little JL, Ramsey MG, Ruble KM, *et al.* Pharmacokinetics of saquinavir after intravenous and oral dosing of saquinavir: Hydroxybutenyl-β-cyclodextrin formulations. *Biomacromolecules.* 2007; 9(1): 305–313.

35. Loftsson T, Brewster ME. Pharmaceutical applications of cyclodextrins. 1. Drug solubilization and stabilization. *Journal of Pharmaceutical Sciences.* 1996; 85(10): 1017–1025.

36. Zia V, Rajewski R, Stella V. Effect of cyclodextrin charge on complexation of neutral and charged substrates: Comparison of (SBE) 7M-β-CD to HP-β-CD. *Pharmaceutical Research.* 2001; 18(5): 667–673.

37. Stella VJ, Rajewski RA. Cyclodextrins: Their future in drug formulation and delivery. *Pharmaceutical Research.* 1997; 14(5): 556–567.

38. Sallas F, Darcy R. Amphiphilic cyclodextrins–advances in synthesis and supramolecular chemistry. *European Journal of Organic Chemistry.* 2008; 2008(6): 957–969.

39. Thompson DO. Cyclodextrins—enabling excipients: their present and future use in pharmaceuticals. *Critical Reviews in Therapeutic Drug Carrier Systems.* 1997; 14(1): 1–104.

40. Bonini M, Rossi S, Karlsson G, Almgren M, Lo Nostro P, Baglioni P. Self-assembly ofβ-cyclodextrin in water. Part 1: Cryo-TEM and dynamic and static light scattering. *Langmuir.* 2006; 22(4): 1478–1484.

41. Loftsson T, Masson M. Cyclodextrins in topical drug formulations: Theory and practice. *International Journal of Pharmaceutics*. 2001; 225(1–2): 15–30.

42. Piel G, Moutard S, Uhoda E, Pilard F, Piérard GE. Skin compatibility of cyclodextrins and their derivatives: A comparative assessment using a corneoxenometry bioassay. *European Journal of Pharmaceutics and Biopharmaceutics*. 2004; 57(3): 479–482.

43. Okamoto H, Komatsu H, Hashida M, Sezaki H. Effects of β-cyclodextrin and di-O-methyl-β-cyclodextrin on the percutaneous absorption of butylparaben, indomethacin and sulfanilic acid. *International Journal of Pharmaceutics*. 1986; 30(1): 35–45.

44. Loftsson T, Vogensen S, Brewster ME, Konráðsdóttir F. Effects of cyclodextrins on drug delivery through biological membranes. *Journal of Pharmaceutical Sciences*. 2007; 96(10): 2532–2546.

45. Bentley M, Vianna RF, Wilson S, Collett JH. Characterization of the influence of some cyclodextrins on the stratum corneum from the hairless mouse. *Journal of Pharmacy and Pharmacology*. 1997; 49(4): 397–402.

46. Korting H, Schäfer-Korting M. Carriers in the topical treatment of skin disease. *Handbook of Experimental Pharmacology*. 2010; 197: 435–468.

47. Kreilgaard M. Influence of microemulsions on cutaneous drug delivery. *Advanced Drug Delivery Reviews*. 2002; 54: S77–S98.

48. Proksch E, Brandner JM. The skin: An indispensable barrier. *Experimental Dermatology*. 2008; 17(12): 1063–1072.

49. Másson M, Loftsson T, Másson G. Cyclodextrins as permeation enhancers: Some theoretical evaluations and *in vitro* testing. *Journal of Controlled Release*. 1999; 59(1): 107–118.

50. Bos JD, Meinardi M. The 500 Dalton rule for the skin penetration of chemical compounds and drugs. *Experimental Dermatology*. 2000; 9(3): 165–169.

51. Barry BW. Transdermal drug delivery. In: Aulton ME, ed. *Pharmaceutics: The science of dosage form design*. Fourth ed. London: Elsevier, Churchill Livingstone 2007; 580–585.

52. Münster U, NakamuraNachname C, Haberland A, Jores K, Mehnert W, Rummel S, *et al.* RU 58841-myristate–prodrug development for topical treatment of acne and androgenetic alopecia. *Die Pharmazie—An International Journal of Pharmaceutical Sciences*. 2005; 60(1): 8–12.

53. Bochot A, Piel G. Applications of cyclodextrins for skin formulation and delivery. In: (Bilensoy E, ed), *Cyclodextrins in Pharmaceutics, Cosmetics, and Biomedicine: Current and Future Industrial Applications. New Jersey;* 2011. pp. 159–176.

54. Lopez RFV, Collett JH, Bentley M. Influence of cyclodextrin complexation on the *in vitro* permeation and skin metabolism of dexamethasone. *International Journal of Pharmaceutics.* 2000; 200(1): 127–132.

55. Ventura CA, Tommasini S, Falcone A. Influence of modified cyclodextrins on solubility and percutaneous absorption of celecoxib through human skin. *International Journal of Pharmaceutics.* 2006; 314(1): 37–45.

56. Ascenso A, Vultos F, Ferrinho D, Salgado A, Guerra Filho S, Ferrari V, *et al.* Effect of tretinoin inclusion in dimethyl-β-cyclodextrins on release rate from a hydrogel formulation. *Journal of Inclusion Phenomena and Macrocyclic Chemistry.* 2012; 73(1–4): 459–465.

57. Larrucea E, Arellano A, Santoyo S, Ygartua P. Study of the complexation behavior of tenoxicam with cyclodextrins in solution: Improved solubility and percutaneous permeability. *Drug Development and Industrial Pharmacy.* 2002; 28(3): 245–252.

58. Irie T, Uekama K. Pharmaceutical applications of cyclodextrins. III. Toxicological issues and safety evaluation. *Journal of Pharmaceutical Sciences.* 1997; 86(2): 147–162.

59. Cal K, Centkowska K. Use of cyclodextrins in topical formulations: Practical aspects. *European Journal of Pharmaceutics and Biopharmaceutics.* 2008; 68(3): 467–478.

60. Arima H, Yunomae K, Hirayama F, Uekama K. Contribution of P-glycoprotein to the enhancing effects of dimethyl-β-cyclodextrin on oral bioavailability of tacrolimus. *Journal of Pharmacology and Experimental Therapeutics.* 2001; 297(2): 547–555.

61. Arima H, Yunomae K, Morikawa T, Hirayama F, Uekama K. Contribution of cholesterol and phospholipids to inhibitory effect of dimethyl-β-cyclodextrin on efflux function of P-glycoprotein and multidrug resistance-associated protein 2 in vinblastine-resistant Caco-2 cell monolayers. *Pharmaceutical Research.* 2004; 21(4): 625–634.

62. Challa R, Ahuja A, Ali J, Khar R. Cyclodextrins in drug delivery: An updated review. *Aaps Pharmscitech.* 2005; 6(2): E329–E357.

63. Uekama K, Irie T, Sunada M, Otagiri M, Arimatsu Y, Nomura S. Alleviation of prochlorperazine-induced primary irritation of

skin by cyclodextrin complexation. *Chemical and Pharmaceutical Bulletin*. 1982; 30(10): 3860–3862.

64. Anadolu RY, Sen T, Tarimci N, Birol A, Erdem C. Improved efficacy and tolerability of retinoic acid in acne vulgaris: A new topical formulation with cyclodextrin complex ψ. *Journal of the European Academy of Dermatology and Venereology*. 2004; 18(4): 416–421.

65. Rode T, Frauen M, Müller B, Düsing H, Schönrock U, Mundt C, *et al.* Complex formation of sericoside with hydrophilic cyclodextrins: Improvement of solubility and skin penetration in topical emulsion based formulations. *European Journal of Pharmaceutics and Biopharmaceutics*. 2003; 55(2): 191–198.

66. Patravale V, Mandawgade S. Novel cosmetic delivery systems: An application update. *International Journal of Cosmetic Science*. 2008; 30(1): 19–33.

67. Shimada K, Ohe Y, Ohguni T, Kawano K, Ishii J, Nakamura T. Emulsifying properties of α-, β-, γ-cyclodextrins. *Nippon Shokuhin Kogyo Gakkaishi*. 1991; 57: 655–656.

68. Laurent S, Serpelloni M, Pioch D. A study of β-cyclodextrin-stabilized paraffin oil/water emulsions. *Journal of the Society of Cosmetic Chemists*. 1999; 50(1): 15–22.

69. Regiert M. Oxidation-stable linoleic acid by inclusion in α-cyclodextrin. *Journal of Inclusion Phenomena and Macrocyclic Chemistry*. 2007; 57(1–4): 471–474.

70. Glomot F, Benkerrour L, Duchêne D, Poelman M-C. Improvement in availability and stability of a dermocorticoid by inclusion in β-cyclodextrin. *International Journal of Pharmaceutics*. 1988; 46(1): 49–55.

71. Caddeo C, Manconi M, Valenti D, Pini E, Sinico C. Photostability and solubility improvement of β-cyclodextrin-included tretinoin. *Journal of Inclusion Phenomena and Macrocyclic Chemistry*. 2007; 59(3–4): 293–300.

72. Ammala A. Biodegradable polymers as encapsulation materials for cosmetics and personal care markets. *International Journal of Cosmetic Science*. 2013; 35(2): 113–124.

73. Arana-Sánchez A, Estarrón-Espinosa M, Obledo-Vázquez E, Padilla-Camberos E, Silva-Vázquez R, Lugo-Cervantes E. Antimicrobial and antioxidant activities of Mexican oregano essential oils (Lippia graveolens HBK) with different composition when microencapsulated in β-cyclodextrin. *Letters in Applied Microbiology*. 2010; 50(6): 585–590.

74. Daletos G, Papaioannou G, Miguel G. In: (Cabral Marques H, eds), *Proceedings of the 14th International Cyclodextrin Symposium*, Kyoto; Japan 2008. Tokyo, Japan: The Society of Cyclodextrins, Japan.
75. Sigmundsdottir H. Improving topical treatments for skin diseases. *Trends in Pharmacological Sciences*. 2010; 31(6): 239–245.
76. Dawson AL, Dellavalle RP, Elston DM. Infectious skin diseases: A review and needs assessment. *Dermatologic Clinics*. 2012; 30(1): 141–151.
77. Ceschel G, Bergamante V, Maffei P, Borgia SL, Calabrese V, Biserni S, *et al.* Solubility and transdermal permeation properties of a dehydroepiandrosterone cyclodextrin complex from hydrophilic and lipophilic vehicles. *Drug Delivery*. 2005; 12(5): 275–280.
78. Loftsson T, Brewster ME. Pharmaceutical applications of cyclodextrins: Basic science and product development. *Journal of Pharmacy and Pharmacology*. 2010; 62(11): 1607–1621.
79. Rigopoulos D, Ioannides D, Kalogeromitros D, Katsambas AD. Comparison of topical retinoids in the treatment of acne. *Clinics in Dermatology*. 2004; 22(5): 408–411.
80. Thevarajah S, Balkrishnan R, Camacho FT, Feldman SR, Fleischer Jr AB. Trends in prescription of acne medication in the US: Shift from antibiotic to non-antibiotic treatment. *Journal of Dermatological Treatment*. 2005; 16(4): 224–228.
81. Ascenso A, Cruz M, Euletério C, Carvalho Fa, Santos NC, Marques HC, *et al.* Novel tretinoin formulations: A drug-in-cyclodextrin-in-liposome approach. *Journal of Liposome Research*. 2013; 23(3): 211–219.
82. Plewig G, Braun-Falco O. Kinetics of epidermis and adnexa following vitamin A acid in the human. *Acta Dermato-Venereologica Supplementum*. 1974; 74: 87–98.
83. Montassier P, Duchêne D, Poelman M-C. Inclusion complexes of tretinoin with cyclodextrins. *International Journal of Pharmaceutics*. 1997; 153(2): 199–209.
84. Amdidouche D, Montassier P, Poelman M, Duchene D. Evaluation by laser Doppler velocimetry of the attenuation of tretinoin induced skin irritation by β-cyclodextrin complexation. *International Journal of Pharmaceutics*. 1994; 111(2): 111–116.
85. Ascenso A, Guedes R, Bernardino R, Diogo H, Carvalho Fa, Santos NC, *et al.* Complexation and full characterization of the tretinoin and dimethyl-β-cyclodextrin complex. *AAPS Pharm Sci Tech*. 2011; 12(2): 553–563.

86. Ascenso A, Duarte A, Silva A, Salgado A, Marques HC. Formulation studies on a topical gel of tretinoin–dimethyl-β-cyclodextrin complex. *Journal of Inclusion Phenomena and Macrocyclic Chemistry*. 2011; 69(3–4): 339–343.

87. At E, Mallard C. inventors; Galderma Research & Development, assignee. Compositions comprising at least one complex composed of a derivative of naphthoic acid and of at least one cyclodextrin and uses thereof patent 20120094954 A1. 2013.

88. Badilli U, Amasya G, Şen T, Tarimci N. Topical emulgel formulation containing inclusion complex of calcipotriol with cyclodextrin. *Journal of Inclusion Phenomena and Macrocyclic Chemistry*. 2014; 78(1–4): 249–255.

89. Feldman SR, Horn EJ, Balkrishnan R, Basra MK, Finlay AY, McCoy D, *et al.* Psoriasis: Improving adherence to topical therapy. *Journal of the American Academy of Dermatology*. 2008; 59(6): 1009–1016.

90. Finlay AY, Coles E. The effect of severe psoriasis on the quality of life of 369 patients. *British Journal of Dermatology*. 1995; 132(2): 236–244.

91. Singh R, Bharti N, Madan J. Characterization of cyclodextrin inclusion complexes—a review. *Journal of Pharmaceutical Science and Technology*. 2010; 2(3); 171–183.

92. Harris RE, Beebe-Donk J, Namboodiri KK. Inverse association of non-steroidal anti-inflammatory drugs and malignant melanoma among women. *Oncology Reports*. 2001; 8(3): 655–657.

93. Godwin DA, Wiley CJ, Felton LA. Using cyclodextrin complexation to enhance secondary photoprotection of topically applied ibuprofen. *European Journal of Pharmaceutics and Biopharmaceutics*. 2006; 62(1): 85–93.

94. Ammar H, Ghorab M, Mahmoud A, Makram T, Noshi S. Topical liquid crystalline gel containing lornoxicam/cyclodextrin complex. *Journal of Inclusion Phenomena and Macrocyclic Chemistry*. 2012; 73(1–4): 161–175.

95. Vega E, Egea MA, Garduño-Ramírez ML, García ML, Sánchez E, Espina M, *et al.* Flurbiprofen PLGA-PEG nanospheres: Role of hydroxy-β-cyclodextrin on *ex vivo* human skin permeation and *in vivo* topical anti-inflammatory efficacy. *Colloids and Surfaces B: Biointerfaces*. 2013; 110: 339–346.

96. Arima H, Miyaji T, Irie T, Hirayama F. Enhancing effect of hydroxypropyl-β-cyclodextrin on cutaneous penetration and activation of ethyl 4-biphenylyl acetate in hairless mouse skin. *European Journal of Pharmaceutical Sciences*. 1998; 6(1): 53–59.

97. Vianna RFL, Bentley M, Ribeiro G. Formation of cyclodextrin inclusion complexes with corticosteroids: Their characterization and stability. *International Journal of Pharmaceutics* 1998; 167(1): 205–213.

98. Scalia S, Villani S, Scatturin A, Vandelli M, Forni F. Complexation of the sunscreen agent, butyl-methoxydibenzoylmethane, with hydroxypropyl-β-cyclodextrin. *International Journal of Pharmaceutics.* 1998; 175(2): 205–213.

99. Cho Y-C, You S-K, Kim HJ, Cho C-W, Lee I-S, Kang BY. Xanthohumol inhibits IL-12 production and reduces chronic allergic contact dermatitis. *International Immunopharmacology.* 2010; 10(5): 556–561.

100. Ono M, Yamaguchi N, Yamaguchi K. inventors; Google Patents, assignee. Xanthohumol compositions and methods for treating skin diseases or disorders patent European Patent EP 2274411 A1. 2009.

101. Chiller K, Selkin BA, Murakawa GJ. Skin microflora and bacterial infections of the skin. *Journal of Investigative Dermatology Symposium Proceedings*; 2001; 6(3): 170–174.

102. Piérard GE, Hermanns-Lê T, Delvenne P, Piérard-Franchimont C. Miconazole, a pharmacological barrier to skin fungal infections. *Expert Opinion on Pharmacotherapy.* 2012; 13(8): 1187–1194.

103. Lai Y, Di Nardo A, Nakatsuji T, Leichtle A, Yang Y, Cogen AL, *et al.* Commensal bacteria regulate Toll-like receptor 3-dependent inflammation after skin injury. *Nature Medicine.* 2009; 15(12): 1377–1382.

104. Ong PY, Ohtake T, Brandt C, Strickland I, Boguniewicz M, Ganz T, *et al.* Endogenous antimicrobial peptides and skin infections in atopic dermatitis. *New England Journal of Medicine.* 2002; 347(15): 1151–1160.

105. Tenjarla S, Puranajoti P, Kasina R, Mandal T. Preparation, characterization, and evaluation of miconazole–cyclodextrin complexes for improved oral and topical delivery. *Journal of Pharmaceutical Sciences.* 1998; 87(4): 425–429.

106. Sammeta SM, Vaka SRK, Murthy SN. Transcutaneous electroporation mediated delivery of doxepin-HPCD complex: A sustained release approach for treatment of postherpetic neuralgia. *Journal of Controlled Release.* 2010; 142(3): 361–367.

107. Ruiz HK, Serrano DR, Dea-Ayuela MA, Bilbao-Ramos PE, Bolás-Fernández F, Torrado JJ, *et al.* New amphotericin B-gamma cyclodextrin formulation for topical use with synergistic activity

against diverse fungal species and Leishmania spp. *International Journal of Pharmaceutics*. 2014; 473(1): 148–157.

108. Liu Y, Petreaca M, Yao M, Martins-Green M. Cell and molecular mechanisms of keratinocyte function stimulated by insulin during wound healing. *BMC Cell Biology*. 2009; 10(1): 1.

109. Zhang Y, Zhu J, Tang Y, Chen X. The preparation and application of pulmonary surfactant nanoparticles as absorption enhancers in insulin dry powder delivery. *Drug Development and Industrial Pharmacy*. 2009; 35(9): 1059–1065.

110. Sajeesh S, Bouchemal K, Marsaud V, Vauthier C, Sharma CP. Cyclodextrin complexed insulin encapsulated hydrogel microparticles: An oral delivery system for insulin. *Journal of Controlled Release*. 2010; 147(3): 377–384.

111. Valentini SR, Nogueira AC, Fenelon VC, Sato F, Medina AN, Santana RG, *et al.* Insulin complexation with hydroxypropyl-beta-cyclodextrin: Spectroscopic evaluation of molecular inclusion and use of the complex in gel for healing of pressure ulcers. *International Journal of Pharmaceutics*. 2015; 490(1–2): 229–239.

112. Lodén M. Role of topical emollients and moisturizers in the treatment of dry skin barrier disorders. *American Journal of Clinical Dermatology*. 2003; 4(11): 771–788.

113. Cheong WK. Gentle cleansing and moisturizing for patients with atopic dermatitis and sensitive skin. *American Journal of Clinical Dermatology*. 2009; 10(1): 13–17.

114. Gelmetti C. Therapeutic moisturizers as adjuvant therapy for psoriasis patients. *American Journal of Clinical Dermatology*. 2009; 10(1): 7–12.

115. Hon KL, Leung AK, Barankin B. Barrier repair therapy in atopic dermatitis: An overview. *American Journal of Clinical Dermatology*. 2013; 14(5): 389–399.

116. Moloney FJ, Collins S, Murphy GM. Sunscreens. *American Journal of Clinical Dermatology*. 2002; 3(3): 185–191.

117. Matsuda H, Arima H. Cyclodextrins in transdermal and rectal delivery. *Advanced Drug Delivery Reviews*. 1999; 36: 81–99.

118. Guskey G, Bacon D, Junneja P, Motley C, Rizzi G. inventors; Deodorant compositions containing cyclodextrin odor controlling agents, patent US 6123932. 2000.

119. Makin SA, Lowry MR. Deodorant ingredients. In: Laden K, ed. *Antiperspirants and Deodorants*. Second ed. New York: Marcel Dekker, Inc. 1999; 169–214.

120. Kilian J, Panzarella F. Comparative studies of samples of perspiration collected from clean and unclean skin of human subjects. *Proceedings of the Scientific Section of the Toilet Goods Association.* 1947; 7: 3.

121. Shelley WB. Apocrine Sweat 1. *Journal of Investigative Dermatology.* 1951; 17(5): 255–256.

122. Hara K, Mikuni K, Hara K, Hashimoto H. Effects of cyclodextrins on deodoration of "Aging Odor". *Journal of Inclusion Phenomena and Macrocyclic Chemistry.* 2002; 44(1–4): 241–246.

123. Lopedota A, Cutrignelli A, Laquintana V, Franco M, Donelli D, Ragni L, *et al.* β-cyclodextrin in personal care formulations: Role on the complexation of malodours causing molecules. *International Journal of Cosmetic Science.* 2015; 37(4): 438–445.

124. Yamagata Y, Yoshibumi S. inventors; (Lion Corp.) assignee. Hair preparations containing cationic surfactants and cyclodextrins. patent JP 62,267,220 (87,267,220). 1986.

125. Kishi T, inventor; Deodorant adhesive tapes containing cyclodextrins for the control of underarm odour. patent Japanese patent 63280013 1988.

126. Maekawa A, inventor; Sunstar Inc., assignee. Antiperspirant aerosol compositions containing cyclodextrin. patent JP 03, 170, 415 (91,170,415), 1989.

127. Matsuda H, Ito K. inventors; Shiseido Co. Ltd., assignee. Body deodorants containing hydroxyalkylated cyclodextrins. Japan1991.

128. Woo R, Trinh T, Cobb D, Schneiderman E, Wolff A, Rosenbalm E, *et al.* Uncomplexed cyclodextrin compositions for odour control. 1998.

129. Trinh J, Dodd T, Bartolo R, Lucas J. inventors; Cyclodextrin based compositions for reducing body odour. 1999.

130. Traupe B, Raschkle T, Schwanke F. inventors; Cosmetic and/or dermatological preparations, having e.g. antimicrobial and/or deodorant activity, comprise cyclodextrin (derivative) inclusion compound of branched carboxylic acid. 2003.

131. Lammerschop O, Schmidt A. inventors; New cyclodextrin derivatives with modified polysiloxane, useful e.g. in textile washing- or cleaning mediums, cosmetic skin care medium such as deodorants, soaps, body lotions, after shaves and perfumes; and as hair caring agents. 2005.

132. Matsuura I, Kimura Y, Sakai Y, Nakatsuji N. inventors; Glutathione-cyclodextrin inclusion complex for cosmetic compositions, patent EP 0442420A1. 1991.

133. Buschmann HJ, Schollmeyer E. Applications of cyclodextrins in cosmetic products: A review. *Journal of Cosmetic Science* 2002; 53(3): 185–191.

134. Lea K, Roh S, Seok J, Seong J. Inventors; Functional seawater for use in bathing application to e.g. treat and prevent skin diseases, contains cyclodextrin and aqueous phytoncide solution. Patent KR2011080064-A. 2011.

135. Numanoğlu U, Şen T, Tarimci N, Kartal M, Koo OMY. Use of cyclodextrins as a cosmetic delivery system for fragrance materials: Linalool and benzyl acetate. *AAPS PharmSciTech*. 2007; 8(4): E85.

136. Chisvert A, Pascual-Martı M, Salvador A. Determination of the UV filters worldwide authorised in sunscreens by high-performance liquid chromatography: Use of cyclodextrins as mobile phase modifier. *Journal of Chromatography A*. 2001; 921(2): 207–215.

137. Coelho GLN, Dornelas CB, Soares KCC, dos Santos EP, Vergnanini AL, dos Santos TC, *et al.* Preparation and evaluation of inclusion complexes of commercial sunscreens in cyclodextrins and montmorillonites: Performance and substantivity studies. *Drug Development and Industrial Pharmacy*. 2008; 34(5): 536–546.

138. Hayden C, Roberts M, Benson H. Sunscreens: Are Australians getting the good oil? *Australian and New Zealand Journal of Medicine*. 1998; 28(5): 639–646.

139. Maier H, Schauberger G, Brunnhofer K, HoÈnigsmann H. Change of ultraviolet absorbance of sunscreens by exposure to solar-simulated radiation. *Journal of Investigative Dermatology*. 2001; 117(2): 256–262.

140. Scalia S, Molinari A, Casolari A, Maldotti A. Complexation of the sunscreen agent, phenylbenzimidazole sulphonic acid with cyclodextrins: Effect on stability and photo-induced free radical formation. *European Journal of Pharmaceutical Sciences*. 2004; 22(4): 241–249.

141. Scalia S, Villani S, Casolari A. Inclusion complexation of the sunscreen agent 2-ethylhexyl-p-dimethylaminobenzoate with hydroxypropyl-β-cyclodextrin: 30 Effect on photostability. *Journal of Pharmacy and Pharmacology*. 1999; 51(12): 1367–1374.

142. Scalia S, Casolari A, Iaconinoto A, Simeoni S. Comparative studies of the influence of cyclodextrins on the stability of the sunscreen agent, 2-ethylhexyl-p-methoxycinnamate. *Journal of Pharmaceutical and Biomedical Analysis.* 2002; 30(4): 1181–1189.

143. Sarveiya V, Templeton JF, Benson HA. Inclusion complexation of the sunscreen 2-hydroxy-4-methoxy benzophenone (oxybenzone) with hydroxypropyl-β-cyclodextrin: Effect on membrane diffusion. *Journal of Inclusion Phenomena and Macrocyclic Chemistry.* 2004; 49(3–4): 275–281.

144. Felton LA, Wiley CJ, Godwin DA. Influence of cyclodextrin complexation on the *in vivo* photoprotective effects of oxybenzone. *Drug Development and Industrial Pharmacy.* 2004; 30(1): 95–102.

145. Simeoni S, Scalia S, Benson HAE. Influence of cyclodextrins on *in vitro* human skin absorption of the sunscreen, butyl-methoxydibenzoylmethane. *International Journal of Pharmaceutics.* 2004; 280(1–2): 163–171.

146. Simeoni S, Scalia S, Tursilli R, Benson H. Influence of cyclodextrin complexation on the *in vitro* human skin penetration and retention of the sunscreen agent, oxybenzone. *Journal of Inclusion Phenomena and Macrocyclic Chemistry.* 2006; 54(3–4): 275–282.

147. Saeki T, Morifuji T. inventors; Sekisui Plastics, Tanaka Narikazu, Japan, assignee. Moisturizer-containing cyclodextrin composite particles for cosmetics patent Japanese patent 08,151,317 (96, 151, 317). 1996.

148. Matsuda H, Ito K. inventors; Shiseido Co. Ltd., assignee. Cosmetics packs containing hydroxyalkylated cyclodextrin patent Japanese patent 03,287,512 (91,287,512). 1991.

149. Imamura K, Tsuchama Y, Tsunakawa H, Okamura K, Okamoto R, Harada K. inventors; Merushan Kk. Kogyo Gijutsuib, assignee. Cosmetics containing water-soluble cyclodextrin polymers as scrubbing particles patent Japanese patent 05,105, 619 (93,105,619). 1991.

150. Tarimci N. Cyclodextrins in the cosmetic field. In: (Bilensoy E, ed), *Cyclodextrins in Pharmaceutics, Cosmetics, and Biomedicine: Current and Future Industrial Applications.* New Jersey: John Wiley & Sons; 2011. pp. 131–144.

151. Farris PK. Cosmeceuticals and clinical practice. *Cosmeceuticals and Cosmetic Practice.* 2014: 1–7.

152. Takagi T, Ramachandran C, Bermejo M, Yamashita S, Yu LX, Amidon GL. A provisional biopharmaceutical classification of the top 200 oral drug products in the United States, Great Britain, Spain, and Japan. *Molecular Pharmaceutics*. 2006; 3(6): 631–643.

153. Conte C, Caldera F, Catanzano O. β-Cyclodextrin Nanosponges as Multifunctional Ingredient in Water-Containing Semisolid Formulations for Skin Delivery. *Journal of Pharmaceutical Sciences*. 2014; 103(12): 3941–3949.

154. Che J, Wu Z, Shao W, Guo P, Lin Y, Pan W, *et al.* Synergetic skin targeting effect of hydroxypropyl-β-cyclodextrin combined with microemulsion for ketoconazole. *European Journal of Pharmaceutics and Biopharmaceutics*. 2015; 93: 136–148.

Chapter 14

Topical Formulations for Onychomycosis: A Review

Barbara S. Gregorí Valdes,[a] Carolina de Carvalho Moore Vilela,[a] Andreia Ascenso,[a] João Moura Bordado,[b] and Helena Ribeiro[a]

[a]*Research Institute for Medicine and Pharmaceutical Science (iMed.UL), Faculty of Pharmacy, University of Lisbon, Portugal*
[b]*CERENA. Instituto Superior Tecnico, University of Lisbon, Portugal*

barbara.valdes@ist.utl.pt

14.1 Introduction

Nail fungal infections frequently addressed as onychomycosis present a considerable prevalence on the global population.

Onychomycosis often causes discoloration, thickening and detachment from the nail bed [1], leading to eventual destruction of the nail bed and inducing important psychosocial and physical detrimental effects. In more extreme cases, affected individuals may feel pain, discomfort and impaired or lost tactile functions that lead to difficulties on walking, fitting shoes, or other exercise [2]. This nail fungal infection counts for about half of all nail diseases and approximately 30% of mycotic infections [3, 4]. It can be caused by dermatophytes, yeasts, or

Carrier-Mediated Dermal Delivery: Applications in the Prevention and Treatment of Skin Disorders
Edited by Andreia Ascenso, Sandra Simões, and Helena Ribeiro

ISBN 978-981-4745-58-1 (Hardcover), 978-1-315-36447-6 (eBook)
www.panstanford.com

non-dermatophytic molds. Approximately 80–90% of all onychomycosis are caused by *Trichophyton rubrum* and *Trichophyton mentagrophytes* dermatophytes. Several non-dermatophytic fungi, such as *Acremonium* spp., *Alternaria* spp., *Aspergillus* spp., *Fusarium* spp., *Scytalidium* spp., and *Scopulariopsis* spp. have been reported to be involved in 2–11% of onychomycosis cases [4]. Regarding yeasts, *Candida* spp. is responsible for 2–10% of fungal infections.

Transmission of dermatophytes is mainly due to moist infected floors, whereas non-dermatophytic fungi are commonly found in traumatized nails in the elderly [4]. Onychomycosis is also associated with tinea pedis in 30–59% of reported cases [5].

Despite not being a life-threatening disease, this infection causes several problems to the infected patients. Unfortunately, limited topical therapy is available, and oral therapy comprises several problems.

This chapter reviews existing literature and report on transungual permeation and topical formulations for onychomycosis treatment.

14.2 Onychomycosis

14.2.1 Epidemiology

Onychomycosis prevalence has been reported to occur from 3% to 26% in the general population [1, 3–6].

Onychomycosis cases have been increasing in the past decades, from a prevalence of 1–2% to 14–15% in a 20-year time window [4]. According to several large studies, it is expected that onychomycosis prevalence will increase to 20% or more in the next decade [4].

Onychomycosis is a serious public health problem leading to detrimental physical and psychological effects. There are specific population groups affected by this disease. About 50% of all psoriasis patients (1–8% of general population) report nail psoriasis and 40–60% are affected by onychomycosis [5, 7, 8]. Onychomycosis also affects about one third of diabetes patients since they are more likely to suffer foot complications that can lead to dermatophytic infections, ulceration, osteomyelitis, cellulitis, and tissue necrosis that may result at worst in amputation [9].

Onychomycosis greatly affects the patients' quality of life even not being a life-threatening condition. Approximately half of patients with onychomycosis report some discomfort due to this condition: pain, difficulty on wearing footwear and walking, emotional embarrassment, and work-related difficulties.

14.2.2 Risk Factors

Onychomycosis is an ailment that has several different risk factors. As for **age**, it is reported that onychomycosis is more prevalent at older age, being the elderly the most affected population with an age/infection correlation of 20% and up to 50% of subjects over 60 and 70 years old, respectively [1, 10, 11]. This correlation is often attributed to poor peripheral circulation, sedentary lifestyle, weaker immune system, diabetes, nail trauma, difficulty in performing foot hygiene and slower growing nails [2, 10, 12, 13]. Children are rarely affected by onychomycosis, with an incidence of just about 0.4%, since this group has smaller nail surface, faster nail growth, less nail injuries, and lower tinea pedis incidence rate besides less contact with infected surfaces [5, 14–16].

Gender is also considered a risk factor being males more often affected by this type of nail infection [2, 11, 17]. Several authors postulated that this may be due to hormone differences between sexes, leading to a difference in the capacity to inhibit the growth of dermatophytes [18]. It is also considered that the use of occlusive footwear and nail injuries may contribute to higher incidence in males [2].

Recent studies have stated that onychomycosis may have a **genetic** basis, with an autosomal dominant pattern of inheritance related to *T. rubrum* infection, and increased susceptibility when at least one parent had onychomycosis [5, 8, 17, 19–22].

The **environment** plays an important role in contracting a fungal infection. Societies that less wear shoes tend to show a lower infection rate [2, 15]. On the contrary, athletes display an increased infection rate due to occlusive footwear, which allows a dark and moist environment for onychomycosis to develop. In addition, athletes contact more with moist infected surfaces, like swimming pools, communal shower rooms, and public toilets [23]. Nail trauma, synthetic clothing (which retains sweat), and tinea pedis are also associated with sports and the incidence of this disease [2, 5, 13, 21, 23–26].

Immunodeficiency in HIV-infected individuals and in transplant recipients is also a risk factor for onychomycosis. These individuals, whose T-lymphocyte count is as low as $400/mm^3$, tend to have a more widespread infection affecting both toe and fingernails [1, 5, 11, 23, 25–29].

Diabetic individuals are three times more likely to have onychomycosis than the general population, being 34% with onychomycosis. Diabetics have decreased peripheral circulation, neuropathy, impaired wound healing, and increased difficulty on foot checkups due to obesity, retinopathy, or cataracts. Therefore, injuries in toenails may go unnoticed due to neuropathy and act as entryways for bacteria, fungi, and other pathogens, leading to serious complications [4, 5, 11, 21, 24, 30].

14.2.3 Clinical Classification

There are five types of onychomycosis established according to the mode and site of invasion by the pathogen: distal and lateral subungual onychomycosis, superficial white onychomycosis, proximal subungual onychomycosis, total dystrophic onychomycosis (Fig. 14.1), and endonyx onychomycosis [1, 2, 5, 29, 31–33].

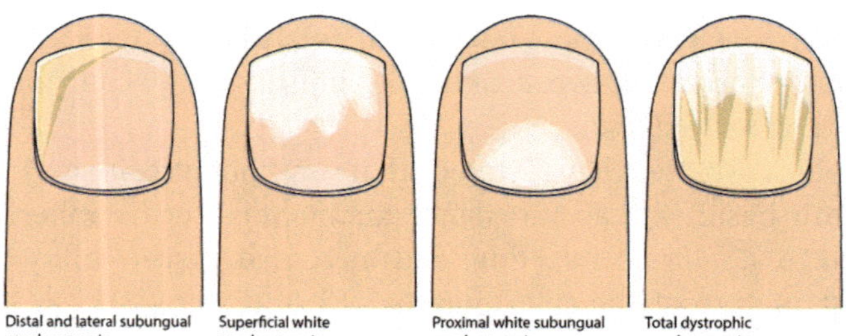

Figure 14.1 Types of onychomycosis according to the mode and site of invasion of the pathogen.

14.2.3.1 Distal and lateral subungual onychomycosis

Distal and lateral subungual onychomycosis (DLSO) is the most common type of onychomycosis [1, 2]. The fungal infection progresses in the nail from the distal to the proximal edge, via the distal-lateral margins or through the lateral nail plate groove, originating from the hyponychium. This type of infection is mainly caused by *Trichophyton* spp. and sometimes by *Scytalidium* spp. and *Candida* spp. [5]. It is frequent to find paronychia in these individuals, leading to subungual hyperkeratosis, nail thickening and onycholysis (nail detachment from the nail bed). The subungual space can be colonized by infectious bacteria and fungi, causing discoloration of the nail plate [1, 2, 5].

14.2.3.2 Proximal subungual onychomycosis

In this type of onychomycosis, fungi invade the area under the nail cuticle and induce the infection of the proximal nail plate. Then, it progresses distally along the nail plate. This clinical manifestation of onychomycosis is mainly prevalent in immunocompromised individuals, being especially common in AIDS patients where it is considered an early HIV infection clinical marker. The fungi responsible for proximal subungual onychomycosis (PSO) are *T. rubrum*, *C. albicans*, *Fusarium* spp., *Aspergillus* spp. and *Scopulariopsis brevicaulis* [5, 25].

14.2.3.3 Superficial white onychomycosis

Superficial white onychomycosis (SWO) presents itself as opaque white patches along the dorsal surface of the nail plate. The upper layers of nail keratin are infected mainly by *Trichophyton mentagrophytes* and *T. rubrum*. Other pathogens responsible for this infection are non-dermatophyte molds such as *Fusarium* spp., *Acremonium* spp. and *Aspergillus* spp. [28].

14.2.3.4 Endonyx onychomycosis

Endonyx onychomycosis (EM) is the most recently described form of onychomycosis, reported as an infection of both superficial and deeper layers of the nail plate. EM does not usually cause nail thickening, detachment, and inflammatory processes, but it leads to lamellar splitting, coarse pitting, and milky white patches within affected nail plates [4]. Usual pathogens are *T. soudanense* and *T. violaceum* [29, 30].

14.2.3.5 Total dystrophic onychomycosis

Total dystrophic onychomycosis (TDO) is usually considered the end stage of any other type of onychomycosis. It can also exist on its own—primary total dystrophic onychomycosis—occurring mostly in patients with chronic mucocutaneous candidiasis. It is described as the nearly complete destruction of the nail plate.

Diagnosing onychomycosis is important in order to guarantee the best possible treatment and with less side effects. Nowadays, the diagnosis is based on a clinical evaluation in which nail samples are collected and analyzed by microscopy and fungal culture [1, 15, 23, 25]. Onychomycosis should be differentiated from other similar conditions, such as psoriasis of the nail, eczema, bacterial infections, contact dermatitis, traumatic onychodystrophies, chronic onycholysis, lichen planus, chronic paronychia, hemorrhage or trauma, onychogryphosis, median canalicular dystrophy, pincer nail, yellow nail syndrome, subungual malignant melanoma, and subungual squamous cell carcinoma [4, 15]. Among these disorders, it is especially difficult to differentiate onychomycosis from nail psoriasis, since they tend to share similar nail morphological changes [34].

14.3 Transungual Delivery

All drugs must reach their target location in order to be more effective. Particularly, the nail unit has different characteristics from the rest of body surface, and consequently, topical formulations may not deliver enough drug amounts to the nail bed. Therefore, it is quite important to clearly understand the nail biology in order to develop more efficient topical formulations.

14.3.1 Nail Structure and Transungual Permeation

The nail plate is the main obstacle in the nail unit for drug permeation. It is hard, thin (0.25–1 mm), slightly elastic, translucent, and convex shaped and consists of 80–90 layers of dead, keratinized, flattened cells tightly bound to each other via intercellular links [35, 36]. The nail is constituted for nail plate, lateral fold, proximal fold, matrix, nail bed and distal groove as showed in Fig. 14.2 [37]. The nail plate can be divided in three macroscopic layers—dorsal, intermediate, and ventral layers—being the dorsal layer the hardest one [37]. In terms of chemical composition, the nail is composed of fibrous proteins, keratins (80% "hard" keratin, remaining "soft" keratin), water and a low amount of lipids (0.1–1%) in contrast to the stratum corneum (~10%). Under normal conditions, the nail plate can contain 7–12% of water, but this content can rise up to 35% [35–37].

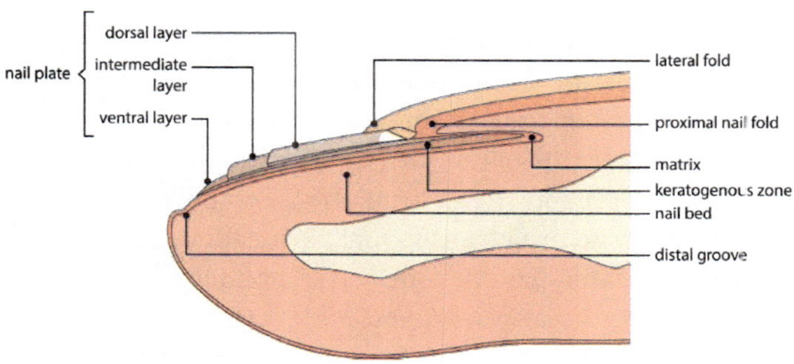

Figure 14.2 Representation of nail structure.

Transungual permeation is dependent on drug properties (molecular weight, log *P*, charge, pKa, etc.), formulation characteristics (pH, water content, etc.), nail plate properties (diseased state, hydration, thickness and keratin content), and drug–keratin interactions.

Accordingly, the most significant properties of the permeant drug are the following:

Molecular weight: It shows an inverse relationship with the drug permeation into the nail plate. Heavier molecules have a higher time permeation through the keratin network on the nail plate [35]. Drugs below 300 $gmol^{-1}$ MW have enhanced transungual permeation rate [37]. Indeed, molecular weight is considered to be the most significant property affecting the nail permeation.

Log *P*: Permeation rate was found to decrease with an increase in carbon-chain length or lipophilicity attributed to the hydrophilic nature of the nail plate [37–40].

Charge: Non-ionic drugs were found to have about 10-fold greater permeability as compared to their ionic counterparts [38]. However, Elsayed *et al.* [37] stated that these results did not take into account the molecular weight of permeant drug, and regarding other studies, drug ionization can have a permeation enhancement effect due to an increase of aqueous solubility maximizing the transungual flux.

pKa and other physicochemical properties: In general, soluble molecules have good permeation across the nail plate. Weak acidic drugs are well permeated at higher pH, while weak basic drugs exhibit better permeation at lower pH values. Ionization can also contribute to increase drug solubility. In addition, if the molecule is sublimable at body temperature, it will be advantageous for permeating through the diseased nail plate, since it is able to overcome air cavities by being sublimed and reach the other side of the cavity [36].

On the other hand, the formulation composition also plays an important role on drug transungual permeation, as follows:

pH: Nail keratin has an isoelectric point of 5.0, turning to a net negative charge at pH 7.4 and net positive charge at pH 2.0. Therefore, some molecules might be repelled due to charging depending on pH. Some studies have stated that antifungal drugs have lower activity in acidic environments [39].

> ***Water content**: Water enhances the diffusion through the nail plate. As water hydrates the nail plate, its volume increases and this swelling causes larger pores, facilitating the permeation of bigger molecules [35].*
>
> *Finally, the nail plate properties influence the drug transungual permeation as well [35, 40]:*
>
> *Diseased nail plate: This state has an enormous influence in permeation as diseased nail plates have uneven thickness, fungi concentration, and onycholysis, which can lead to a detached nail surrounded by non-detached areas. In some cases, this can be beneficial since some formulations can be applied to the detached space [41].*
>
> ***Hydration**: It increases the ungual permeability of polar compounds, as discussed before.*
>
> ***Drug binding to keratin**: It can lead to disappointing results since it reduces the availability of the permeant drug and weakens its concentration gradient. Although this interaction highly influences antifungal activity, it is not frequently considered in many studies [42].*

In summary, all factors that influence the drug transungual permeation are outlined in Fig. 14.3.

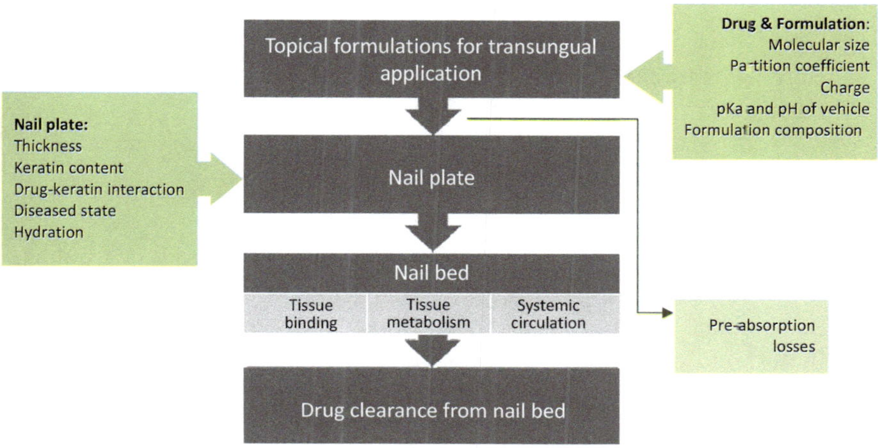

Figure 14.3 Factors influencing nail permeation of topically applied formulations.

14.3.1.1 Mathematical description of nail permeability

Fick's adapted laws of permeability can be used for studying permeation through the nail plate.

Flux (J) occurs when an amount of permeant (Q in mol or g) moves across a membrane with a certain area (A in m^2) during time (t in s) (Eq. 14.1). **Fick's first law** describes the flux where D is the permeant diffusion coefficient (m^2s^{-1}) and $\partial C/\partial x$ is the permeant concentration gradient. The negative sign indicates that the flux is in direction of decreasing concentration (Eq. 14.2).

$$J = \frac{Q}{At} \tag{14.1}$$

$$J = -D\frac{\partial C}{\partial x} \tag{14.2}$$

The permeability coefficient (P) is given by Eq. 14.3:

$$P = \frac{DK}{h}, \tag{14.3}$$

where K is the distribution equilibrium (distribution constant) and h is the membrane thickness (in m) [37, 43].

According to Kobayashi *et al.* [44], the functional dependence of D on molecular weight (MW) of the molecule can be described by Eq. 14.4:

$$D = D^{0(-\beta\cdot\mathrm{MW})} \tag{14.4}$$

Here D^0 represents the diffusivity of a hypothetical molecule having zero molecular weight and β is a constant.

Nail permeability was found to be independent of the drug lipophilicity according to Eq. 14.5 [44]:

$$\log P = \log\left(\frac{D^0}{h}\right) - \beta'\cdot\mathrm{MW}, \quad \text{where } \beta^0 = \beta/2.303 \tag{14.5}$$

As human nails are not always available for studies, it is important to establish a viable *in vitro* model. Animal hooves have been used as a human nail model in the last decades. There are reports of bovine, porcine, and equine hoof use [35]. Saner *et al.* [35] described the relationship between permeability coefficients through human nail plates and hoof membranes can be portrayed by Eq. 14.6:

$$\log P_\mathrm{N} = 3.723 + 1.751 \log P_\mathrm{H}, \tag{14.6}$$

where P_N is the human nail plate permeability coefficient and P_H is the permeability coefficient through bovine hoof membrane obtained experimentally. Recently researchers have used keratin films made from human hair as a model for human nails [36, 45, 46]. Other models used to study nail permeability include wax blocks, nail clippings, cadaver nail plates and excised cadaver toes [37].

14.4 Onychomycosis Topical Therapy

There are several therapeutic options to treat this nail fungal infection: oral, topical, and combined therapy [5, 10, 15, 27, 29, 47–51]. Adhesive patches have been also studied but displayed very low drug permeation. Thus, these pharmaceutical forms have a low representativity in the market [52].

Despite being more effective, oral therapy presents several risks for patients, including adverse reactions and possible drug interactions [10, 53–57]. This is especially relevant in the elderly (the most affected age group) who are polymedicated, and therefore, more susceptible to these reactions [10, 12]. Thus, it would be expected a higher investment on topical therapy.

Currently, most common topical formulations present a low drug deliver to the infection site mainly due to the nail plate characteristics, and the cure rate is lower than oral and combined therapy. Topical formulations are also bothersome to apply if there are several affected nails, reducing patient compliance [58, 59]. Nonetheless, there are several topical formulations available in the market used in monotherapy or in combined therapy. Firstly, it will be reviewed different drug delivery enhancers, antifungal drugs, and finally, several pharmaceutical forms used in topical treatment of onychomycosis.

14.4.1 Drug Delivery Enhancers

The nail plate is a great barrier to drug permeation, as stated before. Thus, it is important to enhance the permeation to guarantee effective drug delivery.

There are several **physical enhancement methods** such as nail plate abrasion, etching of the nail surface with acid, ablation of the nail with pulsed lasers, microporation of the nail plate, application of low frequency ultrasound and electric current through the nail [38]. Although these methods are effective, reduce patient compliance and need the presence of a specialized technician to perform them.

Chemical enhancement of ungual drug delivery can be also useful. This type of enhancement is cheaper, easier to apply, and can be performed by the patient before or concomitantly with drug application. It focuses on breaking chemical and physical bonds that maintain the integrity of the nail plate keratin. Targets for ungual chemical penetration enhancers are disulfide, peptide, hydrogen, and polar bonds. Chemical enhancers are classified according to the targeted bond and respective mechanism [40].

The nail plate is cohesive due to S-S bonds, and these are great targets for enhancers. Enhancers targeting this bond will reduce the disulfide linkage, engaging in nucleophilic attacks as follows:

$$\text{Nail-S-S-Nail} + 2\ \text{R-SH} \Leftrightarrow 2\text{Nail-SH} + \text{R-S-S-R}$$

In this way, enhancers with thiol or ammonia groups with lone pairs of electrons will offer advantage in breaking keratin bonding [43].

14.4.1.1 Disulfide bond cleaving by reducing agents

14.4.1.1.1 Thiols

Thiols are compounds containing sulfhydryl groups (–SH groups) that reduce the disulfide linkage in the keratin matrix of the nail [60]. Examples of thiols used as permeation enhancers include *N*-acetylcysteine, mercaptoethanol, *N*-(2-mercaptopropionyl) glycine (MPG), pyrithione and thioglycolic acid (TGA). Once disulfide bonds are cleaved, they are unlikely to be reformed; thus this alteration is permanent.

14.4.1.1.2 Sulfites

Sodium sulfite is known to cleave disulfide bonds when incubated with proteins and peptides, producing thiols and thiosulfates. It was then found to enhance transungual permeation on both pretreatment and co-application [61].

14.4.1.2 Disulfide bond cleaving by oxidizing agents

In this category of enhancers, hydrogen peroxide has been used alone or in combination with urea, as urea hydrogen peroxide (UHP). Pretreatment of nails with hydrogen peroxide has shown to increase mannitol permeation 3-fold [62]. MedNail® technology consists of pretreating nail with reducing agent TGA, followed by this oxidizing agent UHP. Treatment with this technology increased terbinafine drug flux 18-fold [63].

14.4.1.3 Enhancement by solvents

14.4.1.3.1 Water

The effect of water is recognized in literature as an important enhancer. As nail hydration increases permeability by a mechanism that is thought to be related to swelling and formation of larger pores through the keratin matrix, water content in formulations becomes important to guarantee maximum drug ungual flux. Gunt *et al.* [64] reported a several fold increase related to nail hydration. Kobayashi *et al.* [44] also reported the enhancement of nail permeation by nail hydration. Not with standing, in some cases nail hydration caused by formulation does not seem to improve ungual permeation [38].

14.4.1.3.2 DMSO

DMSO is a transdermal enhancer that interacts with the lipid domains of the stratum corneum, increasing fluidity and promoting partitioning of drugs into the skin. Stüttgen and Bauer [65] reported that DMSO has enhanced the delivery of econazole. However, DMSO has irritating properties at high concentrations.

14.4.1.4 Keratolytic agents

Keratolytic agents disrupt the tertiary structure and hydrogen bonds present in the keratin matrix, "unfolding" keratin and allowing larger molecules to pass through the nail pores [63].

Urea and salicylic acid soften and hydrate the nail plate, enhancing drug permeation. This type of agents weaken and damage the nail plate [38]. Nevertheless, urea cannot induce enhanced flux on its own, but it rather acts by synergizing with other enhancers to increase permeability. For example, urea combined with *N*-acetylcysteine increased the ungual concentration of itraconazole by 94-, 20-, and 49-fold compared with control (no enhancer), only urea and only *N*-acetylcysteine, respectively [38]. It was also reported an increase in permeation for the concomitant use of urea and MPG. This synergy may be explained by easier access of other permeation enhancers during the keratin unfolding by urea.

Urea has also been used to chemically avulse diseased nail at higher concentrations around 40% [66, 67]. It has also been included in nail lacquers [68].

14.4.1.5 Enzymes

Keratinolytic enzymes hydrolyze the keratin matrix of the nail plate, altering its barrier properties and facilitating permeation. Studies have shown that keratinase enzyme markedly enhances nail permeation [61]. This enzyme affects the surface of human nail, causing corneocytes to detach and lift off the plate corroding the surface.

Papain is an endopeptidase containing a highly reactive sulfhydryl group, and has shown promising results as a transungual enhancer [35].

14.4.1.6 Other enhancers

2-*n*-nonyl-1,3-dioxolane (SEPA®), a skin penetration enhancer, has been reported by Hui *et al.* [69] to increase the ungual permeation of econazole. These findings suggest that adding SEPA® to an econazole lacquer can increase drug permeation up to six times, exceeding the MIC necessary to inhibit fungal growth.

Some formulations contain etching agents which are surface modifiers used to disrupt the dorsal surface of the nail to enhance permeation and promote adhesion of films. Phosphoric acid and tartaric acid are two etching agents commonly used as enhancers of transungual permeation [63]. Polyethylene glycols and hydrophobins are other two types of permeation enhancers [70, 71]. Curiously, methanol has also shown some advantages in drug permeation [63].

Table 14.1 overviews the most referenced transungual permeation enhancers.

Table 14.1 Chemical enhancers used for ungual permeation. (structure of enhancers modified from [38])

Enhancer	Structure	Increase in ungual permeation	Experiment setup; permeant (Ref.)
		Thiols/mercaptans	
N-Acetylcysteine (5%)	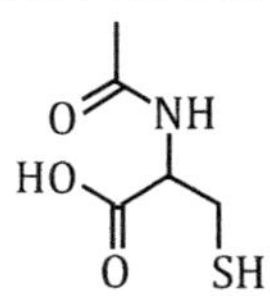	49× increase in nail drug content	Nails immersed in drug solution; Itraconazole
N-Acetylcysteine (15%)		2× increase in mean drug uptake, but not statistically significant. Mean residence time in nail plate increased from 4.2 weeks to 5.5 weeks	Drug was applied twice daily on nail plates of humans for 6 weeks; Oxiconazole
N-Acetylcysteine (3%)		13× increase in flux from an aqueous formulation; 7× increase from lipidic formulation	Diffusion cells; 5 fluoruracilo
N-Acetylcysteine (3%)		Drug measurable in presence but not in absence of enhancer	Diffusion cells; tolnaftate

Enhancer	Structure	Increase in ungual permeation	Experiment setup; permeant (Ref.)
Mercaptoethanol	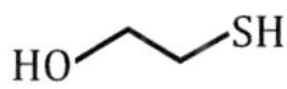	16× increase in flux from an aqueous formulation; 8× increase in flux from lipidic formulation	Diffusion cells; 5 fluoruracil
		Drug measurable in presence but not in absence of enhancer	Diffusion cells; tolnaftate
Thioglycolic acid (TGA)	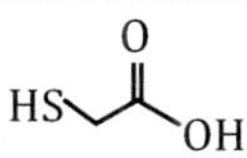	3.8× increase in flux	Diffusion cells; caffeine
Thioglycolic acid (TGA)		2× increase in mannitol concentration in receptor medium	Diffusion cells; mannitol
Cysteine	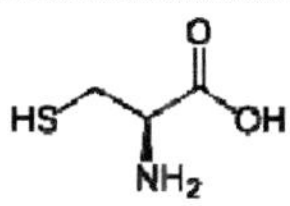	1.7× increase in mannitol concentration in receptor medium	Diffusion cells; mannitol
MPG (*N*-(2-mercaptopropionyl) glycine)	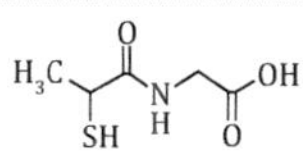	2.5× increase in flux	Diffusion cells; water
		Keratolytic agents	
Hydrogen peroxide	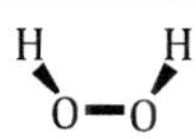	3.2× increase in mannitol concentration in receptor medium	Diffusion cells; mannitol
Urea hydrogen peroxide	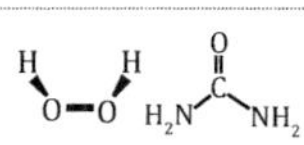	18× increase in flux	Diffusion cells; terbinafine
Urea	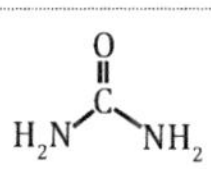	—	Human nail; Fluconazole [68]
Salicylic acid	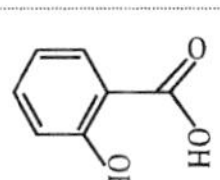	—	[72]
		Sulfites	
Sodium sulfite	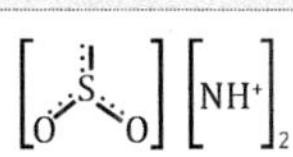	2× increase in permeation through nail clipping	Diffusion cells; 5.6-carboxyfluorescein

(*Continued*)

Table 14.1 (*Continued*)

Enhancer	Structure	Increase in ungual permeation	Experiment setup; permeant (Ref.)
		Enzymes/proteins	
Keratinase		2.3× increase in flux	Diffusion cells; metformin [61]
Hydrophobins		Up to 3.5× enhancement	Diffusion cells; terbinafine [70]
		Etching agents	
Phosphoric acid	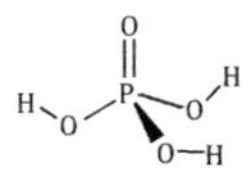	Increase in nail roughness score	Human nail [73]
Tartaric acid	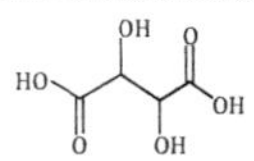	Increase in nail roughness score	Human nail [73]
		Other	
Pyrithione	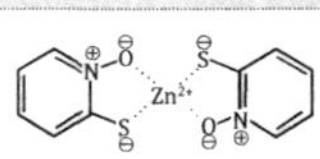	Up to 2.5 × increase in flux	Diffusion cells; water [60]
2-*n*-Nonyl-1,3-dioxolane (SEPA®)		7× increase in concentration, 1.3 × increase in deeper layer flux	Diffusion cell; econazole [69]

14.4.2 Examples of Antifungal Drugs

Table 14.2 represents several antifungal drugs and their main physicochemical characteristics that affect the transungual permeation. It should be noted that permeation coefficient and flux values were not included as reported in the literature considering that there are different methods and protocols to determine them [37]. Since these coefficients are dependent on many factors, it is more appropriate to understand their values in the literature context.

The capability of inhibiting fungal proliferation is usually measured by assessing the MIC (**minimum inhibitory concentration**) of the drug [74, 75]. MIC values for antifungal drugs are summarized in Table 14.3. Although these values do not take into account some characteristics (keratin binding, pH influence), they can serve as guidance [37].

Table 14.2 Antifungal drug properties that influence permeation

Antifungal Drugs	Chemical Structure	MW (g/mol)	Aqueous Solubility (mg/ml)	Log *P*	log n-octanol/ aqueous medium distribution constant	*p*Ka	Ref.
Amorolfine	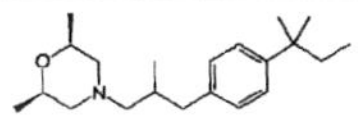	317.5	9.320 (CS, W, T20) 9.995 (CS, W, T32) 8.8 × 10− 3 (CS, pH 7.4 PB, T32)	5.7	0.33 (CS, W, T20)	6.6	[37, 76]
Bifonazole	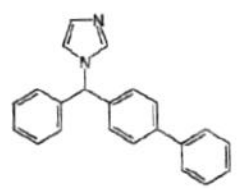	310.4	0.13 × 10− 3 (BS, pH 7.4 PB, T32) 0.51 × 10− 3 (BS, pH 8–10 0.1 M NC, T25) 0.345 (BS, I = 0.02 M pH 7.0 UB, T25)	4.8	4.77 5.2 (BS, 10 mM pH 7.4 PB, RT)	5.85 (I = 0.01 M UB, T25) 5.72 (0.1 M NC, T25)	[37, 76]
Ciclopirox	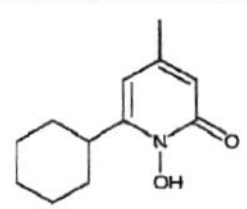	207.3	12.4 (OS, W, RT) 8.590 (OS, W, T32) 1.47 (OS, 5 mM pH 7.0 TB, RT) 1.020 (OS, pH 7.4 PB, T32) 0.22 (OS, 0.1 M HCl, RT)	2	0.53 (OS, 5 mM pH 7.0 TB, RT)	7.2 8.07 ± 0.05 (42% v/v EW, T32)	[37, 76]
Clotrimazole	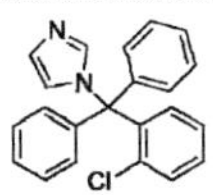	344.8	2.7 × 10− 3 (BS, pH 7.4 PB, T32) 0.39 × 10− 3 (BS, W, T25)	5	4.9 (BS, 10 mM pH 7.4 PB, RT)	6.02 ± 0.05 (0.15 M KC, T25) 4.74 ± 0.04 (42% v/v EW, T32)	[37, 76]
Efinaconazole	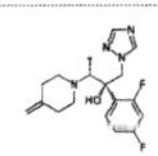	348.4	Predicted: 0.61	2	—	Predicted:7.45-12.7	[37, 76]
Fluconazole	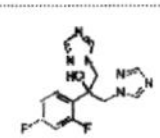	306.3	5 (BS, W, T23) 14 (BS, 0.1 M HCl, T23)	0.4	0.5 (BS, 100 mM pH 7.4 PB)	1.76 ± 0.10 (0.1 M NC, T24)	[37, 76]

(*Continued*)

Table 14.2 (*Continued*)

Antifungal Drugs	Chemical Structure	MW (g/mol)	Aqueous Solubility (mg/ml)	Log *P*	log n-octanol/ aqueous medium distribution constant	*p*Ka	Ref.
Itraconazole	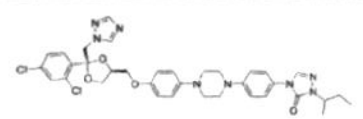	705.63	0.00964	5.7	—	3.70	[37, 76]
Luliconazole	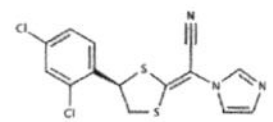	354.3	0.00062 (BS, W)	4	4.34 (pH 7.16 UB, T20) 3.78 (pH 4.00 UB, T20)	4.65	[37, 76]
Tavaborole	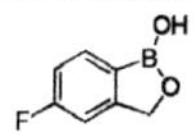	151.9	~1.0	1.24	—	—	[37, 76–78]
Terbinafine	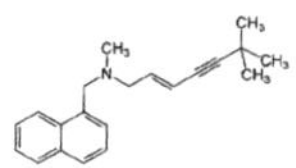	291.43	2.92 (CS, pH 3 UB, T25) 1.57 (CS, pH 3 W, T25) 1.12 (CS, pH 5 W, T25) 0.101 (CS, pH 5 UB, T25) 0.02 (CS, pH 6.8 SIF-P, T37) 1.5 × 10− 3 (CS, pH 9 UB, T25)	5.6	6.0 ± 0.1 (neutral form, 0.15 M KC, T37) [50] 2.3 ± 0.1 (ionized form, 0.15 M KC, T37) 5.5 ± 0.1 (calculated for pH 6.8 from the two above-mention edvalues, T37)	7.05 (0.15 M KC, T37)	[37, 70, 76]
Tioconazole	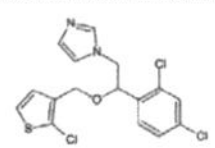	387.7	0.0165	5.3	4.4	6.77	[76]

Source: Adapted from [37].

Note: Measurement conditions, e.g., base/salt used, medium/solvent composition, pH, ionic strength, and temperature are provided between parentheses when available. The base or salt used in measurement is indicated by one of the following abbreviations: BS: base, CS: HCl salt, OS: olamine salt, and NS: nitrate salt. The abbreviations used in the description of aqueous media are: W: in water, NC: in NaCl solution, KC: in KCl solution, UB: in unspecified buffer, PB: in phosphate buffer, TB: in Tris buffer, SIF-P: in simulated intestinal fluid USP XXIII without pancreatin (0.05 M phosphate buffer + 0.01 g/L polysorbate 80), and EW: in hydroethanolic solution. RT is room temperature.

Table 14.3 Minimum inhibitory concentration (MIC) for antifungals used in onychomycosis

MIC (μg/mL)		Amorolfine	Bifonazole	Butenafine	Ciclopirox	Efinaconazole	Itraconazole	Ketoconazole	Tavaborole	Terbinafine
Dermatophytes	*T. rubrum*	0.004–0.015*	≤0.12–1*	0.015–0.12*	0.031–0.5*	≤0.001–0.015*	≤0.015–0.125*	0.06–0.5*	1–2	0.004–0.06*
	T. mentagrophytes	0.004–0.06*	0.5–4*	0.06–0.12*	0.031–0.5*	≤0.001–0.03*	≤0.015–0.5*	0.5–2*	2	0.004–0.5*
	T. tonsurans	0.25	0.5–2*	0.12	0.25	0.016	0.06–0.25*	0.5–2*		0.015–0.06*
	T. verrucosum	0.12	0.25	0.12	0.13	0.0039	0.12	0.12		0.015
	M. canis	0.06–0.25*	1–2*	0.12	0.25	0.13–0.25*	≤0.015–0.03*	0.25–0.5*		0.008–0.03*
	M. gypseum	0.063–0.25*	≤0.12–8*	0.06–0.12*	0.31	0.010	0.031–0.25*	0.5–4*		0.004–0.06*
	E. floccosum	0.13–0.25*	0.25–0.5*	0.06	0.31	0.005	0.03–0.5*	0.25–0.5*	1	0.015–0.06*
	T. interdigitale									0.017
Non-dermatophytes	*Geo. candidum*	1–2*	>64	>0.5			1	2–4*		0.12–0.25*
	Scopulariopsis brevicaulis	0.09	2–4	0.5	0.59	0.25	>4	2–4*		0.12
	Aspergillus fumigatus	>4	2–4*	>0.5	0.42	0.089	0.5–1*	8	0.25	1.4
	Fusarium oxysporum	>4	>64	>0.5	1	1	>4	>8		>2.5
	Fusarium solani	4	>64	>0.5	>4	0.5	>4	>8	2	4
	Fusarium verticillioides	>64	>64	>0.5			2	2		>0.5
	Acremonium potronii	0.26			0.25	0.31	2.5			0.25

(*Continued*)

Table 14.3 (*Continued*)

MIC (µg/mL)		Amorolfine	Bifonazole	Butenafine	Ciclopirox	Efinaconazole	Itraconazole	Ketoconazole	Tavaborole	Terbinafine
	Acremonium sclerotigenum	1			1.4	0.18	4			0.09
	Aspergillus flavus	4			3.4	0.11	0.18			0.11
	Aspergillus niger	4			0.63	0.2	0.63			0.16
	Aspergillus sydowii	4			0.59	0.037	0.3			0.076
	Aspergillus terreus	4			0.5	0.09	0.21			0.13
	Aspergillus nidulans	4			1	0.0078	0.089			0.063
Yeasts	*C. albicans*	0.03–0.5*			0.06–0.5*	0.0005–0.25*	0.004–2*		0.5	0.06–16*
	C. glabrata	4.9			0.13	0.026	0.74			8
	C. krusei	0.27			0.21	0.024	0.38			8
	C. parapsilosis	0.56			0.22	0.0046	0.13			0.28
	C. ropicalis	4			0.5	0.014	0.31			8
	C. guilliermondii	0.25			0.25	0.016	0.13			1
	C. kefyr	0.063			0.13	0.002	0.031			2
	C. lusitaniae	0.5			0.25	0.0039	0.13			4
References		[79, 80]	[79]	[79]	[80, 81]	[80]	[79, 80]	[79]	[77, 78]	[79, 80, 82]

Note: Values displayed in µg/mL.
*MIC range.

Table 14.4 Examples of commercial topical formulation for onychomycosis treatment.

Antifungal	Chemical Group	Commercial name	Approved in	Fungals	Formulation	Posology	Effectiveness	Reference
Amorolfine	Morpholine	Loceryl, Locetar, Sinibal, Unglyfol, Curanail	UK (1991), Switzerland (1991), France (1992), and more than 50 other countries Not approved in USA and Canada	*Candida sp, Trichophyton sp, Microsporum sp, Epidermophyton*	5% lacquer	1–2 weekly, 6–12 months after removal of affected area	50% mycological cure in cases of distal fingernail and toenail onychomycosis	[37, 83–91]
Bifonazole (+40% urea)	Imidazole	Canespor, Canesten, Mycospor, Amypor (Onyster, Canespro)	Australia, Portugal, other countries	*T. rubrum, T. mentagrophytes, T. tonsurans, T. verrucosum, M. canis, M. gypseum, E. floccosum, Scopulariopsis brevicaulis, Aspergillus fumi-gatus*	1% cream (40% urea)	After ablation of the nail with 40% urea cream, application once daily for 4 weeks	33.6% overall cure rates	[37, 66, 67, 77, 88–90]

(*Continued*)

Table 14.4 *(Continued)*

Antifungal	Chemical Group	Commercial name	Approved in	Fungals	Formulation	Posology	Effectiveness	Reference
Ciclopirox	Hydroxypyridone derivative	Mycoster, Niogermos, Batrafen, Kitonail, Onytec, Stieprox, Loprox, Penlac, Ciclopoli, RejuveNail	France (1991), Germany (1992/2008), Austria (1999), Canada (2004), Switzerland (2009) and more than 50 other countries	*Trichophyton sp, S. brevicaulis, Candida sp, Malassezia furfur, Aspergillus sp, Fusarium solani*	8% lacquer	Once daily, 48 weeks	34–46% mycological cure 29–36% overall cure	[37, 77, 83, 88–90, 92–94]
Efinaconazole	Triazole	Jublia	Canada (2013), USA (2014)	*T. rubrum, T. mentagrophytes, Aspergillus sp, Candida albicans*	10% topical solution	Once daily for 48 weeks	53–55% mycological cure rate, 15–18% complete cure	[37, 87–90, 95]
Tavaborole	Oxaborale antifungal	Kerydin	USA (2014)	*T. rubrum, T. mentagrophytes, Aspergillus sp, Fusarium sp, Candida albicans*	5% topical solution	Once daily for 48 weeks	31–36% mycological cure, 6.5–9.5% complete cure	[37, 77, 87–90, 96, 97]
Tioconazole	Imidazole	Trosyl, Trosyd	Portugal, UK, France and 30 more countries	*T. rubrum, T. mentagrophytes*	28% solution	Daily, 3–12 months	22% of patients mycological cure	[3, 37, 83, 88–90, 98]

14.4.3 Examples of Topical Pharmaceutical Forms

Antifungal drugs are formulated in several topical pharmaceutical forms, such as creams, solutions, gels and lacquers (Table 14.4).

14.4.3.1 Cream

Creams are semi-solid topical preparations used for delivery of one or more active substances, or for their emollient or protective action. The base may consist of natural or synthetic substances, being a multiphase preparation with lipophilic and aqueous phases. Lipophilic creams have the lipophilic phase as the continuous one, containing water-in-oil emulsifying agents at higher percentage. The opposite is observed in hydrophilic creams. Creams may also contain other suitable excipients, such as preservatives, antioxidants, stabilizers, emulsifiers, thickeners, and penetration enhancers [99–101].

As reported by Tietz *et al.* [67], a 1% bifonazole cream used after nail ablation by 40% urea paste (Canespor®, Canespro®) was effective in the treatment of onychomycosis. Bifonazole is an imidazole antifungal drug, acting by blocking the conversion of 24-methylendihydrolanosterol to desmethylsterol in fungi as well as inhibiting HMG-CoA (3-hydroxy-3-methylglutaryl-coenzyme A) and compromising the cell membrane. It is active against most dermatophytes, some non-dermatophytes, and molds. This bifonazole cream also contained benzylic alcohol, cetostearyl alcohol, cetyl palmitate, octyldodecanol, polysorbate 60, sorbitan monostearate, and purified water [102]. The urea paste was applied daily with an impermeable bandage until the ablation of the nail was achieved, and then 1% bifonazole cream was applied once daily for 4 weeks. At the end, 33.6% of patients were cured [55, 66, 67, 79, 88–90].

Lahfa *et al.* [66] studied the efficacy of using bifonazole and urea in the same formulation (Amycor Onychoset®).

Other cream formulations in the European market include Selergo®, Loprox®, Ertaczo®, Avage®, Mycoster®, Locetar® and Fougera® [35, 88, 89].

14.4.3.2 Solution

Topical solutions are liquid preparations used for the delivery of one or more active substances that are solubilized in a suitable vehicle. These formulations are transparent and free of visible particles. They may also contain preservatives, antioxidants, stabilizers, and buffers [99–101].

Solutions have some advantages compared to other topical formulations [41, 42, 103]. As a liquid solution, the formula can be applied to the stratum corneum and subungual space as well as on the nail plate surface, and it does not require the patient to posteriorly remove the film as necessary for a nail lacquer [103].

In the US market [42, 96, 104–107], a 10% efinaconazole solution has been recently available. This formulation contains 100 mg of efinaconazole per gram and its inactive ingredients are alcohol, anhydrous citric acid, butylated hydroxytoluene, C12-15 alkyl lactate (co-solvent), cyclomethicone (wetting agent), diisopropyl adipate (co-solvent), disodium edentate, and purified water. This formulation should be applied once daily for 48 weeks. Overall cure rates were reported on 15–18% of patients, mycological cure rates being between 53% and 56% of patients studied. Efinaconazole acts by inhibiting the conversion of lanosterol to ergosterol in the ergosterol synthesis pathway. It possesses a broad spectrum of activity against dermatophytes, non-dermatophytes, and yeasts [16, 23, 37, 77, 80, 83, 87, 108, 109].

5% Tavaborole solution (Kerydin®) is also available in the US market, and it contains ethanol USP, propylene glycol USP, and edetate calcium disodium USP [41, 77]. This solution should be applied once daily for 48 weeks. In conducted clinical trials, 31–36% mycological cure and 6.5–9.5% complete cure was achieved in patients treated with this drug. It shows activity against yeasts, molds, dermatophytes, and filamentous fungi. It was specially developed for onychomycosis treatment [96, 97].

There is a 28% tioconazole topical solution in the European market (Trosyd®, Trosyl®) formulated with undecylenic acid and ethyl acetate [110]. It is applied daily for 3–12 months. The average overall cure rate is 22%. Nowadays, its use is in decline as other more effective antifungal drugs appear in the market [83].

Naumann *et al.* [111] describe several formulations, including a topical solution for the use of EV-086K, a new antifungal with a high lipophilicity which is not beneficial to transungual delivery [111]. This solution was formulated by dissolving 5.7% of EV-086K in 30% ethanol and 63.75% water, besides the presence of Transcutol® P as solubilizer, 0.05% each of citric acid and sodium phosphatemonobasic for pH adjustment and 0.1% of BHA (butylated hydroxytoluene) and EDTA for chemical stability. The study showed that this formulation provided high permeability at the beginning of the penetration process (test in equine hoof), but the maximum drug concentration stabilized at 35% in the acceptor phase after 24 h. These results were not significant compared to other formulations.

Other topical solutions available in the market are Canespor®, Mycoster®, and Lamisil® among others [88].

14.4.3.3 Gel

Gels are formed by suitable gelling agents and can be classified as lipophilic gels or hydrophilic gels. Lipophilic gels (oleogels) usually consist of liquid paraffin with polyethylene or fatty oils gelled with colloidal silica, aluminum, or zinc soaps. Hydrogels (hydrophilic gels) are prepared with water, glycerol, or propylene glycol gelled with gelling agents such as starch, cellulose derivatives, carbomers, and magnesium–aluminum silicates [100].

Although there are some gel formulations in the market (e.g., Lamisil®, a terbinafine formulation), they are mostly used for skin fungal infections instead of onychomycosis.

Along with the cream formulation previously mentioned, Naumann *et al.* [111] also studied a gel formulation for EV-086K delivery. Hydroxyethyl cellulose (HEC) gel was formulated with the following excipients: 10% of EV-086K, 1.5% glycerol, 1% Tylose H 30.000 P2 PHA (HEC) and 87.5% water. Among other tested formulations, the hydrogel had less promising results in equine and bovine hoof penetration test (only 22% and 15% of applied dose was recovered, respectively), and thus it was not evaluated in the human nail.

Kerai *et al.* [53] have recently investigated the use of UV-curable gels (currently used in cosmetics) for the treatment of onychomycosis. The formulated gels contained diurethane dimethacrylate, ethyl methacrylate, 2-hydroxy-2-methylpropiophenone, an antifungal drug (amorolfine HCl or terbinafine HCl) and an organic liquid (ethanol or NMP) as drug solvent. Amorolfine was released at a greater extent than terbinafine, and even not reaching concentrations as high as the comparative nail lacquer, the concentrations were well above the MIC for *T. rubrum*, the main pathogen for onychomycosis. In addition, cured gel formulation could maintain the drug release for longer periods, being a promising candidate for future onychomycosis treatment.

14.4.3.4 Nail lacquer

Nail lacquers have been used for a long time in cosmetics, with both protective and decorative purposes. In the 1990s, nail lacquers started being used to administer drugs to the nail unit. Nail lacquers are **solutions of film-forming polymers**, which leave a polymer film on the nail plate after solvent evaporation. This polymer film can be water resistant or water soluble and acts as a drug reservoir, allowing it to permeate through the nail plate. Depending on the formulation and drug, the lacquer can be removed either mechanically or with organic solvents after a certain time, and it should be reapplied to reestablish the drug pool. Thus, the duration of the film residence in the nail constitutes an important property of a lacquer formulation [112]. Although these formulations are commonly applied with a brush, there are also other types of applicators such as spatulas and sponge tips.

Being occlusive and adhesive to the nail, nail lacquers' films have **advantages** regarding nail hydration, permanence on the nail plate and patient compliance (when compared to creams, gels and solutions).

These preparations usually contain the intended drug, a polymer, a volatile solvent, and suspension agents [56, 91, 113].

The **drug** should present the following properties: (a) low MIC for pathogens causing the disease (dermatophytes, yeasts, etc.), being effective at low concentrations; (b) low molecular weight and volume to better permeate through the keratin matrix pores; (c) water soluble, since water easily permeates through the matrix vehiculating the drug; and (d) affordable and easily obtainable, so that the nail lacquer can be easily produced and obtained by patients [61].

The **polymer** choice can also influence the nail lacquer quality. Hydrophilic polymers have advantages regarding permeation and adhesion to the nail plate, having a soft, flexible, and matte finish, which can improve patient compliance. However, these polymers allow more drug loss and are less occlusive. Hydrophobic polymers have a more durable, harder, and glossier finish. This contributes for a more occlusive lacquer, which can compromise the cumulative amount of permeated drug. Employing both types of polymers seems to be a good approach, since it will combine both good permeation characteristics and higher resistance to environmental drug loss [114].

Effective formulations usually include **permeation enhancers** such as thiols/mercaptans, keratolytic agents, keratinases, etc. Other excipients can be also important to allow better permeation conditions, including hydration, pH, and solubility. This can be achieved by using plasticizers, humectants, solvents, solubilizers, etc. [115].

Colorless and non-glossy medicated nail lacquers are more acceptable by male patients [39]. Despite male patient compliance, it could be considered advantageous for female patients to formulate a colored medicated nail lacquer for the treatment of onychomycosis, since this disease alters the normal appearance of the nail.

After screening for the most suitable formulation components, it is important to check their compatibility and progressively adjust the quantities through pre-formulation studies.

There are several parameters to be evaluated in nail lacquer formulations to assess a higher **quality**, such as drug content, drug permeation studies (permeation coefficient, flow, cumulative amount of permeated drug); gloss; flow; film adhesivity; viscosity; pH; drying time; non-volatile content, etc. According to this quality control, nail lacquers must present the following properties: (a) be physically and chemically stable; (b) release therapeutic levels of drug onto the nail; (c) have a suitable viscosity to freely flow into all the edges and grooves of the nail for easy application; (d) dry quickly (3–5 min) and form an even film once applied; (e) adhere to the nail plate to not come off or flake during daily activities, but at the same time, be easily removed with an enamel remover, and (f) be cosmetically pleasant and well tolerated [38].

14.4.4 Nail Lacquer Formulations for Onychomycosis Treatment

Several examples of nail lacquer formulations for onychomycosis treatment are presented in Table 14.5. However, only nail lacquers containing amorolfine and ciclopirox are currently commercialized [39].

Table 14.5 Nail lacquer formulations

Drug	Commercial name	Polymer	Solvents and other excipients		Ref.
Econazole		Eudragit® RL/PO	2-*n*-nonyl-1,3-dioxolane	Ethanol	[69]
Ciclopirox		Hydroxypropyl chitosan (HPCH)	Cetostearyl alcohol Ethyl alcohol (95°)	Ethyl acetate Purified water	[119]
	Ciclopoli®	Hydroxypropyl chitosan (HPCH)	Ethyl acetate Ethanol 96%	Cetostearyl alcohol water	[114]
		Pluronic F127	Partially methylated β-cyclodextrin	*N*-acetyl-λ-cysteine	[120]
	Mycoster®	Methylvinyl ether	Ethyl acetate Isopropanol	Maleic acid monobutylester	[116]
Fluconazole		Polyvinylpyrroli-done K25	Glycerol triacetate Docusate sodium Urea	Ethanol Demineralized water	[68]
Tavaborole		Poly (vinyl methyl ether alt maleic acid monobutylester)	Ethanol		[78]
		Poly (2-hydroxyethyl methacrylate)	Ethanol Dibutyl sebacate	Water	
		Poly (vinylacetate)	Ethanol Ethyl acetate	Dibutyl sebacate	

Drug	Commercial name	Polymer	Solvents and other excipients		Ref.
Terbinafine		HPMC E-15 PEG 400	Ethanol	Purified water	[121]
		Poly (4-vinyl phenol)	Ethyl acetate	Dibutyl phthalate	
		Cellulose acetate Ethyl cellulose	Triethyl citrate Hydroxypropyl-β-cyclodextrin	Isopropyl alcohol Acetone	[57]
Ketoconazole		Propylene glycol PEG 400	Glycerin Urea hydrogen peroxide	Ethanol Thioglycolic acid	[122]
		Polysilicone-8 Acrylates copolymer	Ethanol Panthenol Tocopheryl acetate Phytantriol	Butylene glycol Benzophe-none-3 Calcium chloride Fragrance	[123]
Miconazole		Nitrocellulose	Urea hydrogen peroxide Propylene glycol	Salicylic acid Ethanol	[124]
EV-086K		Eudragit® E100	Transcutol® P	Ethanol	[111]
Amorolfine	Loceryl®	Methacrylic acid copolymer	Glycerol triacetate Butyl acetate	Ethyl acetate Ethanol	[110]

Amorolfine is a morpholine antifungal agent approved for the treatment of onychomycosis in 1981 [116]. It is usually presented as a nail lacquer containing 5% amorolfine. This active substance inhibits delta14 reductase and delta7-delta8 isomerase, causing ergosterol depletion and ignosterol accumulation in the cytoplasmatic membrane of the fungus cell. Amorolfine is effective against dermatophytes (*Trichophyton* spp., *Microsporum* spp., *Epidermophyton* spp.), yeasts (*Candida* spp., *Cryptococcus* spp., *Malassezia* spp.) and some molds (*Alternaria* spp., *Hendersonula* spp., *Scopulariopsis* spp.). Amorolfine can be present in the nail up to 27% from the nail lacquer formulation after 2 weeks treatment, being this concentration enough to inhibit most fungi. This formulation should be applied once or twice weekly for 6 months. Studies report mycological cures in 52–60% of the patients, and complete cures (both clinical and mycological) up to 44% of patients [54, 77]. Even after apparent cure, some authors recommend the use of amorolfine as prophylaxis [117]. Amorolfine is available in Europe (e.g., Loceryl®, Locetar®, Curanail®), but not in the USA [77, 89–91, 118].

Ciclopiroxolamine (ciclopirox) belongs to the group of hydroxyl-pyridone derivatives, and it has been used as a nail lacquer to treat onychomycosis since 1990. It exerts its antifungal activity by chelating trivalent cations, like Fe^{3+} and Al^{3+}, compromising fungal metal-dependent enzymes and reducing the fungus nutrient intake. Ciclopirox is active against dermatophytes (*Trichophyton* spp., *Microsporum* spp., *Epidermophyton floccosum*), yeasts (*Candida* spp., *Malassezia furfur, Cryptococcus neoformans, Saccharomyces cerevisiae*), molds (*Aspergillus* spp., *Scopulariopsis brevicaulis, Fusarium solani*) and some bacteria, which is advantageous in cases of mixed infection. It is also reported to have some anti-inflammatory activity by inhibiting the local production of prostaglandins and leukotrienes. The nail lacquer available in the market has a concentration of 8% ciclopirox, which increases to 35% after application and evaporation of the volatile solvents. Studies show that it is able to exceed MIC for the three main onychomycosis pathogens—*T. rubrum, T. mentagrophytes* and *Candida albicans*. Common posology is once daily application for 6–12 months over the clean nail plate and slightly over the surrounding skin. However, there are trials stating that once weekly application is effective. Clinical trials have reported complete cure in 29–36% of patients. Ciclopirox is available in both Europe and the USA: Mycoster®, Niogermos®, Batrafen®, Kitonail®, Onytec®, Stieprox®, Loprox®, Penlac®, Ciclopoli®, and RejuveNail® [77, 83, 88–94].

Besides these main drugs, others have been recently studied as well as specific excipients as follows.

Monti *et al.* [119] purposed to evaluate the water-soluble film-forming agent **hydroxypropyl chitosan** (HPCH) included in an experimental nail lacquer (P-3051) containing **ciclopirox**. HPCH is a water-soluble derivative of chitosan. Chitosans are polysaccharides derived from chitin and natural components of the exoskeleton of crustaceans being widely employed in medicine for their wound healing, bacteriostatic, skin moisturizing, and protecting properties. In particular, HPCH was chosen considering its favorable properties, such as high water solubility; high plasticity; affinity to keratin; wound-healing activity; high compatibility with human tissues, etc. P-3051 was composed of 1% HPCH, 1% cetostearyl alcohol, 73% ethyl alcohol (95°), 4% ethyl acetate and 13% purified water. This formulation was compared with a commercial brand (Penlac™) constituted by 8% ciclopirox, ethyl acetate, isopropyl alcohol, butyl monoester of poly(methylvinyl ether/maleic acid) in isopropyl alcohol. Bovine hoof membranes were used as the human nail plate model. Drug concentrations were determined by HPLC (high performance liquid chromatography). Regarding lag times, the respective values obtained were 3.36 ± 0.46 h for P-3051 vs. 12.48 ± 1.31 for Penlac™. The percentage of permeated drug (Q%30h) was also significantly different for the two formulations: 2.58% for P-3051 vs. 1.06% for Penlac™. In fact, a faster drug penetration time might allow the drug to permeate the nail before the hydro soluble film is degraded. Greater efficiency of P-3051 could be attributed to a particular affinity of HPCH to the nail matrix, resulting in an intimate contact and strong adhesion of the lacquer to keratin substrate [119].

An *in vitro* **study** [125] with the formulation previously developed by Monti et al [119] reported an achievement of 13% complete cure rate, which was quite low. It was concluded that ciclopirox formulated in the new hydrolacquer technology is more active and better tolerated than the reference ciclopirox nail lacquer for the long-term treatment of onychomycosis.

In addition, it is much easier to apply without needing any bothersome removal procedures [125].

In 2009, Monti *et al.* compared the transungual permeation of **ciclopirox** with **amorolfine** vehiculated in the same **hydroxypropyl chitosan-based lacquer** and in a **water-insoluble reference** (Loceryl® containing Eudragit® RL100, triacetine, butyl acetate, ethyl acetate, ethyl alcohol and drug). The study was performed on bovine hoof slices and drug concentration was determined by HPLC. Amorolfine experimental lacquer showed higher permeation than the commercial water-insoluble option. In addition, ciclopirox lacquers showed a better performance than amorolfine lacquers [114].

Later in 2012, a **clinical trial** with these formulations was conducted [56]. The results supported *in vitro* data except from day 15 to day 25 in which the nail concentration of these drugs decreased. The authors hypothesized that the steady state had not been reached yet. Moreover, *in vitro* studies do not take into account all the environmental interactions that occur under *in vivo* conditions where both fingers and toenails are constantly exposed to tissues, liquids and blood circulation, resulting in some drug loss. This may be a disadvantage for water soluble formulations since the contact with water can lead to a loss of medication; however, it can be countered by a more frequent application [56].

A **ketoconazole** nail lacquer for onychomycosis treatment was also described by Kiran *et al.* in 2010 [122]. The studied formulations contained ketoconazole, propylene glycol, glycerin, ethanol, polyethylene glycol 400, thioglycolic acid and urea solution in H_2O_2. Several factors were evaluated including non-volatile content, drying time, smoothness of flow, gloss, and permeation studies in human nail plates. Urea hydrogen peroxide enhanced hydration state and thioglycolic acid cleaved keratin bonds, allowing better penetration results, which could be promising upon formulation optimization [122].

In 2011, a novel lacquer formulation for the transungual delivery of **ketoconazole** was reported by Hafeez *et al.* The vehicle used had an anhydrous/alcohol composition with a dual acrylate-silicone hybrid copolymer system that offered film-forming and occlusion properties due to synergistic plasticizing components [123]. It contained ethanol, polysilicone-8, panthenol, acrylates copolymer, tocopheryl acetate, phytantriol, butylene glycol, benzophenone-3, calcium chloride and fragrance, as well as radiolabeled [1-^{14}C]-Ketoconazole. This test formulation was compared with a commercial ketoconazole cream. The study was conducted on human nails placed in a diffusion cell simulating physiological conditions. The formulations were then administered once daily for 7 days. Sampling was made by drilling and radioactivity was measured from the samples obtained. Following the 7-day exposure, the ketoconazole content measured in the ventral/intermediate layers was 0.81 ± 0.39 μg/mg or 535 ± 260 μg/cm^3 for the lacquer formulation and 0.09 ± 0.05 μg/mg or 53 ± 29 μg/cm^3 for the control formulation. The drug concentration attained in the nail was approximately 2140 times higher than the MIC for common dermatophytes, exceeding the MIC for most common onychomycosis pathogens. According to these results, this lacquer formulation can be a potential effective topical treatment for onychomycosis [123].

Hui *et al.* [69] assessed the enhancing properties of **2-*n*-nonyl-1,3-dioxolane** in a lacquer formulation (EcoNail™) to increase the permeation of **econazole**. The test formulation contained 5% (w/w) econazole, Eudragit® RL/PO, ethanol and 18% (w/w) 2-*n*-nonyl-1,3-dioxolane. The assay was performed on healthy human nail plates collected from cadavers and using a diffusion cell. An aliquot of all tested formulations was applied on the nails twice daily for 14 days, washing the nail plates between applications. After incubation, nail samples were obtained with a drill and radioactivity of labeled compounds was measured. The concentration and flux of econazole into the deep layer of human nail in the test group was about 6.3-fold and 7.5-fold higher than the control group, respectively. Econazole concentration was about 15,000 times the MIC for most dermatophytes species and 150 times that for most molds. On the contrary, dioxolane did not penetrate to deeper layers of the nail. These authors hypothesized that besides facilitating diffusion of the drug, dioxolane also functioned as an adhesion promoter and plasticizer for the film-forming polymer, softening the Eudragit film in the lacquer. These results suggest that 2-*n*-nonyl-1,3-dioxolane-enhanced econazole lacquer has the potential to be an effective topical treatment for onychomycosis [69].

The same author performed another study [78] and reported the nail penetration of **tavaborole** (AN2690) from different vehicles and compared with ciclopirox. Four formulations, all containing 10% (w/w) AN2690 were compared for their ability to deliver AN2690 to the deep layers of the nail plate and into the nail bed. The composition (w/w) of these different vehicles was: *formulation A*: 70% ethanol and 20% poly (vinyl methyl ether alt maleic acid monobutylester), a polymer that forms a water-insoluble film, very durable and resistant to damage; *formulation B*: 56% ethanol, 14% water, 15% poly (2-hydroxyethyl methacrylate) and 5% dibutyl sebacate, forming a water soluble film; *formulation C*: 55% ethanol, 15% ethyl acetate, 15% poly (vinylacetate) and 5% dibutyl sebacate, forming a water-insoluble film that can be removed by peeling or scratching the surface; *formulation D*: 20% propylene glycol and 70% ethanol (only solvents). Aliquots of those formulations were applied on human nail plates once daily for 14 days. The ventral/intermediate nail samples were collected at the end of the 14th day dose period, stored at 4°C and analyzed for drug by LC/MS/MS. Considering that any formulation showed a clear advantage over the others, the simplest formulation (D) was chosen for further development [78].

Baran *et al.* [68] evaluated the efficacy of a **fluconazole-urea** nail lacquer in onychomycosis treatment of 13 patients. The lacquer was composed of 1% fluconazole, 20% urea, polyvinylpyrrolidone k25, glycerol triacetate, docusate sodium, ethanol, and demineralized water. After 12–18 months treatment, about 90% of patients had positive results regarding clinical and mycological cure [68].

A **terbinafine bilayered nail lacquer** for onychomycosis was developed by Shivakumar *et al.* in 2010. Since aqueous-based nail lacquers promote nail hydration and drug diffusion through the nail but lack durability, the authors studied an underlying drug-loaded hydrophilic lacquer with an overlying water resistant film. The hydrophilic nail lacquer was composed of 5% (w/v) terbinafine, 6% (w/v) HPMC E-15 (water soluble polymer), 10% (v/v) PEG 400 (penetration enhancer, plasticizer and humectant), 60% (v/v) ethanol and qs 100% purified water. Formulation controls as a hydrophilic nail lacquer of HPMC E-15 containing terbinafine but devoid of PEG 400 and a drug free "placebo" hydrophilic lacquer of HPMCE-15 containing PEG 400 were similarly prepared for comparison. Briefly, terbinafine was dissolved in a mixture of water and ethanol (pH 3.0) by a bath sonicator. HPMC E-15 was soaked overnight in the hydroalcoholic mixture (pH 3.0) and sonicated to ensure complete polymer dissolution. The two hydroalcoholic solutions were mixed and stirred to obtain a clear homogeneous solution to which PEG 400 was added. The hydrophobic nail lacquer was prepared by dissolving poly (4-vinyl phenol) in ethyl acetate at 10% (w/v). Dibutyl phthalate was used as a plasticizer in the lacquer at 4% (v/v). The pH, viscosity and drying time of the hydrophilic nail lacquers was found to be around 4.0, 500 cps, and 300±75 s, respectively. *In vitro* drug permeation studies were performed with cadaver nails. Although therapeutic concentration values were reached in *in vitro* studies, a clinical study on diseased patients should be performed to ascertain clear results [121].

Vipin *et al.* [124] presented in 2014 a **miconazole** nail lacquer. This formulation included a film former (nitrocellulose), permeation enhancers (urea in hydrogen peroxide and propylene glycol), a keratolytic agent (salicylic acid) and an antifungal agent (miconazole nitrate) in ethanol. Several formulations were tested regarding nonvolatile content, gloss, smoothness to flow, drug release (bovine hoof model), drug content (UV spectrophotometry) and antifungal activity. Among ten formulations, the nail lacquer prepared with 2% drug, 3% nitrocellulose, 0.5% ethyl cellulose, 20% salicylic acid, 5% propylene glycol and 5% urea in H_2O_2 exhibited the best results, being a promising formulation for the treatment of *Candida albicans* [124].

Naumann *et al.* [111] studied the controlled delivery of **EV-086K** nail lacquer through the nail plate. 5% Eudragit® E100, dimethylaminoethyl methacrylate, butyl methacrylate, and methyl methacrylate were mixed with 85% ethanol, and stirred until the polymer was completely dissolved. 5% EV-086K and 5% Transcutol® P were stirred in a separate amber glass until a clear solution was obtained. Then, the polymer–ethanol mixture was added. Bovine hoof slices, equine hood slices, and human nails were employed for delivery studies. This nail lacquer penetrated into the nail at higher concentrations than other studied formulations. However, it did not reach the acceptor compartment [111].

14.4.5 Advances in Nail Formulations

14.4.5.1 Colloidal carriers

The literature reports the success of some colloidal carriers for onychomycosis treatment since the drug easily diffuses along the skin tissue to the nail bed. Therefore, application to the skin surrounding the nail could be an effective area to treat for onychomycosis [126–128].

Colloidal carriers or colloidal drug delivery systems (CDDS) are particulate or vesicular dosage forms, having a size range from 1 nm to 0.5 µm (Fig. 14.4). They are essential for successful drug transport and delivery by protecting and maintaining the loaded drug until the site of action is reached.

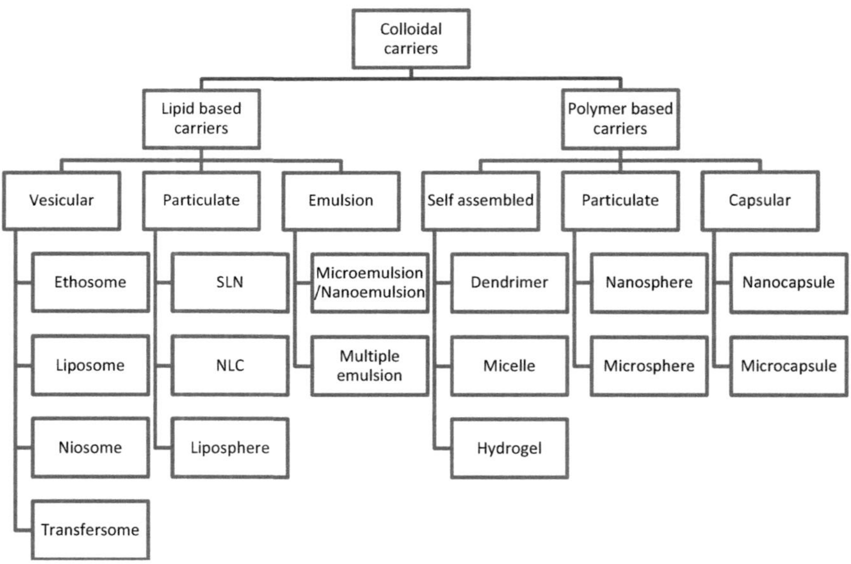

Figure 14.4 Colloidal carriers for drug delivery. Adapted from [126].

Along with gel and solution formulations, Naumann also studied a colloidal carrier formulation for the delivery of **EV-086K**, containing water, propylene glycol, emulsifiers (Tagat® O2V and Synperonic™PE/L 101) and an oil component (Pelemol® BIP). This colloidal carrier system contributed to recover 7.25 ± 0.30% drug from the nail slices. These results were comparable to those obtained with tested solution and nail lacquer [111].

Nogueiras-Nieto *et al.* [120] developed in 2013 an aqueous nail lacquer based on polypseudorotaxanes of Pluronic® F-127 and **ciclopirox** complexed with partially **methylated β-cyclodextrin** (mβ-CD). The obtainment of *in situ* gelling thermosensitive hydrogels was due to the presence of the poloxamer Pluronic ®F127 (PF127), which facilitated drug solubilization via micelle creation forming a gel upon nail application. Partially mβ-CD was also added to further improve drug solubilization. In addition, a penetration enhancer (*N*-acetyl-l-cysteine) alone or in combination with urea was also added to the formulations. Ciclopirox was incorporated into three different vehicles–simple aqueous solutions, thermosensitive hydrogels and polypseudorotaxanes thermosensitive hydrogels. The composition of these vehicles is presented in Table 14.6.

Table 14.6 Vehicles composition of ciclopirox (CPO) nail formulations

Name	Solvent	AC (%)	PF127 (%)	Mβ-CD (%)
CPO–*N*-acetylcysteine solution (CPO–AC)	Water	10	—	—
CPO–thermosensitive hydrogel (CPO–TH)	Water	10	20	—
CPO–polypseudorotaxanes thermosensitive hydrogel (CPO–PPR)	Water	10	20	10

Source: Adapted from [120].

Recently Yang *et al.* [127] have developed a **gel with transfersomes containing terbinafine** prepared by ethanol injection method. The final formulation contained terbinafine, phospholipids, polysorbate 80, sodium cholesteryl sulfate, anhydrous ethanol, sodium benzoate, sodium pyrosulfate, and phosphate buffer (pH 5.0). After 12 h, it was obtained 88.52 ± 4.06 mg·cm^{-2} and 94.38 ± 5.26 mg·cm^{-2} of permeated and penetrated drug, respectively.

Vaghasiya *et al.* [129] studied another formulation composed of **solid lipid nanoparticles** (SLNs) for sustained release and skin targeting of **terbinafine** for the treatment of onychomycosis. The terbinafine loaded SLNs formulation was composed of Compritol® 888 ATO as lipid matrix, Pluronic® F-127 as stabilizer and distilled water as dispersion medium, and prepared by the solvent injection technique. It was observed 40.57 ± 1.76% of the applied drug retained in the skin after 8 h [129].

TDT 067 (terbinafine in transfersome) is a carrier-based liquid spray that has been developed for the delivery of terbinafine to the nail bed to treat onychomycosis. A 2011 study demonstrated that TDT 067 had greater antifungal activity against onychomycosis caused by dermatophytes compared to the free form vehiculated in an oral formulation [130]. TDT 067 formulation has been also tested in clinical trials [131].

There is another study with **terbinafine-loaded liposomes** formulation composed of bioadhesive polymers, pullulan and Eudragit®L100, and prepared by thin film hydration method. This formulation also showed interesting results [132].

Barot *et al.* [133] studied a **microemulsion-based gel containing terbinafine**, oleic acid, Labrasol, Transcutol, water and Carbomer 934P. This formulation was developed in order to be applied between the nail bed and nail plate generated by onycholysis. It was obtained 49.3% ± 4.12% drug retention in the skin, and the total amount of terbinafine permeated after 12 h was 244.65 ±18.43 mg·cm^{-2}. In 2012, Barot also described a microemulsion based antifungal gel incorporating itraconazole [134].

Barot *et al.* [134] developed a gel **microemulsion containing itraconazole** for the topical treatment of onychomycosis. The microemulsion contained benzyl alcohol, isopropyl myristate, Pluronic F68 (surfactant), ethanol, double distilled water, and itraconazole. The microemulsion was incorporated in a gel by adding Carbomer 934P. Nail permeation enhancers like urea and salicylic acid were also used to increase drug penetration through the nail plate. The optimized formulation showed promising results: 92.75% drug entrapment efficacy and a complete drug release in 60 min with a highest nail uptake of 0.386%/mm^2 (39 mg drug) [72].

Angamuthu [135] described the study of **PLGA (poly (lactide-co-glycolide)) microspheres** for the controlled release of **terbinafine** administered by intralesion injection. These microspheres were developed using O/W emulsification and modified solvent extraction/evaporation technique with PVA, methylene chloride and methanol. This formulation achieved controlled release through 30 days and good deposition onto the nail bed and plate.

Chouhan *et al.* [57] studied the influence of a permeation enhancer (hydroxypropyl-β-cyclodextrin or HP-β-CD) in a terbinafine nail lacquer formulation. This formulation was composed of cellulose acetate and ethyl cellulose as film-forming polymers, triethyl citrate as plasticizer, and isopropyl alcohol and acetone as solvents. Formulations containing this enhancer demonstrated a higher flux than the control formulation in *in vitro* studies. The lacquer containing 10% (w/v) HP-β-CD showed maximum flux of 4.586 ± 0.08 μg/mL/cm^2 as compared to the control flux of 0.868 ± 0.06 μg/mL/cm^2, demonstrating its ability to enhance the transungual permeation of poorly soluble drugs [57].

Another **microemulsion-based gel** was studied by Kumar *et al.* [136]. This gel contained **fluconazole** (against *Aspergillus niger*), oleic acid, polysorbate 80, propylene glycol and water. It exhibited an *in vitro* drug release of 72.23% in 7 h.

14.5 Conclusion

Onychomycosis severely affects infected patients' quality of life. Although topical treatments are at the forefront of nail disorders therapy, most of them still have low market representativity. For example, there are few nail lacquers formulations available in the market. Notwithstanding, clinical trials are in progress for new formulations and many research laboratories in different countries (USA, Italy, Spain, Germany, UK, India, Japan) are investigating this issue. In fact, investigation on transungual drug delivery for treatment of onychomycosis is a recent field.

Nail permeation represents the greatest obstacle for transungual drug delivery, and therefore, it is necessary to better understand and study this phenomenon to improve drug bioavailability in the nail matrix. If permeation factors would be clearly elucidated by systematic research, more rational formulations could be developed. Unfortunately, contradicting reports about how drug size, surface charge, and hydrophilicity affect its permeation can be found in literature, and enlighten is urgently needed. Chemical permeation enhancers may also offer many advantages and should be further investigated. Many enhancers currently available show only modest improvement in nail permeation.

Many papers compare commercial products, but not much information is available regarding the formulation and physicochemical properties, possibly due to commercial reasons. Thus, it is complicated to follow a systematic research. In addition, a standardized protocol regarding several procedures should be established in order to obtain a clearer correlation. MIC values determined in aqueous media have been also used to compare antifungal activity in *in vitro* studies. However, this evaluation does not take into account keratin matrix, fungal nutrients and pH characteristics of the nail plate.

In summary, more basic research is still needed in order to improve and develop this interesting field, following standardized protocols to better evaluate and compare different nail formulations for treatment of onychomycosis.

References

1. Lecha M, Effendy, Feuilhade de Chauvin M, Di Chiacchio D, Baran R. Treatment options development of consensus guidelines. *Journal European Academy of Dermatology and Veneorology*. 2005; 19(1): 876–890.
2. Singal A, Khanna D. Onychomycosis: Diagnosis and management. *Indian Journal of Dermatology Venereology and Leprology*. 2011; 77(6): 659–672.
3. Grover C, Khurana A. An update on treatment of onychomycosis. *Mycoses*. 2012; 55(6): 541–551.
4. Thomas J, Jacobson G, Narkowicz CK, Peterson GM, Burnet H, Sharpe C. Toenail onychomycosis: An important global disease burden. *Journal of Clininical Pharmacy Therapeutic*. 2010; 35(5): 497–519.
5. Tchernev G, Penev PK, Nenoff P, Zisova LG, Cardoso JC, Taneva T, *et al.* Onychomycosis: Modern diagnostic and treatment approaches. *Wiener Medizinische Wochenschrift*. 2013; 163(1–2): 1–12.
6. Pierard GE, Pierard-Franchimont C, Arrese JE. The boosted antifungal topical treatment (BATT) for onychomycosis. *Medical Mycology*. 2000; 38: 391–3923.
7. Tan EST, Chong W-S, Tey HL. Nail psoriasis: A review. *American Journal of Clinical Dermatology*. 2012; 13(6): 375–388.
8. Parisi R, Symmons DPM, Griffiths CEM, Ashcroft DM. Global epidemiology of psoriasis: A systematic review of incidence and prevalence. *Journal of Investigative Dermatology*. 2012; 133(2): 377–385.

9. Shi Y, Hu FB. The global implications of diabetes and cancer. *Lancet*. 2014; 383: 1947–1948.
10. Loo DS. Onychomycosis in the elderly: Drug treatment options. *Drugs & Aging*. 2007; 24(4): 293–302.
11. Ghannoum M, Isham N. Fungal Nail Infections (Onychomycosis): A Never-Ending Story? *PLoS Pathogens*. 2014; 10(6): 1–5.
12. Tosti A, Hay R, Arenas-Guzmán R. Patients at risk of onychomycosis-risk factor identification and active prevention. *Journal of European Academy of Dermatology and Venereology*. 2005; 19(1): 13–16.
13. Stone N, Dawber R. Crinkly toenails. Toenail onychomycosis can cause serious problems. *British Medical Journal*. 2000; 320(7232): 448.
14. Gupta AK, Sibbald RG, Lynde CW, Hull PR, Prussick R, Shear NH, *et al.* Onychomycosis in children: Prevalence and treatment strategies. *Journal of the American Academy of Dermatology*. 1997; 36(3): 395–402.
15. Elewski BE. Onychomycosis: Pathogenesis, diagnosis, and management. *Clinical Microbiology Reviews*. 1998; 11(3): 415–429.
16. Feldstein S, Totri C, Friedlander SF. Antifungal therapy for onychomycosis in children. *Clinics in Dermatology*. 2015; 33(3): 333–339.
17. Gupta AK, Gupta MA, Summerbell RC, Cooper EA, Konnikov N, Albreski D, *et al.* The epidemiology of onychomycosis: Possible role of smoking and peripheral arterial disease. *Journal of the European Academy of Dermatology and Venereology*. 2000; 14(6): 466–469.
18. Clemons KV, Schar G, Stover EP, Feldman D, Stevens DA. Dermatophyte-hormone relationships: Characterization of progesterone-binding specificity and growth inhibition in the genera Trichophyton and Microsporum. *Journal of Clinical Microbiology*. 1988; 26(10): 2110–2115.
19. Faergemann J, Correia O, Nowicki R, Ro B-I. Genetic predisposition--understanding underlying mechanisms of onychomycosis. *Journal of European Academy of Dermatology and Venereology*. 2005; 19(1): 17–19.
20. Faergemann J, Baran R. Epidemiology, clinical presentation and diagnosis of onychomycosis. *British Journal Dermatology*. 2003; 149: 1–4.
21. Iorizzo M, Piraccini B, Tosti A. Today's treatments options for onychomycosis. *Journal of the German Society of Dermatology*. 2010; 8: 875–879.

22. Zaias N, Tosti A, Rebell G, Morelli R, Bardazzi F, Bieley H, *et al.* Autosomal dominant pattern of distal subungual onychomycosis caused by Trichophyton rubrum. *Journal of the American Academy of Dermatology*. 1996; 34(2): 302–304.
23. Eisman S, Sinclair R. Fungal nail infection: Diagnosis and management. *British Medical Journal*. 2014; 1800: 1–11.
24. Seebacher C, Brasch J, Abeck D, Cornely O, Effendy I, Ginter-Hanselmayer G, *et al.* Onychomycosis. *Mycoses*. 2007; 50(4): 321–327.
25. Grover C, Khurana A. Onychomycosis: Newer insights in pathogenesis and diagnosis. *Indian Journal of Dermatology, Venereology Leprology*. 2012; 78(3): 263.
26. Baran R, Kaoukhov A. Topical antifungal drugs for the treatment of onychomycosis: An overview of current strategies for monotherapy and combination therapy. *Journal of the European Academy of Dermatology and Venereology*. 2005; 19(1): 21–29.
27. Baran R, Hay RJ, Garduno JI. Review of antifungal therapy and the severity index for assessing onychomycosis: Part I. *Journal of Dermatological Treatment*. 2008; 19(2): 72–81.
28. Repka MA, Mididoddi PK, Stodghill SP. Influence of human nail etching for the assessment of topical onychomycosis therapies. *International Journal of Pharmaceutics*. 2004; 282: 95–106.
29. Baran R, Hay RJ, Garduno JI. Review of antifungal therapy, part II: Treatment rationale, including specific patient populations. *Journal of Dermatological Treatment*. 2008; 19(3): 168–175.
30. Özcan D, Skin D, Demirbilek M. *In vitro* antifungal susceptibility of dermatophyte strains causing tinea pedis and onychomycosis in patients with non-insulin-dependent diabetes mellitus: A case-control study. *Journal of the European Academy of Dermatology and Venereology*. 2010; 24(12): 1442–1446.
31. Baran R, Hay R-J. Nouvelle classification clinique des onychomycoses. *Journal of Medical Mycology*. 2014; 24(4): 247–260.
32. Hay RJ, Baran R. Onychomycosis: A proposed revision of the clinical classification. *Journal of the American Academy of Dermatology*. 2011; 65(6): 1219–1227.
33. Zaias N, Escovar SX, Rebell G. Opportunistic toenail onychomycosis. the fungal colonization of an available nail unit space by non-dermatophytes is produced by the trauma of the closed shoe by an asymmetric gait or other trauma. A plausible theory. *Journal of the European Academy of Dermatology and Venereology*. 2014; 28(8): 1002–1006.

34. Klaassen G, Dulak MG, van de Kerkhof PCM, Pasch MC. The prevalence of onychomycosis in psoriatic patients: A systematic review. *Journal of the European Academy of Dermatology and Venereology*. 2014; 28(5): 533–541.

35. Saner MV, Kulkarni AD, Pardeshi CV. Insights into drug delivery across the nail plate barrier. *Journal of Drug Targeting*. 2014; 2330(9): 1–21.

36. Täuber A, Müller-Goymann CC. *In vitro* permeation and penetration of ciclopirox olamine from poloxamer 407-based formulations-comparison of isolated human stratum corneum, bovine hoof plates and keratin films. *International Journal Pharmaceutics*. 2015; 489(1–2): 73–82.

37. Elsayed MM. Development of topical therapeutics for management of onychomycosis and other nail disorders: A pharmaceutical perspective. *Journal of Controlled Release: Official Journal of the Controlled Release Society*. 2014; 199: 132–144.

38. Murdan S. Enhancing the nail permeability. *Expert Opinion Drug Delivery*. 2008; 5(11): 1267–1282.

39. Murdan S. Drug delivery to the nail following topical application. *International Journal of Pharmaceutics*. 2002; 236(1–2): 1–26.

40. Sugiura K, Sugimoto N, Hosaka S, Katafuchi-Nagashima M, Arakawa Y, Tatsumi Y, *et al.* The low keratin affinity of efinaconazole contributes to its nail penetration and fungicidal activity in topical onychomycosis treatment. *Antimicrobial Agents and Chemotherapy*. 2014; 58(7). 3837–3842.

41. Czaika VA, Phi-Anh L. Trichophyton mentagrophytes cause underestimated contagious zoophilic fungal infection. *Mycoses*. 2013; 56 (1): 33–37.

42. Esteves JA, Baptista AP, Rodrigo FG, Gomes MAM. *Dermatologia*. 3ª Edição ed. Lisboa: Fundação Calouste Gulbenkian; 2005. 1643l.

43. Miron D, Cornelio R, Troleis J, Mariath J, Zimmer AR, Mayorga P, *et al.* Influence of penetration enhancers and molecular weight in antifungals permeation through bovine hoof membranes and prediction of efficacy in human nails. *European Journal of Pharmaceutical Sciences: Official Journal of the European Federation for Pharmaceutical Sciences*. 2014; 51: 20–25.

44. Kobayashi Y, Komatsu T, Sumi M, Numajiri S, Miyamoto M, Kobayashi D, *et al.* *In vitro* permeation of several drugs through the human nail plate: Relationship between physicochemical properties and

nail permeability of drugs. *European Journal of Pharmaceutical Sciences*. 2004; 21(4): 471–477.

45. Lusiana, Reichl S, Müller-Goymann CC. Keratin film made of human hair as a nail plate model for studying drug permeation. *European Journal of Pharmaceutics and Biopharmaceutics*. 2011; 78(3): 432–440.
46. Saner MV, Kulkarni AD, Pardeshi CV. Insights into drug delivery across the nail plate barrier. *Journal of Drug Target*. 2014; 2330(9): 769–789.
47. Gupta AK, Paquet M. Improved efficacy in onychomycosis therapy. *Clinics Dermatology*. 2013; 31(5): 555–563.
48. Cribier BJ, Paul C. Long-term efficacy of antifungals in toenail onychomycosis: A critical review. *British Journal Dermatology*. 2001; 145(3): 446–452.
49. Gupta AK, Paquet M, Simpson FC. Therapies for the treatment of onychomycosis. *Clinics Dermatology*. 2013; 31(5): 544–554.
50. Pajaziti L, Vasili E. Treatment of onychomycosis—a clinical study. *Medical Archives*. 2015; 69(3): 173.
51. Gupta AK, Paquet M. Systemic antifungals to treat onychomycosis in children: A systematic review. *Pediatric Dermatology*. 2013; 30(3): 294–302.
52. Thatai P, Sapra B. Transungual delivery: Deliberations and creeds. *International Journal of Cosmetic Science*. 2014; 36(5): 398–411.
53. Kerai LV, Hilton S, Murdan S. UV-curable gel formulations: Potential drug carriers for the topical treatment of nail diseases. *International Journal of Pharmaceutics*. 2015; 492: 177–190.
54. Van Duyn Graham L, Elewski BE. Recent updates in oral terbinafine: Its use in onychomycosis and tinea capitis in the US. *Mycoses*. 2011; 54(6): 679–685.
55. Schaller M, Borelli C, Berger U, Walker B, Schmidt S, Weindl G, *et al.* Susceptibility testing of amorolfine, bifonazole and ciclopiroxolamine against Trichophyton rubrum in an *in vitro* model of dermatophyte nail infection. *Medical Mycology*. 2009: 753–738.
56. Monti D, Herranz U, Dal Bo L, Subissi A. Nail penetration and predicted mycological efficacy of an innovative hydrosoluble ciclopirox nail lacquer vs. a standard amorolfine lacquer in healthy subjects. *Journal of the European Academy of Dermatology and Venereology*. 2013; 27(2): 153–158.
57. Chouhan P, Saini TR. Hydroxypropyl-β-cyclodextrin: A novel transungual permeation enhancer for development of topical drug

delivery system for onychomycosis. *Journal Drug Delivery*. 2014; 17: 1–7.

58. Evans EGV. Drug synergies and the potential for combination therapy in onychomycosis. *British Journal of Dermatology*. 2003; 149: 11–13.
59. Rosenson RS. The rationale for combination therapy. *American Journal of Cardiology*. 2002; 90(10B): 2K–7K.
60. Malhotra GG, Zatz JL. Investigation of nail permeation enhancement by chemical modification using water as a probe. *Journal of Pharmaceutical Sciences*. 2002; 91(2): 312–323.
61. Mohorčič M, Torkar A, Friedrich J, Kristl J, Murdan S. An investigation into keratinolytic enzymes to enhance ungual drug delivery. *International Journal Pharmaceutical*. 2007; 332(1–2): 196–201.
62. Khengar RH, Jones SA, Turner RB, Forbes B, Brown MB. Nail swelling as a pre-formulation screen for the selection and optimisation of ungual penetration enhancers. *Pharmaceutical Research*. 2007; 24(12): 2207–2212.
63. Shivakumar HN, Repka MA, Narasimha Murthy S. Transungual drug delivery: An update. *Journal of Drug Delivery Science and Technology*. 2014; 24(3): 301–310.
64. Gunt HB, Kasting GB. Effect of hydration on the permeation of ketoconazole through human nail plate *in vitro*. *European Journal Pharmaceutical Sciences*. 2007; 32(4–5): 254–260.
65. Stüttgen G, Bauer E. Bioavailability, skin- and nailpenetration of topically applied antimycotics. *Mykosen*. 1982; 25(2): 74–80.
66. Lahfa M, Bulai-Livideanu C, Baran R, Ortonne JP, Richert B, Tosti A, Piraccini BM, Szepietowski JC, Sibaud V, Coubetergues H, Voisard JJ, Paul C. Efficacy, safety and tolerability of an optimized avulsion technique with Onyster® (40% urea ointment with plastic dressing) ointment compared to bifonazole-urea ointment for removal of the clinically infected nail in toenail onychomycosis: A randomized evaluator-blinded controlled study. *Dermatology*. 2013; 226(1): 5–12. doi: 10.1159/000345105. Epub 2013 Mar 1.
67. Tietz HJ, Hay R, Querner S, Delcker A, Kurka P, Merk HF. Efficacy of 4 weeks topical bifonazole treatment for onychomycosis after nail ablation with 40% urea: A double-blind, randomized, placebo-controlled multicenter study. *Mycoses*. 2013: 414–421.
68. Baran R, Coquard F. Combination of fluconazole and urea in a nail lacquer for treating onychomycosis. *Journal Dermatology Treatment*. 2005; 16(1): 52–55.

69. Hui X, Chan TCK, Barbadillo S, Lee C, Maibach HI, Wester RC. Enhanced econazole penetration into human nail by 2-*n*-nonyl-1,3-dioxolane. *Journal of Pharmaceutical Sciences*. 2003; 92(1): 142–148.
70. Vejnovic I, Huonder C, Betz G. Permeation studies of novel terbinafine formulations containing hydrophobins through human nails *in vitro*. *International Journal of Pharmaceutics*. 2010: 67–76.
71. Vejnovic I, Simmler L, Betz G. Investigation of different formulations for drug delivery through the nail plate. *International Journal Pharmaceutics*. 2010; 386(1–2): 185–194.
72. Pal P, Thakur RS, Ray S, Mazumder B. Design and development of a safer non-invasive transungual drug delivery system for topical treatment of onychomycosis. *Drug Development and Industrial Pharmacy*. 2014: 1–5.
73. Repka MA, O'Haver J, See CH, Gutta K, Munjal M. Nail morphology studies as assessments for onychomycosis treatment modalities. *International Journal of Pharmaceutics*. 2002; 245: 25–36.
74. Andrews JM. Determination of minimum inhibitory concentrations. *Journal of Antimicrobial Chemotherapy*. 2001; 48(l): 5–16.
75. Fothergill AW. Antifungal susceptibility testing: Clinical (CLSI) methods. In: *Interaction of Yeasts, Moulds and Antifungical Agents: How to Detect Resistance* (Hall GS, ed), Chapter 2. XIV. 2012; pp. 65–75.
76. Murdan S, Hinsu D, Guimier M. A few aspects of transonychial water loss (TOWL): Inter-individual, and intra-individual inter-finger, inter-hand and inter-day variabilities, and the influence of nail plate hydration, filing and varnish. *European Journal Pharmaceutics and Biopharmaceutics*. 2008; 70(2): 684–689.
77. Elewski BE, Tosti A. Tavaborole for the treatment of onychomycosis. *Expert Opinion*. 2014; 15(10): 1439–1448.
78. Hui X, Baker SJ, Wester RC, Barbadillo S, Cashmore AK, Sanders V, *et al. In vitro* penetration of a novel oxaborole antifungal (AN2690) into the human nail plate. *Journal of Pharmaceutical Science*. 2007; 96(10): 2622–2631.
79. Tamura T, Asahara M, Yamamoto M, Yamaura M, Matsumura M, Goto K, *et al. In vitro* susceptibility of dermatomycoses agents to six antifungal drugs and evaluation by fractional inhibitory concentration index of combined effects of amorolfine and itraconazole in dermatophytes. *Microbiology and Immunology*. 2014: 1–8.
80. Jo Siu WJ, Tatsumi Y, Senda H, Pillai R, Nakamura T, Sone D, *et al.* Comparison of *in vitro* antifungal activities of efinaconazole and

currently available antifungal agents against a variety of pathogenic fungi associated with onychomycosis. *Antimicrobial Agents and Chemotherapy*. 2013: 1610–1616.

81. Santos DA, Hamdan JS. *In vitro* antifungal oral drug and drug-combination activity against onychomycosis causative dermatophytes. *Medical Mycology*. 2006: 357–362.

82. Carrillo-Muñoz A-J, Giusiano G, Cárdenes D, Hernández-Molina J-M, Eraso E, Quindós G, *et al.* Terbinafine susceptibility patterns for onychomycosis-causative dermatophytes and Scopulariopsis brevicaulis. *International Journal of Antimicrobial Agents*. 2008; 31(6): 540–543.

83. Ameen M, Lear JT, Madan V, Mohd Mustapa MF, Richardson M. British Association of Dermatologists' guidelines for the management of onychomycosis 2014. *British Journal Dermatology*. 2014; 1200–1214.

84. Flagothier C, Piérard-Franchimont C, Piérard GE. New insights into the effect of amorolfine nail lacquer. *Mycoses*. 2005; 48(2): 91–94.

85. Zaug M, Bergstraesser M. Amorolfine in the treatment of onychomycosis and dermatomycoses (an overview). *Clinical and Experimental Dermatology*. 1992; 17(1): 61–70.

86. Crawford F, Hollis S. Topical treatments for fungal infections of the skin and nails of the foot. *The Cochrane Database of Systematic Reviews*. 2007; 3: 2–90.

87. Gupta AK, Daigle D, Foley KA. Topical therapy for toenail onychomycosis: An evidence-based review. *American Journal of Clinical Dermatology*. 2014; 15(6): 489–502.

88. INFARMED. Governo Portugal. (11 July 2015). https://www.infarmed.pt/infomed/inicio.php.

89. FDA. FDA–U.S. Food and Drug Administration. USA. (12 September 2015) http://www.fda.gov/Drugs/default.htm.

90. EMA-European Medicines Agency. European Union. European Union. (2 September 2015). http://www.ema.europa.eu/ema/index.j.

91. Tabara K, Szewczyk AE, Bienias W, Wojciechowska A, Pastuszka M, Oszukowska M, *et al.* Amorolfine vs. ciclopirox-lacquers for the treatment of onychomycosis. *Postpy Dermatologii i Alergol*. 2015; 32(1): 40–45.

92. Bohn M, Kraemer KT. Dermatopharmacology of ciclopirox nail lacquer topical solution 8% in the treatment of onychomycosis. *Journal of the American Academy of Dermatology*. 2000: 57–69.

93. Shemer A, Nathansohn N, Trau H, Amichai B, Grunwald MH. Ciclopirox nail lacquer for the treatment of onychomycosis: An open non-comparative study. *Journal of Dermatology*. 2010; 37(2): 137–139.

94. Gupta AK, Fleckman P, Baran R. Ciclopirox nail lacquer topical solution 8% in the treatment of toenail onychomycosis. *Journal of the American Academy Dermatology*. 2000; 43(4): S70–S80.

95. Elewski BE, Rich P, Pollak R, Pariser DM, Watanabe S, Senda H, *et al.* Efinaconazole 10% solution in the treatment of toenail onychomycosis: Two phase III multicenter, randomized, double-blind studies. *Journal of the American Academy of Dermatology*. 2013: 600–608.

96. Del Rosso JQ, Plattner JJ. From the test tube to the treatment room: Fundamentals of boron-containing compounds and their relevance to dermatology. *Journal of Clinical and Aesthetic Dermatology*. 2014: 13–21.

97. Elewski BE, Aly R, Baldwin SL, González Soto RF, Rich P, Weisfeld M, *et al.* Efficacy and safety of tavaborole topical solution, 5%, a novel boron-based antifungal agent, for the treatment of toenail onychomycosis: Results from 2 randomized phase-III studies. *Journal of the American Academy Dermatology*. 2015; 73(1): 62–69.

98. Hay RJ, Mackie RM, Clayton YM. Tioconazole nail solution--an open study of its efficacy in onychomycosis. *Clinical Experimental Dermatology*. 1985; 10(2): 111–115.

99. Commission BP. British Pharmacopoeia 2009. UK (4 July 2015). https://www.pharmacopoeia.com.

100. Chang R, Raw A, Lionberger R, Yu L. Generic development of topical dermatologic products: formulation development, process development, and testing of topical dermatologic products. *AAPS Journal*. 2013; 15(1): 41–52.

101. Gregorí BS. Estructura y actividad de los antifúngicos. *Revista Cubana Farmacia*. 2005; 39(2): 1–5.

102. Bayer Portugal SA, Resumo das características do medicamento-Canespor. Infarmed. Portugal (4 July 2015). http://www.infarmed.pt/.

103. Santos DA, Hamdan JS. *In vitro* antifungal oral drug and drug-combination activity against onychomycosis causative dermatophytes. *Medical Mycology*. 2006; 44(4): 357–362.

104. Lipner SR, Scher RK. Efinaconazole in the treatment of onychomycosis. *Journal of Infection and Drug Resistance*. 2015; 8: 163–172.

105. Tamura T, Asahara M, Yamamoto M, Yamaura M, Matsumura M, Goto K, *et al. In vitro* susceptibility of dermatomycoses agents to six antifungal drugs and evaluation by fractional inhibitory concentration index of combined effects of amorolfine and itraconazole in dermatophytes. *Microbiology and Immunology*. 2014; 58(1): 1–8.

106. Bhatt V, Pillai R. Efinaconazole Topical Solution, 10%: Formulation development program of a new topical treatment of toenail onychomycosis. *Journal of Pharmaceutical Sciences*. 2015; 104(7): 2177–2182.

107. Del Rosso JQ. The role of topical antifungal therapy for onychomycosis and the emergence of newer agents. *Journal of Clinical and Aesthetic Dermatology*. 2014; 7(7): 10–8.

108. Bahn CA. What is new in fungal pharmacotherapeutics? *Journal of Drug in Dermatology*. 2014; 101(4): 161–165.

109. Gupta AK, Simpson FC. Efinaconazole: A new topical treatment for onychomycosis. *Skin Therapy Letter*. 2014; 19(1).

110. Laboratórios Pfizer. Resumo das Características do Medicamento Trosyd. (3 Janeiro 2015) http://www.medicamentos.com.mx/DocHTM/22353.htm.

111. Naumann S, Meyer JP, Kiesow A, Mrestani Y, Wohlrab J, Neubert RHH. Controlled nail delivery of a novel lipophilic antifungal agent using various modern drug carrier systems as well as *in vitro* and *ex vivo* model systems. *Journal of Controlled Release*. 2014: 60–70.

112. Murdan S, Kerai L, Hossin B. To what extent do *in vitro* tests correctly predict the *in vivo* residence of nail lacquers on the nail plate? *Journal of Drug Delivery Science and Technology*. 2015: 23–28.

113. Joshi M, Sharma V, Pathak K. Matrix based system of isotretinoin as nail lacquer to enhance transungal delivery across human nail plate. *International Journal of Pharmaceutics*. 2014: 268–277.

114. Monti D, Saccomani L, Chetoni P, Burgalassi S, Senesi S, Ghelardi E, *et al.* Hydrosoluble medicated nail lacquers: *In vitro* drug permeation and corresponding antimycotic activity. *British Journal of Dermatology*. 2010; 162(2): 311–317.

115. Monti D, Saccomani L, Chetoni P, Burgalassi S, Tampucci S, Mailland F. Validation of bovine hoof slices as a model for infected human toenails: *In vitro* ciclopirox transungual permeation. *British Journal of Dermatology*. 2011. 99–105.

116. Baraldi A, Jones SA, Guesné S, Traynor MJ, McAuley WJ, Brown MB, *et al.* Human nail plate modifications induced by onychomycosis:

Implications for topical therapy. *Pharaceutical Research*. 2014; 1626–1633.

117. Sigurgeirsson B, Olafsson J, Steinsson J, Kerrouche N, Sidou F. Efficacy of amorolfine nail lacquer for the prophylaxis of onychomycosis over 3 years. *Journal of the European Academy of Dermatology and Venereology*. 2009; 24(8): 910–915.
118. Fiddaman PJ, Rossall S. The production of antifungal volatiles by Bacillus subtilis. *Journal Applied Bacteriology*. 1993; 74(2): 119–126.
119. Monti D, Saccomani L, Chetoni P, Burgalassi S, Saettone MF, Mailland F. *In vitro* transungual permeation of ciclopirox from a hydroxypropyl chitosan-based, water-soluble nail lacquer. *Drug Development and Industrial Pharmacy*. 2005: 11–17.
120. Nogueiras-Nieto L, Begoña Delgado-Charro M, Otero-Espinar FJ. Thermogelling hydrogels of cyclodextrin/poloxamer polypseudorotaxanes as aqueous-based nail lacquers: Application to the delivery of triamcinolone acetonide and ciclopirox olamine. *European Journal of Pharmaceutics and Biopharmaceutics*. 2013: 370–377.
121. Shivakumar HN, Vaka SRK, Madhav NVS, Chandra H, Murthy SN. Bilayered nail lacquer of terbinafine hydrochloride for treatment of onychomycosis. *Journal Pharmaceutical Science*. 2010; 99(10): 4267–4276.
122. Kiran S, Shekar C. Ungual Drug Delivery System of Ketoconazole Nail Lacquer. *In vitro*. *International Journal of Applied Pharmaceutics*. 2010; 2(4): 1–3.
123. Hafeez F, Hui X, Chiang A, Hornby S, Maibach H. Transungual delivery of ketoconazole using novel lacquer formulation. *International Journal Pharmaceutics*. 2013; 456(2): 357–361.
124. Vipin KV, Chandran SC, Augusthy AR, Premaletha K, Kuriakose MR. Formulation and evaluation of an antifungal nail lacquer for onychomycosis. *British Biomedical Bulletin*. 2014; 1(2): 242–248.
125. Baran R, Tosti A, Hartmane I, Altmeyer P, Hercogova J, Koudelkova V, *et al.* An innovative water-soluble biopolymer improves efficacy of ciclopirox nail lacquer in the management of onychomycosis. *Journal of the European Academy of Dermatology and Venereology*. 2009: 773–781.
126. Pradhan M, Singh D, Singh MR. Novel colloidal carriers for psoriasis: Current issues, mechanistic insight and novel delivery approaches. *Journal of Controlled Release*. 2013: 380–395.
127. Yang Y, Ou R, Guan S, Ye X, Hu B, Zhang Y, *et al.* A novel drug delivery gel of terbinafine hydrochloride with high penetration for external use. *Drug Delivery*. 2014; 7544: 1–8.

128. Hamouda T, Flack M, Jr JRB, Mic C. Development of a novel antiviral drug (NB-001) for topical application in humans. *Journal American Academic Dermatology*. 2008; 58(2): 95–96.

129. Vaghasiya H, Kumar A, Sawant K. Development of solid lipid nanoparticles based controlled release system for topical delivery of terbinafine hydrochloride. *European Journal of Pharmaceutical Sciences: Official Journal of the European Federation for Pharmaceutical Sciences*. 2013: 311–322.

130. Ghannoum M, Isham N, Herbert J, Henry W, Yurdakul S. Activity of TDT 067 (terbinafine in transfersome) against agents of onychomycosis, as determined by minimum inhibitory and fungicidal concentrations. *Journal of Clinical Microbiology*. 2011: 1716–1720.

131. Dominicus R, Weidner C, Tate H, Kroon HA. Open-label study of the efficacy and safety of topical treatment with TDT 067 (terbinafine in Transfersome®) in patients with onychomycosis. *British Journal of Dermatology*. 2012: 1360–1362.

132. Tanrýverdi ST, Özer Ö. Novel topical formulations of Terbinafine-HCl for treatment of onychomycosis. *European Journal Pharmaceutical Science*. 2013; 48(4–5): 628–636.

133. Barot BS, Parejiya PB, Patel HK, Gohel MC, Shelat PK. Microemulsion-based gel of terbinafine for the treatment of onychomycosis: Optimization of formulation using D-optimal design. *AAPS Pharm Sci Tech*. 2012: 184–192.

134. Barot BS, Parejiya PB, Patel HK, Mehta DM, Shelat PK. Microemulsion-based antifungal gel delivery to nail for the treatment of onychomycosis: Formulation, optimization, and efficacy studies. *Drug Delivery and Translation Research*. 2012; 2(6): 463–476.

135. Angamuthu M, Nanjappa SH, Raman V, Jo S, Cegu P, Murthy SN. Controlled-release injectable containing terbinafine/PLGA microspheres for onychomycosis treatment. *Journal Pharmaceutical Science*. 2014; 103(4): 1178–1183.

136. Kumar KJR, Muralidharan S, Dhanaraj SA. Antifungal activity of microemulsion based fluconazole gel for onychomycosis against Aspergillus niger. *International Journal of Pharmacy and Pharmaceutical Sciences* 2012; 5(1): 3–9.

Index